RELATIONSHIP BETWEEN MICROBES AND THE ENVIRONMENT FOR SUSTAINABLE ECOSYSTEM SERVICES, VOLUME 1

RELATIONSHIP BETWEEN MICROBES AND THE ENVIRONMENT FOR SUSTAINABLE ECOSYSTEM SERVICES, VOLUME 1

Microbial Products for Sustainable Ecosystem Services

Edited by

JASTIN SAMUEL

Assistant Professor, Waste Valorization Research Lab, Department of Trans-disciplinary Research, Division of Research and Development (DRD), Lovely Professional University, Jalandhar, Punjab, India

AJAY KUMAR

Visiting Scientist, Department of Postharvest Science, Agriculture Research Organization, Volcani center, Rishon Leziyon, Israel

JOGINDER SINGH

Professor, Microbiology Climate Mitigation and Sustainable Agriculture Research lab (CMaSAR)Division of Research and Development Lovely Professional University, Phagwara, Punjab, India

ELSEVIER

Elsevier
Radarweg 29, PO Box 211, 1000 AE Amsterdam, Netherlands
The Boulevard, Langford Lane, Kidlington, Oxford OX5 1GB, United Kingdom
50 Hampshire Street, 5th Floor, Cambridge, MA 02139, United States

Notices
Knowledge and best practice in this field are constantly changing. As new research and experience broaden our understanding, changes in research methods, professional practices, or medical treatment may become necessary.

Practitioners and researchers must always rely on their own experience and knowledge in evaluating and using any information, methods, compounds, or experiments described herein. In using such information or methods they should be mindful of their own safety and the safety of others, including parties for whom they have a professional responsibility.

To the fullest extent of the law, neither the Publisher nor the authors, contributors, or editors, assume any liability for any injury and/or damage to persons or property as a matter of products liability, negligence or otherwise, or from any use or operation of any methods, products, instructions, or ideas contained in the material herein.

ISBN: 978-0-323-89938-3

For information on all Elsevier publications
visit our website at https://www.elsevier.com/books-and-journals

Publisher: Candice Janco
Editorial Project Manager: Howi M. De Ramos
Production Project Manager: Paul Prasad Chandramohan
Cover Designer: Vicky Pearson Esser

Typeset by STRAIVE, India

Contents

Contributors   *xi*

About the editors   *xvii*

Preface   *xix*

1. Microbial food products: A sustainable solution to alleviate hunger   1

Daniela Landa-Acuña, Andi Solorzano-Acosta, Vanessa Sánchez-Ortiz, Edwin Hualpa-Cutipa, Celia Vargas-de-la-Cruz, Bernabé Luis-Alaya, and Eduardo Flores-Juarez

1 Introduction   1

2 General aspects of edible microbial biomass safety   2

3 Safety of edible microbial biomass   6

4 Microorganisms of potential use as food and possible routes to produce edible microbial biomass   13

5 Production of edible microbial biomass   17

6 Biotechnological tools involved in the generation of microbial food products   19

References   22

2. Role of microorganisms in climate-smart agriculture   29

Astha Sinha, Swarnkumar Reddy, and W. Jabez Osborne

1 Introduction   29

2 Effect of pollution on soil and its quality   29

3 Soil pollution and its effects on the quality of soil   31

4 Role of soil microbiome in regulating soil health and plant fertility; A plant-soil-microbial interactions   32

5 Physiology of plant-growth promoting rhizobacteria   34

6 Microbial influence on biogeochemical cycles and its applications   35

7 Role of microbes and enzymes in the restoration and reclamation of soil   38

References   39

3. Microorganisms as biocontrol agents for sustainable agriculture   45

Bhupendra Koul, Manpriya Chopra, and Supriya Lamba

1 Introduction   45

2 Bacteria and fungi as biocontrol agents   46

3 Production of desired product   46

4 Mechanisms of biocontrol agents   48

5 Conclusions   63
References   64

4. Relationship between probiotics and living beings for sustainable life on land   **69**

Celia Vargas-de-la-Cruz, Daniela Landa-Acuña, Md. Shariful Islam, and Eduardo Flores-Juarez

1 Introduction   69
2 Importance of probiotics in animal health   70
3 Feed antibiotics to probiotics like a viable substitute   71
4 Risk assessment protocol established due to probiotics   73
5 Adverse effects because of application for probiotics   74
6 Principle of selection of probiotics   74
7 Competitive exclusion   75
8 Mechanisms of actions of probiotics   76
9 Nutrients and enzymatic attributed to digestion   78
10 Certain impact toward water quality   78
11 Improvement to immune response   79
12 Antiviral effects   79
13 Separation of probiotics   79
14 Method of separation   80
15 Applications of probiotics in aquaculture   80
References   80

5. Microbial adaptation to climate change and its impact on sustainable development   **85**

Srishti Srivastava, Amartya Chakraborty, and K. Suthindhiran

1 Introduction   85
2 Climatic factors   86
3 Future prospective and applications   97
4 Conclusion   99
Acknowledgments   99
References   99
Further reading   105

6. Earthworm-microorganisms interactions for sustainable soil ecosystem and crop productivity   **107**

Sudipti Arora, Sakshi Saraswat, Anamika Verma, and Devanshi Sutaria

1 Introduction   107
2 Earthworms—Nature's ploughman   109

3 Vermicompost as biofertilizer   112

4 Vermicompost on crop growth and productivity   112

5 Effect of vermicompost on nutrient uptake   114

6 Effect of vermicompost on soil physical, chemical and biological properties   115

7 Vermitechnology for co-treatment of OFMSW and wastewater   115

8 Conclusions   117

References   117

7. Avenues of sustainable pollutant bioremediation using microbial biofilms   **121**

Basma A. Omran

1 Introduction   122

2 Environmental pollution   125

3 Biofilms (remarkable biological communities)   128

4 Biofilms and bioremediation   142

5 Applications of biofilms in bioremediation of different pollutants   145

6 Conclusions   149

References   150

8. Endophytic bacteria in a biocontrol perspective   **155**

Riddha Dey and Richa Raghuwanshi

1 Introduction   155

2 Biocontrol activity of endophytes   156

3 Mechanisms of biocontrol   156

4 Strategies to enhance biocontrol efficiency   164

5 Genetic engineering   166

6 Future research prospects   168

References   169

9. Microbiome stimulants and their applications in crop plants   **177**

Shristi Bhandari, Sarvjeet Kukreja, Vijay Kumar, Abhijit Dey, and Umesh Goutam

1 Introduction   177

2 Plant microbiome   178

3 Microbiome: A stimulant in crop plants   180

4 Microbiome stimulants in major crops   181

5 Microbiome stimulants and their application process   185

6 Reduce pathogen infection   190

7 Conclusion   191

References   191

10. Microbes: A sustainable tool for healthy and climate smart agriculture 197

Surojit Bera, Richa Arora, Collins Njie Ateba, and Ajay Kumar

1 Introduction 197
2 Important role of microbes in agriculture 198
3 Combating climate change by microbes 204
4 Conclusions 207
References 208

11. Microbes as biocontrol agent: From crop protection till food security 215

C.R. Vanshree, Muskan Singhal, Mansi Sexena, Mahipal Singh Sankhla, Kapil Parihar, Ekta B. Jadhav, Kumud Kant Awasthi, and Chandra Shekhar Yadav

1 Introduction 215
2 Microbes as biocontrol agents 216
3 Microbial biocontrol for crop protection and food security 217
4 Microbiol biocontrol for postharvest management 223
5 Mechanism of microbial control activity 225
6 Emerging biocontrol strategies 229
7 Conclusion and future prospects 231
References 232

12. Composting process: Fundamental and molecular aspects 239

Ruchi Soni and Sunita Devi

1 Introduction 239
2 Composting: Fundamental aspects 240
3 Types of the composting process (aerobic and anaerobic) 246
4 Phases of composting 248
5 Determination of compost maturity 253
6 Taxonomic and metabolic microbial diversity during composting 256
7 Techniques to analyze microbial diversity in composting 257
8 Role of metagenomics in evaluating microbial diversity in compost 259
9 Conclusion and future outlook 260
References 261

13. Lichenized fungi, a primary bioindicator/biomonitor for bio-mitigation of excessive ambient air nitrogen deposition worldwide 267

Himanshu Rai and Rajan Kumar Gupta

1 Introduction 267
2 Nitrogen assimilation in lichens 269

3 Nitrogen tolerance in lichens: Probable mechanisms   269

4 Lichens are indicators of excessive nitrogen (N) deposition along with multiple scales of diversity dynamics, biochemistry, and ecophysiology   273

5 Lichens, the critical load assessor of N eutrophication   279

6 Long-term lichen air N deposition monitoring: Change in pollution regime   280

7 Stable nitrogen isotope ratio ($\delta^{15}N$ in) lichens function of nitrogen source and their spatial distribution at the landscape level   282

8 Effect of excessive N deposition on lichen food webs of endangered wildlife   293

9 Lichen as a sink for nitrogen in the region of excessive deposition   294

Acknowledgments   294

References   294

Further reading   300

14. Molecular markers and genomics assisted breeding for improving crop plants   **301**

Manish Kumar Vishwakarma, Punam Singh Yadav, Ved Prakash Rai, Uttam Kumar, and Arun Kumar Joshi

1 Introduction   302

2 Trait choice for MAS   303

3 Selection of marker type for better results through MAS   305

4 Types of markers and their utility in plant breeding   306

5 Types of selection and their procedure   311

6 Success stories of MAS and perspectives   321

References   327

15. Nanoherbicides: A sustainable option for field applications   **333**

Vidya Patil-Patankar and Gaurav Sanghvi

1 Introduction   333

2 Weed management   334

3 Herbicides   335

4 Nanoherbicides   337

5 Nanotechnology and its implications in weed management   338

6 Weeds and nanotechnology   340

7 Nano-herbicides and possible actions   340

8 Nanomaterial used in the synthesis   341

9 Nanoherbicides formulations   345

10 Applications of nano-herbicides in farming/agriculture   348

11 Conclusions and future perspectives   349

References   350

16. Intimate coupling of photocatalysis and biodegradation (ICPB): A viable method for removing pesticides from contaminated sites   **355**

Asha S. Raj and Preethy Chandran

1 Introduction   355
2 Environmental distribution of pesticides   356
3 Toxicity of pesticides   357
4 Ecological effect of pesticides   357
5 Pesticides hazardous effects on human health   359
6 Coupling of photocatalysis and biodegradation (ICPB): A new approach   359
7 Mechanism of ICPB   361
8 Different types of photocatalyst used in ICPB   362
9 Different porous carrier materials   362
10 Microorganisms   364
11 Reactors   365
12 Applications of ICPB   365
13 Conclusion   366
References   366

Index   *369*

Contributors

Richa Arora
Department of Microbiology, Punjab Agricultural University, Ludhiana, Punjab, India

Sudipti Arora
Dr. B. Lal Institute of Biotechnology, Jaipur, India

Collins Njie Ateba
Food Security and Safety Niche Area, Faculty of Agriculture, Science and Technology, North West University, Mmabatho, Mafikeng, South Africa

Kumud Kant Awasthi
Department of Life Sciences, Vivekananda Global University, Jaipur, Rajasthan, India

Surojit Bera
Department of Microbiology, School of Bioengineering and Biosciences, Lovely Professional University, Jalandhar, Punjab, India

Shristi Bhandari
School of Bioengineering and Biosciences, Lovely Professional University, Phagwara, Punjab, India

Amartya Chakraborty
Marine Biotechnology and Bio-Products Laboratory, Department of Biomedical Sciences, School of Biosciences and Technology, Vellore Institute of Technology, Vellore, India

Preethy Chandran
School of Environmental Studies, Cochin University of Science and Technology, Kochi, Kerala, India

Manpriya Chopra
Department of Sciences, Bhagwan Mahaveer Public Senior Secondary School, Banga, Punjab, India

Sunita Devi
Department of Basic Sciences, Dr YS Parmar University of Horticulture and Forestry Nauni, Solan, India

Abhijit Dey
Department of Life Sciences, Presidency University, Kolkata, India

Riddha Dey
Department of Botany, Institute of Science, Banaras Hindu University, Varanasi, UP, India

Eduardo Flores-Juarez
Department of Biochemistry, Faculty of Pharmacy and Biochemistry, Universidad Nacional Mayor de San Marcos, Lima, Perú

Umesh Goutam
School of Bioengineering and Biosciences, Lovely Professional University, Phagwara, Punjab, India

Rajan Kumar Gupta
Centre of Advanced Study in Botany, Institute of Science, Banaras Hindu University, Varanasi, Uttar Pradesh, India

Edwin Hualpa-Cutipa
Faculty of Pharmacy and Biochemistry, Microbiology Laboratory, Universidad Nacional Mayor de San Marcos, Lima, Perú

Md. Shariful Islam
Department of Pharmacy, Southeast University, Dhaka, Bangladesh

Ekta B. Jadhav
Government Institute of Forensic Science, Aurangabad, Maharashtra, India

Arun Kumar Joshi
Borlaug Institute for South Asia-CIMMYT, New Delhi, India

Bhupendra Koul
School of Bioengineering and Biosciences, Department of Biotechnology, Lovely Professional University, Phagwara, Punjab, India

Sarvjeet Kukreja
School of Agriculture, Lovely Professional University, Phagwara, Punjab, India

Ajay Kumar
Department of Postharvest Science, Agriculture Research Organization, Volcani center, Rishon Leziyon, Israel; Volcani Center, Rishon LeZion, Israel

Uttam Kumar
Borlaug Institute for South Asia-CIMMYT, New Delhi, India

Vijay Kumar
School of Bioengineering and Biosciences, Lovely Professional University, Phagwara, Punjab, India

Supriya Lamba
Department of Botany, Kamla Nehru College for Women, Phagwara, Punjab, India

Daniela Landa-Acuña
Laboratory of Microbial Ecology and Biotechnology, Department of Biology, Faculty of Sciences, National Agrarian University La Molina (UNALM); Professional Career in Environmental Engineering, Faculty of Engineering, Private University of the North, Los Olivos Campus, Lima, Perú

Bernabé Luis-Alaya
Laboratory of Microbial Ecology and Biotechnology, Department of Biology, Faculty of Sciences, National Agrarian University La Molina (UNALM), Lima, Perú

Basma A. Omran
Petroleum Biotechnology Laboratory, Processes Design and Development Department, Egyptian Petroleum Research Institute (EPRI), Cairo, Egypt

W. Jabez Osborne
Biomolecules Lab, School of Bio Sciences and Technology, Vellore Institute of Technology, Vellore, India

Kapil Parihar
State Forensic Science Laboratory, Jaipur, Rajasthan, India

Vidya Patil-Patankar
Department of Botany, PDEA's Waghire College, Pune, Maharashtra, India

Richa Raghuwanshi
Department of Botany, Mahila Mahavidyalaya, Banaras Hindu University, Varanasi, UP, India

Himanshu Rai
Centre of Advanced Study in Botany, Institute of Science, Banaras Hindu University, Varanasi, Uttar Pradesh, India

Ved Prakash Rai
Agricultural Research Station, Navsari Agricultural University, Tanchha, Bharuch, India

Asha S. Raj
School of Environmental Studies, Cochin University of Science and Technology, Kochi, Kerala, India

Swarnkumar Reddy
Biomolecules Lab, School of Bio Sciences and Technology, Vellore Institute of Technology, Vellore, India

Vanessa Sánchez-Ortiz
Professional Career in Environmental Engineering, Faculty of Engineering, Private University of the North, Los Olivos Campus, Lima, Perú

Gaurav Sanghvi
Department of Microbiology, Marwadi University, Rajkot, Gujarat, India

Mahipal Singh Sankhla
Department of Forensic Science, Vivekananda Global University, Jaipur, Rajasthan, India

Sakshi Saraswat
Institute of Environmental and Occupational Health Sciences, School of Medicine, National Yang Ming University, Taipei, Taiwan

Mansi Sexena
Department of Forensic Science, Vivekananda Global University, Jaipur, Rajasthan, India

Muskan Singhal
Department of Forensic Science, Vivekananda Global University, Jaipur, Rajasthan, India

Astha Sinha
Biomolecules Lab, School of Bio Sciences and Technology, Vellore Institute of Technology, Vellore, India

Andi Solorzano-Acosta
Laboratory of Microbial Ecology and Biotechnology, Department of Biology, Faculty of Sciences, National Agrarian University La Molina (UNALM), Lima, Perú

Ruchi Soni
Regional Centre of Organic Farming (HQ), Ghaziabad, Uttar Pradesh, India

Srishti Srivastava
Marine Biotechnology and Bio-Products Laboratory, Department of Biomedical Sciences, School of Biosciences and Technology, Vellore Institute of Technology, Vellore, India

Devanshi Sutaria
Dr. B. Lal Institute of Biotechnology, Jaipur, India

K. Suthindhiran
Marine Biotechnology and Bio-Products Laboratory, Department of Biomedical Sciences, School of Biosciences and Technology, Vellore Institute of Technology, Vellore, India

C.R. Vanshree
Department of Forensic Science, Institute of Sciences, SAGE University, Indore, Madhya Pradesh, India

Celia Vargas-de-la-Cruz
Department of Pharmacology, Bromatology and Toxicology, Faculty of Pharmacy and Biochemistry, Centro Latinoamericano de Enseñanza e Investigación en Bacteriología Alimentaria-CLEIBA, Universidad Nacional Mayor de San Marcos; E-Health Research Center, Universidad de Ciencias y Humanidades, Lima, Perú

Anamika Verma
Department of Biotechnology and Bioinformatics, Jaypee University of Information Technology, Waknaghat, Himachal Pradesh, India

Manish Kumar Vishwakarma
Borlaug Institute for South Asia–CIMMYT, New Delhi, India

Chandra Shekhar Yadav
School of Forensic Science, NFSU, Gandhinagar, India

Punam Singh Yadav
Department of Genetics and Plant Breeding, Institute of Agricultural Sciences, BHU, Varanasi, India

About the editors

Dr. Jastin Samuel (PhD) is working as assistant professor and leads the waste valorization research lab at Lovely Professional University, Phagwara, Punjab, India. He has more than a decade of experience in R&D. He has been an active part of DST, EU-DBT, and CSIR-funded projects. He was awarded the prestigious CSIR-SRF fellowship in Engineering (ENG42) in 2013. Along with his team, he has developed a lab-scale mine wastewater treatment system for opencast mines at Orissa, India; established an ETP tertiary treatment facility for two industries (sugar industry and distillery industry) with reuse and business proposition in Andhra Pradesh, India, that is approved by CPCB; established a water treatment facility at Pondicherry in PPP mode with the State Government. He is presently working on pathway assessment and mitigation of micro- and nano-plastics.

Dr. Ajay Kumar is currently working at the Agriculture Research Organization, Volcani Center, Rishon LeZion, Israel. He completed his doctoral research from the Department of Botany, Institute of Science, Banaras Hindu University, Varanasi, India, in 2015. In his 9 years of research, he has published more than 55 research articles and book chapters in international and national journals of repute. He has a wide area of research, especially in microbe interactions, endophytes related to medicinal plants and microbial inoculants. He has already edited some books for Elsevier like *PGPR Amelioration in Sustainable Agriculture* and *Climate Change and Agriculture System* and others that are still in production. Recently, the London Journal Press awarded him the Quarterly Franklin Membership (Membership ID#TM89775) for his significant contribution.

Prof. Joginder Singh is presently working as professor in the Department of Microbiology, Lovely Professional University, Punjab, India. He is an active member of various scientific societies and organizations including the Association of Microbiologists of India and the European Federation of Biotechnology. He has published more than 150 research and review articles in peer-reviewed journals, edited 16 books published by Elsevier Science Publishing, Springer International Publishing, and authored/co-authored 75 chapters in edited books. He serves as a reviewer for many prestigious journals. Dr. Singh has attended several international and national seminars, symposia, and conferences; chaired technical sessions; and presented papers at these meetings.

Preface

Microorganisms have always been remarkable and can unequivocally be considered the ancestors of life on earth, as they have been fixing atmospheric nitrogen for all life on earth, which is one of the wonders of this universe. Microbial life is remarkably diverse and it is estimated that there are 100,000,000 times more microbial cells on the planet than the stars in the observable universe. They are present in all parts of the biosphere and cover the planet. Microbes are vital to every ecosystem. Microorganisms participate in most ecological processes, including production, decomposition, and fixation. They have diverse effects on the ecosystem, considered ecological services. Ecosystem services reiterate human reliance on nature and frame the decisions that emphasize the value of nature to our well-being. They are the direct and indirect contributions of ecosystems to human well-being. Directly or indirectly, they support our survival and quality of life. This book, *Relationship Between Microbes and the Environment for Sustainable Ecosystem Services, Volume 1: Microbial Products for Sustainable Ecosystem Services*, presents advances in sustainable solutions, value-added products, human nutrition, and fundamental research on microbes and the environment. The endeavor of this book is to present the state of the art including more advanced and recent descriptions of the use of microbes for sustainable development. It includes the direct and indirect role of bacteria, fungi, actinomycetes, viruses, mycoplasma, and protozoans in developing products contributing toward sustainable ecosystem services. The book covers the latest biotechnological interventions for harnessing microbial biotechnological aspects on a large scale for sustainable development. This book series will be helpful to scientists, experts, and industry professionals working in the field of microbe-based products. This volume presents authoritative information about recent advances in microbial biotechnological approaches providing sustainable options for future endeavors. It covers reference information ranging from describing various microbial applications for sustainability in different aspects of food, energy, environment, industry, and society. The book includes the latest description of the relationship between microbes and the environment, focusing on their impact on ecosystem services. This volume serves as an excellent reference providing a holistic approach to the most recent advances in applying various microbes as a biotechnological tool for a vast range of sustainable applications, modern practices, and exploring futuristic strategies to harness its full potential.

This volume will be as an excellent reference book for microbial science scholars, especially microbiologists, biotechnologists, researchers, technocrats, and agriculture scientists in microbial biotechnology. We are honored that leading scientists with extensive, in-depth experience and expertise in the use of microbial biotechnology for sustainable practices took the time and effort to contribute these excellent chapters.

We thank the Elsevier team for their generous assistance, constant support, and patience in initiating the volume. We are also grateful to our esteemed friends, well-wishers, and faculty colleagues at Lovely Professional University, India, and Agriculture Research Organization, Volcani Center, Rishon LeZion, Israel.

Jastin Samuel[a], Ajay Kumar[b], and Joginder Singh[c]
[a]Lovely Professional University, India
[b]Volcani Center, Rishon LeZion, Israel
[c]Lovely Professional University, India

CHAPTER 1

Microbial food products: A sustainable solution to alleviate hunger

Daniela Landa-Acuña[a,c], Andi Solorzano-Acosta[a], Vanessa Sánchez-Ortiz[c], Edwin Hualpa-Cutipa[d], Celia Vargas-de-la-Cruz[b,f,*], Bernabé Luis-Alaya[a], and Eduardo Flores-Juarez[e]

[a]Laboratory of Microbial Ecology and Biotechnology, Department of Biology, Faculty of Sciences, National Agrarian University La Molina (UNALM), Lima, Perú
[b]Department of Pharmacology, Bromatology and Toxicology, Faculty of Pharmacy and Biochemistry, Centro Latinoamericano de Enseñanza e Investigación en Bacteriología Alimentaria-CLEIBA, Universidad Nacional Mayor de San Marcos, Lima, Perú
[c]Professional Career in Environmental Engineering, Faculty of Engineering, Private University of the North, Los Olivos Campus, Lima, Perú
[d]Faculty of Pharmacy and Biochemistry, Microbiology Laboratory, Universidad Nacional Mayor de San Marcos, Lima, Perú
[e]Department of Biochemistry, Faculty of Pharmacy and Biochemistry, Universidad Nacional Mayor de San Marcos, Lima, Perú
[f]E-Health Research Center, Universidad de Ciencias y Humanidades, Lima, Perú
*Corresponding author: e-mail address: cvargasd@unmsm.edu.pe

1. Introduction

There is currently a worrying situation that has been increasing over the years—the problem of producing enough food to meet the needs of the world population, which is estimated to reach 9 million inhabitants by 2050 (Godfray et al., 2010), especially in tropical and subtropical regions where the population growth rate is increasing faster than in the rest of the world (Rodriguez and Sanders, 2014). Thus, strategies have been generated and implemented globally, due to the need to improve sustainable agricultural activities, to ensure the current and future demand for food, i.e., to guarantee a stable production that is in accordance with the quality of the environment. Among others, the objectives pursued are food security, eradicating poverty, and conserving and protecting the environment and natural resources (Guerra, 2008).

The solution to this problem is focused on increasing the yield and exceeding the current global capacity to produce food by sustainable practices, highlighting the need to develop new technologies and apply the best technologies known for a long time, such as promoting the growth of microbial plant symbionts, in a more effective way (Bennett et al., 2013), the use of microorganisms and their metabolites in food processing that confers exceptional properties and benefit to human health. On the other hand, advances in the detection, study, and characterization of microorganisms have made a great leap in recent decades, molecular techniques have made it possible to know and gather information on the genome of various microorganisms as well as their amino acid sequence and characterization of their metabolites, with the purpose of not only

Relationship Between Microbes and the Environment for Sustainable Ecosystem Services, Volume 1
https://doi.org/10.1016/B978-0-323-89938-3.00001-3

including microorganisms and their metabolites as part of various foods, but also as useful tools in the application of food safety strategies, as well as models that predict the behavior of microorganisms (Havelaar et al., 2010).

The following is a review of the implication of microorganisms in food production, from the associations they establish with plants ensuring nutrient uptake, food safety, and molecular detection techniques.

2. General aspects of edible microbial biomass safety

2.1 Contextual approaches to nutritional value

According to the World Health Organization, "*Nutrition is the intake of food in relation to the body's dietary needs*", a definition that clings to the achievement of the Sustainable Development Goal of zero hunger.

In this perspective, the WHO warns in its latest report that more than 820 million people continue to suffer from hunger in the world and that the most worrisome projection is that even though this scenario has been unabated for three years, obesity continues to grow.

In an inherent analysis, towards economic, social, food production and consumption aspects, availability and access to food have been recorded, at the cost of generating intensive production in large territories; under unsustainable procedures, allowing to diagnose results of negative impacts on receiving bodies and biodiversity. Parallel results have given conformity in the consumption of ultra-processed foods, which include large amounts of sugars, sodium, and fat, favoring the existence and permanence of chronic non-communicable diseases (Soares et al., 2020).

So, how should we focus our efforts to reduce nutritional deficiencies that limit or shorten the well-being and normal functioning of the organism, in a positive scenario, will there be additional alternatives in food improvement that will enhance the capacity of digestion and absorption of nutrients, and according to the condition of obese and stable weight individuals, will there be a difference in the correct assimilation of nutrients? With the intention of answering the questions, let us contemplate the Japanese vision; the intake of quality-food-related to an exercise routine projects the purpose of sustaining a healthy life that addresses the opportunity of longevity in its inhabitants (Iwatani and Yamamoto, 2019).

In terms of proper nutrition, the Estapà (2006) argues that the inclusion of simple and complex carbohydrates, as well as lipids are listed as facilitators of immediate energy providing a total daily energy between (50%–60%) and (30%–35%) to correspond. While proteins, in their structural function of vital organs and complements should not only consider the intake of 0.8 g/kg/day, but that this should also be mixed; that is to say, of animal and vegetable origin in equal proportions. In this environment, vitamins, and essential chemical elements such as phosphorus, calcium, sodium, and potassium should be ingested as recommended since excess could also be a precursor of diseases.

Finally, vegetable fibers, mostly non-absorbable carbohydrates, and lignin, as well as consumables derived from fermentation (saprophytic bacteria) make up the judicious scheme of an adequate diet.

Although the concept of nutrition is still focused on the variables "obtaining energy from food", the considerations on the deficit of macronutrients such as vitamins, minerals, and essential amino acids have encouraged inquiries related to whether what we consume really conserves all that is required (Corio Andújar and Arbonés Fincias, 2009). Namely, from the processes linked to the conservation and/or preservation that food requires; terminologies such as dissociation or piezo-stability for proteins and vitamins respectively, the verification of the percentage of nutritional quality that we can lose in consumables from processes often linked to safety arises as an imperative background (Mor–Mur, 2010).

In attribution to this, functional foods, or nutraceuticals, are postulated as the new initiative of substantial utilization. From providing the content of the food with compounds of nutritional value as a fortifier or subtracting those that are potentially intolerant to the consumer (Ros, 2001).

According to Rocandio Pablo and Arroyo Izaga (2001), the attribution of the term "functional" to what we know as foods does not have a categorically accepted definition. However, the International Food Information Council (IFIC) relates the circumspect term "functional food" to that which provides health benefits through nutrition, while the European Consensus (EC) defined it as that which beneficially affects one or more functions of the organism including the reduction of disease risks.

In these instances, the suggested modifications lean towards biotechnological applications from the most incipient ones such as dairy-derived production and fermentation to the most complex ones such as the use of transgenics (Muñoz, 2008). Similarly, Ros (2001) exemplifies that, although the greatest effort lies in genetic changes aimed at reducing pests, the greatest challenge is to produce plants that synthesize more photochemical of added value for health.

In the evolutionary diligence of biotechnological aspects, according to Olarte (2014) transgenics as a public and global issue are framed in the definition of; modifying, eliminating, or inserting genes inside the DNA of a living being of the same species or of another, a reality that has generated conjectures linked to myths and truths about the optimization of processes and products.

From the published references, the requirement of confirmatory answers attributed by Meléndez Illanes et al. (2013), who in the analysis of cases question whether food is at the service of health or the business of the food industry, is highlighted, and it is that, in attribution to the evidence these must have the ability to be reproducible so that there is no ambiguity in the results of benefits issued in advertising and that in contrast a range of new options for continuous improvement in nutritional aspects is glimpsed.

In this regard, the adequate nutritional value may not be subject to extreme modifications, perhaps the initiatives observe the use of natural components as main mechanisms in the food supply, hence the interaction of physical principles between the stress and deformation of matter "Rheology" (fluid movement) (RAE), may be the link that annexes the participation of microorganisms, as an opportunity to generate food.

In short, we agree with Aranceta-Bartrina (2010) that in the short term, the need to determine nutritional requirements at specific levels will be feasible in terms of results based on gene expression subject to food design models. While, in the field of nutrigenomics, the articulation of functional foods, fortified foods, probiotics, transgenic foods, new foods and pharmacological supplements are ingredients applicable to new research, nanotechnology is projected as the configurator of new reference frameworks of possibilities not yet considered.

2.2 Profit dynamics

Over time and in the context of empirical knowledge, the perception of nutritional gains in foods processed under natural fermentation conditions was an option for consumption in terms of alleviating gastrointestinal problems. In the same direction and under scientific scrutiny, it would later be recognized that the participation of microorganisms not only intervened in the increase of the nutritional contribution of a specific food, but also allowed in collateral scenarios; benefits in the balance of the internal systems of the human being (Olveira and González-Molero, 2016).

Considering the initial identification of lactic acid bacteria (LAB), present in fermented milk at the end of the nineteenth and beginning of the twentieth century, it was demonstrated that *Lactobacillus*, *Leuconostoc*, *Lactococcus* and *Bifidobacterium* actively participate in the production of organic acids, allowing pH stability and product safety. In this sense, it was recorded that the production of metabolites during milk fermentation allowed the integration of a microflora capable of producing changes in flavor, texture, appearance, color, aroma, and nutritional properties of the resource, resulting in a wide variety of products for consumption (Domínguez González et al., 2014).

It is pertinent to emphasize that, the viability of the properties attributed to microorganisms is parameterized in the brief control of the shelf life of food and the analysis of the physicochemical and organoleptic properties of a product also recognized as "probiotics" (Soares et al., 2019).

According to Serra (2016), bifidobacterial strains were able to express induction of immune responses in an anti-inflammatory profile correlated to irritable bowel syndromes. This shows that bacterial flora is rooted in modulatory functions associated with digestion. Of the many guarantees that are attributed to these products, it is feasible to highlight among the most important the regulation of intestinal functions and stimulation of the immune system, allowing the latter to obtain macrophages and antibodies (Ballesta et al., 2008).

2.3 Functional properties: Probiotics-prebiotics-symbiotics

Today, more than 1500 trials on probiotics and about 350 on prebiotics are recognized. However, these publications include different variables in their study and the accumulated evidence supports the opinion that the benefit is measurable in many parameters (Organización Mundial de Gastroenterología, 2011). So, against this background, it is easy to ask: What are probiotics, prebiotics and why are they of interest to research?

In the scope of the WHO scientific experiences, it is mentioned that the term Probiotics, implies referring to live microorganisms that when administered in adequate quantity exert a beneficial effect on the health of the host, while Prebiotics are defined as selectively fermented ingredients that allow changes in the composition and/or activity of the gastrointestinal microbiota, thus providing health benefits to the host (Organización Mundial de Gastroenterología, 2011).

In addition to admitting that probiotics are not toxigenic pathogens and can additionally survive the acidic environment of the stomach and the effect of bile in the duodenum (Dunne et al., 2001), comment that participation has also been observed without displacing the existing native microbiota.

While Ramírez et al. (2018), records that the most studied probiotics are lactic acid producing bacteria, such as *Lactobacillus*, and *Propionibacterium* and *Bifidobacterium* species. Valdovinos-Díaz (2013) refers in terms of molecular biology; it has been identified that the number of bacteria in an individual is 10 times greater than that of human cells and that in the interaction with antibiotics the suppression of all bacterial groups is obtained, in contrast the application of *Saccharomyces boulardii* in antibiotic dosage effectively reduces the changes in the intestinal microbiota.

In the characterization of bacterial species populations, the approximations register from 300 to 400 species, considering that only between 30 and 40 of them coincide for 99% of the population. In correlative aspects, it has been demonstrated that in the newborn the participation of bifidobacteria potentially inhibit the growth of pathogens and are involved in the production of vitamins of group B and folic acid (Narbona López et al., 2014), while Lemale (2014), attributes that the enrichment of milk with *Bifidobacterium lactis* or *Lactobacillus reuteri* register significant reduction of ulceronecrosing enterocolitis in low-weight preterm infants. Thus, according to Bover Cid et al. (2001), from a habitual point of view, feeding does not involve an act of survival, but rather reinforces the need to generate protection against chronic diseases of persistent incidence.

In this regard, it is recorded that the dynamics of work between *Lactobacillus acidophilus* and *B. lactis* (5 × 109 cfu), reduces cases of pollen allergy in children; from the inhibition of access of eosinophils *via* nasal mucosa, while *L. fermentum* (1 × 109 cfu) or placebo associated with probiotics, predisposed Th1 lymphocyte increase to interferon gamma in children with atopic dermatitis (Rueda-Rodríguez et al., 2016; Ouwehand et al., 2009; Prescott et al., 2005).

On the other hand, from the preventive point of view and with the intention of attending to the treatment of inflammatory intestinal diseases, the evaluations of nutrition and interaction of food with the genome give the opportunity to identify advances in the use of prebiotics and probiotics as biological systems of gene stimulation in the control of cancer risks (Peña, 2016).

According to, Valdovinos-García et al. (2018), in a survey of gastroenterologists and nutritionists; one of the characteristics that predisposes the prescription of this element is constituted by the recognition of strains analyzed in a clinical study for the specific symptom or disease. The evaluation connoted the acceptance of 44.33% (141) gastroenterologists and 34.30% (106) nutritionists. Finally, the findings also confirmed that 97% of the participants fully or partially agree that probiotics are safe and confer no health risks.

Likewise, the nature of prebiotics as oligo- or polysaccharides represent the opportunity to be growth-stimulating and opportune substrates for hydrolysis or fermentation by bacteria such as bifidobacteria and lactobacilli (Lemale, 2014), the interactions of both agents are relevant in the continuous improvement of the utilization and assimilation of benefits in the nutrition process.

According to Ramírez and Gordón (2014), in a symbiotic balance of probiotics (saprophytic bacteria and yeasts) and prebiotics (compounds that stimulate the microbiota), the beneficial interaction of resistance to potential pathogens and their inclusion as precursors of the local or systemic immune response, increases the qualitative and quantitative scope of future research. Consequently, for the confirmation of benefits administered independently by probiotics, prebiotics, or symbiotic, it is still required that several researchers involve their efforts in the approach of objectives; based on the confirmation of their preventive benefits, of sustenance in the dietary routine and of possible interactions to the strengthening of recuperative treatments.

3. Safety of edible microbial biomass

As nutritional deficiencies express in the population the need for new strategies available to contribute essential elements and protection against highly adaptable microorganisms, the question prevails in the assumption of whether more than one species of this group can provide nutrients and natural immunity. A clear example of the edible safety of microbial biomass is Kefir, the scopes of research to date present Kefir as a highly nutritious food between proteins, mucopolysaccharides, vitamins B, K, tryptophan, Ca, P and Mg, in addition to a low proportion of fat (12%) (Rosa et al. (2017); Pereira et al., 2017).

Whose versatility in its preparation has been described by research by Pereira et al. (2017), as having a minimum irregular shape (3 cm), besides being an input in the fermentation of milk, which at the end of the process offers us a texture similar to yogurt. The authors highlight the importance of unlimited varieties of cultural media such as cow, goat, sheep, camel, buffalo, peanut, soybean, and rice milk. Then, if it is desired to generate food production based on these microscopic populations, the homemade or artisanal formula includes combining any of the mentioned culture media together with kefir grains in an interval of 8°C to 25°C for 24 h, after which it is possible to obtain derived products such as kefir, Greek yogurt, and buttermilk. According to the authors, the food evidence easily digestible proteins and concentrations between $1220 \pm 88\,\mu g\,mL$ of phosphorus (P) in whey, while for iron (Fe) between $6.7 \pm 0.2\,\mu g\,g$ in kefir and $8.9 \pm 0.2\,\mu g$ in Greek yogurt. Finally, for Greek yogurt they corroborated percentages of calcium (Ca) (21.2%), phosphorus (P) (59.7%), and iron (Fe) (22.3%).

According to Kabakcı et al. (2020), the recognition of digestive dynamics related to cholesterol reduction, obesity, and increased life expectancy, have made kefir over the years, a product with high demand in proactive consumers in reducing gaps related to carcinogenic risks or simply a life with quality limitations. In the quote, the contribution of Vitamins B12, B1, B2, and B6 is highlighted, as well as essential amino acids (phenylalanine, tyrosine, leucine, and glycine) and minerals (magnesium, potassium, and calcium).

Likewise, several investigations conclude that obesity and type 2 diabetes express delayed glucose and lipid absorption through the inhibition of lipid and carbohydrate hydrolyzing enzymes, namely α-amylase and lipases, in the digestive organs (Lee et al., 2013; Tiss et al., 2020; Vieira et al., 2017). The above was a reason for (Tiss et al., 2020), to motivate the comparative in vitro testing between unfermented soy milk and kefir-fermented soy milk to know if these foods could be related to enzymatic activity. On the other hand, although many of the diseases that are currently foreseen are related to microorganisms, according to Lopitz-Otsoa et al. (2006), the positive association of yeasts and bacteria in kefir grains are evidence of probiotic formation that can generate random improvements in resistance to colonization and immunomodulation of the gastrointestinal microbiota in the habit of a balanced diet. Under the symbiotic description of agents such as *Lactobacillus, S. cerevisiae, Kluyveromyces lodderae, K. marxianus,* and *Candida humilis* presented by the authors, this acid fermented milk with light carbonated and alcohol components according to reference, is postulated as an accessible food component that contributes to lactose tolerance, stimulates gastric motor function, and colonizes the intestine.

In a particular mention of the species *Lactobacillus casei,* evidence of survival at the level of the intestinal tract and immunostimulant effect in the significant increase of secretory IgA is reported (Tormo Carnicer et al., 2006). On the other hand, in terms of multiplicity of microbial flora, Wang et al. (2021) report that, in adjudication of the microbiome, kefir gathers varieties of predominant genera interrelated among bacteria such as *Lactobacillus* and yeasts such as *Saccharomyces, Kazachstania, Kluyveromyces,* and *Pichia,* which ratifies the mention made by (Lopitz-Otsoa et al., 2006).

In estimation of the emerging scenario of new viruses and the limitation of antiviral drugs, the appreciation of Hamida et al. (2021), cites the interest of interacting probiotic products with antiviral agents, in the context that kefir enhances the development of antiviral cytokines and dendritic cells derived from human monocytes so that they can be applied as antivirals and anticancer agents. In this sense, the meta-analysis performed by the authors compiled a correspondence between antiviral activity and probiotic agents identified in the composition of kefir.

In the first analysis, the representativeness of the genus Lactobacillus included the following species; *Lactobacillus casei*—Rotavirus, *Lactobacillus brevis*—*Herpes simplex* virus type 2 (HSV-2), *Lactobacillus plantarum*—Echovirus E7 and E19, Influenza virus H1N1, Coxsackie virus, Influenza virus, Seasonal and Avian Influenza viruses, *Lactobacillus acidophilus*—Hepatitis C, Influenza virus, Rotavirus, Coxsackie, *Lactobacillus gasseri*—Influenza A virus, Spiratory syncytial virus (RSV), *Lactobacillus crispatus*—HSV-2, *Lactobacillus amylovorus*—Echovirus E7 and E19, *L. rhamnosus*—Influenza virus, *Herpes simplex* virus type 1, Coxsackie, *L. sakei*—Salmonid viruses, *L. reuteri*—Coxsackievirus A and Enterovirus 71.

The Lactococcus genus was less representative, registering *Lactococcus lactis* subsp. lactis—Feline Calicivirus, norovirus (NV), Herpes simplex virus 1 (HSV-1), Poliovirus (PV-1) and *Lactococcus lactis* subsp. cremoris—Influenza virus.

In retrospect, it is worth noting that the intent of the analysis was structured on the attempt to find a sustainable solution to alleviate hunger in the power of microbial food products. In this episode, the gap analysis on the health consequences in an extreme population (malnourished and obese) exposed to loss of quality of life seems to have a common solution.

Based on what has been mentioned by the authors, kefir and products including probiotics as a symbiotic food with participation of varieties of microorganisms, is postulated as a nutritional alternative that increases defenses and collaborates with the removal of the characteristics of obesity. Including aspects of simplicity in its preparation, the products originated in fermentative activities in correlation to the efficiency of these organisms refer a reliable and nutritious source in the line of defense, preventive and collaborative in the treatment of emerging diseases not always visible or detectable in time.

On the other hand, from another perspective, microorganisms have widely contributed to the generation of integrated ecosystems, closely associated with plants, generating consortia and physicochemical relationships, in benefit of the correct absorption of nutrients, as they possess mechanisms that allow the release of different nutrients that are retained in the soil. For this reason, this microbial versatility has been mentioned below, and its study has allowed discovering new capacities that can be included in the generation of edible biomass:

3.1 Microorganisms and their plant-microorganism-soil relationship

The use of beneficial microorganisms in modern agriculture plays an increasingly significant role in food security, such as plant growth-promoting bacteria, nitrogen-fixing bacteria, phosphate solubilizing microorganisms, arbuscular mycorrhizal fungi, etc., which contribute significantly to maintaining the fertility of agricultural soils (Dohrmann et al., 2013). A simple way to explain the physiology of plant nutrition is that plants transfer the available and dissolved nutrients in the soil through their specialized conduction system (vascular system), carrying these nutrients to all the edible organs of the plant that are used for human nutrition, which is why microorganisms play an important role as they help to capture nutrients from the soil and we call these microorganisms symbionts (Campaña-Olaya et al., 2017).

The available information shows that so far only a minimum percentage, varying between 1% and 10% of all microbial diversity, has been cultivated by in vitro techniques of classical microbiology (de los Santos Villalobos et al., 2018), which is why the study of microorganisms represents a very broad opportunity that needs to be investigated to be used as biotechnological tools to increase the productivity of agricultural crops in an environmentally friendly and cost-effective way (Kalia and Gupta, 2005). A great deal of research has been reported that has demonstrated the important role played by microorganisms in the growth, development, yield tolerance, and protection against plant pathogens at greenhouse and experimental levels, however it is vital to continue with research and above all to standardize and firmly position their use at the level of agricultural fields destined for human consumption to guarantee food security (Landa-Acuña et al., 2020).

In this section, we want to show the importance of the study of microorganisms and how fundamental is the continuous discovery and subsequent application of new microbial species by the scientific community, both at the experimental, teaching, and business levels. These experiences on the biotechnological potential of microorganisms will serve to strengthen and create new highly specialized institutions focused on continuing research, applying new technologies and improving the agrobiotechnology potential of microorganisms to guarantee food security in the most vulnerable countries, in a context of a solid economy, greater scientific-technological development and awareness of the importance of safeguarding microbial communities, with emphasis on native microorganisms.

3.2 Microbial food safety in relation to plant-microorganism-soil

3.2.1 Bacteria in agriculture

Among the best-known roles of bacteria is their link to the biogeochemical cycles (nitrogen, carbon, sulfur, and phosphorus), which participate in the recycling of various bioelements of vital importance for plant nutrition, solubilization of mineral elements such as potassium, calcium, magnesium, etc. On the other hand, bacteria play roles in physiological processes of great importance in plants such as photosynthesis and metabolic processes such as degradation of xenobiotic compounds, control of plant diseases and finally processes related to the maintenance of the structure and function of the soil (Pal and McSpadden, 2006; Pankratova, 2006; Ryan et al., 2008).

For many years it has been reported and is still being investigated and applied on the use of bacteria for agricultural benefit, in this sense, various bacterial genera, such as: Pseudomonas, Bacillus, Pseudomonas, among others, have been worked for their metabolic characteristics of producing phytohormones, involved in root elongation and greater use of nitrogen or in the production of organic acids that contribute to neutralize soil pH, favoring the solubilization of insoluble phosphorus (Kathiresan et al., 1995; Campaña-Olaya et al., 2017).

In this context, strategies have been developed such as the application of bacterial inoculants with diverse metabolic capacities of interest, focused on complementing better agricultural management through the partial or total substitution of synthetic agricultural inputs. Specialized institutions such as the laboratory of microbial ecology and biotechnology of the Universidad Nacional Agraria La Molina have developed several studies focused on the characterization and application of beneficial bacteria in the agriculture of various crops such as potato, quinoa, cocoa, coffee, beans, among others (Ogata-Gutiérrez and Zuñiga, 2020). It is therefore important to focus studies on native spices where edaphoclimatic and crop conditions are better and where favorable results have generally been obtained using microbial inoculants in the agricultural productivity of a nation, thus encouraging the involvement of the productive sector in the use of these microorganisms.

3.2.2 Bacteria and agrochemicals

Agricultural soil is a type of soil that is used for productive activities of various crops, considered of great importance in terms of environmental decontamination since plants, apart from absorbing nutrients such as CO_2, also absorb different polluting compounds from the soil (Soto et al., 2014). To achieve greater productivity, it has been necessary to use agrochemicals for the control, reduction, and elimination of various pests and diseases; however, the consequences of their excessive use bring with them serious problems since they are considered toxic substances, which can have negative effects on soils, human health, and the environment, affecting crop production and consequently food security (Jaramillo Colorado et al., 2016).

More than 50% of agricultural soils are contaminated by the excessive use of agrochemicals, mainly due to the lack of guidance for the management of these compounds used to control pests and plant diseases, generating health problems and a decrease in soil nutrients. It can be deduced that the mismanagement of agrochemicals such as pesticides and fertilizers are detrimental to soil conservation and food security (Durán and Ladera, 2016), which is why in countries such as Venezuela and Thailand they have banned the use of certain agrochemicals due to the harmful effects that this generates (Jara-Peña et al., 2017).

There are several alternatives to reduce the degradation of the affected soil, one of them is the introduction of bacteria that degrade toxic compounds to replace agrochemical compounds (Molina-Montenegro et al., 2016). For this reason, several investigations have been carried out which indicate that bioremediation using bacteria is a promising alternative to

Table 1 Degradation of agrochemicals by different bacteria.

Agrochemical	Bacterial species	Authors
Organophosphates	*Bacillus* sp. *Pantoea agglomerans* *Pseudomonas aeruginosa* *Enterobacter* sp. *Bacillus* sp. *Vibrio metschinkouii* *Pantoea agglomerans* *Serratia ficaria* (A3) *Pseudomonas rhizophila*	Marín and Jaramillo (2015); Jaramillo Colorado et al. (2016); Hassen et al. (2018)
Organochlorines	*Pseudomona* sp. *Klepsiella* sp. *Pseudomona* sp. *Entetobacter* sp. *Pseudomona* sp. *Klepsiella* sp.	Soto et al. (2014); Durán and Ladera (2016)
Chlorothalonil	*Pseudomona* sp. *Klepsiella* sp. *Enterobacter* sp.	Botero et al. (2011); Marín and Jaramillo (2015)
Carbofuran	*Sphingomonas* sp. *Pseudomona* sp. *Streptomyces* spp.	Jaiswal et al. (2017)

recover agricultural soils. Studies on bacteria indicate that their ability to absorb herbicides in agricultural soils shows the positive effect that bacteria of this type have on the degradation of agricultural soils, since they release nutrients and associate minerals with each other, thus regenerating soil nutrients for planting (Akintui et al., 2015).

Four agrochemicals have been reported to be harmful to agricultural soils: organophosphates, organochlorines, chlorothalonil, and carbofurates. Different research on the microorganisms used has shown the positive effects of bacteria for the degradation of agrochemicals (Table 1).

3.2.3 Mycorrhizae in food safety

Among the most common symbiotic organisms at the rhizosphere level, arbuscular mycorrhizal fungi (AMF) are the ones that colonize about 90% of terrestrial plants (Toro et al., 2017). The mycorrhizal symbiosis generated has recently gained interest as a natural biofertilizer due to the multiple benefits it provides to the plant, including tolerance to salinity, drought, heavy metals, and the uptake of phosphorus, a nutrient that is not very mobile in the soil (Landa–Acuña et al., 2020). In leguminous plants it has been demonstrated that mycorrhization increases the capacity of nitrogen fixation in joint association with bacteria of the genus Rhizobium, likewise, when there is a low level of phosphorus, there is a low level of phospholipids in the plant membrane, which leads to a greater root exudation, which brings a better stimulation in mycorrhizal colonization (Liriano et al., 2012).

At the environmental level, arbuscular mycorrhizae contribute to increased crop productivity, regeneration of degraded plant communities, maintenance of ecosystem balance, as well as mitigating the negative effects caused by the presence of heavy metals (HM) in the soil such as Cd, Pb, Zn, and Cu, among others (Audet, 2014). This last contribution has been studied due to the important results in terms of food safety, heavy metals are accumulated in the roots by minimizing their content in the aerial parts of the plant (which are mainly edible) (Hristozkova et al., 2017), or by immobilizing heavy metals in the rhizosphere through soil proteins associated with glomalin, a process known as phytostabilization, which reduces the bioavailability of these elements and thus their toxic effects ((Ferrol et al., 2016); Toro et al., 2017).

Mycorrhizal fungi enrich crops by their contribution of nutrients and provide an acceptance of heavy metals in the terrestrial environment, this is demonstrated in experiments where their good absorption capacity of the pollutant decreases favorably when using these fungi (Ning et al., 2019). Using the information obtained, some fungi involved in bioremediation of soils with crops, which are damaged by heavy metals, were selected (Table 2). It is interpreted by the results that cadmium is the most frequent metal in crops; and the fungus that has been used the most is of the higher ranking Glomus.

At the economic level, mycorrhizal fungi contribute substantially to the efficient use of fertilizers, and at the social level they contribute to integrated rural development, with the use of natural resources (development of native inoculums) at the local scale, thus favoring the establishment of sustainable production agroecosystems (Toro and Andrade, 2020). Successful experiences mention that by using native *Glomeromycota fungi*, phosphorus fertilization can be reduced by up to 33% without affecting cereal production and yields (Cabrales et al., 2019), on the other hand the isolation of native mycorrhizal fungi is suitable

Table 2 Mycorrhizal fungi in heavy metal bioremediation processes.

Mycorrhizal fungi	Heavy metals	References
Fungal Mycorrhizae Arbuscular Formers (HMFA)—natives *Rhizoglomus intraradices* *Glomus macrocarpum*	Cd	Pérez et al. (2019)
Funneleformis mosseae *Rhizophagus intraradices*	Cd	Li et al. (2016)
Glomus intraradices *Glomus mosseae*	Cd and Pb	Bano and Ashfaq (2013)
Glomus intrarradices *Glomus desertícola* *G. mosseae*	Pb	Alvarado et al. (2011)

for phytoremediation purposes and can also serve as a potential biotechnological tool for the successful restoration of degraded ecosystems and can be appropriate to produce *Glomeromycota* fungal inoculate to apply to crop fields at a lower cost (Ferrol et al., 2009).

Due to this background, the use of arbuscular mycorrhizal fungi fits very well in the multiple objectives pursued by sustainable agriculture and together with other mycorrhizal organisms such as bacteria, i.e., mixed consortia show a biotechnological potential that supports food security in current and future times.

4. Microorganisms of potential use as food and possible routes to produce edible microbial biomass

Agriculture constitutes the most important milestone of human technology in our history (Peters and Morgan, 2004), it has made possible the development of culture and population expansion through food security (Johnston and Mellor, 1961); However, the size of the current population and its constant growth have exposed its limitations associated with the management of fresh water, eutrophication, pollution, an increase in plagues and phytopathogenic diseases associated with monoculture and loss of biodiversity, which exert an irreparable environmental footprint on our planet (Kishor et al., 2020; Strokal et al., 2016). In this way a vicious circle has originated in the production of food where each action of man ends up contributing to climate change and the alteration of ecosystems, directly affecting the development of crops that serve us of food, increasing adverse biotic and abiotic factors (Waraich et al., 2011).

Special mention should be made of livestock in obtaining proteins, which according to some authors a dietary transition towards predominantly plant-based foods could significantly alleviate the pressure on the environment; but the change in eating habits is another factor against it (Goldstein et al., 2017; Springmann et al., 2018; Lynch et al., 2018). At this juncture, there is an urgent need to consider alternatives to conventional means of food production (Van Huis and Oonincx, 2017; Sonnino and Marsden, 2006). Some of the alternative strategies include using microorganisms for food, which have been used since at least several centuries ago, however, have received significantly less attention (Thompson et al., 1999). So, this review will present some of the microorganisms that are potential food sources (Caplice and Fitzgerald, 1999).

4.1 Microorganisms of potential use as food

Microbial biomass of potential use in food can be produced from a variety of microorganisms such as bacteria, fungi, and algae (Boze et al., 1995). Many derived products can be developed from the microbial biomass given its compositional variety: carbohydrates, lipids, proteins, nucleic acids, vitamins (Suman et al., 2015). Microbial biomass can be used to develop supplementary products for malnutrition and other applications for its lipid and protein content (Table 3) (Matassa et al., 2016).

Table 3 Uses of components of microbial biomass.

Component	Derivative	Uses
Proteins	Functional protein	Gels, thickeners, and emulsifiers
	Nutritional protein	Food
	Enzymes	Biotransformation
Carbohydrates	Saccharides	Fermentation substrates
	Conversions	Emulsifiers
Lipids	Fats	Food
	Phospholipids	Food
	Carotenoids	Food
	Glycerides	Food and cosmetics
Nucleic acids	Nucleotides	Flavors and gelling agents
	DNA	Drug synthesis precursors
	RNA	

Table 4 Average percentage of nutritional composition on a dry basis of microorganisms.

Microorganisms	Protein	Fat	Minerals	Nucleic acids
Bacteria	72–83	1–3	7–8	8–16
Yeast	45–56	2–6	5–10	6–12
Algae	40–63	7–20	10–15	3–8
Filamentous fungi	30–50	2–8	5–10	7–10

The microbial biomass contains as a major component protein, which is present in percentages greater than 40% on a dry basis, with bacteria having the highest percentage of protein, followed by yeasts and algae (Table 4) (Kuhad et al., 1997). The quality of the protein can be evaluated through its amino acids profile, in this sense the microbial protein shows high levels of lysine and low contents of cysteine and methionine, which makes it like soy protein (Nasseri et al., 2011). Microbial biomass, due to its nutritional composition, is a suitable raw material for the development of a wide variety of products that can be used for human and animal nutrition (Horwath and Paul, 1994).

4.2 Microalgae

Microalgae are unicellular organisms (protists), cenobial (made up of several cells without division among themselves) and thalophytes (unstructured in root, stem, and leaves) that live in fresh or marine waters, and that were provided with assimilation pigments (euglenophytes, chrysophyceae, pyrrophyceae, xanthophyceae, among others) (Tytler et al., 1997). They are a diverse group of photosynthetic microorganisms, this means that they are capable of capturing sunlight to synthesize energy exchange molecules such as APT or ADP, that are used by these organisms to produce other molecules such as carbohydrates,

Table 5 Derivative products and uses of microalgae biomass.

Type	Product/use
Biomass	Food
	Animal feed
	Aquaculture
	Soil conditioner
Colorants	Astaxanthin
	Phycocyanin
	Phycoerythrin
Antioxidants	B–carotene
	Tocopherol
	ARA
	DHA
	Polyunsaturated fatty acids
Special products	Toxins
	Isotopes

lipids, or proteins. Microalgae are expected to develop an important role in the food industry in the future. Photosynthetic microorganisms, such as cyanobacteria and microalgae, can be cultivated for food and feed independently of arable land (Muller-Feuga, 2013).

Microalgae began to be used in the diet since 1950, mainly as dietary supplements (protein and vitamin), in powder, capsules, pills, or tablet. These are usually incorporated in foods such as pasta, cookies, bread, candies, yogurts, or soft drinks (Christaki et al., 2011). Currently today, it is estimated that approximately 30% of the algae produced in the world is used in human nutrition due to its high protein content (Table 5). The most important component of algal biomass are proteins, which can represent up to 50% of the total dry weight and which, added to lipids and carbohydrates, constitute up to 90% of the total dry weight, while minerals, nucleic acids, pigments, and other minor components make up the remaining 10% (Table 6) (Gantar and Svirčev, 2008).

Table 6 Chemical composition of different microalgae as a percentage of the total dry weight.

Species	Protein	Carbohydrates	Lipids
Scenedesmus obliqus	50–56	10–17	12–14
Scenedesmus dimorphus	8–18	21–52	16–40
Clamydomonas rheinhardii	48	17	21
Chlorella vulgaris	51–58	12–17	14–22
Spirulina maxima	60–71	13–16	7
Dunaliella salina	57	32	9
Tetraselmis suecica	39	8	7
Isochrysis galbana	41	5	21

Table 7 Bromatological composition of wild mushrooms.

Species	Humidity	Fats	Minerals	Raw protein	Raw fiber
Agaricus bisporus	91.4	0.3	0.8	1.8	2
Amanita caesarea	93.8	–	0.7	0.8	1
Boletus edulis	90.8	0.5	0.6	1.7	2.1
Pleurotus ostreatus	92	0.4	0.9	1.6	–
Pleurotus spp.	92.4	–	0.6	1.2	1.7
Rmaria botrytis	92.7	–	0.6	1.1	1.7

4.3 Bacteria

These microorganisms present an advantage that they are capable of doubling in a period of 20 to 30 min and have high content of proteins that can reach 85%. Among the genera of bacteria most used to produce protein can be named: *Methylomonas, Pseudomonas, Bacillus* and *Aerobacter* (Bhalla et al., 2007).

4.4 Filamentous fungi

They constitute an attractive source of biomass and microbial protein production, due to their easy harvest and low cost of production; in addition to its ability to produce textured foods, acceptability of consumption and that crop residues are used as substrates to produce these microorganisms (Punt et al., 2002). The filamentous fungi mostly used to produce microbial protein are from the genera *Agaricus, Pleurotus, Lentinula, Auricularia, Volvariella,* and *Flammulina* (Kües and Liu, 2000). Studies on the nutritional properties of some species have increased interest in their consumption, they have a high content of water, protein, and fiber, as well as a great antioxidant power, due to the high content of phenolic compounds (Table 7) (Zied and Pardo-Giménez, 2017).

4.5 Yeast

The first microorganisms used as a source of biomass as unicellular protein were yeasts (Table 8), among them the most representative is *S. cerevisiae*, which represents one of the yeasts of first choice for the industrial production of biomass and ethanol and since

Table 8 Yeast species and substrates used in biomass production.

Yeast	Substrate
Candida utilis	Sulfite liquors and ethanol
Candida novellus	N-alkanes
Candida lipolytica	N-alkanes
Kluyveromyces fragilis	Cheese serum
Saccharomyces cerevisiae	Molasses
Fusarium graminearum	Glucose
Gliocladium deliquescens	Corn liquor

they are lacking b-galactosidase, amylase and glucoamylase activity, budding yeasts are unable to ferment starch (Amata, 2013).

5. Production of edible microbial biomass

Edible microbial biomass derived from bacteria, yeasts, filamentous fungi, or microalgae, is a promising alternative to conventional food sources because they are a source of proteins, vitamins and, in some cases, also lipids (Linder, 2019). Furthermore, due to its ability to use simple organic substrates, it can be produced on an industrial scale and in places that would not compete with agricultural production. However, the substrate used for the growth of the microorganism must be taken into consideration, which will be decisive in its composition (Sankaran et al., 2010).

5.1 Advantages and disadvantages of edible microbial biomass production

The biomass produced by microorganisms, compared to the protein produced conventionally by plants and animals, offers advantages such as that it can be based on substrate from agricultural by-products, which have high levels of protein, have a short generation time, and can be cultivated in a small space, without considering of course that its nutritional composition can be controlled by genetic manipulation (Tables 9 and 10) (Nasseri et al., 2011).

5.2 Possible routes of conversion of CO_2 into microbial biomass

As established by Serrano-Ruiz and Dumesic (2011), the possible routes would be the following:
- Direct conversion of CO_2 into biomass by photoautotrophic microorganisms such as cyanobacteria or eukaryotic microalgae.
- Direct conversion of CO_2 into biomass by H_2 oxidant chemoautotrophic bacteria.

Table 9 Advantages of edible microbial biomass production.

Advantages	
Fast doubling time (Growth rate)	Bacteria 0.3–2 h Yeast 1–3 h Algae 2–6 h Filamentous fungi 4–12 h

High feed conversion rate (grams of protein per kilogram of substrate)
It can be produced in agriculture by-products
Ease of genetic improvement and manipulation
No interference from climatic factors when growing under controlled conditions
Production in reduced and efficient areas
High content of proteins and vitamins in biomass

Table 10 Disadvantages of edible microbial biomass production.

Disadvantages	
There is a cultural rejection to eat microorganisms especially in the West	
Unappetizing organoleptic characteristics	
Possibility of contamination with heavy metals due to the substrates	
High content of nucleic acids	Algae 4%–6%
	Bacteria 10%–16%
	Yeast 6%–10%
	Fungi 2.5%–6%
Slow or no digestion of the cell wall of some microorganisms	
It requires a process of concentration	

- Catalytic conversion of CO_2 to methanol which is then used to grow methylotrophic bacteria or fungi.
- Electrocatalytic conversion of CO_2 to formic acid which is then used to grow methylotrophic bacteria or fungi.
- Biocatalytic conversion of CO_2 into acetic acid using bacteria capable of hydrogenotrophic acetogenesis (synthesis gas fermentation). The resulting acetic acid is used to grow acetotrophic bacteria or fungi.
- Biocatalytic conversion of CO_2 into acetic acid by bacteria capable of electrotrophic acetogenesis. The resulting acetic acid is used to grow acetotrophic bacteria or fungi.
- Biocatalytic conversion of CO_2 into methane using archaea capable of hydrogenotrophic methanogenesis. The resulting methane is used to grow methanotrophic bacteria.
- Biocatalytic conversion of CO_2 into methane using archaea capable of electrotrophic methanogenesis. The resulting methane is used to grow methanotrophic bacteria.

5.3 Microbial biomass production process

The production of microbial biomass consists of generating the cell mass of a microorganism from a substrate. The process in general requires a source of carbon, nitrogen, essential nutrients, and oxygen to ferment, that allows the massive multiplication of the crop. If any of these conditions are deficient in the production medium, it must be supplemented to achieve adequate microbial growth (Fig. 1) (Miltner et al., 2012).

The general fermentation process to produce microbial biomass begins with the selected strain that becomes an inoculum and enters the fermenter, which contains the substrate to be fermented. The substrate must be conditioned by particle size reduction and homogenization, pH adjustment, and sterilization. The fermenter must be designed to function aseptically during the process and have an aeration and agitation system that allows meeting the metabolic needs of the microorganisms. Once the

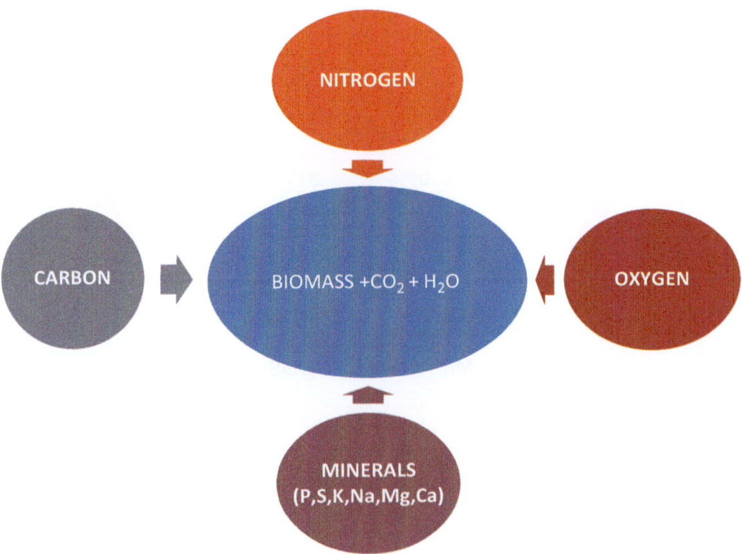

Fig. 1 Biomass production process.

fermentation is finished, the biomass is recovered, generally by centrifugation, drying, and grinding, with other subsequent unit processes depending on the nature of the microbial biomass (Parekh et al., 2000).

6. Biotechnological tools involved in the generation of microbial food products

The nutrients present in food are a good source of basic biomolecules for food, however, their cultivation leads to waste and environmental problems, consequently, it is important to search for new non-conventional food sources that do not depend on soils and that use recycled material. Microorganisms such as bacteria, yeasts, filamentous fungi, or microalgae are a very good source of proteins, vitamins and in some cases contain lipids that benefit human beings.

Biotechnological tools (Fig. 2) play a fundamental role in the development of the biotechnology industry. These tools are used by researchers for the cellular manipulation of the biochemical processes of living organisms to develop new and alternative methods oriented to the production of clean and effective forms of microbial food products that are in harmony with the environment in an eco-friendly and low-cost approach for the company in charge (Stephanopoulos et al., 1998).

6.1 Genetic modification

The use of recombinant DNA approaches and techniques have been applied to obtain different bacterial, fungal, yeast and mold strains, seeking the expression of important

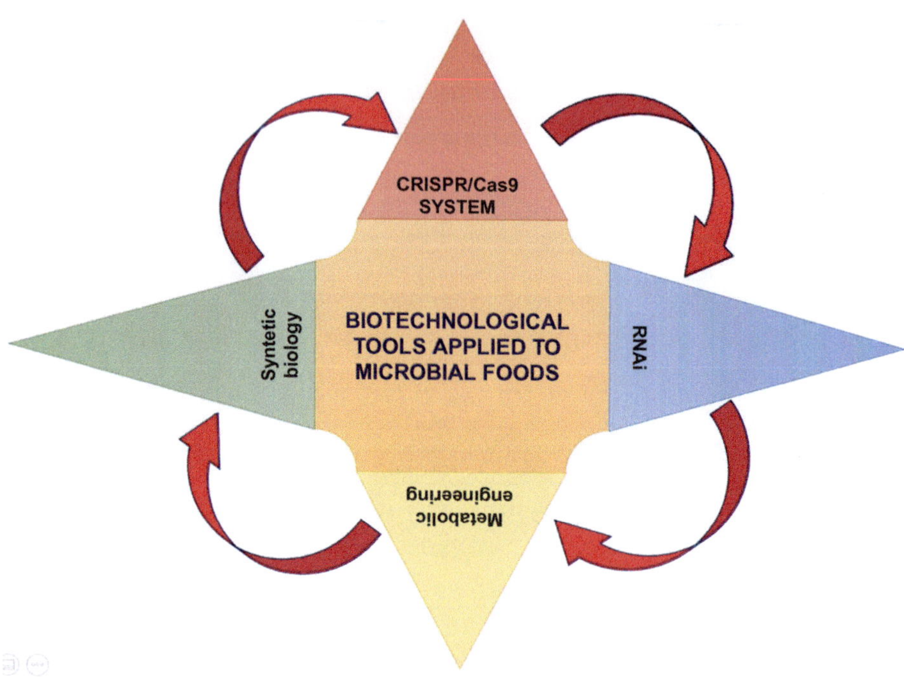

Fig. 2 Molecular approaches and tools oriented to the improvement of microorganisms producing compounds with food properties.

genes through processes such as: induced overexpression of genes of interest, inhibition of non-essential genes, alteration of specific genes or through the inactivation or blocking of specific metabolic pathways. The genetic tools (plasmids as vectors) used for the modification of microorganisms capable of producing microbial food products must not possess antibiotic resistance genes and must be sequences from microorganisms that have been characterized and recognized as safe (GRAS) (Bessin, 2000)

There are several research that used genetic modification as part of the methodology to obtain a food product produced by a microorganism. Studies developed on an oleaginous fungus Mortierella alpine used for the commercial production of Arachidonic Acid (AA), in addition to its great capacity to accumulate up to 50% of polyunsaturated fatty acids (PUFA) in dry weight, which are important for human health Wang et al., 2011; Ho et al., 2007; Ji and Huang, 2019), were carried out through simultaneous overexpression of G6PD (glucose-6-phosphate dehydrogenase), PGD (6-phosphogluconate dehydrogenase) which are genes that play an essential role in the synthesis of fatty acids by the fungus and are part of the pentose phosphate metabolic pathway (Chen et al., 2015; Zhao et al., 2015), and the IDH; (Isocitrate dehydrogenase) gene in which it has been found to play an essential part in fatty acid metabolism in adipocytes and is a potential source of

NADPH (Nicotinamide adenine dinucleotide phosphate) (Zhao et al., 2015). The three genes were overexpressed in M. alpine to evaluate and compare their effect on fatty acid synthesis with the effects of malic enzyme (ME) production (Hao et al., 2014a,b). The authors conclude that the simultaneous overexpression of G6PD and ME represents an effective strategy to improve the production of AA in this fungus, and these assays significantly improve the progress in the production of polyunsaturated acids (PUFA) by this microorganism (Hao et al., 2016).

6.2 RNA interference (RNAi)

RNAi molecules are highly conserved and considered as key elements in epigenetic mechanisms that allow maintaining genome integrity and stability in the fungal group (Billmyre et al., 2013). The function of RNAi is related to the phenomenon of post-transcriptional gene silencing through a double-stranded RNA that allows the degradation of homologous mRNA species, reducing gene expression. The use of RNAi technology is constantly growing, specifically in fungi where it is applied as a reverse genetics tool that allows the modification of gene expression, having demonstrated its efficacy in more than 40 fungal species. In addition to omics technologies (transcriptomics, proteomics, metabolomics, nutrigenomics), RNAi technology can be very useful for the discovery of genes or metabolic pathways, which would allow improving the nutritional value and yield of fungi (Billmyre et al., 2013).

6.3 Cas9 CRISPR system

In recent years, genomic editing through the CRISPR/Cas System has revolutionized the way of genetically modifying living organisms. Due to its versatility, this tool has been adapted to many higher organisms and microorganisms. The CRISPR system comes from the acronym for regularly spaced short palindromic repeats, it is a new genetic tool that allows the development of gene insertions, deletions, and replacements (Cong et al., 2013). This system is a new technology that uses a highly flexible RNA-guided endonuclease and has become a tool that has transformed different fields of study such as genomics, proteomics, gene editing, gene therapy, etc. The CRISPR system associated with a Cas protein in nature comes from the adaptive immune system that is activated against the attack of bacteriophages that try to infect bacteria and archaea (Brouns et al., 2008).

The edible medicinal fungus *Cordyceps militaris* is quite popular in East Asia, with a great potential to produce various bioactive compounds of commercial interest. Previous traditional gene-editing studies have been ineffective and quite complex, so a new tool was required to develop gene editing in this mushroom. The application of the CRISP Cas editing system on this mushroom allowed the optimization of a CAS9 codon with new promoters reported, this assay developed site-specific deletions and insertions demonstrating for the first time the success of the application of the CRISPR–Cas9 system in *C. militaris*, accelerating the reconstruction of the *C. militaris* genome with the aim of generating an accelerated development in the mushroom industry (Chen et al., 2018).

References

Akintui, M., Alcántara, H., Alva, A., Castillo, H., Huayán, J., Llenque, L., 2015. *Pseudomonas fluorescens* from agricultural soils degradator of 2,4 dichlorophenoxyacetic acid herbicide. Revista De Rebiolest 3 (2), s49. Retrieved from https://Revistas.Unitru.Edu.Pe/Index.Php/ECCBB/Article/View/1699/1676.

Alvarado, C., Dasguta, N., Amriz, E., Sanchez, J., Villegas, J., 2011. Hongos micorrízicos arbusculares y la fitorremediación de plomo. Revista internacional de contaminación ambiental 27 (4), 357–364.

Amata, I., 2013. Yeast a single cell protein: characteristics and metabolism. Int. J. Appl. Biol. Pharm. Technol. 4 (2013), 158–170.

Aranceta-Bartrina, J., 2010. Nuevos retos de la nutrición comunitaria. Revista Española de Nutrición Comunitaria 16 (1), 51–55.

Audet, P., 2014. Arbuscular mycorrhizal fungi and metal phytoremediation. In: Emerging Technologies and Management of Crop Stress Tolerance. Elsevier, pp. 133–160.

Ballesta, S., Velasco, C., Victoria Borobio, M., Argüelles, F., Perea, E.J., 2008. Yogures frescos frente a pasteurizados: estudio comparativo de sus efectos sobre los parámetros microbiológicos, inmunológicos y el bienestar gastrointestinal. Enferm. Infecc. Microbiol. Clin. 26 (9), 552–557.

Bano, S., Ashfaq, D., 2013. Role of mycorrhiza to reduce heavy metal stress. Nat. Sci. 5 (12 A), 16–20.

Bennett, A.E., Daniell, T.J., White, P.J., 2013. Benefits of breeding crops for yield response to soil microorganisms. In: Molecular Microbial Ecology of the Rhizosphere, first ed, pp. 17–27, https://doi.org/10.1002/9781118297674.ch3.

Bessin, R.T., 2000. Genetic modified organisms: a consumer perspective. In: NCB Meeting, Minneapolis, MN MO Symposium. North Central Branch. Entomological Society of America, pp. 36–40.

Bhalla, T.C., Sharma, N.N., Sharma, M., 2007. Production of Metabolites, Industrial Enzymes, Amino Acid, Organic Acids, Antibiotics, Vitamins and Single Cell Proteins.

Billmyre, R.B., Calo, S., Feretzaki, M., Wang, X., Heitman, J., 2013. RNAi function, diversity, and loss in the fungal kingdom. Chromosom. Res. 21 (6–7), 561–572.

Botero, et al., 2011. Efecto De La Concentración Del Metil Paratión y El Extracto De Levadura Como Factores De Selección De Microorganismos Degradadores Del Pesticida A Partir De Suelos Contaminados. Revista Ingenierías Universidad de Medellín 10 (19), 13–20.

Bover Cid, S., Vidal Carou, M.C., Mariné Font, A., 2001. ¿Son necesarios los suplementos nutritivos en los alimentos? FMC Formación Médica Continuada En Atención Primaria 8 (9), 620–627.

Boze, H., Moulin, G., Galzy, P., 1995. Production of microbial biomass. Biotechnology 9, 167–220.

Brouns, S.J.J., Jore, M.M., Lundgren, M., Westra, E.R., Slijkhuis, R.J.H., Snijders, A.P.L., Dickman, M.J., Makarova, K.S., Koonin, E.V., Van Der Oost, J., 2008. Small CRISPR RNAs guide antiviral defense in prokaryotes. Science 321 (5891), 960–964.

Cabrales, E., Lopez-Hernández, D., Toro, M., 2019. Effect of inoculation with Glomeromycota fungi and fertilization on maize yield in acid soils. In: Microbial Probiotics for Agricultural Systems, pp. 205–212.

Campaña-Olaya, J., Luis-Alaya, B., Mendez-Farroñan, S., Quezada-Salirrosas, M., Nouchi-Moromizato, E., Munaya-Sáenz, N., Sánchez, H., Prada-Gutierrez, E., Mialhe, E., 2017. Caracterización molecular de la microbiota rizosférica nativa de Opuntia ficus-indica y evaluación de los efectos de cepas microbianas aisladas sobre el desarrollo del cactus. Manglar 14 (1), 3–11.

Caplice, E., Fitzgerald, G.F., 1999. Food fermentations: role of microorganisms in food production and preservation. Int. J. Food Microbiol. 50 (1–2), 131–149.

Chen, B.-X., Wei, T., Ye, Z.-W., Yun, F., Kang, L.-Z., Tang, H.-B., Guo, L.-Q., Lin, J.-F., 2018. Efficient CRISPR-Cas9 gene disruption system in edible-medicinal mushroom cordyceps militaris. Front. Microbiol. 9, 1157.

Chen, H., Hao, G., Wang, L., Wang, H., Gu, Z., Liu, L., Zhang, H., Chen, W., Chen, Y.Q., 2015. Identification of a critical determinant that enables efficient fatty acid synthesis in oleaginous fungi. Sci. Rep. 5, 11247.

Christaki, E., Florou-Paneri, P., Bonos, E., 2011. Microalgae: a novel ingredient in nutrition. Int. J. Food Sci. Nutr. 62 (8), 794–799.

Cong, L., Ran, F.A., Cox, D., Lin, S., Barretto, R., Habib, N., et al., 2013. Multiplex genome engineering using CRISPR/Cas systems. Science 339 (6121), 819–823.

Corio Andújar, R., Arbonés Fincias, L., 2009. Nutrición y salud. SEMERGEN Medicina de Familia 35 (9), 443–449.

de los Santos Villalobos, S., Parra Cota, F.I., Herrera Sepúlveda, A., Valenzuela Aragón, B., Estrada Mora, J.-C., 2018. Colección de microorganismos edáficos y endófitos nativos para contribuir a la seguridad alimentaria nacional. Revista mexicana de ciencias agrícolas 9 (1), 191–202.

Dohrmann, A.B., Ku, M., Ju, S., Jaenicke, S., Schlu, A., Tebbe, C.C., 2013. Importance of rare taxa for bacterial diversity in the rhizosphere of Bt - and conventional maize varieties. ISME J. 7 (1), 37–49.

Domínguez González, K.N., Cruz Guerrero, A.E., Márquez, H.G., Gómez Ruiz, L.C., García-Garibay, M., Rodríguez Serrano, G.M., 2014. El efecto antihipertensivo de las leches fermentadas. Rev. Argent. Microbiol. 46 (1), 58–65.

Dunne, C., O'Mahony, L., Murphy, L., Thornton, G., Morrissey, D., O'Halloran, S., Collins, J.K., 2001. Criterios de selección in vitro para bacterias probióticas de origen humano: correlación con hallazgos in vivo. Am. J. Clin. Nutr. 73 (2), 386s–392s.

Durán, L., Ladera, M., 2016. Biorremediación De Suelos Contaminados Por Organoclorados Mediante La Estimulación De Microorganismos Autóctonos, Utilizando Biosólidos. Nexo Revista Científica 29 (1), 22–28.

Estapà, J.C., 2006. Nutrient and energy intake recommendations. Balanced nutrition; what are RDAs food composition. In: FMC: Continuing Medical Education in Primary Care. vol. 13(10), p. 2.

Ferrol, N., Gonzalez, M., Valderas, A., Benabdellah, K., Azcón-Aguilar, C., 2009. Survival strategies of arbuscular mycorrhizal fungi in cu-polluted environments. Phytochem. Rev. 8, 551–559.

Ferrol, N., Tamayo, E., Vargas, P., 2016. The heavy metal paradox in arbuscular mycorrhizas: from mechanisms to biotechnological applications. J. Exp. Bot. 67 (22), 6253–6265.

Gantar, M., Svirčev, Z., 2008. Microalgae and cyanobacteria: food for thought 1. J. Phycol. 44 (2), 260–268.

Godfray, H.C.J., Beddington, J.R., Crute, I.R., Haddad, L., Lawrence, D., Muir, J.F., et al., 2010. Food security: the challenge of feeding 9 billion people. Science 327, 812–818.

Goldstein, B., Moses, R., Sammons, N., Birkved, M., 2017. Potential to curb the environmental burdens of American beef consumption using a novel plant-based beef substitute. PLoS One 12 (12), e0189029.

Guerra, S., 2008. Micorriza arbuscular. Recurso microbiológico en la agricultura sostenible Tecnología en Marcha 21 (1), 191–201.

Hamida, R.S., Shami, A., Ali, M.A., Almohawes, Z.N., Mohammed, A.E., Bin-Meferij, M.M., 2021. Kefir: a protective dietary supplementation against viral infection. Biomed. Pharmacother. 133, 110974.

Hao, G., Chen, H., Du, K., Huang, X., Song, Y., Gu, Z., Wang, L., Zhang, H., Chen, W., Chen, Y.Q., 2014b. Increased fatty acid unsaturation and production of arachidonic acid by homologous overexpression of the mitochondrial malic enzyme in *Mortierella alpina*. Biotechnol. Lett. 36, 1827–1834.

Hao, G., Chen, H., Gu, Z., Zhang, H., Chen, W., Chen, Y.Q., 2016. Metabolic engineering of Mortierella alpina for enhanced arachidonic acid production through the NADPH-supplying strategy. Appl. Environ. Microbiol. 82, 3280–3288. https://doi.org/10.1128/AEM.00572-16.

Hao, G., Chen, H., Wang, L., Gu, Z., Song, Y., Zhang, H., Chen, W., Chen, Y.Q., 2014a. Role of malic enzyme during fatty acid synthesis in the oleaginous fungus *Mortierella alpina*. Appl. Environ. Microbiol. 80, 2672–2678.

Hassen, W., Neifar, M., Cherif, H., Najjari, A., Chouchane, H., Driouich, R.C., Salah, A., Naili, F., Mosbah, A., Souissi, Y., Raddadi, N., Ouzari, H.I., Fava, F., Cherif, A., 2018. Pseudomonas rhizophila S211, a new plant growth-promoting rhizobacterium with potential in pesticide-bioremediation. Front. Microbiol. 9, 34.

Havelaar, A.H., Brul, S., de Jong, A., de Jonge, R., Zwietering, M.H., ter Kuile, B.H., 2010. Future challenges to microbial food safety. Int. J. Food Microbiol. 139, S79–S94.

Ho, S.-Y., Jiang, Y., Chen, F., 2007. Polyunsaturated fatty acids (PUFAs) content of the fungus *Mortierella alpina* isolated from soil. J. Agric. Food Chem. 55, 3960–3966.

Horwath, W.R., Paul, E.A., 1994. Microbial biomass. In: Methods of Soil Analysis: Part 2 Microbiological and Biochemical Properties. 5, pp. 753–773.

Hristozkova, M., Geneva, M., Stancheva, I., Iliev, I., Azcon-Aguilar, C., 2017. Symbiotic association between golden berry (*Physalis peruviana*) and arbuscular mycorrhizal fungi in heavy metal-contaminated soil. J. Plant Protect. Res. 57 (2), 173–184.

Iwatani, S., Yamamoto, N., 2019. Functional food products in Japan: a review. Food Sci. Human Wellness 8, 96–101.

Jaiswal, S., Kiran, J., Soni, R., Shrivastava, K., 2017. Bioremediation of chlorpyrifos contaminated soil by microorganism. Int. J. Environ. Agric. Biotechnol. 2 (4), 1624–1630.

Jaramillo Colorado, B.E., Bermúdez Tobón, A., Tirado Ballestas, I., 2016. Organophosphorus pesticides degrading bacteria present in contaminated soils. Revista Ciencias Técnicas Agropecuarias 25 (3).

Jara-Peña, E., Gómez, C., José, M.T., Haydeé, S., Tito, T., Liliana, C., Noema, & Dextre, Abigail., 2017. Acumulación de metales pesados en *Calamagrostis rigida* (Kunth) Trin. ex Steud. (Poaceae) y *Myriophyllum quitense* Kunth (Haloragaceae) evaluadas en cuatro humedales altoandinos del Perú. Arnaldoa 24 (2), 583–598.

Ji, X.J., Huang, H., 2019. Engineering Microbes to Produce Polyunsaturated Fatty Acids. Trends in Biotechnology, Vol. 37 Elsevier Ltd., pp. 344–346.

Johnston, B.F., Mellor, J.W., 1961. The role of agriculture in economic development. Am. Econ. Rev. 51 (4), 566–593.

Kabakcı, S.A., Türkyılmaz, M., Özkan, M., 2020. Changes in the quality of kefir fortified with anthocyanin-rich juices during storage. In: Química de los alimentos, p. 126977.

Kalia, A., Gupta, R.P., 2005. Conservation and utilization of microbial diversity. India. NBA Sci. Bull. 9 (7), 307.

Kathiresan, G., Manickam, G., Parameswaran, P., 1995. Efficiency of phosphobacteria addition on cane yield and quality. India. Cooperative Sugar 26, 629–631.

Kishor, D.R., Moses, S., Kumar, M., 2020. Study of important insecticide samples. Agric. Food Newslett. 5.

Kües, U., Liu, Y., 2000. Fruiting body production in basidiomycetes. Appl. Microbiol. Biotechnol. 54 (2), 141–152.

Kuhad, R.C., Singh, A., Tripathi, K.K., Saxena, R.K., Eriksson, K.E.L., 1997. Microorganisms as an alternative source of protein. Nutr. Rev. 55 (3), 65–75.

Landa-Acuña, D., Acosta, R.A.S., Hualpa Cutipa, E., Vargas de la Cruz, C., Luis Alaya, B., 2020. Bioremediation: a low-cost and clean-green technology for environmental management. In: Microbial Bioremediation & Biodegradation, pp. 153–171.

Lee, B.H, Lo, Y.H, Pan, T.M, 2013. Anti-obesity activity of *Lactobacillus* fermented soy milk products. J. Funct. Foods 5 (2), 905–913. https://doi.org/10.1016/j.jff.2013.01.040.

Lemale, J., 2014. Alimentación para lactantes: leches maternizadas y leches de continuación. EMC Pediatría 49 (1), 1–7.

Li, H., Luo, N., Zhang, L., Zhao, H., Li, Y., Cai, Q., Wong, M., Mo, C., 2016. Do arbuscular mycorrhizal fungi affect cadmium uptake kinetics, subcellular distribution and chemical forms in Rice? Sci. Total Environ. 571, 1183–1190.

Linder, T., 2019. Making the case for edible microorganisms as an integral part of a more sustainable and resilient food production system. Food Secur., 1–14.

Liriano, R., Bárbara, D., Barceló, R., 2012. Efecto de la aplicación de Rhizobium y Mycorriza en el crecimiento del frijol (*Phaseolus vulgaris* L) variedad CC-25-9 negro. Centro Agrícola 39 (4), 17–20.

Lopitz-Otsoa, F., Rementeria, A., Elguezabal, N., Garaizar, J., 2006. Kefir: A symbiotic yeasts-bacteria community with alleged healthy capabilities. Rev. Iberoam. Micol. 23 (2), 67–74.

Lynch, H., Johnston, C., Wharton, C., 2018. Plant-based diets: considerations for environmental impact, protein quality, and exercise performance. Nutrients 10 (12), 1841.

Marín, F.L., Jaramillo, B., 2015. Aislamiento de bacterias degradadoras de pesticidas organofosforados encontrados en suelos y en leche bovina. Revista Chilena de Nutrición 42 (2), 179–185.

Matassa, S., Boon, N., Pikaar, I., Verstraete, W., 2016. Microbial protein: future sustainable food supply route with low environmental footprint. Microb. Biotechnol. 9 (5), 568–575.

Meléndez Illanes, L., González Díaz, C., Álvarez-Dardet, C., 2013. Los funcionales a examen: ¿alimentos al servicio de la salud o nuevo negocio para la industria alimentaria? Atención Primaria 45 (6), 287–289.

Miltner, A., Bombach, P., Schmidt-Brücken, B., Kästner, M., 2012. SOM genesis: microbial biomass as a significant source. Biogeochemistry 111 (1–3), 41–55.

Molina-Montenegro, M.A., Oses, R., Atala, C., Torres-Díaz, C., Bolados, G., León-Lobos, P., 2016. Nurse effect and soil microorganisms are key to improve the establishment of native plants in a semiarid community. J. Arid Environ. 126, 54–61.

Mor-Mur, M., 2010. Alimentos tratados por alta presión. Aspectos mutricionales. Actividad Dietética 14 (2), 53–58.

Muller-Feuga, A., 2013. Microalgae for aquaculture: the current global situation and future trends. In: Handbook of Microalgal Cultures: Applied Phycology and Biotechnology, second ed. Wiley Blackwell, West Sussex, pp. 615–627.

Muñoz, E., 2008. Evolución del impacto socioeconómico de las biotecnologías en la salud. Med. Clin. 131, 48–54.

Narbona López, E., Uberos Fernández, J., Armadá Maresca, M.I., Couce Pico, M.L., Rodríguez Martínez, G., Saenz de Pipaon, M., 2014. Grupo de Nutrición y Metabolismo Neonatal, Sociedad Española de Neonatología: recomendaciones y evidencias para la suplementación dietética con probióticos en recién nacidos de muy bajo peso al nacer. Anales de Pediatría 81 (6), 397.e1–397.e8.

Nasseri, A.T., Rasoul-Amini, S., Morowvat, M.H., Ghasemi, Y., 2011. Single cell protein: production and process. Am. J. Food Technol. 6 (2), 103–116.

Ning, C.H., Li, W.B., Xu, Q.K., Li, M., Guo, S.X., 2019. Arbuscular mycorrhizal fungi enhance cadmium uptake of wetland plants in contaminated water. Ying yong sheng tai xue bao = J. Appl. Ecol. 30 (6), 2063–2071.

Ogata-Gutiérrez, K., Zuñiga, D., 2020. Interacciones bacteria-planta: un valor añadido de la inoculación microbiana. Rev. Peru. Biol. 27 (1), 021–025.

Olarte, S.H., 2014. Los alimentos transgénicos como bienes públicos globales. Suma de Negocios 5 (10), 59–66.

Olveira, G., González-Molero, I., 2016. Actualización de probióticos, prebióticos y simbióticos en nutrición clínica. Endocrinol. Nutr. 63 (9), 482–494.

Organización Mundial de Gastroenterología, 2011. Guía Práctica de la Organización Mundial de Gastroenterología. Probióticos y prebióticos. In: Guías Mundiales la WGO Probióticos y prebióticos, p. 29. Retrieved from: https://www.worldgastroenterology.org/UserFiles/file/guidelines/probiotics-spanish-2011.pdf. (Accessed 28 April 2016).

Ouwehand, A.C., Nermes, M., Collado, M.C., Rautonen, N., Salminen, S., Isolauri, E., 2009. Los probióticos específicos alivian la rinitis alérgica durante la temporada de polen de abedul. World J. Gastroenterol. 15 (26), 3261.

Pal, K.K., McSpadden, G.B., 2006. Biological control of plant pathogens. Estados Unidos de América. Plant Health Instructor. 2, 1117–1142.

Pankratova, E.M., 2006. Functioning of cyanobacteria in soil ecosystems. Federación Rusa. Eur. Soil Sci. 39 (1), S118–S127.

Parekh, S., Vinci, V.A., Strobel, R.J., 2000. Improvement of microbial strains and fermentation processes. Appl. Microbiol. Biotechnol. 54 (3), 287–301.

Peña, A.S., 2016. Un nuevo enfoque de la nutrición y nutrigenómica en el tratamiento de las enfermedades inflamatorias intestinales crónicas y en la prevención de malignización. Enfermedad Inflamatoria Intestinal Al Día 15 (1), 1–3.

Pereira, A., Alves, G., Cassiana, S.C., Naozuka, J., 2017. Elemental chemical composition of products derived from kefir fermented milk. J. Food Compos. Anal. 78, 86–90.

Pérez, U., Gómez, M., Serralde, D., Peñaranda, A., Wilches, W., Ramirez, L., Rengifo, G., 2019. Hongos formadores de micorrizas arbusculares (HFMA) como estrategia para reducir la absorción de cadmio en plantas de cacao (*Theobroma cacao*). Terra Latinoamericana 37 (2), 121–130.

Peters, S.J., Morgan, P.A., 2004. The country life commission: reconsidering a milestone in American agricultural history. Agric. Hist., 289–316.

Prescott, S.L., Dunstan, J.A., Hale, J., Breckler, L., Lehmann, H., Weston, S., Richmond, P., 2005. Los efectos clínicos de los probióticos se asocian con un aumento de las respuestas de interferón-γ en niños muy pequeños con dermatitis atópica. Alergia clínica y experimental 35 (12), 1557–1564.

Punt, P.J., van Biezen, N., Conesa, A., Albers, A., Mangnus, J., van den Hondel, C., 2002. Filamentous fungi as cell factories for heterologous protein production. Trends Biotechnol. 20 (5), 200–206.

Ramírez, F.B., Mas, M.T., Núñez, C.G., Jiménez, N.R., Vargas, M.R., 2018. La alimentación en el síndromedel intestino irritabletie. FMC Formación Médica Continuada En Atención Primaria 25 (7), 422–432.

Ramírez, P., Gordón, M., 2014. Microbiota intestinal en el paciente crítico. Una aproximación positiva mediante el aporte de simbióticos. Med. Clin. 143 (4), 161–162.

Rocandio Pablo, A.M., Arroyo Izaga, M., 2001. Alimentos funcionales en Europa. Revista Clínica Española 201 (5), 260–261.

Rodriguez, A., Sanders, I.R., 2014. The role of community and population ecology in applying mycorrhizal fungi for improved food security. ISME J. 9 (5), 1053–1061.

Ros, E., 2001. Introducción a los alimentos funcionales. Med. Clin. 116 (16), 617–619.

Rosa, D.D., Dias, M.M.S, Grześkowiak, Ł.M., Reis, S.A., Conceição, L.L., Peluzio, M.d.C.G, 2017. Milk *kefir*: nutritional, microbiological and health benefits. Nutr. Res. Rev. 30 (1), 82–96. https://doi.org/10.1017/S0954422416000275.

Rueda-Rodríguez, M.C., Arjona, J.S., Soto-Salazar, C., De Castro, M., Castañeda-Cardona, C., Rosselli, D., 2016. Suplementos nutricionales como inmunomoduladores en niños: una revisión de la literatura. Pediatría 49 (4), 103–109.

Ryan, R.P., Germaine, K., Franks, A., Ryan, D.J., Dowling, D.N., 2008. Bacterial endophytes: recent developments and applications. Reino Unido. FEMS Microbiol. Lett. 278 (1), 1–9.

Sankaran, S., Khanal, S.K., Jasti, N., Jin, B., Pometto III, A.L., Van Leeuwen, J.H., 2010. Use of filamentous fungi for wastewater treatment and production of high value fungal byproducts: a review. Crit. Rev. Environ. Sci. Technol. 40 (5), 400–449.

Serra, J., 2016. Microbiota intestinal. Atención Primaria 48 (6), 345–346.

Serrano-Ruiz, J.C., Dumesic, J.A., 2011. Catalytic routes for the conversion of biomass into liquid hydrocarbon transportation fuels. Energy Environ. Sci. 4 (1), 83–99.

Soares, M.B., Martinez, R.C.R., Pereira, E.P.R., Balthazar, C.F., Cruz, A.G., Ranadheera, C.S., Sant'Ana, A.S., 2019. La resistencia de cepas de Bacillus, Bifidobacterium y Lactobacillus con propiedades probióticas declaradas en diferentes matrices alimentarias expuestas a condiciones simuladas del tracto gastrointestinal. Food Res. Int. 108542.

Soares, P., Secci Martinelli, S., Barletto Cavalli, S., Davó-Blanes, M.C., 2020. Methodological proposal to explore the healthy and sustainable food service purchasing. In: G Model, Gaceta 1847. No. of Page 4.

Sonnino, R., Marsden, T., 2006. Beyond the divide: rethinking relationships between alternative and conventional food networks in Europe. J. Econ. Geogr. 6 (2), 181–199.

Soto, L., David, S., Quintanar, V., Isabel, A., Coronado, G., de Lourdes, M., Almada, B., del Carmen, M., Hernández, G., Jaqueline, A.M., Lourdes, M., Cota, G., Patricia, S.G., Isabel, M., Montenegro, M., Mercedes, M., Durán, P., Alejandra, S., García, L., Nepomuceno, G., Gómez, C., Olivia, B., Navarro, V., Paulina, C., 2014. Residuos de plaguicidas organoclorados en suelos agrícolas. Terra Latinoamericana 32 (1), 1–11.

Springmann, M., Wiebe, K., Mason-D'Croz, D., Sulser, T.B., Rayner, M., Scarborough, P., 2018. Health and nutritional aspects of sustainable diet strategies and their association with environmental impacts: a global modelling analysis with country-level detail. Lancet Planet. Health 2 (10), e451–e461.

Stephanopoulos, G., Aristidou, A.A., Nielsen, J.H., 1998. Metabolic Engineering: Principles and Methodologies. Academic Press, San Diego.

Strokal, M., Ma, L., Bai, Z., Luan, S., Kroeze, C., Oenema, O., et al., 2016. Alarming nutrient pollution of Chinese rivers as a result of agricultural transitions. Environ. Res. Lett. 11 (2), 024014.

Suman, G., Nupur, M., Anuradha, S., Pradeep, B., 2015. Single cell protein production: a review. Int. J. Curr. Microbiol. App. Sci. 4 (9), 251–262.

Thompson, F.L., Abreu, P.C., Cavalli, R., 1999. The use of microorganisms as food source for *Penaeus paulensis* larvae. Aquaculture 174 (1–2), 139–153.

Tiss, M., Souiy, Z., Abdeljelil, N., ben, Njima, M., Achour, L. and Hamden, K., 2020. Fermented soy milk prepared using kefir grains prevents and ameliorates obesity, type 2 diabetes, hyperlipidemia and liver-kidney toxicities in HFFD-rats. J. Funct. Foods 67, 103869.

Tormo Carnicer, R., Infante Piña, D., Roselló Mayans, E., Bartolomé Comas, R., 2006. Intake of fermented milk containing *Lactobacillus casei* dn-114 001 and its effect on gut flora. Anales de Pediatría 65 (5), 448–453.

Toro, M., Andrade, G., 2020. Arbuscular mycorrhizae, beneficial microorganisms for sustainable agriculture. In: Leal, F.W., Azul, A., Brandli, L., Lange, S.A., Wall, T. (Eds.), Life on Land. Encyclopedia of the UN Sustainable Development Goals. Springer, Cham, https://doi.org/10.1007/978-3-319-71065-5_122-1.

Toro, M., Gamarra, R., López, L., Infante, C., 2017. Arbuscular mycorrhizal fungi and the remediation of soils contaminated with hydrocarbons. In: Anjum, N. (Ed.), Chemical Pollution Control via Microorganisms. Nova Science Publishers, pp. 79–96.

Tytler, P., Ireland, J., Murray, L., 1997. A study of the assimilation of fluorescent pigments of microalgae *Isochrysis galbana* by the early larval stages of turbot and herring. J. Fish Biol. 50 (5), 999–1009.

Valdovinos-Díaz, M.Á., 2013. Microbiota intestinal en los trastornos digestivos. Probióticos, prebióticos y simbióticos. Rev. Gastroenterol. Mex. 78, 25–27.

Valdovinos-García, L.R., Abreu, A.T., Valdovinos-Díaz, M.A., 2018. Uso de probióticos en la práctica clínica: resultados de una encuesta nacional a gastroenterólogos y nutriólogos. Rev. Gastroenterol. Mex. 84 (3). https://doi.org/10.1016/j.rgmxen.2018.10.001.

Van Huis, A., Oonincx, D.G., 2017. The environmental sustainability of insects as food and feed. A review. Agron. Sustain. Dev. 37 (5), 43.

Wang, H., Sun, X., Song, X., Guo, M., 2021. Effects of kefir grains from different origins on proteolysis and volatile profile of goat milk kefir. Food Chem. 339, 128099.

Vieira, C.P., Cabral, C.C., da Costa Lima, B.R., Paschoalin, V.M.F., Leandro, K.C., Conte-Junior, C.A., 2017. *Lactococcus lactis* ssp. cremoris MRS47, a potential probiotic strain isolated from kefir grains, increases *cis*-9, *trans*-11-CLA and PUFA contents in fermented milk. J. Funct. Foods 31, 172–178.

Wang, L., Chen, W., Feng, Y., Ren, Y., Gu, Z., Chen, H., Wang, H., Thomas, M.J., Zhang, B., Berquin, I.M., Li, Y., Wu, J., Zhang, H., Song, Y., Liu, X., Norris, J.S., Wang, S., Du, P., Shen, J., Wang, N., Yang, Y., Wang, W., Feng, L., Ratledge, C., Chen, Y.Q., 2011. Genome characterization of the oleaginous fungus Mortierella alpina. PLoS One 6 (12), e28319.

Waraich, E.A., Ahmad, R., Ashraf, M.Y., Saifullah, Ahmad, M., 2011. Improving agricultural water use efficiency by nutrient management in crop plants. Acta Agric. Scand. Sect. B: Soil Plant Sci. 61 (4), 291–304.

Zhao, L., Zhang, H., Wang, L., Chen, H., Chen, Y.Q., Chen, W., Song, Y., 2015. ^{13}C-metabolic flux analysis of lipid accumulation in the oleaginous fungus Mucor circinelloides. Bioresour. Technol. 197, 23–29.

Zied, D.C., Pardo-Giménez, A. (Eds.), 2017. Edible and Medicinal Mushrooms: Technology and Applications. John Wiley & Sons.

CHAPTER 2

Role of microorganisms in climate-smart agriculture

Astha Sinha, Swarnkumar Reddy, and W. Jabez Osborne*
Biomolecules Lab, School of Bio Sciences and Technology, Vellore Institute of Technology, Vellore, India
*Corresponding author: e-mail address: jabezosborne@vit.ac.in

1. Introduction

Soil is the one of the most diverse and complex natural habitats encompassing of millions of inhabitants such as fungi, bacteria, and other microorganisms (Bardgett and Van Der Putten, 2014). The various enzymatic actions taking place between the soil and its inhabitants paves an important path in maintaining and sustaining the quality of soil for the ecosystem (Aislabie et al., 2013). Diverse array of microbial communities residing in the soil forms an integral association with the soil biosystem. The soil produces various enzyme as a result of these interactions that triggers various biochemical process such as maintaining soil quality and decomposition of organic matter, sequestration of soil pollutants, regulating nutrient recycling, etc. (Dong et al., 2015; Han et al., 2007; Zhang et al., 2017). Recent unorthodox agricultural practices have altered the biodiversity of soil microbial communities and have rendered the soil composition. Understanding the extent to which soil microbial communities control ecosystem processes is thus critical to establish effective policies to preserve microbial diversity hotspots and the key ecosystem functions and services. In this chapter, we have attempted to discuss about the soil and its quality which influences the role of soil microbiome in directly and indirectly regulating various biochemical processes of the soil.

2. Effect of pollution on soil and its quality

Soil can be defined as loose, organic substances and transformed mineral present at the earth's surface which are capable of supporting plant growth. Healthy soil promotes biological activities and microbial growth, regulates water flow and helps to transform waste organic and inorganic materials and regulate nutrient cycle. Soil function and effectiveness can be monitored by assessing the physical, chemical, and biological qualities of soil (Hillel and Hatfield, 2005). Some of the indicators which helps to measure soil quality includes organic matter content, available nutrients, salinity, rooting depth, etc. The nature and characteristics of soil develop through the complex interactions among climate, biological processes, topography, and parent material (Doran and Zeiss, 2000). Among these the quality of soil is greatly influenced by climate change. The nature and distribution of ecosystems relies on the unique balance between soil and the climate (Karlen et al., 1997).

2.1 Soil quality and its assessment

Soil quality can be defined as "the capacity of a specific kind of soil to function, within natural or managed ecosystem boundaries, to sustain plant and animal productivity, maintain or enhance water and air quality, and support human health and habitation" (Karlen et al., 1997). Maintaining soil quality and soil health is of utmost importance for supporting plant growth. The quality of soil can be assessed based on the type of function it performs within an ecosystem niche (Doran and Zeiss, 2000). Healthy soil quality is a major factor for the growth of plants and availability of nutrients. The term "biological soil quality" indicates the biological function of soil with respect to crop productivity and plant growth. According to Singh et al. (2010) there is an urgent need to boost the food grain production worldwide by 280 million tons in order to fulfill the food demands of the growing population by 2020.

Soil quality cannot be measured directly, it is evaluated with the help of set of indicators. Indicators are set of physical, chemical, and biological measurable properties of soil or plants which suggests about how well the soil can function (Fig. 1).

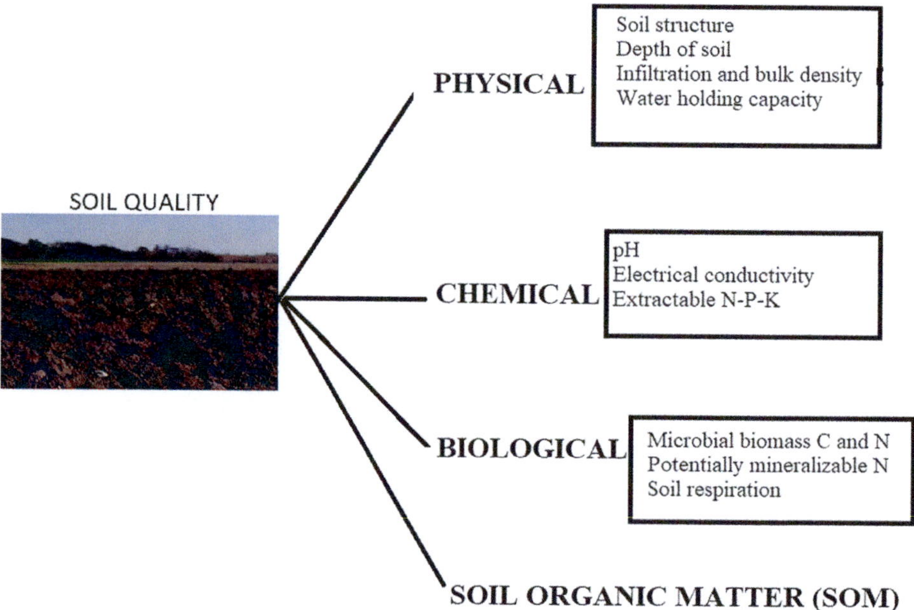

Fig. 1 Soil quality indicators. Indicators are set of physical, chemical, and biological measurable properties of soil or plants.

3. Soil pollution and its effects on the quality of soil

Soil pollution can be defined as a depletion in the productivity of soil and degradation of land due to human anthropogenic activities. Alteration in the physical, chemical, and biological properties of natural soil environment due to the introduction of pollutants such as industrial effluents, sewage, chemicals, plastics, and sewage wastes, fertilizers, pesticides, etc., can also be defined as soil pollution (Reddy and Jabez, 2020). Theses pollutants get mixed in soil and alters its natural chemistry, which imbalances the entire ecosystem. These pollutants also lead to a change in the soil natural pH, which makes it unfit for agricultural purposes (Osaigbovo and Orhue, 2006). According to a study conducted by Saravanamoorthy and Ranjitha (2007), soils when continuously irrigated with wastewater from textile industries, resulted in 0.4 unit increase in pH as compared to the one irrigated with groundwater. Wastewater from sewage, textile, and fertilizer plants, generates huge amount of Ca^{2+} and Mg^{2+} waste which alters and increases the soil pH making them unfit for agriculture (Gupta et al., 2015). Discharge of wastewater from sewage and leather industry when dumped into the ecosystem leads to a considerable decrease in soil electrical conductivity (EC) (Dheri et al., 2007). Soils when comes in contact with polluted wastewater, it traps the nutrients on its surface and subsurface layers, hence increasing the electrical conductivity to a significant level in turn alters the growth and yield of plant (Aghabarati et al., 2008). Soil organic carbon (SOC) level signifies the quality and quantity of nutrients utilized and accumulated by a plant. Disposal of wastewater enriched with many kinds of organic materials, leads to an increase in soil organic carbon and a decrease in decomposition rate (Dheri et al., 2007). According to a study by Rattan et al. (2005) continuous soil irrigation by sewage water had led to an increase in SOC levels from 38% to 79% in subtropical soils of India. The continuous discharge of industrial wastewater into the soil also affects the Nitrogen, Phosphorous, and Potassium levels of the soil (El-Hady, 2007; Mohammad Rusan et al., 2007; Osaigbovo and Orhue, 2006; Yadav et al., 2002). It is also evident from various studies that wastewater from any source has a considerable amount of heavy metals such as Mn, Zn, Fe, Cu, Pb, and Cd, and macro- and micronutrients (Mohammad Rusan et al., 2007). Plants can absorb toxic heavy metals from the soil and deposition them on various parts (Haiyan and Stuanes, 2003). Toxic heavy metals when in the plant, inhibits the root and shoot growth and nutrient uptake. Over usage of chemical fertilizers for getting higher agricultural yields leads to harmful effects on the ecosystem (Reddy and Osborne, 2020). The harmful pesticides and fertilizers get leached into the groundwater resource through surface runoffs (Shivakumar and Srikantaswamy, 2012). Many ecological processes like nitrification, denitrification, phosphorus solubilization, ammonification, and methanogenesis get affected due to the introduction of certain pesticides. Excessive use of strong chemicals and pesticides also affects the microscopic balance of the soil. Increased anthropogenic activities of soil, also affect the micro- and macroscopic flora and fauna affecting soil health and productivity. Increased anthropogenic activities have led to adverse effects on the environment, sustainable agriculture in particular (Fang et al., 2018; Malla et al., 2019).

4. Role of soil microbiome in regulating soil health and plant fertility; A plant-soil-microbial interactions

Microbes are the most diverse group of organism and account for almost 60% of the earth's total biomass (Bar-On et al., 2018). Soil microbial community encompasses a wide array of microorganisms like bacteria, fungi, actinomycetes, algae, protozoa, etc., which plays significant role in the decomposition of organic matter, maintenance of soil health, nutrient recycling, etc. Microbes are responsible for maintaining an equilibrium between the biosphere and its life forms (Vibha and Neelam, 2012). Soil has a complex nature and is perhaps one of the most challenging and exigent of all the natural environments. A complex relationship exists between the plants and their surroundings, in which both have evolved an intimate partnership that enables them to coexist (Nihorimbere et al., 2011). A complex plant-soil-microbial association influences plant health, productivity and the exchange of various substances. It is important to understand how these associations form and how they help the plant for better survival (Morales and Holben, 2011).

Microbes regulates and enhances numerous plant activities such as (1) manipulation of growth hormonal signaling of plants by the production of enzymes and antimicrobial agents prevention against pathogenic microbial strains (Mendes et al., 2013); (2) enhancing the bioavailability of soil-borne nutrients (van der Heijden et al., 2008); (3) microorganisms regulate nutrient cycle of the plant and leads to decomposition of soil organic matters making the soil suitable for all the biotic and abiotic stress (Arrigo, 2005; Conley et al., 2009); (4) promotes plant growth and nutrient availability (Aislabie et al., 2013; Supramaniam et al., 2016; Supramaniam et al., 2016); (5) regulate mineralization, immobilization, and detoxification of contaminants (Arrigo, 2005; Francis et al., 2007; Lugtenberg and Faina, 2009).

4.1 Effect of plants on soil microbiome

Plants and their surroundings interact in a dynamic state where the plants monitor the environmental changes and react to it. Traditionally the root system was considered to be responsible for providing anchorage, support and transport of nutrients and water. Later researchers claimed roots to be a key element in establishing the interactions between the plant and soil (Bais et al., 2006). Unique chemical signals are released by the soil microorganisms, these signals are received and recognized by the complex system of plants. Plant responses to these signals via the release of diverse array of chemical compounds known as root exudates, which create a unique environment in the rhizosphere and consist of sugars, amino acids, flavonoids, aliphatic acids, proteins, and fatty acids (Badri et al., 2009; Sasse et al., 2018). Root exudates are released via different processes such as ion channels, diffusion, and vesicle transport (Bertin et al., 2003).

The secretion, composition, and concentration of root exudates differ in terms of plants species and environment (De-la-Peña et al., 2010; Micallef et al., 2009) and are grouped into low and high molecular weight compounds. In the densely populated rhizosphere, the plant roots compete with neighboring plant species and root-assisted microbes for water, space, and nutrients (Ryan et al., 2001). Root exudates have a significant effect on the colonization of soil microbiome. Strong correlation between the soil microbial communities' structure and host plants was found (Broeckling et al., 2008). According to Uren (2007), rhizodeposition (border cells, root debris, and root exudates) is a major source of organic C for the soil and contributes to 30%–40% of the total C level. These rhizodeposits attract the soil microorganisms which helps in the secretion of growth-promoting hormones, preventing disease, or obtaining nutrients. The rhizosphere acts as a hotspot of ecological richness, with plant roots serving as host for huge array of microbial taxa (Bulgarelli et al., 2013). Although soil factors provide a strong influence on microbial communities, root exudates have been shown to also strongly influence the soil microbial community.

4.2 Role of soil microbes in soil health, plant productivity and health

Plants are associated with a well-structured and organized community of microbes known as phytomicrobiome (Caporaso et al., 2010; Chaparro et al., 2014; Hararuk et al., 2015). Phytomicrobiome along with the plant constitute a holobiont (Berg et al., 2016; Hararuk et al., 2015). The microbial community residing in association with the roots of the plant are called rhizomicrobiome (Zhang et al., 2017). Rhizomicrobiome has a positive effect on soil productivity, plant health and yield (Babalola, 2010). Some of the key roles played by microorganisms for maintaining soil fertility are follows:

- Improving soil structure: Some bacteria and fungi are known to produce chemical substances responsible for integrating soil particles during organic matter disintegration and hence improve the soil structure (Torsvik and Øvreås, 2002)
- Release of nutrient from organic matter: Large sum of nutrient discharge from organic matter takes place due to action of soil microorganism (Arrigo, 2005). They are responsible for the decomposition and degradation of plant and animal waste into organic particles (Shraddha et al., 2011).
- Fixing atmospheric nitrogen and increasing phosphorous availability
- Degrading pesticides and other contaminants in the soil (Sinha and Osborne, 2016)
- Controlling pathogen
- Increasing the bioavailability of soil-borne nutrients (van der Heijden et al., 2008)

Important nutrients such as N, P, and S are present in soil as bound organic molecules and are therefore minimally available for plants. Soil microbes possess metabolic machinery to depolymerize and mineralize organic forms of these nutrients and subsequently release them into the soil, hence liberating inorganic N, P and S that are the preferred nutrient forms for plants (Bonkowski and Roy, 2005).

4.3 Interactions of plant root and microbes

Soil microbiota forms an integral part of soil ecosystem, and plays important role in retention and biotransformation of nutrients, production of hormones known as phytohormones that regulates plant growth, uptake of nutrients and helps the plant in combating stress (Fahad et al., 2015). Plant growth-promoting rhizobacteria produces auxin hormone known as Indole-3-acetic acid, which is involved in plant-microbe interactions and growth (Afzal et al., 2015). Many PGPR's also produce cytokinin's and gibberellins (Gupta et al., 2015; Kumar et al., 2015). Microorganisms in the soil form an association with the soil biosystem and produce soil enzyme-like β-glucosidase and hydrolase responsible for the breakdown of organic matters. Amylase, urease, phosphatase, and sulfatase enzymes take part in nutrient mineralization (Bing-Cheng and Dong-Xia, 2012; Sherene, 2017). Catalase enzymes like dehydrogenase and phosphatase lead to the degradation of heavy metals (Khan et al., 2007). Rhizospheric microbe and their host plants have coevolved together and have developed a series of mechanisms to modulate their communication. Host plants depend on the propensity of their roots to establish a synchronized communication with microbes by the production of root exudates.

5. Physiology of plant-growth promoting rhizobacteria

Recent advances in the area emphasize that introduction of beneficial microbes can enhance the nutrient uptake by plants and has a positive effect on growth (Cummings, 2009), confer resistance to abiotic stress and suppress disease (Selvakumar et al., 2012). Application of free-living or symbiotic soil bacteria known as plant growth-promoting rhizobacteria (PGPR's) has emerged as an effective agricultural treatment technique. PGPRs colonize in the roots rhizosphere region of the plant system and impart beneficial effects. These microorganisms feed on the root exudates derived from the plant system for an enhanced degradation of contaminants. There as two types of PGPR's, symbiotic bacteria (iPGPR) colonize inside the plant cells and produce root nodules that helps in fixation of atmospheric nitrogen fixation, whereas free-living rhizobacteria (ePGPR) tends to grow independently outside the plant cells (Hou et al., 2015). Plants interact with a variety of PGPR's that are capable of enhancing photosynthetic capability of the plant (Xie et al., 2009), plant growth (Hayat et al., 2010), helps in salt and drought tolerance (Dimkpa et al., 2009), enhanced disease suppression, and improved iron acquisition mechanisms (Xie et al., 2009; Zhang et al., 2009). These discoveries have encouraged the potential application of PGPR to enhance agricultural productivity and sustainability. Some of the PGPR's microbial genera include *Bacillus* sp., *Pseudomonas* sp., *Ralstonia* sp., *Burkholderia caribensis*, *Azotobacter* sp., *Rhizobium* sp. Plant growth-promoting rhizobacteria are known to impart the following effects on the plant.

- Direct influence: It comprises the production of growth hormones like auxins, cytokines, and gibberellins, solubilization of insoluble phosphate and nitrogen fixation and the production of indole acetic acid, and release of siderophore molecules
- Indirect influence: it comprises with the production of various secondary metabolites and removal of pathogens

5.1 Signal exchange between plant roots and PGPR's

Rhizospheric soil bacteria excretes hormones that regulate the plant growth and manage plant stress responses. PGPR bacteria produces auxin mostly Indole-3-acetic acid (IAA) in the soil that enhances plant growth and maintains plant-micro interaction (Gupta et al., 2015). Production of auxin by the PGPR bacteria elicits many transcriptional changes in hormone, cell wall-related and defence-related genes in the plant (Spaepen and Vanderleyden, 2011), known to have a positive effect on longer roots and activate auxin response genes that enhance plant growth (Ruzzi and Aroca, 2015), increased root biomass and decrease stomata size (Mesa-Marín et al., 2018). PGPR bacteria provides great support to the plant both in stressed and unstressed condition (Rubin et al., 2017). Ethylene is considered as an important growth hormone responsible for different plant stress responses and developmental process (Nadeem et al., 2014). An increased level of ethylene leads to early ripening of fruits, falling of leaves, and an inhibited root elongation. PGPR bacteria secrete 1-aminocyclopropane-1-carboxylase (ACC) deaminase enzyme that is known to reduce ethylene stress in plants (Glick, 2014; Vejan et al., 2016). ACC is hydrolyzed into α-ketobutyrate and ammonia by the action of ACC deaminase produced by PGPR microorganism and hence promote the plant growth and metabolism. Phosphate is a major macronutrient required by the plant and plays a key role in plant nutrition, growth, and proliferation. However, phosphate is present in an insoluble form in the soil hence limiting its uptake by the plant. PGPR's such as *Pseudomonas* sp., *Bacillus* sp., *Rhizobium* sp., etc., can solubilize the insoluble form of inorganic phosphorus present in the soil and release its soluble form which can be readily taken up by the plant species for its growth and metabolism. Many PGPR's can produce cytokinins and gibberellins enzyme that helps in enhanced plant shoot growth (Gupta et al., 2015; Kumar et al., 2015). The presence of iron is an important growth-limiting factor for the microorganism. PGPR microorganism are known to produce siderophores that are low molecular weight (600–1500 Da) iron chelators that bind to the insoluble ferric iron present in the soil with a very high affinity and release soluble form of iron.

6. Microbial influence on biogeochemical cycles and its applications

Microorganisms residing in the soil, i.e., rhizomicrobia plays a significant role in plant nutrient cycling (Hashem et al., 2017). Microbes are called "Key engines" that drive

Table 1 Influence of soil microbes on various ecosystem processes.

Ecosystem process	Microorganism involved	Contribution to ecosystem	Reference
Carbon cycle	Nitrogen-fixing bacteria, mycorrhizal fungi	0%–50%	Vogelsang et al. (2006)
Decomposition	Bacteria, fungi	Up to 100%	Bardgett, 2005
Phosphorus cycle	Mycorrhizal fungi; P-solubilizing bacteria	0%–90%	Johnson et al. (2002)
Nitrogen cycle	Rhizobia, free-living bacteria, actinomycetes	0%–20%	DeLuca et al. (2002)
Stimulation of plant diversity	Arbuscular mycorrhizal fungi, rhizobia	0%–50%	Scheublin and Van Der Heijden (2006)

and control the Earth's biogeochemical cycles. Microbes are essential for the transformation and breakdown of inaccessible forms of important macro and micronutrients into forms that can be reused and utilized by another microorganism (Table 1).

6.1 Soil carbon and microbial decomposers

Carbon is an essential building block of all organic compounds. Plants utilize carbon and produce compounds such as carbohydrates, lipids, and fatty acids which are in turn used for metabolic activities. Soil carbon (C) level plays a vital role in the regulation of climate and nutrient cycle. The terrestrial carbon cycle works via the balance between photosynthesis and respiration. Carbon fixing autotrophic plants and some microbes synthesize atmospheric carbon dioxide (CO_2) into organic material. Fixed carbon returns to the atmosphere via respiration and decomposition of organic matter thatches place by the action of microorganism. Soil microorganisms contribute as an organic matter decomposers and plant symbiont and help in regulating the nutrient availability and C content in the soil (Bardgett and Van Der Putten, 2014). Studies by Allison et al. (2010) and Hararuk et al. (2015) have provided a helpful insight highlighting the mechanism of soil C transformation by the activity of soil microbial extracellular enzymes. However, any solid empirical evidence correlating the microbial community functional traits, environment context, and the ecosystem processes was lacking. Research by Trivedi et al. (2016) suggested that the soil microbiota composition regulates C degradation via enzymatic activity in the soil and offers a novel evidence of a strong connection between the structure of the soil microbial community and the abundance of genes encoding four different enzymes involved in C degradation cycle. Their study concluded that the variation in microbial community composition has a direct correlation with functional gene abundance, which determines the consequences of enzyme activity regulating C degradation.

6.2 Nitrogen cycle

Nitrogen is an essential for the synthesis of building block of life (DNA, RNA, and amino acids). Nitrogen–fixing bacteria and archaea are the only organism having access to atmospheric nitrogen, other organism relies on reactive forms of N like ammonium and nitrate for their growth. Microbial transformations of atmospheric nitrogen consist of six distinct processes known as nitrogen cycle, where a molecule of dinitrogen gas is first fixed into ammonia, which is assimilated into organic nitrogen. The degradation of organic nitrogen is called ammonification where a molecule of ammonia is released, which is subsequently oxidized to nitrate through nitrification brought about by the action of nitrifying soil bacteria such as members of the genus *Nitrosomonas*, and eventually converted back to a molecule of dinitrogen gas through denitrification process with the help of soil bacteria, such as members of the genera *Pseudomonas* and *Clostridium*. Only when the atmospheric nitrogen is converted from dinitrogen gas into ammonia it can be utilized by primary producers, such as plants. The major transformations of atmospheric nitrogen include processes like nitrogen fixation, nitrification, denitrification, anammox, and ammonification. The transformation of nitrogen into its many oxidation states with the help of microorganisms, such as bacteria, archaea, and fungi are the key to productivity in the biosphere. The process of converting N_2 into biologically available nitrogen by some free-living or symbiotic organism is called nitrogen fixation. Symbiotic association between N-fixing bacteria and plant root is a very complex and specific process. Root exudates of some leguminous plants such as peas, soyabean, pulses, etc., secrete root exudates which acts as signal and attracts specific bacteria of *Rhizobium* family (Howarth, 2008; Johnson et al., 2010; Kuypers et al., 2018). Human activity and process have a profound effect on the amount of released nitrogen into the environment, the use of nitrogen and phosphorus–based fertilizers has led to decreased soil quality and eutrophication (Table 2).

Table 2 Prokaryotes involved in nitrogen fixation.

Genus of bacteria	Lifestyle
Pseudomonas, Azotobacter	Free-living aerobic, chemoorganotrophic
Rhizobium, Frankia	Symbiotic, aerobic, chemoorganotrophic
Chlorobium, Chromatium	Free-living, anaerobic, phototrophic
Thiobacillus, Alcaligenes	Free-living, aerobic, chemolithotrophic
Clostridium	Free-living, anaerobic, chemoorganotrophic

7. Role of microbes and enzymes in the restoration and reclamation of soil

Soil microbiota and their enzymes act as a complete factor in determining soil health. Even minute changes in the structure, composition, and quality of soil affect the distribution and activity of microbes and microbiome population residing in that ecosystem (Gupta et al., 2015; Kumar et al., 2015). The enzymes produced by microorganism exhibits as a useful sensor towards various environmental stress and nutrient availability). The presence of microorganisms in soil is greatly affected by the availability of compound cellulose, where cellulase enzyme plays an important role in global recycling of cellulose. The presence of dehydrogenase enzyme in the soil acts as an indicator of microbial activity in soil and presence of organic matter. Enzymes like cellulase, dehydrogenase, urease, and amidase are also used as soil indicators. Soil enzymes such as oxidoreductases, laccases and hydrolase, etc., are for the breakdown and detoxification of various contaminants from the soil. Researchers have established a set of enzymes produced by bacteria via protease, urease, various phosphatases, and sulfatase linked to high-functioning and boosting of nutrient transfers to plants (García-Rivero and Peralta-Pérez, 2008). Microorganisms play a significant role in the degradation and mineralization of pollutants like volatile organic compounds, benzene, toluene, ethyl benzene, xylene (BTEX) compounds, polycyclic aromatic hydrocarbons (PAHs), textile waste, hydrocarbons, petroleum, heavy metals, polyethylene, etc. (EPA, 2005). Bioremediation is the removal of toxic contaminants with the aid of bacteria, yeast, fungi, plants, or a combination system (Sinha et al., 2018; Sinha and Osborne, 2016). Bioremediation with the help of microorganism is used to convert toxic contaminants into a less toxic form.

7.1 Emerging technologies in understanding plant-microbe responses

Traditional molecular techniques have helped researchers in understanding the compositional patterns of microbial communities, however, very limited insight was available regarding the functional aspects. With the advancement in molecular-omic such as use of metagenomics, metabolomics, transcriptomics, and proteomics, a new wave to explore the host-microbe interactions at a much higher dynamics has begun, enabling a better understanding towards the taxonomic, genetic and functional aspects of the various microbial communities and their correlation (Dubey et al., 2019; Malla et al., 2019). Shortgun metagenomics has replaced 16s rRNA sequencing for understanding the structural and functional composition of the microbial communities within a habitat as the latter yielded information only about the presence and saturation of microbial species. Many bioinformatic software such as MOTHUR (Schloss et al., 2009), QIIME (Caporaso et al., 2010), MG-RAST, MEGAN (Huson and Weber, 2013), etc., have also helped to expand our understanding of plant-microbe interactions. Recently, metagenomics studies have helped to significantly increase our understanding towards the soil microbial diversity, distribution, and functional gene pattern. Incorporation of metagenomic approaches to study the pattern and distribution of soil microbiomes will greatly help in understanding and restoring ecosystem functioning. Next-generation sequencing technologies are far superior and accurate than the conventional sequencing methods. Metagenomic research of different microbial communities has yielded substantial insights into the distribution of gene families across different ecosystems along with the role of precise functional attributes in acclimatizing to varied environmental conditions.

References

Afzal, I., Shinwari, Z.K., Iqrar, I., 2015. Selective isolation and characterization of agriculturally beneficial endophytic bacteria from wild hemp using canola. Pak. J. Bot. 47 (5), 1999–2008. http://www.pakbs. org/pjbot/PDFs/47(5)/47.pdf.

Aghabarati, A., Hosseini, S.M., Maralian, H., 2008. Heavy metal contamination of soil and olive trees (*Olea europaea* L.) in suburban areas of Tehran, Iran. Res. J. Environ. Sci. 2, 323–329. https://doi.org/10.3923/rjes.2008.323.329.

Aislabie, J., Deslippe, J.R., Dymond, J., 2013. Soil microbes and their contribution to soil services. In: Ecosystem Services in New Zealand–Conditions and Trends, pp. 143–161.

Allison, S.D., Wallenstein, M.D., Bradford, M.A., 2010. Soil-carbon response to warming dependent on microbial physiology. Nat. Geosci. 3 (5), 336–340. https://doi.org/10.1038/ngeo846.

Arrigo, K.R., 2005. Marine microorganisms and global nutrient cycles. Nature 437 (7057), 349–355. https://doi.org/10.1038/nature04159.

Babalola, O.O., 2010. Beneficial bacteria of agricultural importance. Biotechnol. Lett. 32 (11), 1559–1570. https://doi.org/10.1007/s10529-010-0347-0.

Badri, D.V., Weir, T.L., van der Lelie, D., Vivanco, J.M., 2009. Rhizosphere chemical dialogues: plant-microbe interactions. Curr. Opin. Biotechnol. 20 (6), 642–650. https://doi.org/10.1016/j.copbio.2009.09.014.

Bais, H.P., Weir, T.L., Perry, L.G., Gilroy, S., Vivanco, J.M., 2006. The role of root exudates in rhizosphere interactions with plants and other organisms. Annu. Rev. Plant Biol. 57, 233–266. https://doi.org/10.1146/annurev.arplant.57.032905.105159.

Bardgett, R., 2005. The Biology of Soil: A Community and Ecosystem Approach.

Bardgett, R.D., Van Der Putten, W.H., 2014. Belowground biodiversity and ecosystem functioning. Nature 515 (7528), 505–511. https://doi.org/10.1038/nature13855.

Bar-On, Y.M., Phillips, R., Milo, R., 2018. The biomass distribution on Earth. Proc. Natl. Acad. Sci. U. S. A. 115 (25), 6506–6511. https://doi.org/10.1073/pnas.1711842115.

Berg, G., Rybakova, D., Grube, M., Köberl, M., 2016. The plant microbiome explored: implications for experimental botany. J. Exp. Bot. 67 (4), 995–1002. https://doi.org/10.1093/jxb/erv466.

Bertin, C., Yang, X., Weston, L.A., 2003. The role of root exudates and allelochemicals in the rhizosphere. Plant Soil 256 (1), 67–83. https://doi.org/10.1023/A:1026290508166.

Bing-Cheng, Y., Dong-Xia, Y., 2012. Soil microbial and enzymatic activities across a chronosequence of Chinese pine plantation development on the loess plateau of China. Pedosphere 22, 1–12. https://doi.org/10.1016/s1002-0160(11)60186-0.

Bonkowski, M., Roy, J., 2005. Soil microbial diversity and soil functioning affect competition among grasses in experimental microcosms. Oecologia 143 (2), 232–240. https://doi.org/10.1007/s00442-004-1790-1.

Broeckling, C.D., Broz, A.K., Bergelson, J., Manter, D.K., Vivanco, J.M., 2008. Root exudates regulate soil fungal community composition and diversity. Appl. Environ. Microbiol. 74 (3), 738–744. https://doi.org/10.1128/AEM.02188-07.

Bulgarelli, D., Schlaeppi, K., Spaepen, S., Van Themaat, E.V.L., Schulze-Lefert, P., 2013. Structure and functions of the bacterial microbiota of plants. Annu. Rev. Plant Biol. 64, 807–838. https://doi.org/10.1146/annurev-arplant-050312-120106.

Caporaso, J.G., Kuczynski, J., Stombaugh, J., Bittinger, K., Bushman, F.D., Costello, E.K., Fierer, N., Pěa, A.G., Goodrich, J.K., Gordon, J.I., Huttley, G.A., Kelley, S.T., Knights, D., Koenig, J.E., Ley, R.E., Lozupone, C.A., McDonald, D., Muegge, B.D., Pirrung, M., et al., 2010. QIIME allows analysis of high-throughput community sequencing data. Nat. Methods 7 (5), 335–336. https://doi.org/10.1038/nmeth.f.303.

Chaparro, J.M., Badri, D.V., Vivanco, J.M., 2014. Rhizosphere microbiome assemblage is affected by plant development. ISME J. 8 (4), 790–803. https://doi.org/10.1038/ismej.2013.196.

Conley, D.J., Paerl, H.W., Howarth, R.W., Boesch, D.F., Seitzinger, S.P., Havens, K.E., Lancelot, C., Likens, G.E., 2009. Ecology: controlling eutrophication: nitrogen and phosphorus. Science 323, 1014–1015. https://doi.org/10.1126/science.1167755.

Cummings, S.P., 2009. The application of plant growth promoting rhizobacteria (PGPR) in low input and organic cultivation of graminaceous crops. Environ. Biotechnol. 5, 43–50.

De-la-Peña, C., Badri, D.V., Lei, Z., Watson, B.S., Branda, M.M., Silva-Filho, M.C., Sumner, L.W., Vivanco, J.M., 2010. Root secretion of defense-related proteins is development-dependent and correlated with flowering time. J. Biol. Chem. 285 (40), 30654–30665. https://doi.org/10.1074/jbc.M110.119040.

DeLuca, T.H., Zackrisson, O., Nilsson, M.-C., Sellstedt, A., 2002. Quantifying nitrogen-fixation in feather moss carpets of boreal forests. Nature 419 (6910), 917–920.

Dheri, G.S., Brar, M.S., Malhi, S.S., 2007. Heavy-metal concentration of sewage-contaminated water and its impact on underground water, soil, and crop plants in alluvial soils of northwestern India. Commun. Soil Sci. Plant Anal. 38 (9–10), 1353–1370. https://doi.org/10.1080/00103620701328743.

Dimkpa, C., Weinand, T., Asch, F., 2009. Plant-rhizobacteria interactions alleviate abiotic stress conditions. Plant Cell Environ. 32 (12), 1682–1694. https://doi.org/10.1111/j.1365-3040.2009.02028.x.

Dong, W.Y., Zhang, X.Y., Liu, X.Y., Fu, X.L., Chen, F.S., Wang, H.M., Sun, X.M., Wen, X.F., 2015. Responses of soil microbial communities and enzyme activities to nitrogen and phosphorus additions in Chinese fir plantations of subtropical China. Biogeosciences 12 (18), 5537–5546. https://doi.org/10.5194/bg-12-5537-2015.

Doran, J.W., Zeiss, M.R., 2000. Soil health and sustainability: managing the biotic component of soil quality. Appl. Soil Ecol. 15 (1), 3–11. https://doi.org/10.1016/S0929-1393(00)00067-6.

Dubey, A., Kumar, A., Abd Allah, E.F., Hashem, A., Khan, M.L., 2019. Growing more with less: breeding and developing drought resilient soybean to improve food security. Ecol. Indic. 105, 425–437. https://doi.org/10.1016/j.ecolind.2018.03.003.

El-Hady, B.A.A., 2007. Compare the effect of polluted and River Nile irrigation water on contents of heavy metals of some soils and plants. Res. J. Agric. Biol. Sci. 3 (4), 287–294.

EPA, 2005. Soil bioremediation-EPA guideline. pp. 1–9.

Fahad, S., Hussain, S., Bano, A., Saud, S., Hassan, S., Shan, D., Khan, F.A., Khan, F., Chen, Y., Wu, C., Tabassum, M.A., Chun, M.X., Afzal, M., Jan, A., Jan, M.T., Huang, J., 2015. Potential role of phytohormones and plant growth-promoting rhizobacteria in abiotic stresses: consequences for changing environment. Environ. Sci. Pollut. Res. 22 (7), 4907–4921. https://doi.org/10.1007/s11356-014-3754-2.

Fang, J., Yu, G., Liu, L., Hu, S., Stuart Chapin, F., 2018. Climate change, human impacts, and carbon sequestration in China. Proc. Natl. Acad. Sci. U. S. A. 115 (16), 4015–4020. https://doi.org/10.1073/pnas.1700304115.

Francis, C., Beman, J., Kuypers, M., 2007. New processes and players in the nitrogen cycle: the microbial ecology of anaerobic and archaeal ammonia oxidation. ISME J. 1 (1), 19–27.

García-Rivero, M., Peralta-Pérez, M.R., 2008. Cometabolismo en la biodegradación de hidrocarburos. Revista Mexicana de Ingeniera Quimica 7 (1), 1–12. http://www.rmiq.org/Pdfs/Vol%207%20No%201/RMIQ_Vol7No1_1.pdf.

Glick, B.R., 2014. Bacteria with ACC deaminase can promote plant growth and help to feed the world. Microbiol. Res. 169 (1), 30–39. https://doi.org/10.1016/j.micres.2013.09.009.

Gupta, G., Parihar, S.S., Ahirwar, N.K., Snehi, S.K., Singh, V., 2015. Plant growth promoting rhizobacteria (PGPR): current and future prospects for development of sustainable agriculture. J. Microb. Biochem. Technol. 7 (2), 96–102.

Haiyan, W., Stuanes, A.O., 2003. Heavy metal pollution in air-water-soil-plant system of Zhuzhou City, Hunan Province, China. Water Air Soil Pollut. 147 (1–4), 79–107. https://doi.org/10.1023/A:1024522111341.

Han, W., Kemmitt, S.J., Brookes, P.C., 2007. Soil microbial biomass and activity in Chinese tea gardens of varying stand age and productivity. Soil Biol. Biochem. 39 (7), 1468–1478. https://doi.org/10.1016/j.soilbio.2006.12.029.

Hararuk, O., Smith, M.J., Luo, Y., 2015. Microbial models with data-driven parameters predict stronger soil carbon responses to climate change. Glob. Chang. Biol. 21 (6), 2439–2453. https://doi.org/10.1111/gcb.12827.

Hashem, A., Abd Allah, E.F., Alqarawi, A.A., Radhakrishnan, R., Kumar, A., 2017. Plant defense approach of *Bacillus subtilis* (Bera 71) against macrophomina phaseolina (tassi) goid in mung bean. J. Plant Interact. 12 (1), 390–401. https://doi.org/10.1080/17429145.2017.1373871.

Hayat, R., Ali, S., Amara, U., Khalid, R., Ahmed, I., 2010. Soil beneficial bacteria and their role in plant growth promotion: A review. Ann. Microbiol. 60 (4), 579–598. https://doi.org/10.1007/s13213-010-0117-1.

Hillel, D., Hatfield, J.L., 2005. Encyclopedia of Soils in the Environment. Vol. 3 Academic Press.

Hou, J., Liu, W., Wang, B., Wang, Q., Luo, Y., Franks, A.E., 2015. PGPR enhanced phytoremediation of petroleum contaminated soil and rhizosphere microbial community response. Chemosphere 138, 592–598. https://doi.org/10.1016/j.chemosphere.2015.07.025.

Howarth, R.W., 2008. Coastal nitrogen pollution: A review of sources and trends globally and regionally. Harmful Algae 8 (1), 14–20. https://doi.org/10.1016/j.hal.2008.08.015.

Huson, D.H., Weber, N., 2013. Microbial community analysis using MEGAN. In: Methods in Enzymology. Vol. 531. Academic Press Inc, pp. 465–485, https://doi.org/10.1016/B978-0-12-407863-5.00021-6.

Johnson, D., Leake, J.R., Read, D., 2002. Transfer of recent photosynthate into mycorrhizal mycelium of an upland grassland: short-term respiratory losses and accumulation of ^{14}C. Soil Biol. Biochem. 34 (10), 1521–1524.

Johnson, P.T.J., Townsend, A.R., Cleveland, C.C., Glibert, P.M., Howarth, R.W., Mckenzie, V.J., Rejmankova, E., Ward, M.H., 2010. Linking environmental nutrient enrichment and disease emergence in humans and wildlife. Ecol. Appl. 20 (1), 16–29. https://doi.org/10.1890/08-0633.1.

Karlen, D.L., Mausbach, M.J., Doran, J.W., Cline, R.G., Harris, R.F., Schuman, G.E., 1997. Soil quality: a concept, definition, and framework for evaluation: (A guest editorial). Soil Sci. Soc. Am. J. 61 (1), 4–10. https://doi.org/10.2136/sssaj1997.03615995006100010001x.

Khan, S., Cao, Q., Hesham, A.E.L., Xia, Y., He, J.Z., 2007. Soil enzymatic activities and microbial community structure with different application rates of Cd and Pb. J. Environ. Sci. 19 (7), 834–840. https://doi.org/10.1016/S1001-0742(07)60139-9.

Kumar, A., Bahadur, I., Maurya, B.R., Raghuwanshi, R., Meena, V.S., Singh, D.K., Dixit, J., 2015. Does a plant growth promoting rhizobacteria enhance agricultural sustainability? J. Pure Appl. Microbiol. 9 (1), 715–724. http://www.microbiologyjournal.org/.

Kuypers, M.M., Marchant, H.K., Kartal, B., 2018. The microbial nitrogen-cycling network. Nat. Rev. Microbiol. 16 (5), 263–276.

Lugtenberg, B., Faina, K., 2009. Plant-growth-promoting rhizobacteria. Annu. Rev. Microbiol. 63, 541–556.

Malla, M.A., Dubey, A., Kumar, A., Yadav, S., Hashem, A., Allah, E.F.A., 2019. Exploring the human microbiome: the potential future role of next-generation sequencing in disease diagnosis and treatment. Front. Immunol. 10. https://doi.org/10.3389/fimmu.2018.02868.

Mendes, R., Garbeva, P., Raaijmakers, J.M., 2013. The rhizosphere microbiome: significance of plant beneficial, plant pathogenic, and human pathogenic microorganisms. FEMS Microbiol. Rev. 37 (5), 634–663. https://doi.org/10.1111/1574-6976.12028.

Mesa-Marín, J., Del-Saz, N.F., Rodríguez-Llorente, I.D., Redondo-Gómez, S., Pajuelo, E., Ribas-Carbó, M., Mateos-Naranjo, E., 2018. PGPR reduce root respiration and oxidative stress enhancing *Spartina maritima* root growth and heavy metal rhizoaccumulation. Front. Plant Sci. 9, 1500.

Micallef, S.A., Shiaris, M.P., Colón-Carmona, A., 2009. Influence of *Arabidopsis thaliana* accessions on rhizobacterial communities and natural variation in root exudates. J. Exp. Bot. 60 (6), 1729–1742. https://doi.org/10.1093/jxb/erp053.

Mohammad Rusan, M.J., Hinnawi, S., Rousan, L., 2007. Long term effect of wastewater irrigation of forage crops on soil and plant quality parameters. Desalination 215 (1–3), 143–152. https://doi.org/10.1016/j.desal.2006.10.032.

Morales, S.E., Holben, W.E., 2011. Linking bacterial identities and ecosystem processes: can "omic" analyses be more than the sum of their parts? FEMS Microbiol. Ecol. 75 (1), 2–16. https://doi.org/10.1111/j.1574-6941.2010.00938.x.

Nadeem, S.M., Ahmad, M., Zahir, Z.A., Javaid, A., Ashraf, M., 2014. The role of mycorrhizae and plant growth promoting rhizobacteria (PGPR) in improving crop productivity under stressful environments. Biotechnol. Adv. 32 (2), 429–448. https://doi.org/10.1016/j.biotechadv.2013.12.005.

Nihorimbere, V., Ongena, M., Smargiassi, M., Thonart, P., 2011. Effet bénéfique de la communauté microbienne de la rhizosphère sur la croissance et la santé des plantes. Biotechnol. Agron. Soc. Environ. 15 (2), 327–337. http://www.pressesagro.be/base/text/v15n2/327.pdf.

Osaigbovo, A.U., Orhue, E.R., 2006. Influence of pharmaceutical effluent on some soil chemical properties and early growth of maize (*Zea mays* L). Afr. J. Biotechnol. 5 (18), 1612–1617. http://www.academicjournals.org/AJB/PDF/pdf2006/18Sep/Osaigbovo%20et%20al.pdf.

Rattan, R.K., Datta, S.P., Chhonkar, P.K., Suribabu, K., Singh, A.K., 2005. Long-term impact of irrigation with sewage effluents on heavy metal content in soils, crops and groundwater—a case study. Agric. Ecosyst. Environ. 109 (3–4), 310–322.

Reddy, S., Jabez, W.O., 2020. Biodegradation and biosorption of Reactive Red 120 dye by immobilized *Pseudomonas guariconensis*: kinetic and toxicity study. Water Environ. Res. 92 (8), 1230–1241.

Reddy, S., Osborne, W.J., 2020. Heavy metal determination and aquatic toxicity evaluation of textile dyes and effluents using *Artemia salina*. Biocatal. Agric. Biotechnol. 25, 101574.

Rubin, R.L., van Groenigen, K.J., Hungate, B.A., 2017. Plant growth promoting rhizobacteria are more effective under drought: a meta-analysis. Plant Soil 416 (1–2), 309–323. https://doi.org/10.1007/s11104-017-3199-8.

Ruzzi, M., Aroca, R., 2015. Plant growth-promoting rhizobacteria act as biostimulants in horticulture. Sci. Hortic. 196, 124–134. https://doi.org/10.1016/j.scienta.2015.08.042.

Ryan, P.R., Delhaize, E., Jones, D.L., 2001. Function and mechanism of organic anion exudation from plant roots. Annu. Rev. Plant Biol. 52, 527–560. https://doi.org/10.1146/annurev.arplant.52.1.527.

Saravanamoorthy, M.D., Ranjitha Kumari, B.D., 2007. Effect of textile waste water on morphophysiology and yield on two varieties of peanut (*Arachis hypogaea* L.). J. Agric. Technol. 3 (2), 335–343.

Sasse, J., Martinoia, E., Northen, T., 2018. Feed your friends: do plant exudates shape the root microbiome? Trends Plant Sci. 23 (1), 25–41. https://doi.org/10.1016/j.tplants.2017.09.003.

Scheublin, T.R., Van Der Heijden, M.G.A., 2006. Arbuscular mycorrhizal fungi colonize nonfixing root nodules of several legume species. New Phytol. 172 (4), 732–738.

Schloss, P.D., Westcott, S.L., Ryabin, T., Hall, J.R., Hartmann, M., Hollister, E.B., Lesniewski, R.A., Oakley, B.B., Parks, D.H., Robinson, C.J., Sahl, J.W., Stres, B., Thallinger, G.G., Van Horn, D.J., Weber, C.F., 2009. Introducing mothur: open-source, platform-independent, community-supported software for describing and comparing microbial communities. Appl. Environ. Microbiol. 75 (23), 7537–7541. https://doi.org/10.1128/AEM.01541-09.

Selvakumar, G., Panneerselvam, P., Ganeshamurthy, A.N., 2012. Bacterial mediated alleviation of abiotic stress in crops. In: Bacteria in Agrobiology: Stress Management. Vol. 9783642234651. Springer-Verlag, Berlin, Heidelberg, pp. 205–224, https://doi.org/10.1007/978-3-642-23465-1_10.

Sherene, T., 2017. Role of soil enzymes in nutrient transformation: a review. Bio Bull 3 (1), 109–131.

Shivakumar, D., Srikantaswamy, S., 2012. Heavy metals pollution assessment in industrial area soil of Mysore city, Karnataka, India. Int. J. Appl. Sci. Eng. Res. 1, 604–611. https://doi.org/10.6088/ijaser.0020101062.

Shraddha, Shekher, R., Sehgal, S., Kamthania, M., Kumar, A., 2011. Laccase: microbial sources, production, purification, and potential biotechnological applications. Enzyme Res. 2011 (1). https://doi.org/10.4061/2011/217861, 217861.

Singh, B.K., Bardgett, R.D., Smith, P., Reay, D.S., 2010. Microorganisms and climate change: terrestrial feedbacks and mitigation options. Nat. Rev. Microbiol. 8 (11), 779–790. https://doi.org/10.1038/nrmicro2439.

Sinha, A., Osborne, W.J., 2016. Biodegradation of reactive green dye (RGD) by indigenous fungal strain VITAF-1. Int. Biodeterior. Biodegrad. 114, 176–183. https://doi.org/10.1016/j.ibiod.2016.06.016.

Sinha, A., Lulu, S., Vino, S., Banerjee, S., Acharjee, S., Osborne, W.J., 2018. Degradation of reactive green dye and textile effluent by *Candida* sp. VITJASS isolated from wetland paddy rhizosphere soil. J. Environ. Chem. Eng. 6 (4), 5150–5159.

Spaepen, S., Vanderleyden, J., 2011. Auxin and plant-microbe interactions. In: Cold Spring Harbor Perspectives in Biology. Vol. 3.

Supramaniam, Y., Chong, C.W., Silvaraj, S., Tan, I.K.P., 2016. Effect of short term variation in temperature and water content on the bacterial community in a tropical soil. Appl. Soil Ecol. 107, 279–289. https://doi.org/10.1016/j.apsoil.2016.07.003.

Torsvik, V., Øvreås, L., 2002. Microbial diversity and function in soil: from genes to ecosystems. Curr. Opin. Microbiol. 5 (3), 240–245. https://doi.org/10.1016/S1369-5274(02)00324-7.

Trivedi, P., Delgado-Baquerizo, M., Trivedi, C., Hu, H., Anderson, I.C., Jeffries, T.C., Zhou, J., Singh., B.K., 2016. Microbial regulation of the soil carbon cycle: evidence from gene–enzyme relationships. ISME J. 10 (11), 2593–2604.

Uren, N.C., 2007. Types, amounts, and possible functions of compounds released into the rhizosphere of soil-grown plants. In: Pinton, R., Varanini, Z., Nannipieri, P. (Eds.), The Rhizosphere: Biochemistry and Organic Substances at the Soil-Plant Interface. Marcel Dekker, New York, pp. 19–40.

van der Heijden, M.G., Bardgett, R.D., van Straalen, N.M., 2008. The unseen majority: soil microbes as drivers of plant diversity and productivity in terrestrial ecosystems. Ecol. Lett. 11, 296–310. https://doi.org/10.1111/j.1461-0248.2007.01139.x.

Vejan, P., Abdullah, R., Khadiran, T., Ismail, S., Nasrulhaq Boyce, A., 2016. Role of plant growth promoting rhizobacteria in agricultural sustainability – a review. Molecules 21 (5). https://doi.org/10.3390/molecules21050573.

Vibha, B., Neelam, G., 2012. Importance of exploration of microbial biodiversity. Int. Res. J. Biol. Sci. 1 (3), 78–83.

Vogelsang, K.M., Reynolds, H.L., Bever, J.D., 2006. Mycorrhizal fungal identity and richness determine the diversity and productivity of a tallgrass prairie system. New Phytol. 172 (3), 554–562.

Xie, X., Zhang, H., Paré, P.W., 2009. Sustained growth promotion in Arabidopsis with long-erm exposure to the beneficial soil bacterium *Bacillus subtilis* (GB03). Plant Signal. Behav. 4 (10), 948–953. https://doi.org/10.4161/psb.4.10.9709.

Yadav, R.K., Goyal, B., Sharma, R.K., Dubey, S.K., Minhas, P.S., 2002. Post-irrigation impact of domestic sewage effluent on composition of soils, crops and ground water—a case study. Environ. Int. 28 (6), 481–486. https://doi.org/10.1016/S0160-4120(02)00070-3.

Zhang, H., Sun, Y., Xie, X., Kim, M.S., Dowd, S.E., Paré, P.W., 2009. A soil bacterium regulates plant acquisition of iron via deficiency-inducible mechanisms. Plant J. 58 (4), 568–577. https://doi.org/10.1111/j.1365-313X.2009.03803.x.

Zhang, W., Lu, Z., Yang, K., Zhu, J., 2017. Impacts of conversion from secondary forests to larch plantations on the structure and function of microbial communities. Appl. Soil Ecol. 111, 73–83. https://doi.org/10.1016/j.apsoil.2016.11.019.

CHAPTER 3

Microorganisms as biocontrol agents for sustainable agriculture

Bhupendra Koul[a],*, **Manpriya Chopra[b]**, **and Supriya Lamba[c]**
[a]School of Bioengineering and Biosciences, Department of Biotechnology, Lovely Professional University, Phagwara, Punjab, India
[b]Department of Sciences, Bhagwan Mahaveer Public Senior Secondary School, Banga, Punjab, India
[c]Department of Botany, Kamla Nehru College for Women, Phagwara, Punjab, India
*Corresponding author: e-mail address:

1. Introduction

Biocontrol agents are the organisms or microorganisms that control the plant pathogen. They play a key role in resisting pest population and prevent the crop from various diseases like leaf wilt, curling of disease, root rot disease, crown gall disease, etc. They kill the pest population before they spread the disease (Corn et al., 2014). Biocontrol agents are used at large scales as they do not cause any harm to the main crop and kill the pest population naturally. Many bacteria (*Bacillus* sp., *Pseudomonas* sp., etc.) and fungi (*Trichoderma* sp., *Candida* sp. etc.) are used as biocontrol agents. The major benefit of these biocontrol agents is that the hosts do not consume them, and thus host organism remains unaffected (Savita and Anuradha, 2019).

Fertilizers and pesticides are used to prevent the crop from various pests. These pesticides and chemical fertilizers effectively increase the crop yield, but it also shows a negative impact on human health and atmosphere. The overdose of these fertilizers interrupts the ecological balance of biotic and abiotic components of the ecosystem (Bale et al., 2008). If a person is accidentally exposed to any chemical fertilizer or pesticide, muscular or nerve disorder may appear in the person. Regular contact with such chemicals may lead to sweating, headache, muscle weakness, and protracted exposure may lead to alteration in work of excretory organ, central nervous system, TSH, bladder, and liver (Colnot and Dekant, 2016). Long-term exposure to such pesticides may weaken the immune system, affect cellular respiration, and cause skin cancer and skin sensitization (Anderson and Meade, 2014). Continuous use of fertilizers and pesticides in the field also pollute the soil and decrease the growth of beneficial microbes in the soil. The government has banned many hazardous pesticides in many countries (Shakir et al., 2016).

Researchers are developing alternatives to reduce the effect of such pesticides and fertilizers. Biocontrol agents are one of the alternatives to control pests in various crops (Bale et al., 2008). This term by first used by Harry Scott Smith. Biocontrol agents are used in both entomology and plant pathology to reduce the growth of microbes, predatory insects, entomopathogenic nematodes (Wratten, 2008). The microbes used in both (entomology and plant pathology) to kill or control the pest population are named 'biocontrol agents' or 'biological control agents'. Moreover, the extracts or fermented products are taken from various natural sources can also act as biocontrol agents. These products or extracts can be used in any form (pure or mixture), having various effects on target microbes or pathogens (Usta, 2013; Siegwart et al., 2015). According to USRC (US Research Council), biocontrol agents are natural or modified organisms, genes or a gene product used to suppress the growth of the unwanted organism (pest/pathogen) and to enhance the growth and yield of the wanted organism (crops, microorganism, or insects) (Sandhu et al., 2012). Biological control agents reduce the activity of pathogens to damage the target organism, hence called 'natural enemies' (Heydari and Pessarakli, 2010). The utilization of fungi and bacteria is done for its various profit-oriented products to kill the pests, weeds, and to control diseases (Butt et al., 2001; Raaijmakers et al., 2002; De Silva et al., 2019). It is found that their association enhances the productivity of the plant, and contributes to animal and human health (Savita and Anuradha, 2019). Thus, bacteria and fungi are extensively used as biocontrol agents to prevent the crop from diseases and enhance both the quality and quantity of food (Singh et al., 2019).

2. Bacteria and fungi as biocontrol agents

Several bacteria and fungi are known for their biocontrol activity. The first biocontrol activity of bacteria was reported in 1979 by bacterium *Agrobacterium radiobacter* strain K 84 against crown gall disease. This was registered with USEPA (US Environmental Protection Agency), and the foremost fungus used was *Trichoderma harzianumi* ATCC 20476 was used to control the diseases of plants. Till 2013, only 14 bacteria and 12 fungi species have been noticed and registered with EPA that control pest population or kill pathogens (Fravel, 2005). It was patented by the environmental protection agency (EPA). One cannot use bacteria and fungi directly as a biocontrol agent. Researchers have developed products using such bacteria and fungi to sway away from the pest population. Production of products is a long process involving different steps.

3. Production of desired product

The production of such biocontrol agents is a long and multi-step process that includes a wide range of activities. Activities involved in the commercialization of such biocontrol agents or their products are:

3.1 Isolation of desired microorganism from natural environment

The success of the biocontrol agent depends upon how the searching screening processes have been carried out. Both processes depend upon the pathogen to be targeted, the crop and cropping system with which the pest is related. The selection of isolates with desirable traits is made only after a good screening process. These isolates are screened for their biocontrol activity against the pathogen in both conditions (*in vitro* and glasshouse conditions). The main focus is to isolate, screen, and select those fungi and bacteria that have the ability to produce the biocontrol activity against the pathogens *in vitro* or *in vivo* conditions. They are screened using various media in a way to regulate the activity of fungi by the formation of clear zones (Berg et al., 2005).

3.2 Evaluation of the biocontrol agent in both conditions (*in vitro* and glasshouse)

The isolates collected are then evaluated for their efficiency to kill pathogens in both lab conditions as well as in greenhouse conditions. If an isolate does not show an effective result in field, then before rejecting it is again subjected to lab conditions to evaluate the reason for its inefficiency.

3.3 Testing of isolates in field conditions

Those isolates that show effective results in both *in vitro* and greenhouse conditions are then exposed to field conditions. Their activity is observed regularly. Once the isolate provides good results, it is then sent for its mass production. Otherwise, work was done on it in lab conditions to increase its efficiency in fields. Only then biocontrol agents are exposed to natural settings of the environment when the isolate shows positive results in the natural environment, then its moves further for mass production (Samish et al., 2014).

3.4 Mass production

Mass production is one of the biggest obstacles in the commercialization of any biocontrol agent. It is a time-consuming step. Mass production is a major step of commercialization. It can be done liquid, solid-state, etc., type of fermentation to produce the isolate in great amount for its biocontrol activity (De Cal et al., 2010).

3.5 Formulation of desired product

The formulation is a process in which active ingredients are blended with inert material (diluents or surfactants) in order to change the physical characters of microorganisms to a better form. The formulation is used to stabilize the fungi isolates during production. Proper distribution and storage of products are done. It assists in the proper handling and application of the product. This process should keep in mind to protect the agent from unhealthy environmental factors, and thus it enhances the activity of organisms (Wraight et al., 2001). A complete, final formulation product must be:
- Easy to handle.
- Stable at different temperature conditions (-5 to $35°C$).
- Minimum shelf life of 2 y (at room temperature).

3.6 Registration

It means that the product from the biocontrol agent got patented in the literature and got ready to be released in the market. It should keep in mind some major things while registering are the proper name of the biocontrol agent, its morphological description, its habitat of living, toxic percentage to animals and humans. It involves two essential factors of biocontrol agents. These are:

- Environmental fate
- Toxicity

In India, total 12 biopesticides are registered under Insecticide Act, 1968. Various include neem-based pesticides, *Bacillus thuringiensis*, *Trichoderma*, NPV are major biocontrol agents used in India (Montesinos, 2003). Registration of the biocontrol agent or any bio-pest is done only after having certain required information which is provided under section 9(3) of the pesticide act of India, 1968. This information includes:

- Common and systematic name of biocontrol/bio pest agent
- Natural habitat
- Morphological description
- Detailing of the manufacturing process
- Toxicity level in mammals
- Toxicity level to the environment
- Residual analysis

Table 1 shows different products manufactured using different strains of bacteria and fungi (Junaid et al., 2013). Bacteria and fungi are excessively used as biocontrol agents in controlling the pest population without harming the main crop. Following is the table showing the potential of different microorganisms in controlling the pest or pathogen population in various fields. Table 2 summarizes the reports on the efficiency of bacteria and fungi against specific microbe/pathogens.

Like bacteria, various fungal BCAs are also used where the chemical fertilizers are totally prohibited or terminated (Thambugala et al., 2020). The antagonistic property of various biocontrol agents like *Trichoderma viride, Beauveria bassiana, Fusarium oxysporium* makes them useful for the control of pathogens in crops like rice, tomato, sunflower, pepper, cabbage, etc. (Rajeshkumar, 2019). Table 3 summarizes the reports on the use of fungi as biocontrol agents providing different levels of protection.

4. Mechanisms of biocontrol agents

The mechanism of biocontrol agents has been studied to know how it works and under what conditions it gives effective results. It provides optimization of control and helps screen and isolate more effective strains of biocontrol agents. Biocontrol agents include many different mechanisms that help in the control of the pathogen or pest population.

Table 1 List of commercialized products used as biocontrol agents.

Bio control agents	Strain	Pathogen/disease	Host	Product formed	Manufacturer
Bacteria					
Agrobacterium radiobacter	**(a)** 84 **(b)** K 1026	*A. tumefaciens*	Nuts, fruits ornamental	Galtrol Nagol	Agbiochem, Bio-care, UK
Bacillus subtilis	**(a)** GB34 **(b)** GB03	*Rhizoctonia fussarium* *R. Aspergillus*	Soyabean, wheat, barely, peas	GB34, Companion, Kodiac	Gustafon, Growth products USA
Pseudomonas aurofaciens	TX-1	*R. solani* *Phythium* spp.	vegetables	Bio-jet, spot less	EcoSoil, Canada
P. flurorescence	A506	Bunch rot, fire blight	Potato, tomato	Frostban	PHT (Plant health technologies), USA
Streptomycinegriseoviridis	–	Soil borne pathogen	Tree seedlings	Mycostop	Kemira Agro Oy, Finland
Trichoderma harzianum	**(a)** T-22 **(b)** T-39	Soil borne pathogens *B. cinerea*	Food crops	Root shield, trichodex	Bio Works, USA
A. quisquallis	M-10	Powdery mildew	Vegetables, fruits	Trichodex	Ecogen, USA
Pseudomonas protegens	AS15	*A. flavus*	cotton	Alfa guard	Circleonegloba, USA
G. catenulatum	JI446	Soil borne pathogens	Spices	Prima stop soil guard	Kemira Agro Oy, Finland
G. virens	GL-21	Nematodes (parasitic)	Fibers	Prima stop soil guard	Kemira Agro Oy, Finland
Fungi					
Trichoderma harzianum	T-22	Soil borne pathogens	Green house nurseries	Root shield Plant shield	Bio works, USA
Aspergillus flavus	AF36	*Aspergillus flavus*	Cotton	Alfa guard	Circleonegloba, USA
Ampelomycesquisqualis	M-10	Powdery mildew	Vegetables, fruits	Trichodex	Ecogen, USA
Gliocladiumcatrnulatum	JI446	Soil borne pathogens	Spices	Prima stop soil guard	Kemira Agro Oy, Finland
Trichoderma harzianum	T-39	*Botrytis cinerea*	Most of food crops	Trichodex	Bio works, USA
Gliocladium virens	GL-21	Nematodes (parasitic)	Fibers	Prima stop soil guard	Kemira Agro Oy, Finland

Table 2 Reports on bacteria as bio control agents against different pathogens.

Bacteria	Class	Pathogen	Host	Level of protection	References
Lactic Acid Bacteria (LAB)	Bacilli	*Listeria monocytogenes*	Cheese, Alfaalfa, raddish	Two traditional Sicilian cheeses PS (*Pecorino siciliano*) and VD (Vastedda della valle) were used in combination with LAB to control L. *monocytogenes*. • PS + LAB = slightly reduction in bacteria. • VD + LAB = complete elimination of bacteria is observed.	Scatassa et al. (2017)
Lactic Acid Bacteria (LAB)	Bacilli	*Pythium ultimum*	Roots of tomato	Reduced the growth of *Pythium* by 60%.	Lutz et al. (2012)
B. cereus UW85	Bacilli	*Phytophthora medicaginis*	Alfaalfa	Reduced seedling mortality to 0%. Healthy plants were produced.	Handelsman et al. (1990)
Bacillus amyloliquefaciens	Bacilli	*Aspergillus flavus* *Aspergillus parasiticus*	Grains Legumes Pistachio	• Controlled growth of *Aspergillus falvus* to 72%–75%. • Degraded 4 alfatoxins during first three days after inoculation in *Aspergillus parasiticus*. • Inhibited the growth of bacteria.	Lahlali and Jijaki (2007)
Bacillus subtilis	Bacilli	*Phytophthora medicaginis*	Alfaalfa, Chickpea	Reduced the growth of bacteria by inhibiting alfatoxin production.	Federici (2007)

Bacillus thuringiensis	Bacilli	Larva of moth, flies and beetles, cabbage loopers, tomato hornworm, mosquitoes	Cotton Corn	• showed its maximum effect when applied to caterpillars during 1st and 2nd instar stage. • 95%–100% control was done by Bt-i in cabbage loopers, caterpillers • Bt-k controlled mosquito growth rate to an appropriate level.	Federici (2007)
Bacillus sphaericus	Bacilli	Mosquito: Culex, Anopheles	Ornamental plants	• Released toxin in gut after 15 min of ingestion. • Controlled pest population to 92%–99%.	Federici (2007)
Paenibacilluspopilliae	Bacilli	Japanese bettle	Ornamental plants	• 92%–100% pest population was controlled after implication.	Federici (2007)
*PaenibacillusjamilaeHS-*26	Bacilli	*Fusarumoxysporum, Botryosphaeria ribis, Rhizoctonia solani , Bipolarissorokiniana*	Banana Cucumber	• 42%–55% population of pest was controlled. • HS-26 strain showed 82%–85% antagonistic effect.	Wang et al. (2019)
Rhizobium etli strain G12 + *Fusarium oxysporum strain* Fo162	Alphaproteobacteria	Root knot nematods (*Meloidogyne* spp.)	Tomato	• Reduction in root penetration by *M. incognita.* • Reduction in development and reproduction of *M. incognita.*	Martinuz et al. (2013)
Burkholderiacepacia BC11	Betaproteobacteria	*Rhizoctonia solani*	Potato Cotton	• Showed antifungal activities. • Effectively controlled damping off disease in cotton (89%–92%).	Lahlali and Jijaki (2007)

Continued

Table 2 Reports on bacteria as bio control agents against different pathogens—cont'd

Bacteria	Class	Pathogen	Host	Level of protection	References
Pseudomonas fluorescence	Gammaproteobacteria	*Rhizoctonia solani, Phythium* sp., *Erwinia carotovora, F. Glycinia, P. ultimum, Sarocladiumoryzae, Gaeumannomycesgraminis var tritici*, Growth of virulent bacteria	Cotton Sugarbeet Potato Soyabean Rice Wheat	• Controlled the growth of pathogen (*Rhizoctonia solani*) by 79%. • Enhanceed growth of crops by 26.3–52.6%. • Reduced disease rate to 60.0%. • Increased crop/grain yield to 33.8%.	De Souza et al. (2003) Shaikh and Sayyed (2015)
Pseudomonas fluorescens CHAO	Gammaproteobacteria	*Thielaviopsis basicola Pythium ultimum*	Pea Tobacco	• Increased shoot weight by 35%, root weight by 52%. • Increased number of lateral roots.	Berg et al. (2005)
Serratia entomophila	Gammaproteobacteria	*Costelystrazealandica*	Grass grub	• Treated infested pastures in New Zealand effectively.	Federici (2007)
Pseudomonas putida	Gammaproteobacteria	*Fusaria* spp., *E. carotovora*	Potato Beans Cucumber Radish Tomato	• Reduced wilt in main crop. • C7 strain-controlledroot rot of tomato and fusarium crown. • About 78%–83% of plant growth was enhanced.	Shaikh and Sayyed (2015)
Pseudomonas cepacia	Gammaproteobacteria	*Colletotrichum gloeosporiodes, Bipolarismaydis, F. oxysporum*	Mango, Maize, Onion, Tomato	• More than 90% colonization of the pathogen has reduced. • Decreased disease rate by 30%–65%.	Shaikh and Sayyed (2015)

Table 3 Reports on fungi as biocontrol agents against different pathogens.

Fungal species	Class	Pathogen	Host	Level of protection	References
Verticillium lecanii	Sordariomycetes	*Penicillium digitatum*	Citrus fruits	Prevented colonization of the pathogen in the exocarp and mesocarp.	Benhamou (2004)
Pichia anomala Strain K	Saccharomycetes	*Botrytis cinerea*	Apples	Significant protection from *B. cinerea*.	Gisbert (2007) and Grevesse et al. (2003)
Trichoderma harzianum	Sordariomycetes	*Rhizoctonia solani*, *Pythium ultimum*	Alfaalfa, cocunut	Increased survival of plants by 12%.	Limon et al. (1999), Woo et al. (1999)
Trichoderma longibrachiatum	Sordariomycetes	*Pythium ultimum*	Cucumber seedlings.	*T. longibrachiatum* transformants (having extra copy of *eglI* gene) suppressed the growth of *P. ultimum*. (biocontrol).	Migheli et al. (1998)
Trichoderma virens	Sordariomycetes	*Sclerotium rolfsii*, *Rhizoctonia solani*	Grapevine	Controlled 15 plant pathogens. Provided 82.2% protection against *Aspergillusniger*.	Mukherjee et al. (2004)
Acremonium byssoides	Sordariomycetes	*Oidium heveae*	Water hyacinth	56%–65% pathogen get destroyed.	Hawksworth (1981)
Cercosporarodmanii	Dothideomycetes	*Eichhornia crassipes*	Vegetables Fruits	Spray of fungicide at high pressure (100 psi) and high water volume(15 to 20 gal/ac) gives good result.	Butt et al. (2001)
Ampelomycesquisqualis	Dothideomycetes	*Erysiphales*	Cucumber	Controlled 59%–89% of pest population.	Butt et al. (2001)

Continued

Table 3 Reports on fungi as biocontrol agents against different pathogens—cont'd

Fungal species	Class	Pathogen	Host	Level of protection	References
C.gloeosporioides	Dothideomycetes	*Cuscutachinesis, C. australisin* soybeans	Ornamental vegetables	Extract with 2.5% chitosan showed good results.	Butt et al. (2001)
Metarhizium anisopliae	Sordariomycetes	*Schistocerca gregaria*	Food crops	Affected the life cycle of pest (5th nymphal instar).	Shairra et al. (2015)
Paecilomyces fumosoroseus	Eurotiomycetes	Nymphs of *Aleyrodidae*	Rice	30% of the pest population get destroyed.	Kim et al. (2008)
Paecilomyces lilacinus	Eurotiomycetes	Cyst nematodes and root-knot	Tomato	Pre-plant application is sufficient to control *Meloidogyne incognita.*	Kiewnick and Sikora (2006)
Puccinia chondrillina	Pucciniomycetes	Rush skeleton weed of *Chondrillajuncea*	Rice	At 0.05 mM fungus control the growth of pest.	Hasan (1973)
Gliocladium virens	Sordariomycetes	Root pathogens and damping-off	Ground nut	73–76% control of the pest population.	Butt et al. (2001)
Nomuraearileyi	Sordariomycetes	*Spodoptera litura*	Cash crops	Enhanced crop production by 43%.	Ignoffo and Clemente (1985)
Peniophora gigantea	Agaricomycetes	*Heterobasidionannossum*	Annosum	Controlled pest population before infection spreads.	Butt et al. (2001)
Beauveria bassiana, Paecilomyces fumosoroseus, Metarhizium anisopliae, Pythium oligandrum	Sordariomycetes	*Helicoverpaarmigera*	Lab investigation	Offered 87% protection to the plant. Biocontrol of larvae and pupae.	Nguyen et al. (2007)
Pythium oligandrum	Oomycete	*Pythium ultimum*	Sugarbeet	100% inhibit the growth of pest.	Butt et al. (2001)
Verticillium lecanii	Sordariomycetes	*Trileurodesvaporariorum, Myzuspersicae*	Ornamental	Before infection killed 98–99% of the larval population.	Chandrashekharaiah et al. (2013)

There are various different mechanisms used to control the pathogens that cause diseases. Different interactions occur between the microorganisms, which control the pathogens (Islam et al., 2005). Biological control occurs due to interactions occurring in different environmental conditions, which results in mutual relations between microorganisms and plant hosts (Bull et al., 2002). Broadly two types of mechanisms are involved: Indirect antagonisms and direct antagonism. A mixed path antagonism mechanism is also observed in some cases.

Direct antagonism in the bacterial population arises due to direct contact of bacteria with the pathogen or pest, while indirect antagonism is due to the result of competition by the biocontrol agent for the binding site at which pathogen or pest attacks. These mechanisms involve the following interactions:

- Hyperparasitism
- Competition
- Secretion of Lytic enzymes
- Plant growth promotion
- Induced systemic host resistance
- Antibiosis

4.1 Hyperparasitism

It is the most efficient form of direct antagonism (Pal and Gardener, 2006). In hyperparasites, firstly, biocontrol agent shows tropic growth towards the target (pest or population) organism, then coils around it, attack it, and then dissolves the cell membrane or cell wall of the target organism by enzymatic activity (Tiwari, 1996). In bacteria, hyperparasitism is rare. Only a few bacteria of Bacillus species show this phenomenon. In this case, a particular biocontrol agent attacks the pathogen directly that will kill the pathogen or its propagules. There are four major groups of hyperparasites that are hypoparasites, predators, facultative parasites, and obligate parasites. It includes various fungal hypoparasites which attacks the sclerotia, e.g., *Coniothyrium minitans*, attacks hyphae of fungi, e.g., *Pythium oligandrum*, fungi which attack pathogens of powdery mildew, e.g., *Gliocladium virens*, the sole pathogen of fungus ferociously attacked by numerous hyperparasites, e.g., *Cladosporium* (Milgroom and Cortesi, 2004). There are fungi like *Trichoderma,* which shows the predatory action under nutrient-restricted conditions and forms diverse enzymes that act against the cell walls of various fungi, which are pathogenic in nature (Benhamou and Chet, 1997).

4.2 Competition

In this mechanism, both pathogen and microorganism used as biocontrol agents compete for food, shelter, and nutrients. It is a type of indirect mechanism as biocontrol agents do not directly interact with the pathogen but indirectly kill it or reduce its population by competing with it for space and nutrients without harming the host. They kill or suppress the pest population before they spread disease as these microbes depend upon certain substances (fatty acids, acetaldehyde, ethanol) to overcome their dormant stage, and biocontrol agents exert competition for these substances or stimulants and thus decrease their disease-causing ability. An important factor in biocontrol is the competition occurring between pathogens and non-pathogens. The most effective BCAs competes for nutrients and thus leads to show protection of plants from phyopathogens (Jiang et al., 2019).

4.3 Antibiosis

It is a process of biological interaction that occurs between two or more organisms that are damaging to at least one of them. Many microbes release various compounds which show activity that is antibiotic in nature (Shahrak et al., 2009). Antibiotics help in repressing the magnification of the pathogen to be targeted *in situ* or *in vitro*. The effectual antibiotic should be generated in adequate doses close to the pathogen (Thomashow et al., 2002). Mainly the isolates of *Gliocladium* and *Trichoderma* show antibiotic activity. *Gliocladium virens* produce the gliotoxin that biocontrols *Rhizoctonia solani* and *Pythium ultimum* (Whilhite et al., 1994; Howell and Stipanovic, 1995).

4.4 Secretion of lytic enzymes

Several bacteria secrete metabolites that interfere with pest/pathogen growth and their activities. Bacteria and many other microorganisms secrete lytic enzymes, which hydrolyse many polymeric compounds (proteins, chitin, hemicellulose, cellulose and DNA). These secretions suppress the activity of pathogens. It is a type of mixed path antagonism. The other microbial by-products also decrease the pathogen's growth. For example, HCN production by *Fluorescent pseudomonads* suppresses the root pathogens.

4.5 Metabolite production

It means the formation of a metabolite which is the translational or final product of metabolism. It is usually the small molecules. There are two types, i.e., primary and secondary. The primary metabolites are related to growth, development and production, and secondary metabolites are produced by bacteria, fungi, or plants, which are not directly related to growth or development (Berdy, 2005). Biocontrol agents lead to the production of various metabolites that leads to interference of growth of pathogen activities. Metabolites like lytic enzymes help in breaking down various polymeric compounds like proteins, cellulose, hemicelluloses, chitin, DNA (Wilhite et al., 2001). For example, *Serratia marcescens* shows chitinase activity to control *Sclerotium rolfsii*, the root rot generated by *Fusarium oxysporum* (radicis-lycopersici in tomato) put down by chitosan (Lafontaine and Benhamou, 1996).

4.6 Plant growth promotion

Bacterial biocontrol agents control plant diseases by enhancing the growth of plants through following ways:

- Increase solubilization of nutrients
- Enhance root growth to increase nutrient uptake
- Nutrient sequestrations (Chaube et al., 2003)

Biocontrol agents also produce plant growth-promoting hormones, including cytokinins, gibberellins, auxins that promote the growth of the plant and reduce the effect of the pathogen.

4.7 Induced systemic host resistance

It is an indirect form of antagonism. It is the mechanism of resistance that is activated by infection. It has two modes of interaction, i.e., SAR pathway and the ISR pathway. Resistance induced may be systemic or local. *Avr* genes (avirulence genes, a group of proteins that provide defense mechanism in plants against certain diseases. They function as pathogen-specific elicitors of defense-related responses in the plant, which include corresponding R-gene (Resistance gene). It includes SAR (systematic acquired resistance) and ISR (induced systemic resistance). SAR is salicylic mediated, while ISR is mediated by ethylene or jasmonic acid, which are produced by the following applications of some non-pathogenic rhizobacteria. In SAR pathway elicit a hypersensitive response in which the pathogen is limited to a small area. Salicyclic acid plays a major role in the SAR pathway. In ISR pathway is induced by jasmonic acid. It includes *F. oxysporum*, *Trichoderma*, which shows ISR pathway and they are basically plant growth-promoting fungi (PGPF) that increase plant growth and crop yield (Pieterse et al., 2014).

4.8 Antibiosis

Antibiosis is the process of producing low molecular weight compounds (antibiotics) by microorganisms that directly affect the growth of pest or pathogenic populations (Weller, 1998). Antibiotics are produced by using several biocontrol agents. It is difficult to measure their effective quantities due to the production of small quantities compared to other, less toxic compounds available in the phytosphere. A biocontrol agent is said to be effective if it produces enough quantity of antibiotics in the vicinity of pathogen present in the plant (Chaube et al., 2003).

Different bacteria and fungi produce different antibiotics, and their mode of action differs in controlling the pest/pathogen population. It was estimated that by 2010, around 24,000–27,000 microbial antibiotics are known, but all among them, only a few works with efficiency. Biocontrol agents produce three kinds of antibiotics, i.e., volatile/non-polar, non-volatile/polar, water-soluble. Among these antibiotics, non-polar/volatile antibiotics are considered best as they are effective in the regions away from their site of production. Bacteria can produce more than one antibiotic and thus suppress the growth of more than one pathogen, as observed in *Bacillus cereus* spp. strain UW85 which produces two antibiotics kanosamine and zwittermycin. It increases the biological control efficiency of the bacteria. Table 4 and Table 5 show some bacterial and fungal determinants of ISR and SAR in different plants. Tables 6 and 7 show some antibiotics-antifungal produced by different bacteria-fungi to control pathogenic populations. Table 8 shows some of the antibiotics that are clinically useful.

Table 4 Bacterial determinants of ISR and SAR in different plants.

Bacteria	Strain	Host	Determinant	References
ISR (induced systemic resistance)				
B. mycoides	Bac J	Sugarbeet	Chitinase; peroxidise; beta 1,3 gluconase	Barbagus et al. (2004)
B. subtilis	GB03 IN937a	*Arabidopsis*	2,3 butanediol	Ryu et al. (2004)
P. fluorescence	WCS374 WCS417	Raddish *Arabidopsis* Tomato	Antibiotics; siderophores; Fe regulating factors; lipopolysaccharides.	Van Wees et al. (1997); Leeman et al. (1995)
SAR (systemic acquired resistance)				
P. fluorescence	CHAO	Tabacco	Siderophores	Maurhofer et al. (1994)

Table 5 Fungal determinants of ISR and SAR in different plants (Mandal and Ray, 2011).

Fungi	Strain	Host	Determinant	Type of resistance
Trichoderma harzianum	T-39	Cucumber	Chitinase, β-1,3 glucanase	ISR
Fusarium oxysporum	npF0	Cucumber	Peroxidase, Phenylalanine	SAR
Pythium oligandrum	–	Tomato	Oligandrin, elictin like protein	SAR
Penicillumoxalicum	–	Tomato, Tobacco	Glycoprotein	SAR

Table 6 Antibiotics produced by different bacteria.

Bacteria	Strain	Antibiotic	Pathogen	Disease	References
Agrobacterium radiobacter	–	Agrocin 84	*A. tumefaciens*	Crown gall	Kerr (1980)
Bacillus subtilus	AU195	Bacilomycin D	*Aspergillus flavus*	Reduced effect of alfatoxin	Moyne et al. (2001)
Bacillus amyloliquefaciens	FZB42	Bacilomycin D	*F. oxysporium*	Wilt	Koumoutsi et al. (2004)
B. cereus	UW85	Zwitermycin A.	*P. aphanidermatum*	Damping off	Smith et al. (1993)
B. subtilus	QST713	Iturin	*Botrytis, R. solani*	Damping off	Paulitz and Belanger, 2001
Lycobacter spp.	K88	Xanthobacin A	*A. cochliodes*	Damping off	Islam et al. (2005)
Pseudomonas fluorescence	F113	2,4 dicetylpholoroglucinol	*Pythium* spp	Damping off	Shanahan et al. (1992)
Pseudomonas aeruginosa	–	Fluoroquinolones carbapenems	*Pythium*	Damping off	Shanahan et al. (1992)
Pantoea spp.	C91	Herbicolin	*E. amylovora*	Fire blight disease	Sandra et al. (2001)
B. cepacea	–	Pyrrolnitrin	*P. oryzae*	Rice blast	Verma et al. (2020)
Saccharotrix	SA198	A4, A5	*Aspergillus*	Wilt,rot	Moyne et al. (2001)
Streptomycese	NTK937	Caboxamycin	*Anti-tumor*	Crown gall	Kerr (1980)
	NTK963	Lysolipin	*Anti bacterial*	Damping off	
Verrucosispora	AB 18–032	Abyssomicin C	*Gram positive bavcteria*	Daming off	Kerr (1980)

Table 7 Antibiotics produced by different fungi.

Fungi	Antifungal	Pathogen	Disease	References
Candida auris	Fluconazole, Anidulafungin	Mouse	Bloodstream, ear and eye infection	Ostrowsky et al. (2020)
Candida glabrata	Azole	Higher animals	Infection effects heart, brain, eyes	Pappas et al. (2016)
Aspergillus	Fluconazole	Animals	Bad germs in gut	Patterson et al. (2016)
Cryptococcosis	Cryptococcal injection	Virus	HIV/AIDS, Lungs and Nervous system disorder	Lortholary et al. (2011)
Trichoderma virens	Gliotoxin	*R. solani*	Root rot disease in plants	Wilhite et al. (2001)

Table 8 Antibiotics that are commercial biocontrol agents.

Biocontrol agents	Antibiotic	Pathogen/activity	Mode/site of action
Bacillus polymyxa	Polymyxin B	Gram negative bacteria	Cell membrane
B. subtilis	Bacitracin	Gram positive bacteria	Wall synthesis
Ephalosporium acremonium	Cephalosporin	Broad spectrum	Wall synthesis
Microspora purpurea	Gentamicin	Broad spectrum	Protein synthesis
Streptomyces spp.	Amphotericin B	Fungi	Protein synthesis
	Streptomycin	Broad spectrum	Cell membrane
	Vancomycin	Gram negative bacteria	
	Erythromycin	Gram positive bacteria	
	Tertacyclin		
	Neomycin		
Penicillium griseofulvum	Griseofulvin	Dermatophytic fungi	Act upon microtubules
Penicillium chrysogenum	Penicillin Oxacillin Dicloacillin Amoxicillin Ampicillin	Gram positive bacteria	Wall synthesis, antipseudomonal penicillins

There is a number of antibiotics available, having different classes and different modes of action. They show their peculiar activity towards different pathogens. Nowadays, even though antibiotics are widely used in the treatment of various diseases, they simultaneously show side effects on health. For example, streptomycin may cause tuberculosis, doripenem and meropenem are responsible for nausea, headache, allergic reactions, roxithromycin may cause jaundice, etc.

4.9 Actinomycetes

Actinomycetes are gram-positive bacteria that are terrestrial or aquatic in nature. They play a great role in maintaining the soil ecology. They are known to produce enzymes that degrades the lignin, chitin and other organic material. They usually form the hyphae during their development stages (Goodfellow and Williams, 1983).

Actinomycetes have the potential to be used as biocontrol agents in sustainable agriculture (Table 9). It can be used to control soil-borne diseases, foliar diseases, which may promote plant growth (Shimizu et al., 2000). To suppress the growth of soil pathogens, various actinomycetes help get rid of those pathogens. The *Gaeumannomyces graminis* a root disease eradicated by *Pseudomonas fluorescens* 2-79 (Thomashow and Weller, 1988). The clubroot of Chinese cabbage is the disease caused by *Plasmodiophora brassicae*, which

Table 9 Reports on actinomycetes as biocontrol agents.

Actinomycete species	Host plant species	Level of protection	References
Stretomyces, Nocardia, Microbispora	*Azadirachta indica*	• Showed significant antimicrobial activity against *Pythium* and *Phytophthora* sp.	Verma et al. (2009)
Glycomyces	*Sambucus adnata*	• Recovered from stem. • Provided 70% of protection.	Gu et al. (2007)
Saharoploysppora, Actinomadura	*Maytenus austroyunnanensis*	• 78% enhanced the growth of plant. • *Actinomadura* was recovered from stems and provide 19% protection from pest.	Qin et al. (2008)
Micromonospora, Nocardia, Stretomyces, Microbispora	*Alpinia galanga*	• Resisted the growth of *C. albicans* and *Colletotrichum* sp.	Taechowisan et al. (2008)
Streptomyces	*Thottea grandiflora*	• Inhibited growth of pathogenic bacteria (*Bacilus* spp., *Pleisiomonas* spp., *Pseudomonas*). • Showed antifungal activity, i.e., inhibited growth of *Aspergillus fumigates* by 44%, *Phytophthora* spp. by 23%, *Fusarium solani* by 62%, *G. candidum* by 23% and *Pythium ultimum* by 20%.	Ghadin et al. (2008)

Continued

Table 9 Reports on actinomycetes as biocontrol agents—cont'd

Actinomycete species	Host plant species	Level of protection	References
Stretomyces, Nocardioides	*Oryza sativa*	• *Streptomyces* controlled pathogen (fungi-*Pyricularia* sp.) that cause rice blast disease. • Showed maximum of 87.3% resistance to plants (*in vitro*). • *Streptomyces* acts as biofertilizers.	Tian et al. (2007)
Streptomyces, Micromonospora	*Triticum aestivum*	• Streptomyces affected the growth of *Fusarium graminearum* causing agent of FHB (Fusarium head blight). • Increased the growth of the crop.	Coombs and Franco (2003)
Streptosporangium, Streptomyces	*Zea mays*	• Total 53 isolates were recovered, 22 from roots and 31 from leaves. • 43.4% of isolated actinomycetes showed their antimicrobial activity against yeast and some species of bacteria.	De Araujo et al. (2000)
Microbispora, Micromonospora	*Brassica campestris*	• Enhanced the growth of main crop. • Showed antifungal activity.	Lee et al. (2008)
Actinoplanes	*Lupinus termis*	• It possessed antifungal activity. • Reduced germination of spores($P < .05$).	El-Tarabily (2003)
Actinoallimurus	*Acacia auriculiformis*	• Provided defense mechanism to plants. • Enhanced growth of a plant.	Thamchaipenet et al. (2010)
Nocardia, Pseudonocardia, Nonomuraea	*Aquilaria crassna*	• About 10 strains were isolated from roots and shoots. • Produced siderophores and Indole Acetic Acid (IAA). • Possessed antimicrobial, immunosuppressive, antitumor and some other pharmaceutical compounds.	Taechowisan et al. (2008)
Streptomyces	*Kalmia latifolia*	• Exhibited antifungal activity. • Protected the plant from the infection caused by *P. sydowiana*.	Nishimura et al. (2002)

Table 9 Reports on actinomycetes as biocontrol agents—cont'd

Actinomycete species	Host plant species	Level of protection	References
Streptomyces, Kitasatospora, Actinoplanes	*Taxus*	• Produced hydrolytic enzymes and thus growth of phytopathogens get suppressed. • Possessed antifungal activities.	Caruso et al. (2000)
Streptomyces	*Rhododendron*	• Showed less resistance against pathogen causing Pestalotia disease.	Shimizu et al. (2000)

are eliminated by *Microbispora rosea* and *Streptomyces olivochromogenes* (Lee et al., 2008). In control of foliar diseases like cucumber plant, cucumber anthracnose disease is caused by *Colletotrichum orbiculare* is controlled by *Streptomyces* sp. strain MBCu-56 (Shimizu et al., 2009). The actinomycetes, which show plant growth-promoting activity, release substances as indole acetic acid and siderophores. The *Streptomyces* sp.MBR-52 produces Indole acetic acid and Indole-3-pyruvic acid, which increases the growth and elongation of the plant (Meguro et al., 2006). Actinomycetes are considered biofertilizers as they enhanced soil fertility and the growth of beneficial microorganisms in the soil. Table 9 shows the use of different actinomycete species for medicinal plants, crop plants, and woody plants.

5. Conclusions

Biological control agents are considered best in controlling pest or pathogen populations as compared to insecticides and pesticides used in fields. They control the disease–causing agents without having any effect on soil and the environment. Thus, biocontrol agents are safe, effective, economical, and eco-friendly alternatives to synthetic agricultural pesticides. More biocontrol agents need to be formulated/discovered to enhance growth and curtail the pest-induced yield penalty. However, to get better results, one must have proper knowledge about the specificity/activity spectrum of the biocontrol agent. Suppose the governments are inclined towards sustainable agriculture. In that case, the joint efforts of agricultural scientists, private stakeholders, and volunteer farmers must be towards encouraging organic farming and integrated pest management using biocontrol agents.

References

Anderson, S.E., Meade, B.J., 2014. Potential health effects associated with dermal exposure to occupational chemicals. Env. Health Insights 2014, 8.

Bale, J.S., Van Lenteren, J.C., Bigler, F., 2008. Biological control and sustainable food production. Philos. Trans. R. Soc. B. 3631492, 761–776.

Barbagus, R.L., Zidack, N.K., Sherwood, J.E., Jacobsen, B.J., 2004. Screening for the identification of potential biological control agents that induce systemic acquired resistance in sugar beet. J. Biol. Control 302, 342–350.

Benhamou, N., 2004. Potential of the mycoparasite, *Verticillium lecanii*, to protect citrus fruit-against *Penicillium digitatum*, the causal agent of green mold, a comparison with-the effect of chitosan. Phytopathology 94, 693–705.

Benhamou, N., Chet, I., 1997. Cellular and molecular mechanisms involved in the interaction between *Trichoderma harzianum* and *Pythium ultimum*. Appl. Environ. Microbiol. 63 (5), 2095–2099.

Berdy, J., 2005. Bioactive microbial metabolites. J. Antibiot. 58, 1–26.

Berg, G., Krechel, A., Ditz, M., Sikora, R.A., Ulrich, A., Hallmann, J., 2005. Endophytic and ectophytic potato-associated bacterial communities differ in structure and antagonistic function against plant pathogenic fungi. FEMS Microbiol. Ecol. 51, 215–229.

Bull, C.T., Shetty, K.G., Subbarao, K.V., 2002. Interactions between myxobacteria, plant pathogenic fungi and biocontrol agents. Plant Dis. 86, 889–896.

Butt, T.M., Jackson, C., Magan, N. (Eds.), 2001. Fungi as Biocontrol Agents: Progress, Problems and Potential. CABI, Wallingford, UK, pp. 1–390.

Caruso, M., Colombo, A.L., Fedeli, L., Pavesi, A., Quaroni, S., Saracchi, M., Ventrella, G., 2000. Isolation of endophytic fungi and actinomycetes taxane producers. Ann. Microbiol. 50, 3–13.

Chandrashekharaiah, Gowda, G.B., Kumar, A., Chakravarthy, A.K., 2013. Bio–efficacy of *Verticillium lecanii* against white fly on greengram. Insect Environ. 192, 83–86.

Chaube, H.S., Mishra, D.S., Varshney, S., Singh, U.S., 2003. Biocontrol of plant pathogens by fungal antagonists, a historical background, present status and future prospects. Ann Rev. Plant Pathol. 2, 1–42.

Colnot, T., Dekant, W., 2016. Approaches for grouping of pesticides into cumulative assessment groups for risk assessment of pesticides residues in food. Regul. Toxicol. Pharmacol. 83.

Coombs, J.T., Franco, C.M.M., 2003. Isolation and identification of actinobacteria from surface sterilized wheat roots. Appl. Environ. Microbiol. 69, 5603–5608.

Corn, M.J.P., Christos, Z., Roeland, L.B., David, M.W., Van Wees, S.C.M., Peter, A.H.M., 2014. Induced systemic resistance by beneficial microbes. Ann. Rev. Phytopathol. 52, 347–375.

De Araujo, J.M., da Silva, A.C., Azevedo, J.L., 2000. Isolation of endophytic actinomycetes from roots and leaves of maize *Zea mays* L. Braz. Arc. Biol. Technol. 43, 447–451.

De Cal, A., Lrena, I., Guijarro, B., Melgarejo, P., 2010. Mass production of conidia of *Penicillium frequentans*, a biocontrol agent against brown rot of stone fruits. Biocontrol Sci. Technol. 126, 715–725.

De Silva, N.I., Brooks, S., Lumyong, S., Hyde, K.D., 2019. Use of endophytes as biocontrol agents. Fungal Biol. Rev. 33 (2), 133–148.

De Souza, J.T., Boer, M., Waard, P., Beek, T.A., Raaijmakers, J.M., 2003. Biochemical, genetic and zoosporicidal properties of cyclic lipopeptide surfactants produced by *Pseudomonas fluorescens*. J. Appl. Environ. Microbial. 6912, 7161–7172.

El-Tarabily, K.A., 2003. An endophytic chitinase–producing isolate of Actinoplanesmissouriensis, with potential for biological control of root rot of lupin caused by *Plectosporiumtabacinum*. Aust. J. Bot. 51, 257–266.

Federici, B.A., 2007. Bacteria as biological control agents for insects: economics, engineering, and environmental safety. In Vurro, M., Gressel, J. (Eds.), Novel Biotechnologies for Biocontrol Agent Enhancement and Management. In: NATO Security through Science Series, Springer, Dordrecht, pp. 25–51.

Fravel, D.R., 2005. Commercialization and implementation of biocontrol. Annu. Rev. Phytophthol. 43, 337–359.

Ghadin, N., Zin, N.M., Sabaratnam, V., Badya, N., Basri, D.F., Lian, H.H., Sidik, N.M., 2008. Isolation and characterization of a novel endophytic Streptomyces SUK 06 with antimicrobial activity from Malaysian plant. Asian J. Plant Sci. 7, 189–194.

Gisbert, Z., 2007. Review on safety of the entomopathogenic fungi *Beauveria bassiana* and *Beauveria brong-niartii*. Biocontrol Sci. Technol. 17 (6), 553–596.

Goodfellow, M., Williams, S.T., 1983. Ecology of actinomycetes. Annu. Rev. Microbiol. 37 (1), 189–216.

Grevesse, C., Lepoivre, P., Jijakli, M.H., 2003. Characterization of the exoglucanase-encoding gene PaEXG2 and study of its role in the biocontrol activity of *Pichia anomala* strain K. Phytopathology 93 (9), 1145–1152.

Gu, Q., Zhen, W., Huang, Y., 2007. *Glycomyces sambucus* sp. nov., and endophytic actinomycetes isolated from the stem of Sambucus adnata wall. Int. J. Syst. Evol. Microbiol. 57, 1995–1998.

Handelsman, J., Raffel, S., Mester, E.H., Wunderlich, L., Grau, C.R., 1990. Biological control of damping-off of alfalfa seedlings with *Bacillus cereus* UW85. Appl. Environ. Microbiol. 56 (3), 713–718.

Hasan, S., 1973. The biology of *Puccinia chondrillina* a potential biological control agent of skeleton weed. Ann. Appl. Biol. 74, 325–332.

Hawksworth, D.L., 1981. A survey of the fungicolous conidial fungi. In: Cole, G.T., Kendrick, B. (Eds.), Biology of Conidial Fungi. Vol. I. Aca Press, New York, pp. 171–235.

Heydari, A., Pessarakli, M., 2010. A review on biological control of fungal plant pathogens using microbial antagonists. J. Biol. Sci. 104, 273–290.

Howell, C.R., Stipanovic, R.D., 1995. Mechanisms in the biocontrol of *Rhizoctonia solani*-induced cotton seedling disease by *Gliocladium virens*: antibiosis. Phytopathology 85 (4), 469–472.

Ignoffo, C.M., Clemente, G., 1985. Host spectrum and relative virulence of an Ecuadoran and a Mississippian biotype of *Nomuraearileyi*. J. Inverteber. Pathol. 453, 346–352.

Islam, T.M., Hashidoko, Y., Deora, A., Ito, T., Tahara, S., 2005. Suppression of damping off disease in host plants by the rhizoplane bacterium Lysobacter sp. StrainSB–K88 is linked to plant colonisation and antibiosis against soilborne *Perenosporomycetes*. J. Appl Environ. Microbiol. 71, 3786–3796.

Jiang, L., Jeong, J.C., Lee, J.S., Park, J.M., Yang, J.W., Lee, M.H., Choi, S.H., Kim, C.Y., Kim, D.-H., Kim, S.W., Lee, J., 2019. Potential of *Pantoea dispersa* as an effective biocontrol agent for black rot in sweet potato. Sci. Rep. 9 (1), 1–13.

Junaid, M.J., Dar, N.H., Bhat, T.A., Bhat, A.H., Bhat, M.A., 2013. Commercial biocontrol agents and their mechanism of action in the management of plant pathogens. Int. J. Mod. Plant Anim. Sci. 12, 39–57.

Kerr, A., 1980. Biological control of crown gall through production of Agrocin 84. J. Plant Dis. 64, 25–30.

Kiewnick, S., Sikora, R.A., 2006. Biological control of the root-knot nematode *Meloidogyne incognita* by *Paecilomyces lilacinus* strain 251. Biol. Control 38 (2), 179–187.

Kim, J.S., Roh, J.Y., Choi, J.Y., Shin, S.C., Jeon, M.J., Je, Y.H., 2008. Insecticidal activity of *Paecilomyces fumosoroseus* SFP-198 as a multi-targeting biological control agent against the greenhouse whitefly and the two-spotted spider mite. Int. J. Ind. Entomol. 17 (2), 181–187.

Koumoutsi, A., Chen, X.H., Henne, A., Liesegang, H., Gabriele, H., Franke, P., Vater, J., Borris, R., 2004. Structural and functional characterization of gene clusters directing non ribosomal synthesis of bioactive lipopeptides in *Bacillus amyloliquefaciens* strain FZB42. J. Bacteriol. 86, 1084–1096.

Lafontaine, P.J., Benhamou, N., 1996. Chitosan treatment, an emerging strategy for enhancing resistance of greenhouse tomato plants to infection by *Fusarium oxysporum* f.sp. *radicis–lycopersici*. Biocontrol Sci. Tech. 6, 111–124.

Lahlali, R., Jijaki, M.H., 2007. Isolation and evaluation of bacteria and fungi as biological control agents against *Rhizoctonia solani*. Commun. Agric. Appl. Biol. Sci. 72 (4), 973–982.

Lee, S.O., Choi, G.J., Choi, Y.H., Jang, K.S., Park, D.J., Kim, C.J., Kim, J.C., 2008. Isolation and characterization of endophytic actinomycetes from Chinese cabbage roots as antagonists to *Plasmodiophora brassicae*. J. Microbiol. Biotechnol. 18, 1741–1746.

Leeman, M., Pelt, V.J.A., Ouden, F.M.D., Heinsbroek, M., Bakker, P.A.H.M., Schippers, B., 1995. Induction of systemic resistance by *Pseudomonas fluorescens* in radish cultivars differing in susceptibility to *Fusarium* wilt, using a novel bioassay. Eur. J. Plant Pathol. 1016, 655–664.

Limon, M.C., Pintor-Toro, J.A., Benitez, T., 1999. Increased antifungal activity of *Trichoderma harzianum* transformants that over express a 33–kDa chitinase. Phytopathology 89, 254–261.

Lortholary, O., Desnos-Ollivier, M., Sitbon, K., Fontanet, Bretagne, S., Dromer, F., 2011. Recent exposure to caspofungin or fluconazole influences the epidemiology of candidemia, a prospective multicentre study involving 2,441 patients. Antimicrob Agents Chemother 55, 532–538.

Lutz, M.P., Michel, V., Martinez, C., Camps, C., 2012. Lactic acid bacteria as biocontrol agents of soil-borne pathogens. Biological Control of Fungal and Bacterial Plant Pathogens. IOBC-WPRS Bull. 78, 285–288.

Mandal, S., Ray, R.C., 2011. Induced systematic resistance in biocontrol of plant disease. In: Bioaugmentation, Biostimulation and Biocontrol, pp. 241–260.

Martinuz, A., Schouten, A., Sikora, R.A., 2013. Post-infection development of *Meloidogyne incognita* on tomato treated with the endophytes *Fusarium oxysporum* strain Fo162 and *Rhizobium etli* strain G12. Bio-Control 58 (1), 95–104.

Maurhofer, M., Keel, C., Haas, D., Defago, G., 1994. Pyoluteorin production by *Pseudomonas fluorescens* strain CHA0 is involved in the suppression of *Pythium* damping–off of cress but not of cucumber. Eur. J. Plant Pathol. 1003–4, 221–232.

Meguro, A., Ohmura, Y., Hasegawa, S., Shimizu, M., Nishimura, T., Kunoh, H., 2006. An endophytic actinomycete, *Streptomyces* sp. MBR–52, that accelerates emergence and elongation of plant adventitious roots. Actinomycetologica 20, 1–9.

Migheli, Q., Gonzales-Candelas, L., Dealessi, L., Camponogara, A., Ramon-Vidal, D., 1998. Transformants of *Trichoderma langibrachaitum* over–expressing the beta–1–4–endoglucanase gene agl1 show enhanced biocontrol of *Pythium ultimum* on cucumber. Phytopathology 88, 673–677.

Milgroom, M.G., Cortesi, P., 2004. Biological control of chestnut blight with hypovirulence, a critical analysis. Annu. Rev. Phytopathol. 42, 311–338.

Montesinos, E., 2003. Development, registration and commercialization of microbial pesticides for plant protection. Int. Microbiol. 6 (4), 245–252.

Moyne, A.L., Shelby, R., Cleveland, T.E., Tuzun, S., 2001. Bacillomycin D, an Iiturin with antifungal activity against *Aspergillus flavus*. J. Appl. Microbiol. 90, 622–629.

Mukherjee, P., Latha, J., Hadar, R., Horwitz, B.A., 2004. Role of two G–protein alpha subunits, TgaA and TgaB, in the antagonism of plant pathology by *Trichoderma virens*. Appl. Environ. Microbiol. 70, 542–549.

Nguyen, N.T., Borgemeister, C., Poehling, H.M., Zimmermann, G., 2007. Laboratory investigations on the potential of entomopathogenic fungi for biocontrol of *Helicoverpa armigera* (Lepidoptera: Noctuidae) larvae and pupae. Biocontrol Sci. Technol. 17 (8), 853–864.

Nishimura, T., Meguro, A., Hasegawa, S., Nakagawa, Y., Shimizu, M., Kunoh, H., 2002. An endophytic actinomycete, Streptomyces sp. AOK–30, isolated from mountain laurel and its antifungal activity. JGPP 68, 390–397.

Ostrowsky, B., Greenko, J., Adams, E., Quinn, M., O'Brien, B., Chaturvedi, V., 2020. *Candida auris* isolates resistant to three classes of antifungal medications–New York, 2019. MMWR 69, 6–9.

Pal, K.K., Gardener, B.M., 2006. Biological control of plant pathogens. Plant Health Instr. 1–25. https://doi.org/10.1094/PHI-A-2006-1117-02.

Pappas, P.G., Kauffman, C.A., Andes, D.R., Clancy, C.J., Marr, K.A., Ostrosky-Zeichner, L., 2016. Clinical practice guideline for the management of candidiasis, 2016 update by the infectious diseases' society of America. Clin. Infect. Dis. 62, 1–50.

Patterson, T.F., Thompson, G.R., Denning, D.W., Fishman, J.A., Hadley, S., Herbrecht, R., 2016. Practice guidelines for the diagnosis and management of aspergillosis, 2016 update by the infectious diseases' society of America. Clin. Infect. Dis. 63, 1–60.

Paulitz, T.C., Belanger, R.R., 2001. Biological control in greenhouse systems. Annu. Rev. Phytopathol. 391, 103–133.

Pieterse, C.M.J., Zomioudis, C., Berendsen, R.L., Weller, D.M., Wees, S.C.M.V., Bakker, P.A.H.M., 2014. Induced systemic resistance by beneficial microbes. Annu. Rev. Phytopathol. 52, 347–375.

Qin, S., Li, J., Zhao, G.Z., Chen, H.H., Xu, L.H., Li, W.J., 2008. *Saccharopolyspora endophytica* sp. nov., an endophytic actinomycetes isolated from the root of *Maytenus austroyunnanensis*. Syst. Appl. Microbiol. 31, 352–357.

Raaijmakers, J.M., Vlami, M., De Souza, J.T., 2002. Antibiotic production by bacterial biocontrol agents. Antonie van leeuwenhoek 81 (1), 537–547.

Rajeshkumar, S., 2019. Antifungal impact of nanoparticles against different plant pathogenic fungi. Mycobiology 2, 197–217.

Ryu, C.M., Kloepper, J.W., Zhang, S., 2004. Induced systemic resistance and promotion of plant growth by *Bacillus* spp. J. Phytopathol. 94, 1259–1266.

Samish, M., Rot, A., Ment, D., Barel, S., Glazer, I., Gindin, G., 2014. Efficacy of the emtomopathogenic fungus *Metarhizium brunneum* in controlling the tick *Rhipicephalus annulatus* under field conditions. Vet. Parasitol. 206, 258–266.

Sandhu, S.S., Sharma, A.K., Beniwal, V., Goel, G., Batra, P., Kumar, A., Jaglan, S., 2012. Myco biocontrol of insect pests, factors involved mechanism and regulation. J. Pathog. 2012. https://doi.org/10.1155/2012/126819, 126819.

Sandra, A.I., Wright, C.H., Zumoff, L.S., Steven, V.B., 2001. *Pantoea agglomerans* strain EH318 produces two antibiotics that inhibits *Erwinia amylovora* in vitro. J. Appl. Environ. Microbiol. 67, 282–292.

Savita, S., Anuradha, A., 2019. Fungi as biocontrol agents. In: Biofertilizers for Sustainable Agriculture and Environment, pp. 395–411.

Scatassa, M.L., Gaglio, R., Cardamone, C., Macaluso, G., Arcuri, L., Todaro, M., Mancuso, I., 2017. Anti-Listeria activity of lactic acid bacteria in two traditional Sicilian cheeses. Ital. J. Food Safety 6 (1), 6191.

Shahrak, M., Heydari, A., Hassanzadeh, N., 2009. Investigation of antibiotic, siderophore and volatile metabolites production by *Bacillus* and *Pseudomonas* bacteria. Iran J. Biol. 22, 71–85.

Shaikh, S.S., Sayyed, R.Z., 2015. Role of plant growth–promoting *rhizobacteria* and their formulation in biocontrol of plant diseases. In: Plant Microbes Symbiosis: Applied Facets. Springer, New Delhi, pp. 337–351.

Shairra, S.A., Dorrah, M.A., Hassan, H.M.S., Nabeel, S.M., 2015. Increasing efficacy of the fungus, *Metarhizium anisopliae* on the desert locust, *Schistocerca gregaria* (Forskal) by adding some bio-products. Egypt. J. Biol. Pest Control 25 (3).

Shakir, S., Kanwal, M., Murad, W., Azizullah, A., 2016. Effects of some commonly used pesticides on seed germination, biomass production and photosynthetic pigments in tomato. Ecotoxicology 25 (2).

Shanahan, P., O'Sullivan, D.J., Simpson, P., Glennon, J.D., O'Gara, F., 1992. Isolation of 2,4–diacetylpholoroglucinol from a fluorescent pseudomonads and investigation of physiological parameters influencing its production. J. Appl. Environ. Microbiol. 58, 353–358.

Shimizu, M., Nakagawa, Y., Sato, Y., Furumai, T., Igarashi, Y., Onaka, H., Yoshida, R., Kunoh, H., 2000. Studies on endophytic actinomycetes I. Streptomyces sp. isolated from rhododendron and its antifungal activity. J. Gen. Plant Pathol. 66 (4), 360–366.

Shimizu, M., Yazawa, S., Ushijima, Y., 2009. A promising strain of endophytic *streptomyces* sp. for biological control of cucumber anthracnose. J. Gen. Plant Pathol. 75, 27–36.

Siegwart, M., Gralliot, B., Lopez, C.B., Besse, S., Bardin, M., Philippe, C., 2015. Resistance to bioinsecticides or how to enhance their suistainability, a review. Front. Plant Sci. 6, 381.

Singh, M., Singh, D., Gupta, A., Pandey, K.D., Singh, P.K., Kumar, A., 2019. Plant growth promoting rhizobacteria, application in biofertilizers and biocontrol of phytopathogens. In: PGPR Amelioration in Sustainable Agriculture, pp. 41–66.

Smith, K.P., Havey, M.J., Handelsman, J., 1993. Suppression of cottony leak of cucumber with *Bacillus cereus* strain UW85. J. Plant. Dis. 77, 139–142.

Taechowisan, T., Chuaychot, N., Chanaphat, S., Wanbanjob, A., Shen, Y., 2008. Biological activity of chemical constituents isolated from *Streptomyces* sp. Tc052, and endophyte in *Alpinia galanga*. Int. J. Pharmacol. 4, 95–101.

Thambugala, K.M., Daranagama, D.A., Phillips, A.J., Kannangara, S.D., Promputtha, I., 2020. Fungi vs. fungi in biocontrol: an overview of fungal antagonists applied against fungal plant pathogens. Front. Cell. Infect. Microbiol. 718.

Thamchaipenet, A., Indananda, C., Bunyoo, C., Duangmai, K., Matsumoto, A., Takahashi, Y., 2010. *Actinoallomurus acaciae* sp. nov., an endophytic actinomycetes isolated from Acaciauriculiformis A. Cunn. Ex Benth. Int. J. Syst. Evol. Microbio. 60, 554–559.

Thomashow, L.S., Bonsall, R.F., Weller, D.M., 2002. Antibiotic Production by Soil and Rhizosphere Microbes *in situ*, second ed. ASM Press, Washington, DC, pp. 638–647.

Thomashow, L.S., Weller, D.M., 1988. Role of a phenazine antibiotic from *Pseudomonas fluorescens* in biocontrol of *Gaeumannomyces graminis* var. tritici. J. Bacteriol. 170, 3499–3500.

Tian, X.L., Cao, L.X., Tan, H.M., Han, W.Q., Chen, M., Liu, Y.H., Zhou, S., 2007. Diversity of cultivated and uncultivated actinobacterial endophytes in the stems and roots of rice. Microbiol. Ecol. 53, 700–707.

Tiwari, A.K., 1996. Biological Control of Chick Pea Wilt Complex Using Different Formulations of *Gliocladium virens* through Seed Treatment (Ph.D. thesis). G.B Plant University of Agriculture and Technology, Pantnagar, India, p. 167.

Usta, C., 2013. Microorganisms in biological pest control – a review bacterial toxin application and effect of environmental factors. In: Current Progress in Biological Research, pp. 287–317.

Van Wees, S.C.M., Pieterse, C.M.J., Triissenaar, A., Westende, A.M., Hartog, F., Van Loon, L.C., 1997. Differential induction of systemic resistance in *Arabidopsis* by biocontrol bacteria. J. Mol. Plant-Microbe Interact. 106, 716–724.

Verma, V.C., Gond, S.K., Kumar, A., Mishra, A., Kharwar, R.N., Gange, A.C., 2009. Endophytic actinomycetes from *Azadirachta indica* A. Juss.: isolation, diversity, and anti-microbial activity. Microb. Ecol. 57 (4), 749–756.

Verma, C., Jandaik, S., Gupta, B.K., Kashyap, N., Suryaprakash, V.S., Kashyap, S., Kerketta, A., 2020. Microbial metabolites in plant disease management: review on biological approach. Int. J. Chem. Stud. 8 (4), 2570–2581.

Wang, X., Li, Q., Sui, J., Zhang, J., Liu, Z., Du, J., Xu, R., Zhou, Y., Liu, X., 2019. Isolation and characterization of antagonistic bacteria *Paenibacillus jamilae* HS-26 and their effects on plant growth. BioMed Res. Int. 2019.

Whilhite, S.E., Lumsden, R.D., Straney, D.C., 1994. Mutational analysis of gliotoxin production by the biocontrol fungus *Gliocladium virens* in relation to suppression of *Pythium* damping-off. Phytopathology 84 (8), 816–821.

Wilhite, S.E., Lunsden, R.D., Strancy, D.C., 2001. Peptide synthetase gene in *Trichoderma virens*. J. Appl. Environ. Microbiol. 67, 5055–5062.

Woo, S., Donzelli, B., Scala, F.R.M., Harman, G., Kubicek, C., Del Sorbo, G., Lorito, M., 1999. Disruption of the ech42 endochitinase-encoding gene affects biocontrol activity in *Trichoderma harzianum* P1. Mol. Plant-Microbe Interact. 12, 419–429.

Wraight, S.P., Jackson, M.A., de Kock, S.L., 2001. Production, Stabilization and Formulation of Fungal Biocontrol Agents. 514, pp. 1–390.

Wratten, M.L., 2008. Therapeutic approaches to reduce systemic inflammation in septic associated neurologic complications. Eur. J. Anaesthesiol. Suppl. 25, 1–7.

CHAPTER 4

Relationship between probiotics and living beings for sustainable life on land

Celia Vargas-de-la-Cruz[a,f,*], **Daniela Landa-Acuña**[b,c], **Md. Shariful Islam**[d], **and Eduardo Flores-Juarez**[e]

[a]Department of Pharmacology, Bromatology and Toxicology, Faculty of Pharmacy and Biochemistry, Centro Latinoamericano de Enseñanza e Investigación en Bacteriología Alimentaria-CLEIBA, Universidad Nacional Mayor de San Marcos, Lima, Perú
[b]Laboratory of Microbial Ecology and Biotechnology, Department of Biology, Faculty of Sciences, National Agrarian University La Molina (UNALM), Lima, Perú
[c]Professional Career in Environmental Engineering, Faculty of Engineering, Private University of the North, Los Olivos Campus, Lima, Perú
[d]Department of Pharmacy, Southeast University, Dhaka, Bangladesh
[e]Department of Biochemistry, Faculty of Pharmacy and Biochemistry, Universidad Nacional Mayor de San Marcos, Lima, Perú
[f]E-Health Research Center, Universidad de Ciencias y Humanidades, Lima, Perú
*Corresponding author. e-mail address: cvargasd@unmsm.edu.pe

1. Introduction

Probiotics have been studied extensively in both humans and livestock due to their ability to regulate gut microbiota and the immune process, as well as its use in veterinary preventive therapeutic treatment and for clinical therapeutic purposes (Celiberto et al., 2017; Sun et al., 2018). Consequently, probiotics are known to be viable and safe supplements; furthermore probiotics in animal feed, promotes development, growth and nutrient digestibility in animals (Hai et al., 2007). For this reason, probiotics are used in dairy animals, which decreases serum cholesterol and decreases diarrhea outbreaks (Cavalheiro et al., 2015; Lee et al., 2017). Put another way, probiotics have elevated aerobic conditions in the gastrointestinal (GI) environment by reducing oxygen-scavenging compounds, for example, nitrates. Probiotics have also been demonstrated to excrete hydrolytic enzymes that fight bacterial toxins by disabling toxin receptors, as a result, moderating the occurrence of poison-negotiated pathogenesis infections in livestock (Hossain et al., 2017). There is no doubt that animal feed is significant in livestock cultivation, therefore there are different ongoing investigations into its capacity in feed additives (Chen et al., 2017). It is clear that the European legislation's ban of in-feed antibiotics in 2006 consequently paved the way for the outbreak of genes in the gut microflora of European pigs (Zhao and Kim, 2015). In general, it is affirmed that the use of antibiotics as an enhancer to improve the diet of cattle faced generalized prohibition in several countries (Abd El-Tawab et al., 2016). Likewise, the improvement of different functional feeds for animals guarantees health and safety. Fermented food products

Relationship Between Microbes and the Environment for Sustainable Ecosystem Services, Volume 1
https://doi.org/10.1016/B978-0-323-89938-3.00004-9

applying probiotics are generally accepted additives concentrated in various countries (Johnson et al., 2019; Hwang et al., 2013).

Probiotics are live microorganisms that benefit health in a host when administered in sufficient dosage. Numerous species belonging to the genera of *Streptococcus*, *Lactobacillus*, *Bifidobacterium*, and *Lactococcus* remain the most remarkable probiotic materials to date (Plessas et al., 2017). Zootechnical additives classified as beneficial agents are used at a regulatory level (Bernardeau and Vernoux, 2009). This is necessary for a probiotic candidate in order to estimate its lowest performance characteristics prior to being certified (Bernardeau and Vernoux, 2013).

However, the positive characteristics of probiotics can add modulation to the host's immune and uncertain physiological process, in addition to the introduction of extremely severe markers to various pathogens, the prevention and treatment of infectious and inflammatory diseases, in addition to acting as a biocontrol factor, especially in stopping deterioration, among other benefits (Diaz-Vergara et al., 2017).

2. Importance of probiotics in animal health

Via the formulation of volatile fatty acid digestibility and the stimulation of lactic acid-dependent protozoa, the prosperity of probiotic performance was proven. It is a common opinion that probiotics have also been applied to enhance the potency of effective feed use; it also escalates milk formulation and diminishes diarrhea in cattle and pigs, but it yet unproven to treat the Salmonella colonization of chicken intestinal tracts. Moreover, commercial in-feed probiotics improved immunomodulatory capacity and, besides Lavipan, dramatically minimized *Campylobacter* spp. in the GI tracts of poultry birds, consequent inhibiting pathogenic contaminants and escalating disinfection in poultry conditions (Song et al., 2019). Roselli et al. (2017) investigated the probiotics commonly fed to weaned piglets as well crops produced as pragmatic consequents through flourishing gut health via balanced microbiota, escalating immunological and physiological systems also prohibiting GI disorders (Roselli et al., 2017). Extensive feedback was investigated as to the rapid alterations of the GI microbial ecosystem, which resisted the progress of the nearby pathogens, combined in conjunction with the fermentation of pleasing production products (Sakata, 1990).

Hanczakowska et al. (2016) in a related work discovered that *Enterococcus faecium*, when fed to piglets as a feed supplement, demonstrated a preventive effect opposed to *Clostridium perfringens*. In addition, the microflora in the GI of animals is generally known to be an effective metabolic organ owing to its biodiversity. Therefore, this is significant for managing the intestinal microbiota useful for targeting pathogens in cattle (Fang et al., 2015). It is a common opinion that lactic acid bacteria, including probiotic potential organic acids that increase the acidity of the GI environment, lower the risk of pathogen infestation, while at the same time regulate the gut habitat as well microbial ecosystem (Shokryazdan et al., 2017).

3. Feed antibiotics to probiotics like a viable substitute

Over the decades, antibiotics have been widely applied as prophylactic as well as growth-promoting agents in livestock. Furthermore, it is major contributor to the unregulated growth in multidrug-resistant pathogens. It has, for this reason, minimized remedial options, in both veterinary and human clinics, as well used minimally in clinical prosperity, formerly for remediable diseases and as a consequence can result in an extended hospital stay (Sarrazin et al., 2019). Particularly, in Europe, feed antimicrobials are the best general route of drug administration in pig farming (Seal et al., 2018). Therefore, drug administration in this form is not used in healthy animals to avoid unnecessary antimicrobials when feeding affected ones, thus consequently escalating the danger of reverberating bacteria. Furthermore, escalating tendency of the execution, which led to antimicrobial-resistant genes, prompted the United States Animal Agriculture Sector to inhibit the use of subtherapeutic Antibiotic Growth Promoters (AGPs) in 2017 via execution of the Veterinary Feed Directive (VED) (Helm et al., 2019; FAO, 2002). Additionally, antimicrobial consumption through livestock was evaluated to be up 240,000 metric tons yearly across continents, although a few countries are experiencing a substantial reduce in the sales of antimicrobials for food-producing animals (Vine et al., 2004). Therefore, resistance associated with the use of antibiotics in agricultural practices, as well as the potential transfer of antibiotic-resistant bacteria from food-producing animals to humans, is also a health concern (Cattaneo et al., 2009).

Antibiotics administered to farm animals are eliminated in urine and feces into the environment and consequently affecting the microorganisms in the environment in a bid for survival (Ponce et al., 2011). In fact, the use of antibiotics in food-producing animals is under severe perusal owing to the risk of zoonotic move of resistant pathogens between the human populace (Coyne et al., 2018). One example is that β-lactam antibiotics, namely penicillin, carbapenem, and cephalosporin, are no longer related to human clinical therapy (Tesfaye and Hailu, 2019). An increased degree of usefulness of colistin, as opposed to plasmid mediated mcr-1 and carbapenemase-producing *Enterobacteriaceae*, was reported in 2015 and also was later evaluated in more than 50 various countries, including seven mcr-1 gene variations (Wang et al., 2000, 2017). Colistin was used for years in China in the livestock industries and was also used to research innovation that advocated mcr-1 developed from animals prior to use in the human population (Suresh et al., 2013). In 2017, Chinese authorities banned the agricultural usage of colistin as growth promoter (Xiao et al., 2016). In addition, in Europe countries, vancomycin-resistant *Enterococci* were forbidden in the use of glycopeptide antibiotics such as avoparcin in animal feed (Sornplang and Piyadeatsoontorn, 2016). The animal Health Authority in

France in 2016 accepted an edict controlling the administration of critically significant antibiotics, namely third- and fourth-generation cephalosporins, macrolides, and fluoroquinolones. It was planned to sponsor the rational administration of antibiotics and to promote the use of substitutes in the veterinary clinical process (Bourély et al., 2018). Furthermore, in the UK, there was an attempt to find a substitute approach to prophylactics that compared to conventional antimicrobials. Also various farmers demonstrated scrutiny when declining the administration of antibiotics. Consequently, advice on sponsoring discreet scripts, supplying suggestions of other procedures to administer, and prohibiting diseases in pigs also were gathered by the Medicines in Agriculture Alliance in 2003 as well the Pig veterinary Society in 2014. Burrow and co-workers evaluated the risk kinematics of *E. coli* resistance in pigs with significant clinical antimicrobials and concluded that depletion of antibiotic resistance in pigs can result in a low level of beta-lactam-resistant *E. coli* between their progeny. According to Wang (2007), Pacific white shrimp feed was supplemented with sulfonamide and ciprofloxacin to estimate the microbial community by selecting the V4 region of 16S rRNA genes. The extent of ciprofloxacin- and sulfonamide-resistant genes was respectively investigated. It is generally thought that, in clinical protection as well as conservation of treatment options opposed to infectious diseases, there is a need for the restriction of antibiotics application to inevitable instances. So, the elimination of in-feed antibiotics in animal farming significantly increased around the world. As well, it has prompted different investigations on growing, developing, and promoting substitute additives for probiotics, prebiotics, acidifiers, plant secondary metabolites (De Llano et al., 2016), enzymes, bacteriophages, and bacteriocins (Servin, 2004; Cowieson and Kluenter, 2019).

Connected to probiotics and exogenous enzymes, when administered as feed substitutes, they have an advantageous influence on extension representation and also on the weight-obtained feed ration in calves (Prieur et al., 1990; Lan et al., 2017). Therefore, the results of two sufficient market-feed substitutes containing lactic acid bacteria, *Lactobacillus* fermentation products and plant-sourced enzymes, in comparison to in-feed antibiotics were widely estimated based on outgrowth acting, as well as remaining features and blood metabolites of steers through Ran and fellow co-workers (Rengpipat et al., 2000; Qiao et al., 2018). In summary, the investigation revealed which supplements developed mean daily weight gains and feed ability through the early part of the increasing phase; in addition, steers supplemented with these products need a small amount of therapeutic antimicrobials, so probiotic supplements are a useful substitute to AGPs in escalating steers. Besides that, a diet supplemented with *Astyanax bimaculatus* including *Lactobacillus* spp. increased the amounts of leucocytes in the circulatory system, yielding greater resistance to gut pathogens. For this reason, higher survival rates were associated with elevated feed ability (De Moraes et al., 2018). One more study was carried out by Vase-Khavari et al. (2019). Probiotics in poultry bird feed demonstrated an important effect toward the progress of the broilers, including laudable responses on total cholesterol, triglyceride levels, as well immunological parameters, along with a notable depletion in colon pathogens (Verschuere et al., 2000).

4. Risk assessment protocol established due to probiotics

Despite the clinical productivity of probiotics, protection against the presence of microorganisms, should be safe. It is advised that the estimation of risk factors that may be caused by any particular strain should be taken seriously. It goes without saying that the most probiotic strains have gained a "Generally Recognized as Safe" status owing to their successful history of use, although each and every novel probiotic cannot be assumed to share historical safety, including conventional strains (Tovar et al., 2002; Donohue, 2006). Other probiotics may control unpleasant characteristics, namely virulence factors, hemolytic potential, transferable antimicrobial resistance, and toxic biochemicals production (Lilly and Stillwell, 1965). One example is that the potential risk of a transfer of antimicrobial-resistant genes in a probiotic strain to other bacteria in the microbiota of the host's GI environment, as well as the rare event of additional infectious diseases such as endocarditis, pneumoniae, bacteraemia, meningitis, and septic arthritis connected with *Lactobacillus* and *Enterococcus* strains that have been passed to immunocompromised patients (Vankerckhoven et al., 2008).

Academic and industrial scientists estimated the biosafety of lactic acid bacteria as probiotics in a 2006 two-day workshop at the University of Antwerp, Belgium. This was preferred, as opposed to any strong probiotic strain not associated to the wild-type distribution of related antimicrobials, as well as virulence determinants, should be avoided as upcoming functional products for mankind consumption.

The WHO/FAO expert consultation, in a joint report in Rome (Vase-Khavari et al., 2019), mentioned *Enterococcus* because one species of the bacteria has very strong virulence characteristics despite lactic acid bacteria. Contributors therefore were disheartened by the application of *Enterococcus* as a probiotic in which the genus generally expresses a higher level of vancomycin-resistant genes; for example, its resistance could be moved to nearby pathogens, thus increasing their pathogenicity. Of course, this includes whatever strains of vancomycin-resistant *Enterococci* are often connected to nosocomial infections. Moreover, the evaluation of the strain identification was combined with the mechanism to identify strains that assume sufficient consciousness. More, the transition potent of the strain beyond the GI barrier must be examined carefully also (Roselli et al., 2017; Huys et al., 2013). Other potential risk factors to watch for are the ability to move obtained antimicrobial resistance, probability of sponsoring detrimental metabolic consequence, and excessive immune stimulation, as well as virulence determinants, toxigenicity of the strain, and level of cleanliness of the product, colonization, and stability of the strain over time (Suzer et al., 2008).

The working group report (FAO) (Srinivas et al., 2017) indiscriminately suggests that potential probiotic strains should be screened owing to unwanted secondary medical effects to end consumers, as well as virulence in animal models including compromised immunity and adverse disadvantageous effects regarding end users. According to Kim et al. (2018), due to the security evaluation and ruling of probiotics, the EU Union Scientific Committee regarding Animal Nutrition mentions taxonomical definition of the strains and gathering of considerable report publishing data such as a history of use, human interference, industrial applications, ecological niche, description of conclusion consumers, and exclusion of pathogenicity (Kamei et al., 1988).

5. Adverse effects because of application for probiotics

Clearly the safety of probiotics applies to both farm animals and mankind. The peculiar functions of a probiotic are more important than the source of the isolate. This is significant in that the usefulness of a probiotic agent remains workable over a substantial period at the site of action; this is very critical to determine the source of the organisms in the GI tract as the origin of the intestinal microbiota has not been well established. Further, the probability of probiotics used for animal feed cross-contaminating the human food chain cannot be downplayed, even though there is very little available information regarding the risk to human food owing to contamination from in-feed probiotics (Kocyigit et al., 2016). The adverse effects of probiotics also depend on the host's predominant physiological and immunological condition. It was felt that there is an insufficiency of evaluation and systematic description regarding adverse effects owing to probiotic interference and that interference is unsatisfactorily documented (Bajagai et al., 2016).

6. Principle of selection of probiotics

Verschuere et al. (2000) extensively researched which probiotics should be considered, and furthermore these characteristics help in developing new, useful, and safe products. As per Merrifield et al. (2010), the characteristics include:

- Both the general aquatic and host species organisms should not be pathogenic.
- They should be obstructive because of bile salts.
- They must be able to survive and cultivate well in the intestinal mucosa.
- They must show convenient development features.
- They must demonstrate antagonistic features to a single or else more principal pathogens.
- They must continue to be durable according to standard storage conditions and remain alive during feed production procedures.

Fig. 1 presents information on a selection of probiotics, including screening of healthy animals in the time of outbreak, in which good healthy animals are screened out. It is

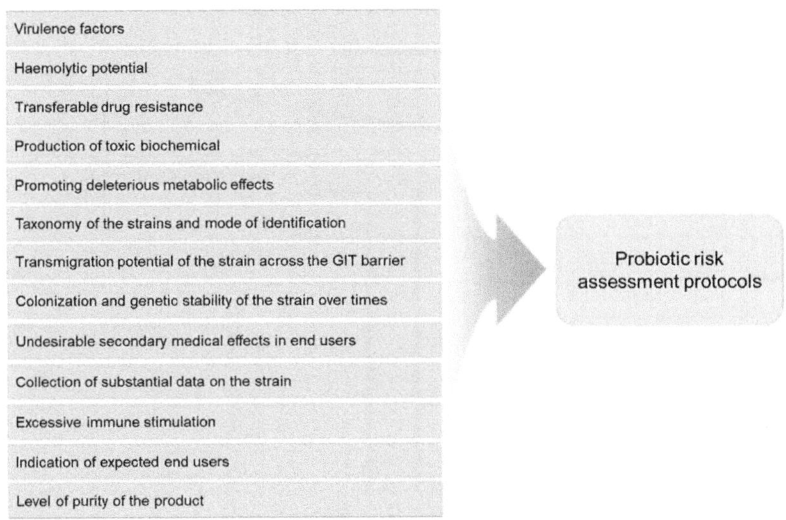

Fig. 1 Established safety assessment protocols for a probiotic candidate.

pursued through separation of microbial strains because of the animal's skin, gill, and culture medium. However, the separated strands can be evaluated by the diseased organisms, likewise *in vivo* and *in vitro*. Consequently, if the strain is unable to colonize and overcome the pathogen, the strains can later be tested for its economic durability and also put into a control body for testing prior to release as a commercial product (Fig. 2).

According to these selection criteria, several microorganisms can be used as probiotics; a clear example is aquaculture, an activity that uses a large amount of chemicals to maintain optimal conditions, which have studied nonpathogenic strains of *Vibrio, Aeromonas, Pseudomonas,* and *Alteromonas* with antagonistic properties to pathogenic strains that can be used as probiotics, significantly reducing the use of contaminating compounds (Hai et al., 2007). These strains must be carefully selected to guarantee their specificity and appropriate density for use (Balcázar et al., 2006). In addition, the efficiency of antimicrobial activity is one of the main mechanisms through which probiotic organisms function (Lü et al., 2014). Another important criterion for the selection of probiotic strains is the ability of certain animals to adhere to the intestinal epithelium, as reported by Meidong et al. (2017), where strains of *Lactobacillus plantarum* adhere strongly to porcine gastric mucin. This adhesion is a prerequisite for bacterial colonization of the host gut (Vankerckhoven et al., 2008).

7. Competitive exclusion

Bacterial antagonism is a common phenomenon in nature; consequently, microbial interactions contribute a significant role to the equilibrium among contending

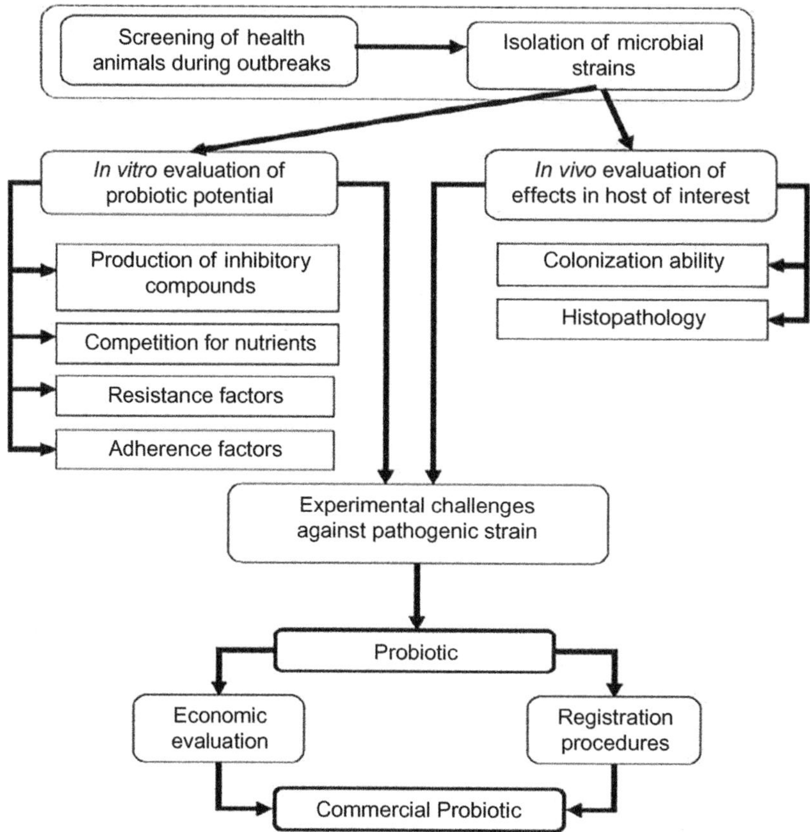

Fig. 2 Illustration of selection of probiotics as biocontrol agents.

pathogenic microorganisms, but the constitution of microbial communities can be changed through farming implementation and conditions of the environment, which bring to mind the expansion of chosen bacterial species. However, it is commonly known that the microbiota of the GI tract of aquatic animals could be modified, for example, through ingestion of other microorganisms; whereas, microbial manipulation constitutes a viable tool to reduce the incidence of chosen pathogens (Morelli, 2000).

8. Mechanisms of actions of probiotics

Improvement of colonization resistance and direct inhibitory effects in opposition to pathogens are the most significant factors where probiotics have declined the incidence of prolongation of diseases. Furthermore, probiotic strains have been demonstrated to prohibit pathogenic bacteria between *in vitro* and *in vivo* by different mechanisms (Balcázar, 2002). Also, the methodological and ethical limitations of evaluate animals make it difficult to understand the action of probiotics, and limited clarification is available. Despite that, some benefits connected to the administering of probiotics have been recognized:

(a) Reasonable exclusion of pathogenic bacteria.

(b) Nutrients and enzymatic performance to digestion.

(c) Bacteria supported to direct uptake of dissolved organic material mediated.

(d) Magnification of the immune response in opposition to pathogenic microorganisms.

(e) Antiviral effects.

These mechanisms are not only limited to the intestinal region, but the benefits can be extrapolated to other forms of action, as stated in Fuller's review, whose original concept was applied to protozoa that favored the growth of other protozoa (Lilly and Stillwell, 1965). This concept establishes one of the main requirements for the colonization of the GI tract, whose mechanism of competitive exclusion of fixation sites can represent a significant advantage during the initial stages of fertilization of some individuals, as in the case of the larval stage of some marine organisms (Irianto and Austin, 2002). Likewise, the acquisition of iron is another necessary mechanism for some pathogenic bacteria for adequate growth, which is limited both in tissues and body fluids of animals in its $Fe+3$ form, which is insoluble, so the siderophores are considered as binding agents of iron for its acquisition, thus consolidating the siderophores as important mechanisms of virulence in some pathogens. Some probiotic microorganisms have that capacity (Kesarcodi-Watson et al., 2008).

According to Ajuwon (2015), in the poultry industry, the use of antibiotics that counteract the action of pathogens is important. However, studies have shown that various mechanisms of probiotics generate benefits in poultry, such as (Abd El-Tawab et al., 2016) maintaining normal intestinal microflora by competitive exclusion and antagonism; (Abushelaibi et al., 2017) altering metabolism by increasing the activity of digestive enzymes and decreasing the activity of bacterial enzymes and ammonia production; (Ajuwon, 2015) stimulating the immune system; and (Bajagai et al., 2016) improving feed intake and digestion. However, the results are still debatable due to inconsistencies in the expected effects on performance and health of the poultry.

Likewise, *in vitro* studies on animals have shown that the production of a restrictive environment, with respect to pH, redox potential, and hydrogen sulfide production, are the mechanisms by which the intestinal flora resists colonization by pathogenic bacteria. This has shown that probiotic bacteria possess the ability to lower the pH in UC (ulcerative colitis) patients after ingestion of a probiotic preparation (Venturi et al., 1999). Also, the production of antimicrobial compounds known as bacteriocins, produced by probiotic bacteria, also beneficially contribute; this inhibitory activity varies, some inhibiting other lactobacilli or taxonomically related Gram-positive bacteria, and some being active against a much wider range of Gram-positive and Gram-negative bacteria, as well as yeasts and molds (Ng et al., 2009).

9. Nutrients and enzymatic attributed to digestion

Researchers recommended that microorganisms have a beneficial effect in digestive procedure of aquatic animals. However, in fish, it has been indicated that *Bacteroides* sp. and *Clostridium* sp. contribute to the host's nutrition, particularly by providing fatty acids and vitamins (Balcázar et al., 2006). Besides, various bacteria may assist in the digestion systems of bivalves through manufacturing extracellular enzymes like kinds of proteases, with lipases also supplying important growth factors (Sanders et al., 2010). Indistinguishable examination has been demonstrated for the microbial flora of adult penaeid shrimp, where an addition of enzymes for digestion was included to the animal (Ran et al., 2019).

There are several *in vitro* studies that show the activity of probiotics on lipases, proteins, and carbohydrates. Tovar et al. (2002) found that the inclusion of *Debaryomyces hansenii* in concentrations of 7×105 cells/Kg improves the lipidic activity of *Dicentrarchus labrax*, as well as later trials in fish, evidenced that the addition of *Lactobacillus* spp., in the diet of *Sparus aurata* L and *Cyprinus carpio* improves the activity of lipase (Suzer et al., 2008). On the other hand, studies have shown that the addition of probiotic bacteria (*Bacillus* sp.) in shrimp diets significantly improves protease activity in 28 days (Wang, 2007). Likewise, an increase of the protease activity in juvenile fish was evidenced when they were fed a diet with probiotics containing strains of *Lactobacillus*, *Saccharomyces*, photosynthetic bacteria, *Cusuanjun*, *B. natto*, *Actinobacteria*, *Sulfolobus acidocaldarius*, and *Streptococcus faecium*, under defined formulations of these and their combinations, in a period of 80 days (Fang et al., 2015).

On the other hand, strains of *Bacillus subtilis*, in different concentrations, were tested in white shrimp feed, promoting amylase activity after 8 weeks, as well as probiotics of *B. mojavensis* and a mixture of *B. mojavensis* and *Virgibacillus proomii*. Amylase activity was evidenced after 20 days after hatching of the fry and with low amylase activity between 40 and 60 days. This finding may indicate that the activity of digestive enzymes can be used as an indicator of food acceptance by the larvae, as well as for digestive capacity in relation to the type of feeding offered, as well as improved growth performance, survival, and healthy intestinal microenvironment of the host (Hoseinifar et al., 2017).

10. Certain impact toward water quality

Enhancement of the water standard has particularly been linked with *Bacillus* sp. The rationale is that Gram–positive bacteria are much more greater converters of organic matter back to CO_2 than Gram-negative bacteria. In the time of the production cycle, higher level of Gram-positive bacteria could diminish the escalation of dissolved organic carbon. Nevertheless, the use of *Bacillus* sp. developed water standard or quality, and in addition, survival and growth rates furthermore increased the health of status of juvenile *Penaeus monodon* and moreover declined the pathogenic vibrios (Xia et al., 2019).

11. Improvement to immune response

Probiotics can be used to stimulate the nonspecific immune process. Dalmin et al. (2001) and FAO (2006) examined the use of *Bacillus* sp. (strain S11) dispensed in tiger shrimp. Another study (Balcázar et al., 2006) revealed that the administration of a mixture of bacterial trains like *Bacillus* sp. and *Vibrio* spp. impacted the extension and development of juvenile white shrimp, which also represented a defensive effect in opposition to the pathogens *Vibrio harvevi* and white spot syndrome virus. Additionally, this defense was due to a stimulation of the immune system through escalating phagocytosis and antibacterial activity.

12. Antiviral effects

Various bacteria administered, namely candidate probiotics, have antiviral effects even though the proper mechanism of these bacteria is unknown. Laboratory tests on the activation of viruses that can occur through not only chemical but also biological substances, for example, as extracts from marine algae and extracellular agents of bacteria, demonstrated that strains of *Vibrio* spp., *Pseudomonas* sp., *Aeromonas* spp., and groups of coryneform separated from salmonid hatcheries revealed antiviral activity in opposition to infectious hematopoietic necrosis virus (IHNV) including more than 50% plaque depletion (Kesarcodi-Watson et al., 2008). According to Girones et al. (1989), a marine bacterium, tentatively categorized under the genus *Moraxella*, also showed antiviral properties including higher particular for poliovirus.

13. Separation of probiotics

It is generally thought that probiotic organisms found on aquatic animals are also in the ambient condition or environment, called the culture medium. However, it is commonly thought that probiotics can be separated from the skin, gill, and the culture medium.

To do this, consider that probiotic properties are specific to each strain, condition, and dose, so it is uncommon to find two unique probiotic strains belonging to the same species with similar characteristics (Shokryazdan et al., 2017). This suggests that probiotic products should contain strains with very well-defined characteristics and with studies that ratify their potential probiotic properties in living organisms (Morelli, 2000).

Likewise, FAO/WHO (2002) considered a series of general *in vitro* tests for the selection and characterization of potential probiotic strains, among which are:

(a) Resistance to gastric acidity

(b) Bile salt hydrolase (BSH) activity and resistance to bile salt

(c) Adhesion to human mucus and/or epithelial cells and cell lines

(d) Antimicrobial and antagonistic activity against potentially pathogenic bacteria

14. Method of separation

A neat and clean piece of cotton wool can be rubbed over the skin of the aquatic animal or a little amount of water can be taken from the culture medium. Therefore, the clean cotton wool or water is placed into a petri dish, which is also diluted, prior to being observed, featured, recognized, and counted under a microscope. In a nutrient medium, a little amount of the organism is inoculated. It is important to think that this agar medium depends on the probiotic species that you want to cultivate, since several organisms are spread in a varied culture medium; one example is Man Ragosa and Sharp (MRS) medium to be used for *Lactobacillus* sp., while Yeast Extract Agar (YEA) is used for *Saccharomyces cerevisiae*. For this reason, the culture species is left in a suitable environment for several hours/days as needed for the escalate of the cultured organism. After that, a preestablished quantity of the cultured organisms can be used for the fish *in vivo/in vitro*.

15. Applications of probiotics in aquaculture

In different ways, probiotics can be provided to the host or added to its aquatic environment; *in vivo*, direct application to the host is known while application to the culture environment is known *in vitro*. Meidong et al. (2017) reported the means of application as follows: addition via live food, bathing, or addition to culture water. Furthermore, commercial diet or live food can be classified under the crucial headings *in vitro* or *in vivo*, while also bathing addition to culture water can be classified under *in vitro*. In all these procedures used by different researchers, positive results were acquired.

References

Abd El-Tawab, M.M., Youssef, I.M., Bakr, H.A., Fthenakis, G.C., Giadinis, N.D., 2016. Role of probiotics in nutrition and health of small ruminants. Pol. J. Vet. Sci. 19, 893–906.

Abushelaibi, A., Al-mahadin, S., El-tarabily, K., Shah, N.P., Ayyash, M., 2017. Characterization of potential probiotic lactic acid bacteria isolated from camel milk. LWT Food Sci. Technol. 79, 316–325.

Ajuwon, K.M., 2015. Toward a better understanding of mechanisms of probiotics and prebiotics action in poultry species. J. Appl. Poult. Res. 25 (2), 277–283.

Bajagai, Y.S., Klieve, A.V., Dart, P.J., Bryden, W.L., 2016. In: Harinder, P.S. (Ed.), Probiotics in Animal Nutrition–Production, Impact and Regulationby FAO Animal Production and Health Paper No. 179. FAO, Rome, Italy, ISBN: 978-92-5-109333-7.

Balcázar, J.L., 2002. Use of probiotics in aquaculture: general aspects. In: de Blas, I. (Ed.), Memorias del Primer Congreso Iberoamericano Virtual de Acuicultura. Zaragoza, Spain, pp. 877–881.

Balcázar, J.L., de Blas, I., Ruiz-Zarzuela, I., Cunningham, D., Vendrell, D., Muzquiz, J.L., 2006. The role of probiotics in aquaculture. Vet. Mircrobiol. 144, 173–186.

Bernardeau, M., Vernoux, J., 2009. Overview of the use of probiotics in the feed/food chain. Probiotics: production, evaluation and uses in animal feed, Kerala India. Biotechnol. Appl. Biochem. 2009, 15–45.

Bernardeau, M., Vernoux, J.P., 2013. Overview of differences between microbial feed additives and probiotics for food regarding regulation, growth promotion effects and health properties and consequences for extrapolation of farm animal results to humans. Clin. Microbiol. Infect. 19, 321–330.

Bourély, C., Fortanéd, N., Calavas, D., Leblond, A., Gay, É., 2018. Why do veterinarians ask for antimicrobial susceptibility testing? A qualitative study exploring determinants and evaluating the impact of antibiotic reduction policy. Prev. Vet. Med. 159, 123–134.

Cattaneo, A.A., Wilson, R., Doohan, D., Lejeune, J.T., 2009. Bovine veterinarians' knowledge, beliefs, and practices regarding antibiotic resistance on Ohio dairy farms. J. Dairy Sci. 92, 3494–3502.

Cavalheiro, C.P., Ruiz-Capillas, C., Herrero, A.M., Jiménez-Colmenero, F., Ragagnin, M.C., Martins, F.-L.L., 2015. Application of probiotic delivery systems in meat products. Trends Food Sci. Technol. 46, 120–131.

Celiberto, L.S., Bedani, R., Rossi, E.A., Cavallini, D.C., 2017. Probiotics: the scientific evidence in the context of inflammatory bowel disease. Crit. Rev. Food Sci. Nutr. 57, 1759–1768.

Chen, J., Wang, Q., Liu, C.M., Gong, J., 2017. Issues deserve attention in encapsulating probiotics: critical review of existing literature. Crit. Rev. Food Sci. Nutr. 57, 1228–1238.

Cowieson, A.J., Kluenter, A.M., 2019. Contribution of exogenous enzymes to potentiate the removal of antibiotic growth promoters in poultry production. Anim. Feed Sci. Technol. 250, 81–92.

Coyne, L.A., Latham, S.M., Dawson, S., Donald, I.J., Pearson, R.B., Smith, R.F., Williams, N.J., Pinchbeck, G.L., 2018. Antimicrobial use practices, attitudes and responsibilities in UK farm animal veterinary surgeons. Prev. Vet. Med. 161, 115–126.

Dalmin, G., Kathiresan, K., Purushothaman, A., 2001. Effect of probiotics on bacterial population and health status of shrimp in culture pond ecosystem. Indian J. Exp. Biol. 39, 939–942.

De Llano, D.G., Gil-sánchez, I., Esteban-fernández, A., Ramos, A.M., Fernández-díaz, M., Cueva, C., Moreno-arribas, M.V., Bartolomé, B., 2016. Reciprocal beneficial effects between wine polyphenols and probiotics: an exploratory study. Eur. Food Res. Technol. 243, 531–538.

De Moraes, A.V., Pereira, M.D.O., Moraes, K.N., Rodrigues-Soares, J.P., Jesus, G.F., 2018. Autochthonous probiotic as growth promoter and immunomodulator for Astyanax bimaculatus cultured in water recirculation system. Aquac. Res. 49, 2808–2814.

Diaz-Vergara, L., Pereyra, C.M., Montenegro, M., Pena, G.A., Aminahuel, C.A., Cavaglieri, L.R., 2017. Encapsulated whey-native yeast Kluyveromyces marxianus as a feed additive for animal production. Food Addit. Contam. Part A Chem. Anal. Control Expo. Risk Assess. 34, 750–759.

Donohue, D.C., 2006. Safety of probiotics. Asia Pac. J. Clin. Nutr. 15, 563–569.

Fang, C., Ma, M., Ji, H., Ren, T., Mims, S.D., 2015. Alterations of digestive enzyme activities, intestinal morphology, and microbiota in juvenile paddlefish, *Polyodon spathula*, fed dietary probiotics. Fish Physiol. Biochem. 41, 91–105.

FAO, 2002. Joint FAO/WHO Working Group Report on Drafting Guidelines for the Evaluation of Probiotics in Food, 30 April and 1 May 2002. Food and Agriculture Organization of the United Nations, Rome, Italy. Available from: https://www.who.int/foodsafety/fs_management/en/probiotic_guidelines.pdf. (Accessed 6 April 2019).

FAO, 2006. Report of a Joint FAO/WHO Expert Consultation on Evaluation of Health and Nutritional Properties of Probiotics in Food Including Powder Milk with Live. In Health and Nutrition Properties of Probiotics in Food Including Powder Milk with Live Lactic Acid Bacteria; FAO Food and Nutrition Paper 85. FAO, Rome, Italy.

Food and Agriculture Organization/World Health Organization, 2002. Report of a Joint FAO/WHO Working Group on Drafting Guidelines for the Evaluation of Probiotics in Food. Food and Agriculture Organization/World Health Organization, London, Ontario, Canada.

Girones, R., Jofre, J.T., Bosch, A., 1989. Isolation of marine bacteria with antiviral properties. Can. J. Microbiol. 35, 1015–1021.

Hai, N.V., Fotedar, R., Buller, N., 2007. Selection of probiotics by various inhibition test methods for use in the culture of western king prawns, *Penaeus latisulcatus* (Kishinouye). Aquaculture 272 (1–4), 231–239.

Hanczakowska, E., Swiatkiewicz, M., Natonek-Wisniewska, M., Okon, K., 2016. Medium chain fatty acids (MCFA) and/or probiotic *Enterococcus faecium* as a feed supplement for piglets. Livest. Sci. 192, 1–7.

Helm, E.T., Curry, S., Trachsel, J.M., Schroyen, M., Gabler, N.K., 2019. Evaluating nursery pig responses to in-feed sub-therapeutic antibiotics. PLoS One 14, e0216070.

Hoseinifar, S.H., Dadar, M., Ringø, E., 2017. Modulation of nutrient digestibility and digestive enzyme activities in aquatic animals: the functional feed additives scenario. Aquac. Res. 48 (8), 3987–4000.

Hossain, M.I., Sadekuzzaman, M., Ha, S.D., 2017. Probiotics as potential alternative biocontrol agents in the agriculture and food industries: a review. Food Res. Int. 100, 63–73.

Huys, G., Botteldoorn, N., Delvigne, F., De Vuyst, L., Heyndrickx, M., Pot, B., Dubois, J.J., Daube, G., 2013. Microbial characterization of probiotics-advisory report of the working group "8651 probiotics" of the Belgian Superior Health Council (SHC). Mol. Nutr. Food Res. 57, 1479–1504.

Hwang, C.E., Seo, W.T., Cho, K.M., 2013. Enhanced antioxidant effect of black soybean by cheongguk-jang with potential probiotic *Bacillus subtilis* CSY191. Korean J. Microbiol. 49, 391–397.

Irianto, A., Austin, B., 2002. Probiotics in aquaculture. J. Fish Dis. 25, 633–642.

Johnson, T.A., Sylte, M.J., Looft, T., 2019. In-feed bacitracin methylene disalicylate modulates the Turkey microbiota and metabolome in a dose-dependent manner. Sci. Rep. 9, 8212.

Kamei, Y., Yoshimizu, M., Ezura, Y., Kimura, T., 1988. Screening of bacteria with antiviral activity from fresh water salmonid hatcheries. Microbiol. Immunol. 32 (1), 67–73.

Kesarcodi-Watson, A., Kaspar, H., Lategan, M.J., Gibson, L., 2008. Probiotics in aquaculture: the need, principles and mechanisms of action and screening processes. Aquaculture 274 (1), 1–14.

Kim, M.J., Ku, S., Kim, S.Y., Lee, H.H., Jin, H., Kang, S., Li, R., Johnston, T.V., Park, M.S., Ji, G.E., 2018. Safety evaluations of *Bifidobacterium bifidum* BGN4 and *Bifidobacterium longum* BORI. Int. J. Mol. Sci. 19, 1422.

Kocyigit, R., Aydin, R., Yanar, M., Diler, A., Avci, M., Ozyurek, S., 2016. The effect of direct-fed Micro-bials plus exogenous feed enzyme supplements on the growth, feed efficiency ratio and some behavioural traits of Brown Swiss x eastern Anatolian red F1 calves. Pak. J. Zool. 48, 1389–1393.

Lan, R., Tran, H., Kim, I., 2017. Effects of probiotic supplementation in different nutrient density diets on growth performance, nutrient digestibility, blood profiles, fecal microflora and noxious gas emission in weaning pig. J. Sci. Food Agric. 97, 1335–1341.

Lee, S., Lee, J., Jin, Y.-I., Jeong, J.-C., Chang, Y.H., Lee, Y., Jeong, Y., Kim, M., 2017. Probiotic char-acteristics of Bacillus strains isolated from Korean traditional soy sauce. LWT Food Sci. Technol. 79, 518–524.

Lilly, D.M., Stillwell, R.H., 1965. Probiotics: growth promoting factors produced by microorganisms. Science 147, 747–748.

Lü, X., Yi, L., Dang, J., Dang, Y., Liu, B., 2014. Purification of novel bacteriocin produced by *Lactobacillus coryniformis* MXJ 32 for inhibiting bacterial foodborne pathogens including antibiotic-resistant microor-ganisms. Food Control 46, 264–271.

Meidong, R., Doolgindachbaporn, S., Sakai, K., Tongpim, S., 2017. Isolation and selection of lactic acid bacteria from thai indigenous fermented foods for use as probiotics in tilapia fish oreochromis niloticus. AACL Bioflux 10 (2), 455–463.

Merrifield, D.L., Dimitroglou, A., Foey, A., Davies, J.S., Baker, T.M.R., Bøgwald, J., Castex, M., Ringø, E., 2010. The current status and future focus of probiotic and prebiotic applications for salmonids. Aqua-culture 302, 1–18.

Morelli, L., 2000. In vitro selection of probiotic lactobacilli: a critical appraisal. Curr. Issues Intest. Microbiol. 1, 59–67.

Ng, S.C., Hart, A.L., Kamm, M.A., Stagg, A.J., Knight, S.C., 2009. Mechanisms of action of probiotics: recent advances. Inflamm. Bowel Dis. 15 (2), 300–310.

Plessas, S., Nouska, C., Karapetsas, A., Kazakos, S., Alexopoulos, A., Mantzourani, I., Chondrou, P., Fournomiti, M., Galanis, A., Bezirtzoglou, E., 2017. Isolation, characterization and evaluation of the probiotic potential of a novel Lactobacillus strain isolated from feta-type cheese. Food Chem. 226, 102–108.

Ponce, C.H., Dilorenzo, N., Quinn, M.J., Smith, D.R., May, M.L., Galyean, M.L., 2011. Case study: effects of a directfed microbial on finishing beef cattle performance, carcass characteristics, and in vitro fermentation. Prof. Anim. Sci. 27, 276–281.

Prieur, G., Nicolas, J.L., Plusquellec, A., Vigneulle, M., 1990. Interactions between bivalves mollusks and bacteria in the marine environment. Oceanogr. Mar. Biol. Annu. Rev. 28, 227–352.

Qiao, M., Ying, G.-G., Singer, A.C., Zhu, Y.-G., 2018. Review of antibiotic resistance in China and its environment. Environ. Int. 110, 160–172.

Ran, T., Gomaa, W.M.S., Shen, Y.Z., Saleem, A.M., Yang, W.Z., McAllister, T.A., 2019. Use of naturally sourced feed additives (lactobacillus fermentation products and enzymes) in growing and finishing steers:

effects on performance, carcass characteristics and blood metabolites. Anim. Feed Sci. Technol. 254, 114190.

Rengpipat, S., Rukpratanporn, S., Piyatiratitivorakul, S., Menasaveta, P., 2000. Immunity enhancement in black tiger shrimp (Penaeus monodon) by aprobiont bacterium (*Bacillus* S11). Aquaculture 191, 271–288.

Roselli, M., Pieper, R., Rogel-Gaillard, C., De Vries, H., Bailey, M., Smidt, H., Lauridsen, C., 2017. Immunomodulating effects of probiotics for microbiota modulation, gut health and disease resistance in pigs. Anim. Feed Sci. Technol. 233, 104–119.

Sakata, T., 1990. Microflora in the digestive tract of fish and shellfish. In: Lesel, R. (Ed.), Microbiology in Poecilotherms. Elsevier, Amsterdam, pp. 171–176.

Sanders, M.E., Akkermans, L.M., Haller, D., Hammerman, C., Heimbach, J., Hormannsperger, G., Huys, G., Levy, D.D., Lutgendorff, F., Mack, D., et al., 2010. Safety assessment of probiotics for human use. Gut Microbes 1, 164–185.

Sarrazin, S., Joosten, P., Gompel, L.V., Luiken, R.E.C., Mevius, D.J., Wagenaar, J.A., Heederik, D.J.J., 2019. Quantitative and qualitative analysis of antimicrobial usage patterns in 180 selected farrow-to-finish pig farms from nine European countries based on single batch and purchase data. J. Antimicrob. Chemother. 74, 807–816.

Seal, B.S., Drider, D., Oakley, B.B., Brüssow, H., Bikard, D., Rich, J.O., Miller, S., Devillard, E., Kwan, J., Bertin, G., et al., 2018. Microbial-derived products as potential new antimicrobials. Vet. Res. 49, 66.

Servin, A.L., 2004. Antagonistic activities of lactobacilli and bifidobacteria against microbial pathogens. FEMS Microbiol. Rev. 28, 405–440.

Shokryazdan, P., Faseleh Jahromi, M., Liang, J.B., Ho, Y.W., 2017. Probiotics: from isolation to application. J. Am. Coll. Nutr. 36 (8), 666–676.

Song, X., Huang, Q., Zhang, Y., Zhang, M., Xie, J., He, L., 2019. Rapid multiresidue analysis of authorized/banned cyclopolypeptide antibiotics in feed by liquid chromatography–tandem mass spectrometry based on dispersive solid-phase extraction. J. Pharm. Biomed. Anal. 170, 234–242.

Sornplang, P., Piyadeatsoontorn, S., 2016. Probiotic isolates from unconventional sources: a review. J. Anim. Sci. Technol. 58, 26.

Srinivas, B., Rani, G.S., Kumar, B.K., Chandrasekhar, B., Krishna, K.V., Devi, T.A., Bhima, B., 2017. Evaluating the probiotic and therapeutic potentials of *Saccharomyces cerevisiae* strain (OBS2) isolated from fermented nectar of toddy palm. AMB Express 7, 2.

Sun, J., Zhang, H., Liu, Y.-H., Feng, Y., 2018. Towards understanding MCR-like Colistin resistance. Trends Microbiol. 26, 794–808.

Suresh, K., Srinath, K., Pravesh, B., 2013. Safety concerns of probiotic use: a review. IOSR J. Dent. Med. Sci. 12, 56–60.

Suzer, C., Coban, D., Kamaci, H.O., Saka, S., Firat, K., Otgucuoglu, O., Kucuksari, H., 2008. Lactobacillus spp. bacteria as probiotics in gilthead sea bream (*Sparus aurata* L.) larvae: effects on growth performance and digestive enzyme activities. Aquaculture 280, 140–145.

Tesfaye, A., Hailu, Y., 2019. The effects of probiotics supplementation on milk yield and composition of lactating dairy cows. J. Phytopharmacol. 8, 12–17.

Tovar, D., Zambonino, J., Cahu, C., Gatesoupe, F., Vazquez-Juarez, R., Lesel, R., 2002. Effect of live yeast incorporation in compound diet on digestive enzyme activity in sea bass (*Dicentrarchus labrax*) larvae. Aquaculture 204, 113–123.

Vankerckhoven, V., Huys, G., Vancanneyt, M., Vael, C., Klare, I., Romond, M., Entenza, J.M., Moreillon, P., Wind, R.D., Knol, J., et al., 2008. Biosafety assessment of probiotics used for human consumption: recommendations from the EU-PROSAFE project. Trends Food Sci. Technol. 19, 102–114.

Vase-Khavari, K., Mortezavi, S.-H., Rasouli, B., Khusro, A., Salem, A.Z.M., 2019. The effect of three tropical medicinal plants and superzist probiotic on growth performance, carcass characteristics, blood constitutes, immune response, and gut microflora of broiler. Trop. Anim. Health Prod. 51, 33–42.

Venturi, A., Gionchetti, P., Rizzello, F., et al., 1999. Impact on the composition of the faecal flora by a new probiotic preparation: preliminary data on maintenance treatment of patients with ulcerative colitis. Aliment. Pharmacol. Ther. 13, 1103–1108.

Verschuere, L., Rombault, G., Sorgeloos, P., Verstraete, W., 2000. Probiotics bacteria as biological control agents in aquaculture. Microbiol. Mol. Biol. Rev. 64, 655–671.

Vine, N.G., Leukes, W.D., Horst. Kaiser., 2004. In vitro growth characteristics of five candidate aquaculture probiotics and two fish pathogens grown in fish intestinal mucus. FEMS Microbiol. Lett. 231 (1), 145–152.

Wang, Q., Huang, S.Q., Li, C.Q., Xu, Q., Zeng, Q.P., 2017. *Akkermansia muciniphila* May determine chondroitin sulfate ameliorating or aggravating osteoarthritis. Front. Microbiol. 8, 1955.

Wang, X., Li, H., Zhang, X., Li, Y., Ji, W., Xu, H., 2000. Microbiol flora in the digestives tract of adult penaeid shrimp (*Penaeus chinensis*). J. Ocean. Univ. Qingdao 30, 493–498.

Wang, Y.-B., 2007. Effect of probiotics on growth performance and digestive enzyme activity of the shrimp *Penaeus vannamei*. Aquaculture 269, 259–264.

Xia, X., Wang, Z., Fu, Y., Du, X.-D., Gao, B., Zhou, Y., He, J., Wang, Y., Shen, J., Jiang, H., et al., 2019. Association of colistin residues and manure treatment with the abundance of mcr-1 gene in swine feedlots. Environ. Int. 127, 361–370.

Xiao, L., Estellé, J., Kiilerich, P., Ramayo-Caldas, Y., Xia, Z., Feng, Q.S., Pedersen, A.Ø., Kjeldsen, N.J., Liu, C., 2016. A reference gene catalogue of the pig gut microbiome. Nat. Microbiol. 1, 16161.

Zhao, P., Kim, I., 2015. Effect of direct-fed microbial on growth performance, nutrient digestibility, fecal noxious gas emission, fecal microbial flora, and diarrhea score in weanling pigs. Anim. Feed Sci. Technol. 200, 86–92.

Microbial adaptation to climate change and its impact on sustainable development

Srishti Srivastava, Amartya Chakraborty, and K. Suthindhiran*
Marine Biotechnology and Bio-Products Laboratory, Department of Biomedical Sciences, School of Biosciences and Technology, Vellore Institute of Technology, Vellore, India
*Corresponding author: e-mail address: sudhindhira@gmail.com

1. Introduction

The microbial community, though not studied on a large scale, is crucial for the maintenance and health of any ecosystem. Microbial metabolism and physiology are directly affected by the same environmental disturbances that alter host metabolism, and these changes will alter the interaction between the host and the microbial community within each microhabitat. Documenting the complexity of microbes, their holobiont responses to environmental stress, and their subsequent role in macroecological change, is therefore fundamental for predicting change due to climatic factors (Cavicchioli et al., 2019; Flemming and Wuertz, 2019). Whether it is the role in maintaining the balance of pH in the soil or regulating biogeochemical cycles, microbes play a vital role. Some microbes change with the changing surroundings which may be compromising for the host and some microbiomes remain the same as they adapt themselves in such situations hence protecting the host. The anthropogenic activities in the past few decades have contributed a lot to the increase in climate changes or industrial accidents, for instance "El Niño," which ultimately results in high environmental stress (Cavicchioli et al., 2019; Damjanovic et al., 2017). The microbial species are known for their ability to release certain secondary metabolites which reduces the effects of the xenobiotics or other stress factors in the environment. Some of the instances discussed further in the paper include the following. Growth and regulation of many plants are via the beneficial microbes (rhizobacteria) in the soil which are associated with the functioning (Mayak et al., 2004; Paul and Nair, 2008). Bioremediation technique of microorganisms is another helpful measure to clean the polluted environments. They possess the ability to metabolically mineralize or transform organic contaminants into less detrimental substances which then can be integrated into the biogeochemical cycles of nature (Gibson and Harwood, 2002; Wackett, 2000). The coral-associated bacteria scavenge on the residual nutrients that are sometimes released by the corals and in return they

Relationship Between Microbes and the Environment for Sustainable Ecosystem Services, Volume 1
https://doi.org/10.1016/B978-0-323-89938-3.00005-0

provide essential metabolites and aid in nitrogen and carbon fixation. The microbes are known to aid corals with a high range of functional roles like—Oxygen Cycling, Nutrient Cycling, Antimicrobials, Nitrogen Fixation (by nitrogenase enzyme) (Capone and Carpenter, 1982). Further, the prokaryotic microbiome is very diverse and is occupied with several types of microbes and it was reported that the colonization of one type of microbe prevents the colonization of any pathogenic microbe, mainly by competitive exclusion as sometimes they secrete the antimicrobial peptides which inhibit the growth of opportunistic pathogens (Gibson and Harwood, 2002). Similarly, the health and maintenance of the coral reefs are largely because of the presence of the microbial community in the coral's cavity or sponges. It is therefore essential for the microbes to develop or inherit resistance and adapt to the continuously changing conditions around the environment. The microbes in order to adapt, follow certain measures to help maintain the biodiversity thriving. The current changes in the oceans lead to acidification. Eutrophication or depletion of oxygen threatens the health of the microbial community as well as the corals and other species diversity in the niche (Fishelson, 1977). We focus on some of the adaptive strategies such as Microbiome Mediated Transgenerational Acclimatization (MMTA) followed by species present in the niche. The species *Syncchococcus* is one absolute example of adaptation (Webster and Reusch, 2017). We further focus on the soil ecosystem challenges such as the droughts, the increasing levels of CO_2, the drastic changing parameters, and how the microorganisms survive and maintain balance by certain measures. One such example discussed here is the plant species *Brassica rapa* and its adaptation along with its surrounding plant biota (Lau and Lennon, 2012). Agriculture and plant diversity are a major part of any ecosystem and hence its health becomes a necessity. The microbes help in the maintenance and help to resist the plants against any kind of drastic soil change, one such example being species *Ralstonia eutropha* discussed here having the carbon sequestration ability which helps in survival amidst stress (Rossi et al., 2015). Generally, these studies are conducted in small-scale laboratory facilities called the mesocosm which technically mimics the natural coral habitat and is significantly useful in analyzing the small-scale changes in the environment around the corals and their responses to it (Jokiel et al., 2008; Odum, 1984). They have been used as the most useful lab equipment to understand the dynamics of the corals under environmental stress and how these effects can be directed by the symbiotic microbiome. These changes can be monitored artificially both under control and experimental conditions (Odum, 1984; Stewart et al., 2013).

2. Climatic factors

Biodiversity can be affected by a number of factors including climatic changes such as natural disasters, drought, increased levels of carbon dioxide, or increased pH in the oceans that can lead to acidification and eutrophication. These external factors have a major impact on the health of the species and that of the biome. Microbiome communities associated with the species tend to get adapted in order to protect and preserve biodiversity in numerous ways. Some examples are cited as follows (Table 1).

2.1 Temperature elevation and thermotolerance

There are traces of adaptation via genetic mutations as well due to the meiosis-specific genes present in the species responsible for the thermotolerance and promoting adaptation (Webster and Reusch, 2017). The Horizontal gene transfer is yet another approach for rapid adaptation and works on mechanisms such as conjugation and transformation. The mobile elements are transferred through a transduction mechanism as well which involves viruses transferring beneficial genes to species. One of the examples involves the spherulin encoding gene essential for skeletogenesis which is transferred from

Table 1 List of extreme environmental conditions and their consequences on organisms.

S. No	Types of extreme environmental conditions	Description	Causes of the extreme conditions	Their consequences on the organism
1.	Temperature Elevation	Temperature elevation occurs as a consequence of global warming wherein the temperature is estimated to increase to an extent causing disturbance in the aqua ecosystem. It becomes practically asperse for the sea organisms to survive. The extent of this environmental stress is felt most in the ocean floor dwelling species of organisms, especially the coral species.[a]	Global warming, release of greenhouse gases, industrial development, reduction in forests, dumping of industrial effluents in sea waters.[b]	Loss of species and habitats in marine ecosystem, extinction of plants and animals, reduction in their resilience and response, eutrophication of their habitats leading to destruction of habitats.[c]

Continued

Table 1 List of extreme environmental conditions and their consequences on organisms—cont'd

S. No	Types of extreme environmental conditions	Description	Causes of the extreme conditions	Their consequences on the organism
2.	Drought Stress	Drought stress results from the decreased availability of water in the environment. This stimulus evokes a drought stress response in plants and animals.[d]	Climatic changes, heat waves, irregular rainfall patterns, unavailability of water resources.[e]	Reduces plant growth and development, reduction in crop yielding, loss of agricultural product, inactivation of stress response in some plants.[f]
3.	Increased CO_2 in Lands	The increase in the amount of CO_2 emitted in the environment is linked to a direct decline in biodiversity.[g]	Industrial release of gases, increased burning, and usage of fossil fuels, shoot up in greenhouse gases production.[h]	Chemical alterations in the plants chemicals and composition, an increase in the rate of photosynthetic carbon fixation by leaves, damages the plant physiology.
4.	Ocean Acidification	The diffusion of atmospheric CO_2 is found and accumulated in the ocean and has ultimately reduced the pH to 0.1. This reduction in pH and increase in the acidic condition is termed as ocean acidification.[i]	Burning of fossil fuels, increased levels of CO_2 emission in atmosphere, ineffective disposal of wastes and incorrect measures of land waste management, use of pesticides and fertilizers.[j]	Corals get most affected as they are very sensitive to these changes leading to reduction in coral population, affects the skeletal structure of marine species, causes an increase in the CO_2 values of the ocean environment and a subsequent increase in the concentration of carbonate ions, poor growth and development of marine corals as

Table 1 List of extreme environmental conditions and their consequences on organisms—cont'd

S. No	Types of extreme environmental conditions	Description	Causes of the extreme conditions	Their consequences on the organism
5.	Water Soluble Oil Fractions	The occurrence of hydrocarbons in the ocean ecosystem caused by anthropogenic activities like industrial oil spills, pipeline expulsions, etc., in the form of what is known as water-soluble oil fractions results in the education of the survival rate of the marine species, including corals. In some cases, it causes bioaccumulation.[l]	Release of industrial oil spills, accidents resulting in crude oil release, ineffective dumping of industrial effluents in rivers and oceans, burning of coal.[m]	hampers the Skeletogenesis process.[k] WSF leads to reduced growth, retraction of the coral polyps, extrusion of symbiotic species of zooxanthellae, the increase of the Ca45 deposition, a decrease in the rates of calcification rate in some species, alters the configuration of marine organisms in terms of nutritional uptake and contraction for digestive mixing.[n]

[a]Kaniewska et al. (2015); Luter et al. (2020).
[b]Fragoso Ados Santos et al. (2015); Luter et al. (2020).
[c]Buckley and Huey (2016); Jokiel and Coles (1977); Saint-martin (2003).
[d]Lau and Lennon (2012); Pineda et al. (2013).
[e]Lei et al. (2014).
[f]Fragoso Ados Santos et al. (2015); Zoppellari et al. (2014).
[g]Biagi et al. (2020); Schlüter et al. (2014).
[h]Schlüter et al. (2014).
[i]Gruber (2011); Guinotte and Fabry et al. (2008); Luter et al. (2020).
[j]De Goeij et al. (2013); Silva et al. (2019).
[k]Comeau et al. (2019); Gutner-Hoch et al. (2017); Mollica et al. (2018); O'Brien et al. (2016).
[l]Dubinsky and Stambler (1996); Jackson (2016); Loya and Rinkevich (1980); White et al. (2012).
[m]Pineda et al. (2013); Wackett (2000).
[n]Bak and Elgershuizen (1976); Fragoso Ados Santos et al. (2015); Michel and Fitt (1984); Shafir et al. (2007).

microbes to sponges or the eukaryotic-like protein from host to symbionts in order to regulate phagocytosis of host (Luter et al., 2020). The tolerance of corals, as a matter of fact, can increase enormously if there is a shift in symbiotic microbes rather than species themselves. The MMTA being a powerful approach, will only take place if the

Table 2 List of microorganisms that help withstand and tolerate the extreme conditions for species survival.

S. No.	Extreme environmental conditions	Microorganisms that help withstand and tolerate the extreme conditions for species survival
1.	Temperature Elevation	*Flavobacteria, Rhodobacterales*[a], *Alteromonadales*[b], *Rhizobiales, Vibrionales, Bacteroidetes*[c], *Sphingobacteria, Desulfovibrionales, Bacteriovorax, Vibrio harveyi, Caulobacter, Rhodobacterales*[d]
2.	Drought Stress	*Brassica rapa*[e], *rhizobacterial strain, Bacillus megaterium BOFC15*[f], *Enterobacter* sp.[g], *Bacillus thuringiensis*[h], *Bacillus* sp., *Serratia* sp., *Pseudomonas* sp., *Azospirillum*[i].
3.	Increased CO_2 in Lands	*Ralstonia eutropha*[j], *Arthrospira*[k], *Clostridium jundiai*[l], *Chlorella vulgaris, Bacillus, Acinetobacter, Solibacillus, Pontibacter, Achromobacter, Ensifer, Fusarium, Melanotaenium, Wardomycopsis*[m].
4.	Ocean Acidification	*Rhodobacteraceae (Alphaproteobacteria)*[n], *Gammaproteobacteria, Flavobacteriaceae*[o], *and Campylobacteraceae*[p], *Pseudomonas, Exiguobacterium*[q], *Halotalea, Micrococcus, Halomonas meridiana*[r], *Arthrobacter*[s], *Pseudoalteromonas*[t], *Paracoccus*[u], *and Vibrio*[v].
5.	Water Soluble Oil Fractions	*Erythrobacter citreus*[w], *Acinetobacter calcoaceticus*[x], *Pseudomonas stutzeri*[y], *Bacillus cereus*[z], *Paracoccus homiensis*[aa], *P. amogawaensis*[ab], *P. marcusii*[ac], *Psychrobacter sp*[ad], *Gordonia lacunae*[ae], *Vibrio alginolyticus*[af], *Vibrio natriegens*[ag], *Vibrio fortis*[ah], *V. azureus*[ai], *Bacillus krulwichiae*[aj], *Dietzia maris*[ak], *Kocuria flava, Micrococcus luteus*[al], *Alteromonas macleodii*[am], *Pseudoalteromonas prydzensis*[an].

[a]Wong et al. (2017).
[b]Shiu et al. (2017).
[c]McDevitt-Irwin et al. (2017); Rypien et al. (2010).
[d]Thurber et al. (2009).
[e]Pineda et al. (2013).
[f]Zhou et al. (2016).
[g]Nguyen (2018).
[h]Dash et al. (2014).
[i]Gupta (2012).
[j]Lau and Lennon (2012).
[k]Freitas et al. (2012); Rossi et al. (2015).
[l]Damjanovic et al. (2017).
[m]Liu et al. (2020).
[n]Meron et al. (2012).
[o]Biagi et al. (2020).
[p]Meron et al. (2011).
[q]Das and Mangwani (2015).
[r]James et al. (1990); Mata et al. (2002).
[s]Mohapatra and Bapuji (1998).
[t]Longeon et al. (2004).
[u]Liu et al. (2008).
[v]Das and Mangwani (2015); Dash et al. (2014); Thirugnanasambandam et al. (2019); Lee et al. (2003).
[w]Denner et al. (2002); Zhuang et al. (2015).
[x]Lal and Khanna (1996).
[y]Celik et al. (2008).
[z]Da Cunha et al. (2006).

[aa]Kim et al. (2006).
[ab]López-Cortés et al. (2010).
[ac]Antoniou et al. (2015).
[ad]Moghadam et al. (2016).
[ae]Al-Awadhi et al. (2012).
[af]X. Li et al. (2019).
[ag]Hamdan and Fulmer (2011).
[ah]Ali et al. (2016).
[ai]Bayat et al. (2016).
[aj]S. W. Li et al. (2019).
[ak]W. Chen et al. (2017).
[al]Bayat et al. (2016).
[am]Jin et al. (2012).
[an]Al-Awadhi et al. (2012); Al-Dahash and Mahmoud (2013); Gupta (2012); Varjani (2017).

microbiome is passed through generations vertically or horizontally. Further, MMTA is where the acclimatization is performed and regulated through mechanisms by the microbial–associated species in the niche. The reef microbiomes are expected to have great impacts on the health, tolerance levels, and plasticity for the associated holobiont but they might also be involved in the maintenance of homeostasis for example, in model cnidarian *Nematostella vectensis* (Webster and Reusch, 2017) (Table 2).

2.2 Drought stress

The plant fitness is found to be largely influenced by the rapid responses to climatic change by the soil microbes and their adaptation (Ashraf, 2004). One of the referred articles explains the experiment conducted, a multigeneration selection experiment (selection in a controlled environment experiment) for establishing the relevance of adaptation of plants to drought stress especially when they are associated with neighboring microbial communities. The plant species *B. rapa* was chosen for the experiment to find out the microbial community's population and diversity and how they can affect fitness. The fitness traits were taken as response variables, histories and contemporary environment were taken as fixed factors and mesocosms and interactions are taken as random factors (Lau and Lennon, 2012). The plant fitness was found to be largely influenced by the rapid responses to climatic change by the soil microbes, the male and female parts were exceptionally more productive which refers to the flower and fruit production, respectively. When the changed precipitation regimes were considered, it was noted that in usual situations, the drought would cause a negative effect on the plant fitness and the plant history was not much useful. But when subjected to microbial communities, it showed 20% decrease in fruit production when associated with dry adapted microbial communities (Lau and Lennon, 2012). There are indirect benefits to the adaptive strategies used by the microorganisms. One, the shifts in the diversity of these associated microorganisms might result in the alteration of biogeochemical cycles which are linked to the resources essential to the plants, for example, nitrogen (Zoppellari et al., 2014). Two, the drought stress can change the population of mutualists and pathogens which benefits the plant as it increases the fitness of the hosts (Zhou et al., 2016). Fitness can be influenced either by evolutionary changes in plants or rapid changes by the microbes. But there are some suggestions regarding the beneficial interactions between the plant and the surrounding biota. The fruit production and biomass were observed to increase when there was a mismatch of history of microbes and history of plants (Lau and Lennon, 2012). Some of the

Table 3 List of few microorganisms along with their known and identified mechanisms that help reverting extreme environmental conditions to normal conditions.

S. No.	Extreme environmental conditions	Microorganisms that revert back to normal conditions	Mechanism of action of mentioned microorganisms
1.	Increased CO_2 in lands	*Bacillus, Acinetobacter, OLB13, Solibacillus, Pontibacter, Achromobacter, and Ensifer, Fusarium, Melanotaenium, and Wardomycopsis.*[a]	These microorganisms (bacteria and fungi) help in CO_2 sequestration and reduction and fixation of atmospheric CO_2 by a mechanism called as biomineralization.[b] The bacterial and fungal cell walls are highly electronegative and hence contribute to the absorption of the metal ions like Ca^{2+} and Mg^{2+} which results in the formation of carbonates from SIC (soil inorganic carbon). These carbonaceous minerals can be found in deserts and other rich carbon lands in order to revert to normal concentrations of carbon.[c]
2.	Drought stress	*Serratia* sp., *Pseudomonas stutzeri*[d]	These plant growth-promoting bacteria (PGPB) provide drought resistance to the plants along with better production and physiology development.[e] The mechanisms involve the direct approach of regulation and production of phytohormones such as auxins and the indirect approach involves production of antimicrobial metabolites and

Table 3 List of few microorganisms along with their known and identified mechanisms that help reverting extreme environmental conditions to normal conditions—cont'd

S. No.	Extreme environmental conditions	Microorganisms that revert back to normal conditions	Mechanism of action of mentioned microorganisms
3.	Osmotic stress	*Achromobacter, Acidovorax, Alcaligenes, Bacillus, Enterobacter, Klebsiella, Methylobacterium, Pseudomonas, Rhizobium, and Variovorax*[g].	hydrolytic enzymes to protect from toxins and pathogens.[f] The rhizobacteria species help in regulating osmotic stress in plant species by ACC deaminase enzyme. The enzyme present is responsible for hydrolysis of the ACC compound into ammonia. This reduces the ethylene. The enzymes also provide resistance to high concentration of salinity, assisting in their normal growth and development.[h]
4.	Water soluble oil fractions	*P. stutzeri.*[i]	This is an oleophilic stain of microbe that is known for its degrading property of O–Xylene, in which the compound is converted to 2,3-dimethylphenol by the enzyme activity of mono-oxygenase and can further be converted to 3,4-dimethyl catechol by the same. The reaction can also proceed by a different pathway where O–Xylene can be converted to 3,4-dimethyldihydrothiol by the enzyme reaction of dehydrogenase which gets converted to the product of 3,4-dimethyl catechol. This reaction can further go on for an extra diol cleavage product.[j]

[a]Liu et al. (2020).
[b]Liu et al. (2020).
[c]Liu et al. (2020); Wolfe et al. (2008).
[d]Wang et al. (2014).
[e]Wang et al. (2014).
[f]Lau and Lennon (2012); S. Begum et al. (2021).
[g]Nguyen (2018); Siddikee et al. (2011).
[h]Nguyen (2018); Siddikee et al. (2011).
[i]El Mahdi et al. (2016).
[j]J.E. Chen et al. (2017); El Mahdi et al. (2016); Sweet and Bythell (2017).

other organisms as cited in Table 3, depict the direct and indirect mechanisms to assist the survival of associated plants.

2.3 Increased levels of CO_2 in lands

Carbon dioxide represents 63% under the Greenhouse Gas (GHGs) and possesses a threat of increasing exponential rate due to the coal demands and running industries. These volumes of carbon in the environment affect climate change and lead to desertification with an ultimate decline in biodiversity (Doney et al., 2009; Rossi et al., 2015). The ability of the microorganisms such as cyanobacteria, autotrophic bacteria, and other green algae to fix the carbon dioxide and its remnants in the atmosphere more efficiently than any of the higher green plants (Allison et al., 2010). In arid and semiarid environments, the sustainable mechanism of soil carbon sequestration (CO_2 fixation) by cyanobacteria and biological soil crusts is believed to increase soil restoration (Rossi et al., 2015). There is also the possibility of performing carbon dioxide sequestration using microalgae which also can result in bioproducts used up in the industries. Some other examples would involve particular chemotrophic bacteria, such as *R. eutropha* (or picketii) and related organisms which too are witnessed to have the carbon sequestration ability (Rossi et al., 2015). Chemotrophic bacteria that can fix the CO_2 up to certain amounts along with the cyanobacteria species tend to have 50-folds more efficiency due to their fast reproduction. The cyanobacteria also tend to exist in the deserted areas as a natural solution called biological soil crusts which are responsible for the photosynthesis that takes place there, hence a source of the carbon reduction from the surroundings. The microbiotic crusts act as a neutralizer for the abiotic stresses which arise from climate change. The potential of the microalgae is also viewed as the most effective method for reducing global warming through CO_2 emission (Rossi et al., 2015). The *Chlorella* sp., and *Spirulina platensis* have been studied for experiments for CO_2 sequestration. The mechanism of bacteria such as *Arthrospira* as mentioned in the paper highlights the role of these in the biogeochemical cycles of the uptake and removal of electrons. Yet another example is mentioned that of the *Clostridium jundiai* for the CO_2 fermentation, that is, it converts the CO to C_2 molecules in the environment (Rossi et al., 2015).

2.4 Ocean acidification

After the industrial revolution, the atmospheric CO_2 was seen to have increased gradually through the years. One major impact of CO_2 was the increase in the concentration of the bicarbonate ions in the ocean water level which further resulted in the reduction of the pH of the ocean water level by 0.1 unit which led to ocean acidification (Allison et al., 2010). Today the conclusions from a wide range of studies suggest that the impact of ocean acidification is known to hamper a lot of normal biological processes in the corals like the net energy productivity, maintenance of the internal physiological condition along with maintaining the $CaCO_3$ balance (Turley and Findlay, 2016). There are reports of elevation in the levels of resistance factors to pH stress and particular secondary metabolites at low pH conditions which indicate that the corals do exhibit stress responses to reduced pH (Kaniewska et al., 2015; Meron et al., 2011, 2012; Thurber et al., 2009). This results in the decrease of the levels of the $CaCO_3$ utilization by corals which hampers skeletal structure formation and hence inhibits the coral growth (Doo et al., 2019; Nakamura and Nakamori, 2007; Saint-martin, 2003; Venn et al., 2011). The structure and physiology of the calciferous corals are affected by the imbalance in the CO_2 (Fabry et al., 2008). The coral-associated microbial communities have a major role in the functioning of Carbon, Nitrogen, and Sulfur cycle as a metabolic process in corals (Beman et al., 2011; Rädecker et al., 2015; Ziegler et al., 2016). The microbial species that play a part in the nitrogen cycle with the host are pervasive in all scenarios and are accordant participants of the microbiome (Lau and Lennon, 2012; Rossi et al., 2015). The nitrogen-fixing bacteria associated with corals are responsible for internal nitrogen on reefs (Fiore et al., 2010; O'Brien et al., 2016). Whether driven by pH-induced reductions in host taurine production or through a reduced potential for microbial dissimilation, the reduced capacity for sulfur metabolism could contribute to the poor performance of *S. flabelliformis* in acidified waters. One of the papers had concluded from experiments that the increased temperature and reduced ocean water pH have a strong synergistic effect on polyp mortality rate in corals (Prada et al., 2017). In an experiment, it was noted that coral species *C. singaporensis* showed enhanced survival efficiency. The cause can be traced back to the presence of the symbiotic microbiota and its associated functions integrated with the metabolism of the corals contribute to the better survival of the species in acidified environments (Botté et al., 2019). For understanding a possible solution to reduce the impact of ocean acidification, the probiotics approach was used. From the various coral species that were exposed, the samples were collected and the microbial species were isolated. After the sequencing was done, the data revealed that bacterial communities of many key calcifying coral reef taxa are capable of withstanding short-term exposure to elevated or reduced pH (Cavalcanti et al., 2018; Ribes et al., 2016).

2.5 Water soluble oil fractions

Among all the other factors that affect coral physiology and functioning, the Water-Soluble Oil Fraction (WSF) is one of the major anthropogenic causes that leads to disintegration of the aquatic ecosystems. Acute oil spills cause extensive death of corals and prolonged cessation of growth and reproduction. Following the greater demand of petroleum and its associated products throughout the world, there is a need of transporting through the waters which sometimes lead to leakage of the same (Silva et al., 2019). The extent to which the coral holobiont is getting benefitted by the symbiosis of the microbes present in them is still not clear to us but we do know that microorganisms enhance the adaptation of the corals under such an environmental stress (Fragoso Ados Santos et al., 2015). The petroleum WSF consists of mainly three components which are known to affect the ecosystem. They are Total Petroleum Hydrocarbons (TPH), n-alkanes and Polycyclic Aromatic Hydrocarbons (PAH) (Fragoso Ados Santos et al., 2015). Years of research have concluded that the corals exposed to these petroleum hydrocarbons show severe loss of associated symbionts especially the zooxanthellae and the bacteria leading to coral death (Jackson, 2016; Michel and Fitt, 1984; Simister et al., 2016). The Polycyclic Aromatic Hydrocarbons (PAH) and Produced Formation Water (PFW) are known to inhibit the photosynthetic ability and reduction in photochemical efficiency of the zooxanthellae symbionts which ultimately hamper the health of its partner coral (Jokiel et al., 2008; Jones and Heyward, 2003; Odum, 1984; Ondrasek et al., 2021; Yang et al., 2019, 2020). Study and research have shown that corals have their own innate protective response to the stress related to oil exposure (Bak and Elgershuizen, 1976; May et al., 2020; Renegar et al., 2017; White et al., 2012) but in severe cases, the corals are known to implicate special response via the symbiotic bacterial strains associated with them (Fragoso Ados Santos et al., 2015). There have been reports that there are various strains of hydrocarbon-degrading bacteria that possess special genes that help in degrading the oil concentrations in the surroundings and ultimately help the coals in adapting to the stress (Fragoso Ados Santos et al., 2015). Numerous in situ or ex situ experiments have been carried out throughout the years on the impacts of oil spills and petroleum hydrocarbons on coral holobiont functioning, that actually documents the role of microbial and algal species in mitigating the impact of oil and hydrocarbons on corals.(Almeda et al., 2013; Ozhan et al., 2014; Paniagua-Michel and Fathepure, 2019; Suni et al., 2007). There are research approaches proposed by numerous scientists throughout the globe but in case of quantifying the impacts of the anthropogenic stress of oil and WSF, the mesocosm system tends to be the most effective (Silva et al., 2019; Suni et al., 2007). The mesocosm system has various possibilities to mimic the external ocean environment when the coral microbiome is exposed to stress and helps us monitor the same with all the measures and factors possible (Silva et al., 2019; Teira et al., 2007). In addition, some of the other systematic tools are used today for the prediction of the impact of hydrocarbons on corals (Ozhan et al., 2014). The prediction of the same was on the grounds of three major factors that are: (1) Analyzing the reduction of the hydrocarbons—PAH (analyzing by Liquid Chromatography), TPH (analyzation by Gas Chromatography), and n-alkanes (analyzing by Gas Chromatography) by the introduced bacterial consortia, (2) Chlorophyll Fluorescence (Analyzed by the Pulse Fluorescence), (3) Biochemical Markers (Fragoso Ados Santos et al., 2015). Research has shown that the bacteria associated with coral species can help in the adaptation and is today a new target approach for research to aid the coral holobiont in adapting to this anthropogenic impact factor.

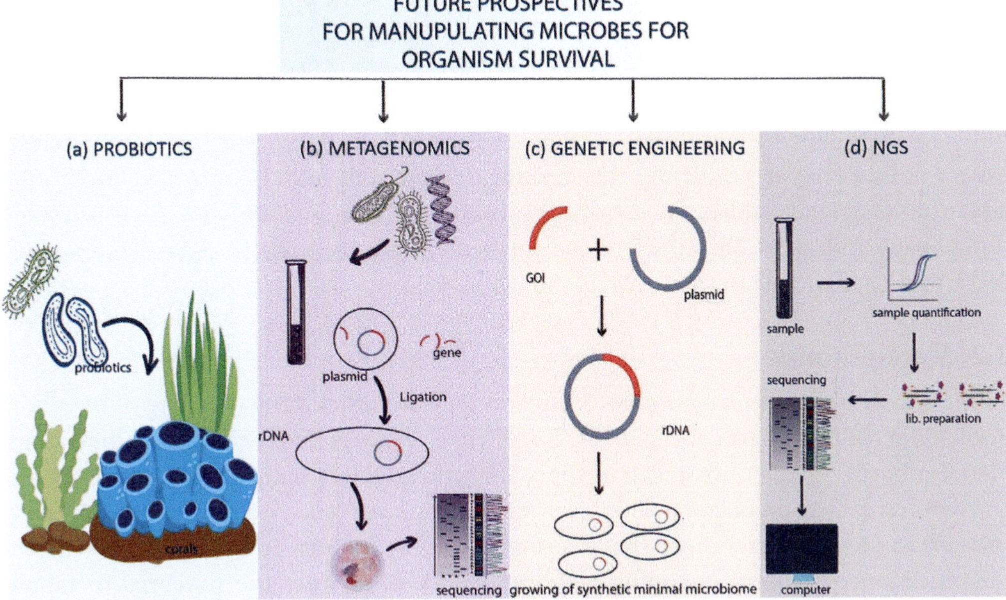

Fig. 1 Future prospective for manipulating microbes; (a) use of probiotics to improve coral health and flexibility; (b) Metagenomic approach to study the microbiome and their changes; (c) Genetic engineering to incorporate stress-resistant genes for better survival; and (d) NGS technique to study the microbial gene expression on the stressful environment.

3. Future prospective and applications

The gradual exposure of various anthropogenic influences like elevated seawater temperature, ocean acidification, elevated concentrations of CO_2 in the ocean, sudden exposure to various crude oils, hydrocarbons, heavy metals and toxic effluents, etc., make the list of factors that are responsible for degradation of ecosystems, be it terrestrial or aquatic (Cianelli et al., 2011; Comeau et al., 2019; May et al., 2020; Yang et al., 2019, 2020). This pressing issue is sustaining since post-industrial revolution (Turley and Findlay, 2016). Today, we have different approaches to read, understand, and manipulate the functional genome of the symbiotic microbial species that will allow us to configure stress tolerance to the holobiont along with some advantageous industrial applications. Some of them are as follows (Fig. 1).

3.1 Probiotics

The most potential scientific procedure that has been suggested today is the use of probiotics to improve the aquatic ecosystems and prevention of degraded corals. Probiotic can be defined as a group of selected collection of bacteria that are provided to the host wherein the microbes are allowed to grow and infer some specific functions, done by both, artificially or naturally (Fragoso Ados Santos et al., 2015). The research says that the coral–associated microbiome can be modulated artificially to improve coral health and flexibility in the changing climatic and other environmental scenarios; known as the "Coral Probiotic Hypothesis" (CPH) (Reshef et al., 2006). We can also selectively use a coral mucus for simulating the survival of the corals under controlled conditions. Today, it is focused on the concept of the minimal microbiome which is the smallest but functionally indispensable subset of the total microbiome (Fragoso Ados Santos et al., 2015; van Oppen and Blackall, 2019).

3.2 Metagenomics

Bioinformatics aided metagenomics are helping us understand the diversity of the entire pool of microbes (Littman et al., 2011). For eliciting a better understanding of these complex symbiotic interactions we can imply high throughput sequencing to characterize entire microbial assemblages or metagenomes within a given system using phylogenetic markers to provide insight into the diversity of the community in combination with the identification of protein-encoding gene sequences to identify the potential metabolic pathways available within the community (Thurber et al., 2009). Our ultimate goal here is to identify the various phylogenetic markers from the collected library of microbes to compare and classify each species and understand the specific phenotypic function associated with it.

3.3 Genetic engineering

Today, the modern genetic engineering tools have paved a clear way to enhance the characteristics of the microbial strains which are generally isolated via highly specific cultures (Goomer and Kunkel, 1992). It is possible to engineer and modify the genome or the plasmid of the bacteria by using the recombinant DNA technology to introduce newer characteristics for adaptation to a particular stress (van Oppen and Blackall, 2019). This modified pool is again collected back to a functional pile of genetically engineered microbes called synthetic minimal microbiome (van Oppen and Blackall, 2019).

3.4 Next-generation sequencing and others

Today with the advent of NGS platforms, we have sequenced the genome of *Symbiodinium*, and found out several genes that are expressed when a particular stress is given to a particular microbiome. By expression profiling, we know which genes are widely expressed during stress that led to the breaking of the symbiotic association with the host. Either by gene knockout which provides insight into the role and criticality of a gene by assessing the effect of its absence, or by the introduction of a new gene while the cell is repairing its genome by non-homologous end joining or by homology-directed repair, research is going on now to alter the genes using the CRISPR Cas-9 technology (J.E. Chen et al., 2017; Cleves et al., 2018; Levin et al., 2017). Other novel strategies encompass the use of Phage therapy, which uses the bacteriophage to insert new genes to the host of the symbiotic bacteria for some new traits.

4. Conclusion

In this chapter, we appraised some of the discoveries on how microorganisms associated with the species assist in the survival and adaptation in the changing climatic conditions. We have focused on the research done on the stress factors such as temperature elevation, drought stress, ocean acidification, and increasing levels of CO_2 and water-soluble fractions. In our understanding, the microbes possess the ability to withstand or evolve as per the changing niche and habitats. With the upcoming modern-day approaches, today the technology stands on the verge of a huge breakthrough which will pave a new way in the conservation of the ecological biosphere. We need to have deeper insight of the molecular mechanisms which will help to sustain a particular microbial community associated with the holobiont and trace our understanding to the level of gene expression, cellular metabolism, and chemical interactions and understand how these changes are influencing the ecosystem and its functioning.

Acknowledgments

The authors wish to thank the management of VIT for providing the necessary facilities.

References

Al-Awadhi, H., Al-Mailem, D., Dashti, N., Hakam, L., Eliyas, M., Radwan, S., 2012. The abundant occurrence of hydrocarbon-utilizing bacteria in the phyllospheres of cultivated and wild plants in Kuwait. Int. Biodeterior. Biodegrad. 73, 73–79.

Al-Dahash, L.M., Mahmoud, H.M., 2013. Harboring oil-degrading bacteria: a potential mechanism of adaptation and survival in corals inhabiting oil-contaminated reefs. Mar. Pollut. Bull. 72 (2), 364–374.

Ali, N., Dashti, N., Salamah, S., Sorkhoh, N., Al-Awadhi, H., Radwan, S., 2016. Dynamics of bacterial populations during bench-scale bioremediation of oily seawater and desert soil bioaugmented with coastal microbial mats. Microb. Biotechnol. 9 (2), 157–171.

Allison, S.D., Wallenstein, M.D., Bradford, M.A., 2010. Soil-carbon response to warming dependent on microbial physiology. Nat. Geosci. 3 (5), 336–340. https://doi.org/10.1038/ngeo846.

Almeda, R., Wambaugh, Z., Wang, Z., Hyatt, C., Liu, Z., Buskey, E.J., 2013. Interactions between zooplankton and crude oil: toxic effects and bioaccumulation of polycyclic aromatic hydrocarbons. PLoS One 8 (6). https://doi.org/10.1371/journal.pone.0067212.

Antoniou, E., Fodelianakis, S., Korkakaki, E., Kalogerakis, N., 2015. Biosurfactant production from marine hydrocarbon-degrading consortia and pure bacterial strains using crude oil as carbon source. Front. Microbiol. 6, 274.

Ashraf, M., 2004. Some important physiological selection criteria for salt tolerance in plants. Flora 199 (5), 361–376. https://doi.org/10.1078/0367-2530-00165.

Bak, R.P.M., Elgershuizen, J.H.B.W., 1976. Patterns of oil-sediment rejection in corals. Mar. Biol. 37 (2), 105–113. https://doi.org/10.1007/BF00389121.

Bayat, Z., Hassanshahian, M., Hesni, M.A., 2016. Study the symbiotic crude oil-degrading bacteria in the mussel *Mactra stultorum* collected from the Persian Gulf. Mar. Pollut. Bull. 105 (1), 120–124.

Begum, N., Wang, L., Ahmad, H., Akhtar, K., Roy, R., Khan, M.I., Zhao, T., 2021. Co-inoculation of arbuscular mycorrhizal fungi and the plant growth-promoting rhizobacteria improve growth and photosynthesis in tobacco under drought stress by up-regulating antioxidant and mineral nutrition metabolism. Microb. Ecol., 1–18.

Beman, J.M., Chow, C.E., King, A.L., Feng, Y., Fuhrman, J.A., Andersson, A., Bates, N.R., Popp, B.N., Hutchins, D.A., 2011. Global declines in oceanic nitrification rates as a consequence of ocean acidification. Proc. Natl. Acad. Sci. U. S. A. 108 (1), 208–213. https://doi.org/10.1073/pnas.1011053108.

Biagi, E., Caroselli, E., Barone, M., Pezzimenti, M., Teixido, N., Soverini, M., Rampelli, S., Turroni, S., Gambi, M.C., Brigidi, P., Goffredo, S., 2020. Patterns in microbiome composition differ with ocean acidification in anatomic compartments of the Mediterranean coral Astroides calycularis living at CO_2 vents. Sci. Total Environ. 724, 138048.

Botté, E.S., Nielsen, S., Abdul Wahab, M.A., Webster, J., Robbins, S., Thomas, T., Webster, N.S., 2019. Changes in the metabolic potential of the sponge microbiome under ocean acidification. Nat. Commun. 10 (1). https://doi.org/10.1038/s41467-019-12156-y.

Buckley, L.B., Huey, R.B., 2016. How extreme temperatures impact organisms and the evolution of their thermal tolerance. Integr. Comp. Biol. 56 (1), 98–109.

Capone, D.G., Carpenter, E.J., 1982. Nitrogen fixation in the marine environment. Science 217 (4565), 1140–1142. https://doi.org/10.1126/science.217.4565.1140.

Cavalcanti, G.S., Shukla, P., Morris, M., Ribeiro, B., Foley, M., Doane, M.P., Thompson, C.C., Edwards, M.S., Dinsdale, E.A., Thompson, F.L., 2018. Rhodoliths holobionts in a changing ocean: host-microbes interactions mediate coralline algae resilience under ocean acidification. BMC Genomics 19 (1), 701. https://doi.org/10.1186/s12864-018-5064-4.

Cavicchioli, R., Ripple, W.J., Timmis, K.N., Azam, F., Bakken, L.R., Baylis, M., Behrenfeld, M.J., Boetius, A., Boyd, P.W., Classen, A.T., Crowther, T.W., Danovaro, R., Foreman, C.M., Huisman, J., Hutchins, D.A., Jansson, J.K., Karl, D.M., Koskella, B., Mark Welch, D.B., et al., 2019. Scientists' warning to humanity: microorganisms and climate change. Nat. Rev. Microbiol. 17 (9), 569–586. https://doi.org/10.1038/s41579-019-0222-5.

Celik, G.Y., Aslim, B., Beyatli, Y., 2008. Enhanced crude oil biodegradation and rhamnolipid production by *Pseudomonas stutzeri* strain G11 in the presence of Tween-80 and Triton X-100. J. Environ. Biol. 29, 867–870.

Chen, J.E., Barbrook, A.C., Cui, G., Howe, C.J., Aranda, M., 2017. The genetic intractability of *Symbiodinium microadriaticum* to standard algal transformation methods. BioRxiv. https://doi.org/10.1101/140616.

Chen, W., Li, J., Sun, X., Min, J., Hu, X., 2017. High efficiency degradation of alkanes and crude oil by a salt-tolerant bacterium *Dietzia* species CN-3. Int. Biodeter. Biodegr. 118, 110–118.

Cianelli, D., Manfra, L., Zambianchi, E., Maggi, C., Cicero, A., 2011. Modelling and observations of produced formation waters (PFW) at sea. In: Fluid Waste Disposal, pp. 113–135.

Cleves, P.A., Strader, M.E., Bay, L.K., Pringle, J.R., Matz, M.V., 2018. CRISPR/Cas9-mediated genome editing in a reef-building coral. Proc. Natl. Acad. Sci. U. S. A. 115 (20), 5235–5240. https://doi.org/10.1073/pnas.1722151115.

Comeau, S., Cornwall, C.E., Pupier, C.A., DeCarlo, T.M., Alessi, C., Trehern, R., McCulloch, M.T., 2019. Flow-driven micro-scale pH variability affects the physiology of corals and coralline algae under ocean acidification. Sci. Rep. 9 (1), 12829. https://doi.org/10.1038/s41598-019-49044-w.

Da Cunha, C.D., Rosado, A.S., Sebastián, G.V., Seldin, L., Von Der Weid, I., 2006. Oil biodegradation by *Bacillus* strains isolated from the rock of an oil reservoir located in a deep-water production basin in Brazil. Appl. Microbiol. Biotechnol. 73 (4), 949–959.

Damjanovic, K., Blackall, L.L., Webster, N.S., van Oppen, M.J.H., 2017. The contribution of microbial biotechnology to mitigating coral reef degradation. Microb. Biotechnol. 10 (5), 1236–1243. https://doi.org/10.1111/1751-7915.12769.

Das, S., Mangwani, N., 2015. Ocean acidification and marine microorganisms: responses and consequences. Oceanologia 57 (4), 349–361.

Dash, H.R., Mangwani, N., Das, S., 2014. Characterization and potential application in mercury bioremediation of highly mercury-resistant marine bacterium *Bacillus thuringiensis* PW-05. Environ. Sci. Pollut. Res. 21 (4), 2642–2653.

De Goeij, J.M., Van Oevelen, D., Vermeij, M.J., Osinga, R., Middelburg, J.J., De Goeij, A.F.P.M., Admiraal, W., 2013. Surviving in a marine desert: the sponge loop retains resources within coral reefs. Science 342 (6154), 108–110.

Denner, E.B., Vybiral, D., Koblízek, M., Kämpfer, P., Busse, H.J., Velimirov, B., 2002. *Erythrobacter citreus* sp. nov., a yellow-pigmented bacterium that lacks bacteriochlorophyll a, isolated from the western Mediterranean Sea. Int. J. Syst. Evol. Microbiol. 52 (5), 1655–1661.

Doney, S.C., Fabry, V.J., Feely, R.A., Kleypas, J.A., 2009. Ocean acidification: the other CO_2 problem. Annu. Rev. Mar. Sci. 1, 169–192. https://doi.org/10.1146/annurev.marine.010908.163834.

Doo, S.S., Edmunds, P.J., Carpenter, R.C., 2019. Ocean acidification effects on in situ coral reef metabolism. Sci. Rep. 9 (1), 12067. https://doi.org/10.1038/s41598-019-48407-7.

Dubinsky, Z.V.Y., Stambler, N., 1996. Marine pollution and coral reefs. Glob. Change Biol. 2 (6), 511–526.

El Mahdi, A.M., Aziz, H.A., Amr, S.S.A., El-Gendy, N.S., Nassar, H.N., 2016. Isolation and characterization of *Pseudomonas* sp. NAF1 and its application in biodegradation of crude oil. Environ. Earth Sci. 75 (5), 380.

Fabry, V.J., Seibel, B.A., Feely, R.A., Orr, J.C., 2008. Impacts of ocean acidification on marine fauna and ecosystem processes. ICES J. Mar. Sci. 65 (3), 414–432. https://doi.org/10.1093/icesjms/fsn048.

Fiore, C.L., Jarett, J.K., Olson, N.D., Lesser, M.P., 2010. Nitrogen fixation and nitrogen transformations in marine symbioses. Trends Microbiol. 18 (10), 455–463. https://doi.org/10.1016/j.tim.2010.07.001.

Fishelson, L., 1977. Stability and instability of marine ecosystems, illustrated by examples from the Red Sea. Helgoländer Meeresun. 30 (1–4), 18–29. https://doi.org/10.1007/BF02207822.

Flemming, H.-C., Wuertz, S., 2019. Bacteria and archaea on earth and their abundance in biofilms. Nat. Rev. Microbiol. 17 (4), 247–260. https://doi.org/10.1038/s41579-019-0158-9.

Fragoso Ados Santos, H., Duarte, G.A.S., Rachid, C.T.D.C., Chaloub, R.M., Calderon, E.N., Marangoni, L.F.D.B., Bianchini, A., Nudi, A.H., Do Carmo, F.L., Van Elsas, J.D., Rosado, A.S., Castro, C.B.E., Peixoto, R.S., 2015. Impact of oil spills on coral reefs can be reduced by bioremediation using probiotic microbiota. Sci. Rep. 5, 18268. https://doi.org/10.1038/srep18268.

Freitas, A.C., Rodrigues, D., Rocha-Santos, T.A., Gomes, A.M., Duarte, A.C., 2012. Marine biotechnology advances towards applications in new functional foods. Biotechnol. Adv. 30 (6), 1506–1515.

Gibson, J., Harwood, C.S., 2002. Metabolic diversity in aromatic compound utilization by anaerobic microbes. Annu. Rev. Microbiol. 56 (1), 345–369. https://doi.org/10.1146/annurev.micro.56.012302.160749.

Goomer, R., Kunkel, G., 1992. The transcriptional start site for a human U6 small nuclear RNA gene is dictated by a compound promoter element consisting of the PSE and the TATA box. Nucleic Acids Res. 20 (18), 4903–4912. https://doi.org/10.1093/nar/20.18.4903.

Gruber, N., 2011. Warming up, turning sour, losing breath: ocean biogeochemistry under global change. Philos. Trans. R. Soc. A: Math. Phys. Eng. Sci. 369 (1943), 1980–1996.

Gupta, V.V., 2012. Beneficial microorganisms for sustainable agriculture. Microbiol. Aust. 33 (3), 113–115.

Gutner-Hoch, E., Ben-Asher, H.W., Yam, R., Shemesh, A., Levy, O., 2017. Identifying genes and regulatory pathways associated with the scleractinian coral calcification process. PeerJ 5, e3590.

Hamdan, L.J., Fulmer, P.A., 2011. Effects of COREXIT® EC9500A on bacteria from a beach oiled by the Deepwater Horizon spill. Aquat. Microb. Ecol. 63 (2), 101–109.

Jackson, A.J.B.C., 2016. Articles Ecological Effects of a Major Oil Spill on Panamanian Coastal Marine Communties. Vol. 243, pp. 37–44.

James, S.R., Dobson, S.J., Franzmann, P.D., McMeekin, T.A., 1990. *Halomonas meridiana*, a new species of extremely halotolerant bacteria isolated from Antarctic saline lakes. Syst. Appl. Microbiol. 13 (3), 270–278.

Jin, H.M., Kim, J.M., Lee, H.J., Madsen, E.L., Jeon, C.O., 2012. Alteromonas as a key agent of polycyclic aromatic hydrocarbon biodegradation in crude oil-contaminated coastal sediment. Environ. Sci. Technol. 46 (14), 7731–7740.

Jokiel, P.L., Coles, S.L., 1977. Effects of temperature on the mortality and growth of Hawaiian reef corals. Mar. Biol. 43 (3), 201–208.

Jokiel, P.L., Rodgers, K.S., Kuffner, I.B., Andersson, A.J., Cox, E.F., Mackenzie, F.T., 2008. Ocean acidification and calcifying reef organisms: a mesocosm investigation. Coral Reefs 27 (3), 473–483. https://doi.org/10.1007/s00338-008-0380-9.

Jones, R.J., Heyward, A.J., 2003. The effects of produced formation water (PFW) on coral and isolated symbiotic dinoflagellates of coral. Mar. Freshw. Res. 54 (2), 153–162. https://doi.org/10.1071/MF02108.

Kaniewska, P., Chan, C.K.K., Kline, D., Ling, E.Y.S., Rosic, N., Edwards, D., Hoegh-Guldberg, O., Dove, S., 2015. Transcriptomic changes in coral holobionts provide insights into physiological challenges of future climate and ocean change. PLoS One 10 (10), e0139223. https://doi.org/10.1371/journal.pone.0139223.

Kim, B.Y., Weon, H.Y., Yoo, S.H., Kwon, S.W., Cho, Y.H., Stackebrandt, E., Go, S.J., 2006. *Paracoccus homiensis* sp. nov., isolated from a sea-sand sample. Int. J. Syst. Evol. Microbiol. 56 (10), 2387–2390.

Lal, B., Khanna, S., 1996. Degradation of crude oil by *Acinetobacter calcoaceticus* and *Alcaligenes odorans*. J. Appl. Bacteriol. 81 (4), 355–362.

Lau, J.A., Lennon, J.T., 2012. Rapid responses of soil microorganisms improve plant fitness in novel environments. Proc. Natl. Acad. Sci. U. S. A. 109 (35), 14058–14062. https://doi.org/10.1073/pnas.1202319109.

Lee, Y.K., Kwon, K.K., Cho, K.H., Kim, H.W., Park, J.H., Lee, H.K., 2003. Culture and identification of bacteria from marine biofilms. J. Microbiol. 41 (3), 183–188.

Lei, Y., Yang, K., Wang, B., Sheng, Y., Bird, B.W., Zhang, G., Tian, L., 2014. Response of inland lake dynamics over the Tibetan Plateau to climate change. Clim. Change 125 (2), 281–290.

Levin, R.A., Voolstra, C.R., Agrawal, S., Steinberg, P.D., Suggett, D.J., van Oppen, M.J.H., 2017. Engineering strategies to decode and enhance the genomes of coral symbionts. Front. Microbiol. 8, 1220. https://doi.org/10.3389/fmicb.2017.01220.

Li, S.W., Liu, M.Y., Yang, R.Q., 2019. Comparative genome characterization of a petroleum-degrading *Bacillus subtilis* strain DM2. Int. J. Genomics 2019.

Li, X., Zheng, R., Zhang, X., Liu, Z., Zhu, R., Zhang, X., Gao, D., 2019. A novel exoelectrogen from microbial fuel cell: bioremediation of marine petroleum hydrocarbon pollutants. J. Environ. Manage. 235, 70–76.

Littman, R., Willis, B.L., Bourne, D.G., 2011. Metagenomic analysis of the coral holobiont during a natural bleaching event on the great barrier reef. Environ. Microbiol. Rep. 3 (6), 651–660. https://doi.org/10.1111/j.1758-2229.2010.00234.x.

Liu, Z., Sun, Y., Zhang, Y., Qin, S., Sun, Y., Mao, H., Miao, L., 2020. Desert soil sequesters atmospheric CO_2 by microbial mineral formation. Geoderma 361, 114104.

Liu, Z.P., Wang, B.J., Liu, X.Y., Dai, X., Liu, Y.H., Liu, S.J., 2008. *Paracoccus halophilus* sp. nov., isolated from marine sediment of the South China Sea, China, and emended description of genus *Paracoccus* Davis 1969. Int. J. Syst. Evol. Microbiol. 58 (1), 257–261.

Longeon, A., Peduzzi, J., Barthelemy, M., Corre, S., Nicolas, J.L., Guyot, M., 2004. Purification and partial identification of novel antimicrobial protein from marine bacterium *Pseudoalteromonas* species strain X153. Mar. Biotechnol. 6 (6), 633–641.

López-Cortés, A., Rodríguez-Fernández, O., Latisnere-Barragán, H., Mejía-Ruíz, H.C., González-Gutiérrez, G., Lomelí-Ortega, C., 2010. Characterization of polyhydroxyalkanoate and the phaC gene of *Paracoccus seriniphilus* E71 strain isolated from a polluted marine microbial mat. World J. Microbiol. Biotechnol. 26 (1), 109–118.

Loya, Y., Rinkevich, B., 1980. Effects of oil pollution on coral reef communities. Mar. Ecol. Prog. Ser. 3 (16), 180.

Luter, H.M., Andersen, M., Versteegen, E., Laffy, P., Uthicke, S., Bell, J.J., Webster, N.S., 2020. Cross-generational effects of climate change on the microbiome of a photosynthetic sponge. Environ. Microbiol. 22 (11), 4732–4744. https://doi.org/10.1111/1462-2920.15222.

Mata, J.A., Martínez-Cánovas, J., Quesada, E., Béjar, V., 2002. A detailed phenotypic characterisation of the type strains of *Halomonas* species. Syst. Appl. Microbiol. 25 (3), 360–375.

May, L.A., Burnett, A.R., Miller, C.V., Pisarski, E., Webster, L.F., Moffitt, Z.J., Pennington, P., Wirth, E., Baker, G., Ricker, R., Woodley, C.M., 2020. Effect of Louisiana sweet crude oil on a Pacific coral, *Pocillopora damicornis*. Aquat. Toxicol. 222. https://doi.org/10.1016/j.aquatox.2020.105454, 105454.

Mayak, S., Tirosh, T., Glick, B.R., 2004. Plant growth-promoting bacteria confer resistance in tomato plants to salt stress. Plant Physiol. Biochem. 42 (6), 565–572. https://doi.org/10.1016/j.plaphy.2004.05.009.

McDevitt-Irwin, J.M., Baum, J.K., Garren, M., Vega Thurber, R.L., 2017. Responses of coral-associated bacterial communities to local and global stressors. Front. Mar. Sci. 4, 262.

Meron, D., Atias, E., Iasur Kruh, L., Elifantz, H., Minz, D., Fine, M., Banin, E., 2011. The impact of reduced pH on the microbial community of the coral *Acropora eurystoma*. ISME J. 5 (1), 51–60. https://doi.org/10.1038/ismej.2010.102.

Meron, D., Rodolfo-Metalpa, R., Cunning, R., Baker, A.C., Fine, M., Banin, E., 2012. Changes in coral microbial communities in response to a natural pH gradient. ISME J. 6 (9), 1775–1785. https://doi.org/10.1038/ismej.2012.19.

Michel, W.C., Fitt, W.K., 1984. Effects of a water-soluble fraction of a crude oil on the coral reef hydroid *Myrionema hargitti*: feeding, growth and algal symbionts. Mar. Biol. 84 (2), 143–154. https://doi.org/10.1007/BF00392999.

Moghadam, M.S., Albersmeier, A., Winkler, A., Cimmino, L., Rise, K., Hohmann-Marriott, M.F., Kalinowski, J., Rückert, C., Wentzel, A., Lale, R., 2016. Isolation and genome sequencing of four Arctic marine *Psychrobacter* strains exhibiting multicopper oxidase activity. BMC Genomics 17 (1), 1–14.

Mohapatra, B.R., Bapuji, M., 1998. Characterization of acetylcholinesterase from *Arthrobacter ilicis* associated with the marine sponge (*Spirastrella* sp.). J. Appl. Microbiol. 84 (3), 393–398.

Mollica, N.R., Guo, W., Cohen, A.L., Huang, K.F., Foster, G.L., Donald, H.K., Solow, A.R., 2018. Ocean acidification affects coral growth by reducing skeletal density. Proc. Natl. Acad. Sci. 115 (8), 1754–1759.

Nakamura, T., Nakamori, T., 2007. A geochemical model for coral reef formation. Coral Reefs 26 (4), 741–755. https://doi.org/10.1007/s00338-007-0262-6.

Nguyen, M.L., 2018. Biostimulant Effects of Rhizobacteria on Wheat Growth and Nutrient Uptake Under Contrasted N Supplies (Doctoral dissertation). Université de Liège, Liège, Belgique.

O'Brien, P.A., Morrow, K.M., Willis, B.L., Bourne, D.G., 2016. Implications of ocean acidification for marine microorganisms from the free-living to the host-associated. Front. Mar. Sci. 3. https://doi.org/10.3389/fmars.2016.00047.

Odum, E.P., 1984. The mesocosm. Bioscience 34, 558–562. https://doi.org/10.2307/1309598.

Ondrasek, G., Kranjčec, F., Filipović, L., Filipović, V., Bubalo Kovačić, M., Badovinac, I.J., Peter, R., Petravić, M., Macan, J., Rengel, Z., 2021. Biomass bottom ash & dolomite similarly ameliorate an acidic low-nutrient soil, improve phytonutrition and growth, but increase Cd accumulation in radish. Sci. Total Environ. 753. https://doi.org/10.1016/j.scitotenv.2020.141902, 141902.

Ozhan, K., Parsons, M.L., Bargu, S., 2014. How were phytoplankton affected by the deepwater horizon oil spill? Bioscience 64 (9), 829–836. https://doi.org/10.1093/biosci/biu117.

Paniagua-Michel, J., Fathepure, B.Z., 2019. Microbial consortia and biodegradation of petroleum hydrocarbons in marine environments. In: Microbial Action on Hydrocarbons. Springer, Singapore, pp. 1–20, https://doi.org/10.1007/978-981-13-1840-5_1.

Paul, D., Nair, S., 2008. Stress adaptations in a plant growth promoting rhizobacterium (PGPR) with increasing salinity in the coastal agricultural soils. J. Basic Microbiol. 48 (5), 378–384. https://doi.org/10.1002/jobm.200700365.

Pineda, A., Dicke, M., Pieterse, C.M., Pozo, M.J., 2013. Beneficial microbes in a changing environment: are they always helping plants to deal with insects? Funct. Ecol. 27 (3), 574–586.

Prada, F., Caroselli, E., Mengoli, S., Brizi, L., Fantazzini, P., Capaccioni, B., Pasquini, L., Fabricius, K.E., Dubinsky, Z., Falini, G., Goffredo, S., 2017. Ocean warming and acidification synergistically increase coral mortality. Sci. Rep. 7, 40842. https://doi.org/10.1038/srep40842.

Rädecker, N., Pogoreutz, C., Voolstra, C.R., Wiedenmann, J., Wild, C., 2015. Nitrogen cycling in corals: the key to understanding holobiont functioning? Trends Microbiol. 23 (8), 490–497. https://doi.org/10.1016/j.tim.2015.03.008.

Renegar, D.A., Turner, N.R., Riegl, B.M., Dodge, R.E., Knap, A.H., Schuler, P.A., 2017. Acute and subacute toxicity of the polycyclic aromatic hydrocarbon 1-methylnaphthalene to the shallow-water coral *Porites divaricata*: application of a novel exposure protocol. Environ. Toxicol. Chem. 36 (1), 212–219. https://doi.org/10.1002/etc.3530.

Reshef, L., Koren, O., Loya, Y., Zilber-Rosenberg, I., Rosenberg, E., 2006. The coral probiotic hypothesis. Environ. Microbiol. 8 (12), 2068–2073. https://doi.org/10.1111/j.1462-2920.2006.01148.x.

Ribes, M., Calvo, E., Movilla, J., Logares, R., Coma, R., Pelejero, C., 2016. Restructuring of the sponge microbiome favors tolerance to ocean acidification. Environ. Microbiol. Rep. 8 (4), 536–544. https://doi.org/10.1111/1758-2229.12430.

Rossi, F., Olguín, E.J., Diels, L., De Philippis, R., 2015. Microbial fixation of CO_2 in water bodies and in drylands to combat climate change, soil loss and desertification. New Biotechnol. 32 (1), 109–120. https://doi.org/10.1016/j.nbt.2013.12.002.

Rypien, K.L., Ward, J.R., Azam, F., 2010. Antagonistic interactions among coral-associated bacteria. Environ. Microbiol. 12 (1), 28–39.

Saint-martin, A., 2003. Interacting effects of CO_2 partial pressure and temperature on photosynthesis and calcification in a scleractinian coral. Glob. Change Biol. 9, 1660–1668. https://doi.org/10.1046/j.1529-8817.2003.00678.x.

Schlüter, L., Lohbeck, K.T., Gutowska, M.A., Gröger, J.P., Riebesell, U., Reusch, T.B., 2014. Adaptation of a globally important coccolithophore to ocean warming and acidification. Nat. Clim. Change 4 (11), 1024–1030.

Shafir, S., Van Rijn, Rinkevich, B., 2007. Short and long term toxicity of crude oil and oil dispersants to two representative coral species. Environ. Sci. Technol. 41 (15), 5571–5574.

Shiu, J.H., Keshavmurthy, S., Chiang, P.W., Chen, H.J., Lou, S.P., Tseng, C.H., Hsieh, H.J., Chen, C.A., Tang, S.L., 2017. Dynamics of coral-associated bacterial communities acclimated to temperature stress based on recent thermal history. Sci. Rep. 7 (1), 1–13.

Siddikee, M.A., Glick, B.R., Chauhan, P.S., jong Yim, Sa, T., 2011. Enhancement of growth and salt tolerance of red pepper seedlings (*Capsicum annuum* L.) by regulating stress ethylene synthesis with halotolerant bacteria containing 1-aminocyclopropane-1-carboxylic acid deaminase activity. Plant Physiol. Biochem. 49 (4), 427–434.

Silva, D.P., Duarte, G., Villela, H.D.M., Santos, H.F., Rosado, P.M., Rosado, J.G., Rosado, A.S., Ferreira, E.M., Soriano, A.U., Peixoto, R.S., 2019. Adaptable mesocosm facility to study oil spill impacts on corals. Ecol. Evol. 9 (9), 5172–5185. https://doi.org/10.1002/ece3.5095.

Simister, R.L., Antzis, E.W., White, H.K., 2016. Examining the diversity of microbes in a deep-sea coral community impacted by the deepwater horizon oil spill. Deep Sea Res. Part II: Top. Stud. Oceanogr. 129, 157–166. https://doi.org/10.1016/j.dsr2.2015.01.010.

Stewart, R.I.A., Dossena, M., Bohan, D.A., Jeppesen, E., Kordas, R.L., Ledger, M.E., Meerhoff, M., Moss, B., Mulder, C., Shurin, J.B., Suttle, B., Thompson, R., Trimmer, M., Woodward, G., 2013. Mesocosm experiments as a tool for ecological climate-change research. In: Advances in Ecological Research. Vol. 48. Academic Press Inc, pp. 71–181, https://doi.org/10.1016/B978-0-12-417199-2.00002-1.

Suni, S., Koskinen, K., Kauppi, S., Hannula, E., Ryynänen, T., Aalto, A., Jäänheimo, J., Ikävalko, J., Romantschuk, M., 2007. Removal by sorption and in situ biodegradation of oil spills limits damage to marine biota: a laboratory simulation. Ambio 36 (2–3), 173–179. https://doi.org/10.1579/0044-7447(2007)36[173:RBSAIS]2.0.CO;2.

Sweet, M., Bythell, J., 2017. The role of viruses in coral health and disease. J. invertebr. Pathol. 147, 136–144.

Teira, E., Lekunberri, I., Gasol, J.M., Nieto-Cid, M., Álvarez-Salgado, X.A., Figueiras, F.G., 2007. Dynamics of the hydrocarbon-degrading Cycloclasticus bacteria during mesocosm-simulated oil spills. Environ. Microbiol. 9 (10), 2551–2562. https://doi.org/10.1111/j.1462-2920.2007.01373.x.

Thirugnanasambandam, R., Inbakandan, D., Kumar, C., Subashni, B., Vasantharaja, R., Abraham, L.S., Ayyadurai, N., Murthy, P.S., Kirubagaran, R., Khan, S.A., Balasubramanian, T., 2019. Genomic insights of *Vibrio harveyi* RT-6 strain, from infected "Whiteleg shrimp" (*Litopenaeus vannamei*) using Illumina platform. Mol. Phylogenet. Evol. 130, 35–44.

Thurber, R.V., Willner-Hall, D., Rodriguez-Mueller, B., Desnues, C., Edwards, R.A., Angly, F., Dinsdale, E., Kelly, L., Rohwer, F., 2009. Metagenomic analysis of stressed coral holobionts. Environ. Microbiol. 11 (8), 2148–2163. https://doi.org/10.1111/j.1462-2920.2009.01935.x.

Turley, C., Findlay, H.S., 2016. Ocean acidification. In: Climate Change: Observed Impacts on Planet Earth, second ed. Elsevier Inc, pp. 271–293, https://doi.org/10.1016/B978-0-444-63524-2.00018-X.

van Oppen, M.J.H., Blackall, L.L., 2019. Coral microbiome dynamics, functions and design in a changing world. Nat. Rev. Microbiol. 17 (9), 557–567. https://doi.org/10.1038/s41579-019-0223-4.

Varjani, S.J., 2017. Microbial degradation of petroleum hydrocarbons. Bioresour. Technol. 223, 277–286.

Venn, A., Tambutté, E., Holcomb, M., Allemand, D., Tambutté, S., 2011. Live tissue imaging shows reef corals elevate pH under their calcifying tissue relative to seawater. PLoS One 6 (5), e20013. https://doi.org/10.1371/journal.pone.0020013.

Wackett, L.P., 2000. Environmental biotechnology. Trends Biotechnol. 18 (1), 19–21. https://doi.org/10.1016/S0167-7799(99)01399-2.

Wang, S., Ouyang, L., Ju, X., Zhang, L., Zhang, Q., Li, Y., 2014. Survey of plant drought-resistance promoting bacteria from *Populus euphratica* tree living in arid area. Indian J. Microbiol. 54 (4), 419–426.

Webster, N.S., Reusch, T.B.H., 2017. Microbial contributions to the persistence of coral reefs. ISME J. 11 (10), 2167–2174. https://doi.org/10.1038/ismej.2017.66.

White, H.K., Hsing, P.Y., Cho, W., Shank, T.M., Cordes, E.E., Quattrini, A.M., Nelson, R.K., Camilli, R., Demopoulos, A.W.J., German, C.R., Brooks, J.M., Roberts, H.H., Shedd, W., Reddy, C.-M., Fisher, C.R., 2012. Impact of the deepwater horizon oil spill on a deep-water coral community in the Gulf of Mexico. Proc. Natl. Acad. Sci. U. S. A. 109 (50), 20303–20308. https://doi.org/10.1073/pnas.1118029109.

Wolfe, D.W., Ziska, L., Petzoldt, C., Seaman, A., Chase, L., Hayhoe, K., 2008. Projected change in climate thresholds in the Northeastern US: implications for crops, pests, livestock, and farmers. Mitig. Adapt. Strateg. Glob. Change 13 (5), 555–575.

Wong, S.K., Park, S., Lee, J.S., Lee, K.C., Ogura, Y., Hayashi, T., Chiura, H.X., Yoshizawa, S., Hamasaki, K., 2017. *Algibacter aquaticus* sp. nov., a slightly alkaliphilic marine Flavobacterium isolated from coastal surface water. Int. J. Syst. Evol. Microbiol. 67 (7), 2199–2204.

Yang, T., Cheng, H., Wang, H., Drews, M., Li, S., Huang, W., Zhou, H., Chen, C.M., Diao, X., 2019. Comparative study of polycyclic aromatic hydrocarbons (PAHs) and heavy metals (HMs) in corals, surrounding sediments and surface water at the Dazhou Island, China. Chemosphere 218, 157–168. https://doi.org/10.1016/j.chemosphere.2018.11.063.

Yang, T., Diao, X., Cheng, H., Wang, H., Zhou, H., Zhao, H., Chen, C.M., 2020. Comparative study of polycyclic aromatic hydrocarbons (PAHs) and heavy metals (HMs) in corals, sediments and seawater from coral reefs of Hainan, China. Environ. Pollut. 264, 114719. https://doi.org/10.1016/j.envpol.2020.114719.

Zhou, C., Ma, Z., Zhu, L., Xiao, X., Xie, Y., Zhu, J., Wang, J., 2016. Rhizobacterial strain *Bacillus megaterium* BOFC15 induces cellular polyamine changes that improve plant growth and drought resistance. Int. J. Mol. Sci. 17 (6). https://doi.org/10.3390/ijms17060976.

Zhuang, L., Liu, Y., Wang, L., Wang, W., Shao, Z., 2015. *Erythrobacter atlanticus* sp. nov., a bacterium from ocean sediment able to degrade polycyclic aromatic hydrocarbons. Int. J. Syst. Evol. Microbiol. 65 (Pt_10), 3714–3719.

Ziegler, M., Roik, A., Porter, A., Zubier, K., Mudarris, M.S., Ormond, R., Voolstra, C.R., 2016. Coral microbial community dynamics in response to anthropogenic impacts near a major city in the central Red Sea. Mar. Pollut. Bull. 105 (2), 629–640. https://doi.org/10.1016/j.marpolbul.2015.12.045.

Zoppellari, F., Malusà, E., Chitarra, W., Lovisolo, C., Spanna, F., Bardi, L., 2014. Improvement of drought tolerance in maize (*Zea mays* L.) by selected rhizospheric microorganisms. Italian J. Agrometeorol. 1, 5–18. http://www.patroneditore.com/includes/ddownloads_fascicoli.php?fascicolo=2019.

Further reading

Malvezzi, A.J., Murray, C.S., Feldheim, K.A., DiBattista, J.D., Garant, D., Gobler, C.J., Chapman, D.D., Baumann, H., 2015. A quantitative genetic approach to assess the evolutionary potential of a coastal marine fish to ocean acidification. Evol. Appl. 8 (4), 352–362.

CHAPTER 6

Earthworm-microorganisms interactions for sustainable soil ecosystem and crop productivity

Sudipti Arora[a],*, Sakshi Saraswat[b], Anamika Verma[c], and Devanshi Sutaria[a]
[a]Dr. B. Lal Institute of Biotechnology, Jaipur, India
[b]Institute of Environmental and Occupational Health Sciences, School of Medicine, National Yang Ming University, Taipei, Taiwan
[c]Department of Biotechnology and Bioinformatics, Jaypee University of Information Technology, Waknaghat, Himachal Pradesh, India
*Corresponding author: e-mail address: sudiptiarora@gmail.com

1. Introduction

India holds the second-largest agricultural land in the world, with 20 agro-climatic regions and 157.35 million hectares of land under cultivation (Eliazer Nelson et al., 2019). Thus, agriculture plays a vital role with 58% of rural households depending on it. Although India is self-sufficient in food production, it has significantly decreased from the 1940–1960s, closely associated with the occurrence of famine-like the Bengal famine of 1943. Therefore, the Green Revolution was initiated in the 1960s to increase food production, alleviate extreme poverty and malnourishment in the country, and to feed millions. The major crops cultivated in the era preceding the Green Revolution were rice, millets, sorghum, wheat, maize, and barley. In addition to this, the area under irrigation in India increased substantially between 1960 and 1990 because of the availability of mechanized pumps. By 1990, almost 50% of the irrigated area in the country was served with groundwater. The productivity of the crops was increased by the use of fertilisers, pesticides, and groundwater resources. The amount of chemical fertilisers used post-advent of the Green Revolution was quite high, and a steep increase was observed in the consumption of chemical fertilisers for the cultivation of the crop. This adopted intensive agriculture practice might have nourished most of the country's inhabitants, but the opposite happened to the agricultural land itself. It rendered the land infertile, led to extensive water consumption and aggravated groundwater loss. Its dependence on singular crops, heavy ploughing machinery, fossil-fuel-based fertilisers, and pesticides is degrading our soil's wildlife and nutrient cycles and contributing to an increase in the global carbon footprint.

In agricultural areas, mismanagement and overuse of chemical fertilisers, pesticides, and lack of crop rotation have caused the soil to become infertile over the years, and groundwater depletion has become a frequent occurrence. By changing the natural microflora and

Relationship Between Microbes and the Environment for Sustainable Ecosystem Services, Volume 1
https://doi.org/10.1016/B978-0-323-89938-3.00006-2

increasing the alkalinity and salinity of the soil, the gorging of chemical fertilisers to achieve high yield caused physical and chemical soil degradation. The widespread use of pumps and the extraction of groundwater led to water depletion in hard rock and arid regions of India and further contributed to land degradation. In addition, climate change and global warming have changed the consistency of the soil and raised the risk of erosion. The temperature rise is expected to increase the rate of mineral production and decrease the content of soil organic matter. The rate of microbial decomposition of organic matter could be increased by high temperatures, thereby adversely affecting soil fertility in the long run. The increase in global temperature results in an increase in the volume of seawater. As sea levels rise, flooding, shoreline submergence and water table salinity could impact agriculture by inundating low-lying areas. Future climate shifts will impact the supply of water for agriculture. India relies on rivers that emanate from the Himalayas for water-resource production, apart from monsoon rains. The rise in temperature will increase snowmelt as a result of global warming, and thus, snow cover will decrease. In the short term, snowmelt in many rivers may increase the flow of water, which in turn may increase the frequency of floods. However, the receding snow line could decrease water flow in these rivers in the long run. The arrival of the summer monsoon across India will become more unpredictable in climate-change scenarios and may be delayed. This will not only affect the rain-fed crops but also the storage of water in irrigated areas. Uncertain rainfall patterns are followed by these thermal shifts. Increased rainfall leads to increased leaching of minerals, especially nitrates, in regions that are already moist. This contributes to the nutrient levels of the cultivable soil being reduced. Again, the use of chemical fertilisers, pesticides and, in recent years, even antibiotics has met these needs. Pests and weeds account for about 40% of crop losses worldwide. Where pests and diseases flourish, climate change is changing, making it harder for farmers to remain resilient. Owing to their harmful impact on the wellbeing of humans and animals, many widely used herbicides, pesticides, and fungicides are now also fighting to be prohibited. Increasing resistance to their behavior, even though they are not, makes controlling weeds, pests, and diseases an increasing challenge for farmers and researchers (MacLaren et al., 2020). The spread of antimicrobial resistance (AMR) through agricultural lands is another emerging concern. On the other side, where pesticides are agrochemicals used to protect plants from insect attacks, antibiotics are medicines used to treat human bacterial diseases. They are not prescribed to be used as pesticides and only after a bacterial disease has been diagnosed in a crop can they be used under professional supervision. In reality, the indiscriminate use of essential antibiotics such as streptomycin and tetracycline is troubling. Unspent antibiotics make their way into the surrounding ecosystem if used in crops. Microorganisms in soil and water exposed to this growing load of antibiotics may develop resistance to it, which then spreads through the transfer of genes to other bacteria. When such resistant microorganisms infect humans or livestock, their diagnosis becomes difficult as well as costly. There is also a possibility that traces of antibiotics remain in edible parts of the plant long after they are sprayed, affecting human health or making them resistant to the antibiotics (Kumar et al., 2005).

Thus, it is the need of the hour to strictly regulate sustainable agricultural practices and food systems which provide sufficient and nutritious food for all, while minimizing environmental impact. Throughout the world, the search is on for sustainable approaches for the management of irrigation water and for policies to improve the performance of irrigated agriculture. There is a global demand to shift to sustainable farming systems, such as zero-budget natural farming (ZBNF). India also introduced ZBNF in its Union Budget 2019–20 (Gupta et al., 2020). As the name suggests, it is the adaptation of an ancient practice that reduces farmers' direct cost and encourages them to use natural inputs, such as cow dung and cow urine. The inputs help manage soil nutrition, fertility, pests, and seeds. But the ways to implement and achieve such sustainable solutions is intensely debated by intellectuals via two narratives: incremental steps to improve efficiency in conventional agriculture while reducing negative externalities, versus transformative redesign of farming systems based on agroecological principles (Eyhorn et al., 2019). Transformative systems such as organic farming have proven sustainability benefits, including improved soil quality, enhanced biodiversity, reduced pollution, and increased farm incomes (Seufert and Ramankutty, 2017), but in many contexts result in lower yields so that their sustainability per unit product is sometimes questioned (Searchinger et al., 2018). On the other hand, complex traditional systems such as precision farming and reduced labor may be productive but have major adverse externalities, including loss of biodiversity, soil degradation, emissions, decreased human health, and low farm incomes. Therefore, we need to concentrate on certain mitigation methods that are both highly productive and beneficial to farmers and do not have any detrimental effect on the soil's biodiversity and its environment.

2. Earthworms—Nature's ploughman

The large secondary decomposers in the soil are earthworms. They make a very significant contribution to the composition of the soil, deeply altering the physical, chemical, and biological properties of the soil through their feeding, burrowing, and casting. These ecosystem engineers' burrowing, in particular the underground network of tunnels and galleries containing termites and ant nests, enhances soil porosity by providing sufficient aeration and water holding capacity below ground, promotes root penetration, and avoids surface crusting and topsoil erosion. One of the most significant contributions is its role in the formation of soil aggregates. Hong et al. (2011) have shown that aggregates formed by earthworms have higher water stability. The activities of earthworms also result in the formation of many organic mineral complexes, which increase the content of water-stable aggregates with $>1000\,\mu m$ diameter, carbon is combined into stable soil aggregates with activities of earthworm, thereby increasing the stability of aggregates and making soil organic matter easier to decompose. The activities of earthworm have also been reported to significantly change the physical and chemical properties of soil which further enhanced the mineralization of organic matter and stabilize soil nutrient cycles (Bertrand et al., 2015). Studies have also demonstrated that the soils refined by earthworms often have higher values of organic matter, total nitrogen, Calcium, Magnesium, and Potassium, available Nitrogen and Phosphorous, etc. (Lemtiri et al., 2014). For example, the contents of mineral nitrogen (NO_3-N^+,NH_{4+}-N), total carbon, nitrogen, and microbial biomass in stalk-returning soils with rotation of paddy and wheat are significantly enhanced after the soils were inoculated with earthworms, indicating that earthworms have dual functions of increasing soil microbial biomass and promoting the mineralization of organic nitrogen (Domínguez et al., 2004).

Earthworms as soil engineers play a key role in soil organic matter turnover, soil structure, mineralization of nutrients, heavy metals bioavailability, and land remediation. Recently the vermicomposting and vermifiltration technology has proven an environmentally sustainable, economically viable, and socially acceptable method for wastewater management and land remediation. The use of vermicompost has also been shown to significantly improve plants growths and yields (Blouin et al., 2019).

2.1 Role of earthworms and microflora in sustainable soil ecosystem
2.1.1 Role of drilosphere

Another recent approach to science and researchers influencing microbial behaviors in the soil is the notion of an earthworm sphere. Earthworms change the properties of soil and make it acceptable for microbes on this basis, the vermic horizons identified by soil scientists in which earthworms affect the dynamic properties of the soil (Brown et al., 2000). Bouché, coined the term "drilisphere" to the area up to 2 mm around earthworm burrow walls, while Haminton and Dindal used another word vermisphere, which is a soil structure and volume produced by earthworm, i.e., midden (Bouché et al., 1997). The drilosphere was expanded later on Lavelle (1988), including earthworm populations along with soil volume, microbial populations. Latter on Lavelle expanded the drilosphere, including earthworm populations along with soil volume, microbial and invertebrate populations which is influenced by their activities. This definition has five main components including following points and as shown in Fig. 1.

- Internal micro-environment of the earthworm gut
- External environment where earthworm attracts with soil
- Above and belowground environment
- Earthworm originated structure, i.e., middens
- Earthworms created burrows

The earthworm body secretes mucus and nitrogenous excretion in their habitat which is useful for soil ecology because the microbiota functions have been strengthened in soil which followed biodegradation of organic matters and may be function as microbial primers. Earthworm develops castings a by-product of gut passage which are primarily in globular and granular forms; these are features of different sizes, high stability and periods also in microbial activities variation. Globular casts are large in size and more

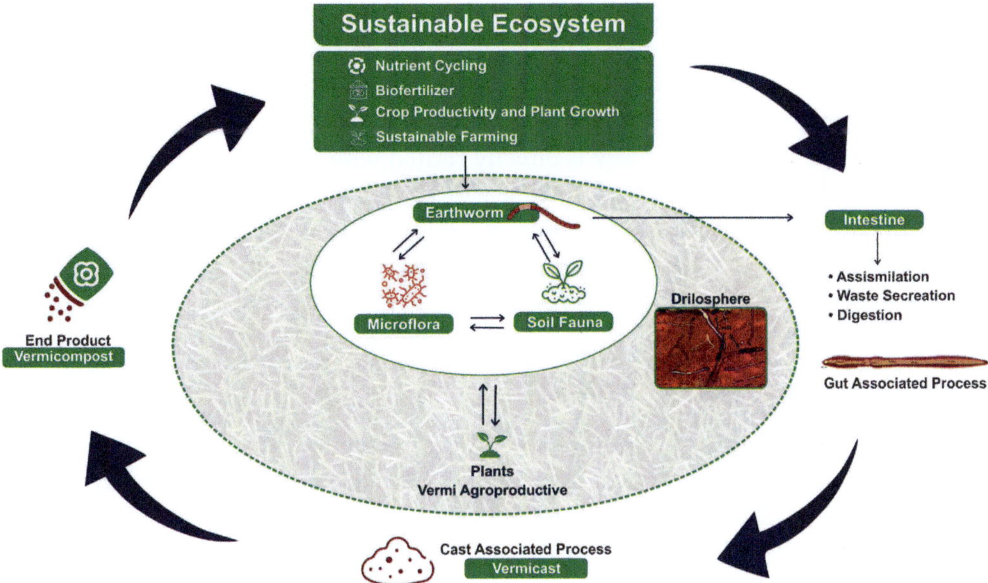

Fig. 1 Representation of sustainable ecosystem between earthworms and microflora.

stable as compared with granular casts. Due to more stability, globular casts are more important and useful for mineralization.

2.1.2 Role of rhizosphere

The rhizosphere (a zone around root system) is directly and indirectly affected by the earthworms that follow the growth and development of the plant. The earthworm feed and behavior in rhizosphere are the direct effects of the earthworm while indirectly influencing the soil texture, physiological, physiochemical, and biological mechanisms that affect the growth of the root and plant. Different visual findings, i.e., root volume, duration, and quantity can be easily accessed by earthworm activities in the rhizosphere; mycorrhizal and soil bacterial activities, etc. (Beven and Germann, 1982; Zaller et al., 2011; Liu et al., 2014). Recently organic farming has great importance than the conventional system because of soil organic matters in organic system. Soil bulk density and porosity are correlated to each other; porosity influence by the addition of organic matters which directly affects the bulk density, i.e., reduction in density; this effect is in organic farming systems and is important for root growth and nutrient availability. The higher soil respiration is always directly proportional to higher microbial activity in the organic farming system because of continuous addition of organic matters; which is most suitable for microbes and follows the high biological activity and rapid decomposition. Because of soil organic matters in the organic system, organic farming has been of great significance lately than the conventional system. Soil bulk density and porosity are related to each other; porosity influence through the addition of organic matter that directly affects bulk density, i.e., density reduction; this effect is important for root growth and nutrient availability in organic farming systems. Higher soil respiration is often directly related to higher microbial activity in the organic farming system because of the continuous inclusion of organic matter that is most appropriate for microbes and follows elevated biological activity and rapid decomposition. Thus we can say the porosity is directly related with soil organic matters and indirectly with the soil bulk density (Cui et al., 2013; Liu et al., 2014; Yan et al., 2018; Esmaeilzadeh and Ahangar, 2014). Hence, decreasing soil bulk density always enhances the porosity due to increase in pore size and numbers which promotes the soil microbial activities following root growth and productivity.

3. Vermicompost as biofertilizer

The flow of energy through soil food web occurs through fungal dominated energy channel in the absence of earthworm which is a slow process. On the other hand, earthworms favour a faster bacterial-dominated energy channel. This may be due to the burrowing activity which likely disrupts the mycelium networks of fungal species. Also, mixing and homogenization of the soil reduce the growth of fungal hyphae. Second, the ingestion, fragmentation, and mixing of residues with soil by earthworms accelerated the release of labile organic substrates that can be metabolized by bacteria, resulting in a higher ratio of carbon to nitrogen-processing bacteria.

Epigenic species of earthworms such as *Eisenia fetida* and *Eudrilus eugeniae* live in organic horizon and feed on organic wastes (vegetable wastes), they can eat as much as half of their weight on a daily basis, and produce castings called as vermicasts. Vermicompost consists of vermicasts and can be described as high-quality organic composts produced from degradation of organic matter present in the waste. It is a complex mixture of earthworm excreta, humified organic matter and microorganisms. The use of vermicompost as a biofertilizer has been reported to be beneficial for plant growth and soil productivity. It is rich in essential microflora required for plant growth like nitrogen fixers, phosphorous solubilizers and free from pathogens, harmful chemicals, weed seeds and bad odor makes it a good substituent of liberally used chemical fertilisers. Additionally, it improves soil structure, texture, aeration, and water holding capacity and prevents soil erosion which in turn increases soil productivity by increasing germination, growth, fruit production of wide varieties of plants.

4. Vermicompost on crop growth and productivity

Sustainable agriculture is the need of the hour and vermicompost can be a potential input in it. Vermicompost can entice a hormonal induced like response from plants on application as it contains high levels of nutrients, humic acids, and humates. The thorough analysis on various crops and plants, a positive influence on vegetative growth, shoot growth, and root development has been documented by various researchers as described in Table 1.

Table 1 A review on positive effects of earthworms on various crops.

Plants	Benefits	References
Ladies finger (*Abelmoschus esculentus*), cucumber (*Cucumis sativus*), and greengram (*Vigna radiata*)	Germination success and promoted the morphological growth and biochemical content of the plant species	Hussain et al. (2018)
Cherry tomato (*Lycopersicon esculentun*)	Improved the quality and productivity of tomato	Truong et al. (2018)
Capsicum (*Capsicum annum*)	Improved plant nutrition, growth, photosynthesis, and chlorophyll content of the leaves	Rekha et al. (2018)
Tomato (*Solanum lycopersicum*)	Works as a protective agent for improvement of salt stress resistance in tomato	Benazzouk et al. (2018)
Strawberry (*Fragaria* × *ananassa*)	Increases in photosynthesis rate, free radical scavenging, and soil enzymatic activity	Zuo et al. (2018)
Marigold (*Tagetes erecta*), Strawberry (*Fragaria* × *ananasa*)	Growth augmentation, increased overall nitrate-nitrogen concentrations in leaf tissues at flowering stage, improves soil fertility	Atiyeh et al. (2001)
Tomatoes (*Lycopersicum esculentum*), Maize (*Zea mays*)	Better plant growth hormones and carbohydrate accumulation	Gutiérrez-Miceli et al. (2007)
Banana (*Musa acuminata*), Cassava (*Manihot esculenta*), Cowpea (*Vigna unguiculata*)	Improvements in yield and biometric characteristics	Padmavathiamma et al. (2008)
Okra (*A. esculentus*)	Increase pod yield and soil fertility	Oroka (2015)
Strawberry (*Fragaria* × *ananassa*)	Improvements in yield and accelerate fruits quality	Singh et al. (2010a,b)
Bhut Jolokia (*Capsicum assamicum*)	Vermiwash significantly affected growth and nutrient utilization	Khan et al. (2014)
Cowpea (*Vigna unguiculata*)	Increases seed yield, straw yield, biological yield, total root nodules and leghaemoglobin content, protein content in seeds	Khan et al. (2015)
Hyacinth bean (*Lablab purpureous*)	Increased growth and yield characteristics	Karmegam and Daniel (2008)
Mulberry (*Morus indica*)	Increases luxuriant growth of bacteria-treated rhizosphere and other micronutrients and NPK (Nitrogen, Phosphorus and Potassium)	Louis Mary et al. (2015)
Rice (*Oriza sativa*)	Increased growth, yield, nutrient uptake, and soil characteristics	Jayakumar et al. (2011)
Soybean (*Glycine max*)	Increase in seed germination, seedling survival, shoot length, and root length	Sheikh (2015)
Tea (*Camellia sinensis*)	A higher radical scavenging activity in tea leaf extracts	Bagchi (2015)
Tomato (*Solanum lycopersicon*)	Increases juiciness, titratable acidity, ascorbic acid content, shelf life	Suge et al. (2011)

Continued

Table 1 A review on positive effects of earthworms on various crops—cont'd

Plants	Benefits	References
Ornamental plant (*Codiaeum variegatum*)	Layering increased root numbers, length and biomass	Karmegam and Daniel (2009)
Mungbean (*Vigna radiate*)	Higher germination (93%), growth and yield.	Zaller (2007)
Maize (*Zea mays*)	Significantly higher plemule length of maize seedlings	Shouche et al. (2014)
Petunia	Stimulated seed germination	Zaller (2007)
Paddy (*Oryza sativa*)	Significant vegetative growth like shoot weight, root weight, root and shoot length	Kumar et al. (2018)
Wheat (*Triticum*)	Increase in quality due to increased gluten content	Gopinath et al. (2008)

5. Effect of vermicompost on nutrient uptake

Vermicompost has dominantly shown its effect on plant growth and productivity, this effect can be attributed to the presence of nutrients required by the plants for growth and development. It is rich in soluble plant nutrient mineral, high nutrient fraction in terms of organic matter and existence of diverse microflora which increases the rate of mineralization and thus bioavailability of nutrients; all these factors contribute to increase productivity. Humus which is abundantly present in the vermicast has also been reported to be responsible for mineral uptake.

The soil aggregates formed due to vermicompost capture the nutrients immobilizing them and releasing them slowly as and when required by the plants, thus, protecting them from leaching. The capture of minerals and ions in certain organic structures reduces their bioavailability to plants. However, remobilization of these elements is possible due to change in pH, soil redox potential, or release of root exudates that increases bioavailability of these nutrients. With both nodular and free living N_2 fixing bacteria, vermicompost helps to increase N_2 fixation and is a proven cheaper form of nitrogen and other essential components for better nodulation and yield, particularly in legumes. Likewise, the introduction of vermicompost to rice plants in association with fertiliser exhibited the greatest absorption by the plant of nitrogen, magnesium, potassium, and phosphorus and vermicompost enhancement with rock phosphate gave cowpea plants superior results in terms of major nutrient absorption, N, P, K, S, Ca, and Mg. However, the application and impact of vermicompost often depend on the form of substrate that is being replaced by vermicompost. As substrates with a relatively high level of plant-available mineral nutrients (as with chemical fertilisers at the optimum level) are supplemented by increasing doses of vermicompost with a relatively lower level of mineral nutrients, the adverse effects on plant growth may be caused by a gradual decrease in the overall supply of minerals and by substrates with a relatively low level of plant-available mineral nutrients.

6. Effect of vermicompost on soil physical, chemical and biological properties

The earthworm burrowing, feeding, and casting operations modify the soil's physical, chemical, and biological properties. Physical property changes such as aeration, lower density of bulk, drainage, erosion, porosity, and infiltration are frequently observed. Due to the presence of organic matter and mucus from the gut of the earthworm present in vermicasts, which improved soil aggregation, soils have increased water retention ability. Previous studies (Bacq-Labreuil et al., 2019; Pathma and Sakthivel, 2015) also show that macropore space increases from 50 to 500 µm, which strengthens the air–water relationship in the soil that favors plant growth. There is a high percentage of humus in vermicasts, which helps to accumulate soil particles, which in turn increases soil aeration and water holding capacity. Humus contains humic acid that binds to many nutrients from plants, viz. potassium, calcium, iron, sulphur, and phosphorus, which are readily released if plants require them, are retained and stored. Vermicompost has a beneficial impact on the chemical properties of the soil, such as pH and soil enzyme activity, due to the presence of various microorganisms that secrete different enzymes into the soil. The synergistic activity of earthworms and microorganisms on chemical species is also reduced by water-soluble chemical matters present in the species that cause environmental pollution. In terms of total biomass and diversity, Earthworm's induces complex shifts in the soil microbiota community, its presence in the soil greatly favors bacterial growth compared to fungal growth. The biomarkers used for the study of the microbial population of PLFAs (phospholipid fatty acid) suggest an exclusive rise in soil gram-negative bacterial population (Willers et al., 2015). These microbes generate different enzymes required for the breakdown of complex organics and the secretion into the soil of different enzymes and nutrients that are taken up for growth and development by the plant.

7. Vermitechnology for co-treatment of OFMSW and wastewater

Vermitechnolgy incorporates traditional vermicomposting for solid waste management into an effortless wastewater treatment process through vermifiltration. Vermifiltration has been widely utilized over the years due to low energy consumption, less operation and maintenance costs, chemical-free method for the treatment of domestic and industrial wastewater as compared to conventional wastewater treatment techniques. Vermicomposting is the mesospheric waste bio-oxidation and stabilization process where earthworms and diverse microbial communities jointly process organic wastes under the aerobic condition to produce nutrient-rich and pathogen-free vermicompost as show in Fig. 2 (Rodríguez-Canché et al., 2010). This process occurs simultaneously along with the wastewater treatment through vermifiltration. Just like the vermicompost, the treated wastewater is rich in organic nutrients and is free from pathogens. Lab-scale studies of this

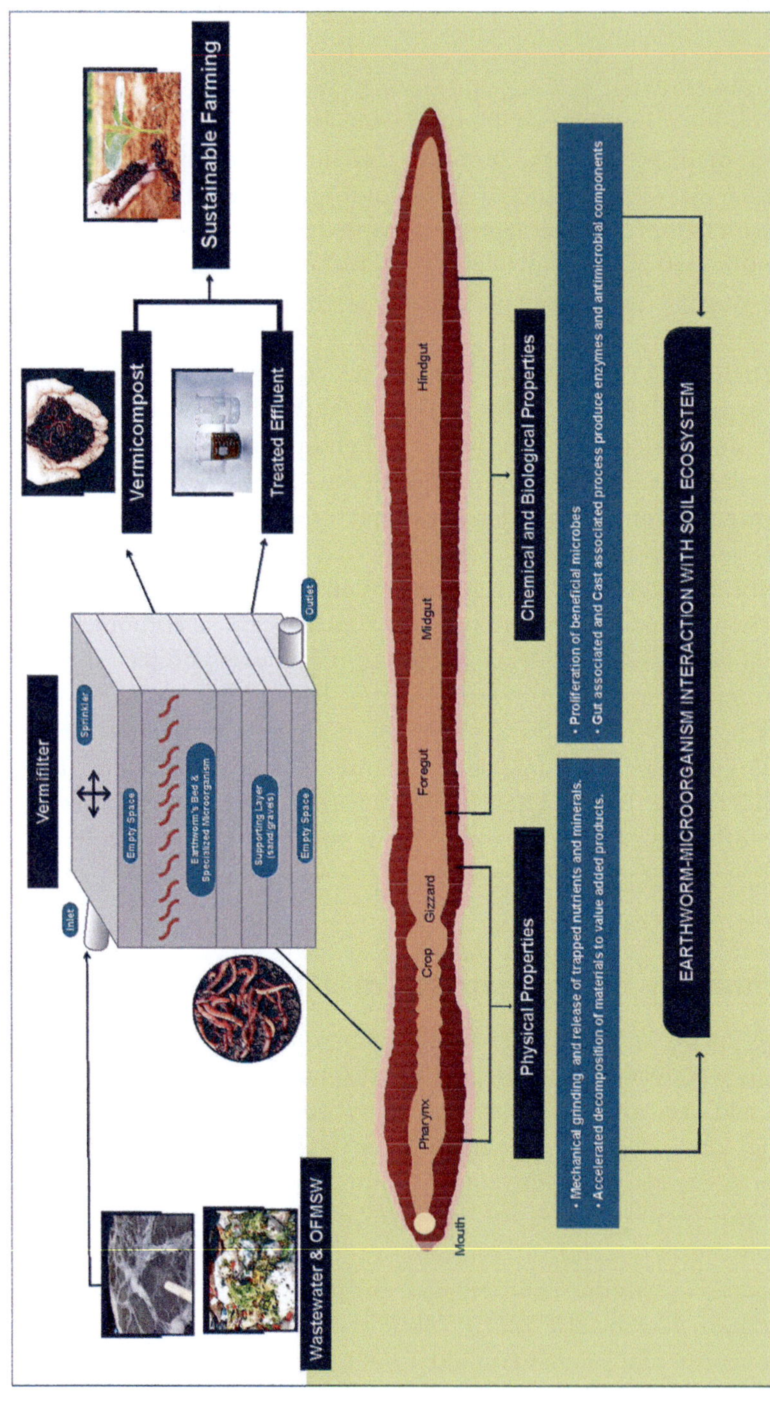

Fig. 2 Earthworm microorganism interaction for the treatment of organic fraction municipal solid waste (OFMSW) and waste water through vermifilter.

integrated approach have also shown positive results for organic fraction of municipal solid waste (OFMSW) along with wastewater, (Rajpal et al., 2014; Xing et al., 2011) with the reusable and ingenious, nutrient-rich end products.

8. Conclusions

Overall, the effects of earthworms on soil fertility and plant production are positive. They enhance soil structure and within their casts stabilize SOM fractions. They increase mineralization in the short term, which makes mineral nutrients accessible to plants. Earthworm causes the release of phytohormonestud analog molecules that aim to enhance the growth of plants. Tillage is normally harmful to earthworms, although practises that increase SOM content have a positive effect on earthworm populations. By transforming biodegradable and organic waste into nutrient-rich products, earthworms preserve the physicochemical properties of the soil, emerging from their burrows to deposit the fecal matter (vermicast) on the surface. Earthworms promote microbial activity, combine the soil, soil water content, and water holding capacity, and aggregate them. The mutual activity of earthworms and microbes brings faster decomposition as the state of the earthworm, aerates, fragments, and microbial action increases the surface area of the organic matter. They also increase the decomposition of litter, the dynamics of soil organic matter, nutrient cycles, encourage plant growth, and minimize some soil-borne diseases. Taken collectively, it is very likely that the use of earthworm services in cropping systems will lead to increasing agricultural sustainability. However, in order to better understand the effect of earthworms on crop production, additional long-term field studies would be necessary.

References

Atiyeh, R.M., Edwards, C.A., Subler, S., Metzger, J.D., 2001. Pig manure vermicompost as a component of a horticultural bedding plant medium: effects on physicochemical properties and plant growth. Bioresour. Technol. 78, 11–20. https://doi.org/10.1016/S0960-8524(00)00172-3.

Bacq-Labreuil, A., Crawford, J., Mooney, S.J., Neal, A.L., Ritz, K., 2019. Cover crop species have contrasting influence upon soil structural genesis and microbial community phenotype. Sci. Rep. 9, 7473. https://doi.org/10.1038/s41598-019-43937-6.

Bagchi, A., 2015. Organic farming practice for quality improvement of tea and its anti parkinsonism effect on health defense. J. Phys. Chem. Biophys. 5 (2). https://doi.org/10.4172/2161-0398.1000178.

Benazzouk, S., Djazouli, Z.E., Lutts, S., 2018. Assessment of the preventive effect of vermicompost on salinity resistance in tomato (*Solanum lycopersicum* cv. Ailsa Craig). Acta Physiol. Plantarum 40, 121. https://doi.org/10.1007/s11738-018-2696-6.

Bertrand, M., Barot, S., Blouin, M., Whalen, J., de Oliveira, T., Roger-Estrade, J., 2015. Earthworm services for cropping systems. A review. In: Agronomy for Sustainable Development., https://doi.org/10.1007/s13593-014-0269-7.

Beven, K., Germann, P., 1982. Macropores and water flow in soils. Water Resour. Res. 18 (5), 1311–1325.

Blouin, M., Barrere, J., Meyer, N., Lartigue, S., Barot, S., Mathieu, J., 2019. Vermicompost significantly affects plant growth. A meta-analysis. In: Agronomy for Sustainable Development., https://doi.org/10.1007/s13593-019-0579-x.

Bouché, M.B., Al-Addan, F., Cortez, J., Hammed, R., Heidet, J.C., Ferrière, G., et al., 1997. Role of earthworms in the N cycle: a falsifiable assessment. Soil Biol. Biochem. 29 (3–4), 375–380.

Brown, G.G., Barois, I., Lavelle, P., 2000. Regulation of soil organic matter dynamics and microbial activityin the drilosphere and the role of interactionswith other edaphic functional domains. Eur. J. Soil Biol. 36 (3–4), 177–198.

Cui, J., Zhang, R., Bu, N., Zhang, H., Tang, B., Li, Z., et al., 2013. Changes in soil carbon sequestration and soil respiration following afforestation on paddy fields in north subtropical China. J. Plant Ecol. 6 (3), 240–252.

Domínguez, J., Bohlen, P.J., Parmelee, R.W., 2004. Earthworms increase nitrogen leaching to greater soil depths in row crop agroecosystems. Ecosystems 7 (6), 672–685. https://doi.org/10.1007/s10021-004-0150-7.

Eliazer Nelson, A.R.L., Ravichandran, K., Antony, U., 2019. The impact of the green revolution on indigenous crops of India. J. Ethnic Foods 6, 8. https://doi.org/10.1186/s42779-019-0011-9.

Esmaeilzadeh, J., Ahangar, A.G., 2014. Influence of soil organic matter content on soil physical, chemical and biological properties. Int. J. Plant Anim. Environ. Sci. 4 (4), 244–252.

Eyhorn, F., Muller, A., Reganold, J.P., Frison, E., Herren, H.R., Luttikholt, L., Mueller, A., Sanders, J., Scialabba, N.E.H., Seufert, V., Smith, P., 2019. Sustainability in global agriculture driven by organic farming. In: Nature Sustainability., https://doi.org/10.1038/s41893-019-0266-6.

Gopinath, K.A., Saha, S., Mina, B.L., Pande, H., Kundu, S., Gupta, H.S., 2008. Influence of organic amendments on growth, yield and quality of wheat and on soil properties during transition to organic production. Nutr. Cycl. Agroecosyst. 82, 51–60. https://doi.org/10.1007/s10705-008-9168-0.

Gupta, N., Tripathi, S., Dholakia, H.H., 2020. Can zero budget natural farming save input costs and fertiliser subsidies? January https://www.ceew.in/sites/default/files/can-zero-budget-natural-farming-save-input-costs-and-fertilizer-subsidies.pdf.

Gutiérrez-Miceli, F.A., Santiago-Borraz, J., Montes Molina, J.A., Nafate, C.C., Abud-Archila, M., Oliva Llaven, M.A., Rincón-Rosales, R., Dendooven, L., 2007. Vermicompost as a soil supplement to improve growth, yield and fruit quality of tomato (*Lycopersicum esculentum*). Bioresour. Technol. 98, 2781–2786. https://doi.org/10.1016/j.biortech.2006.02.032.

Hong, H.N., Rumpel, C., Des Tureaux, T.H., Bardoux, G., Billou, D., Tran Duc, T., Jouquet, P., 2011. How do earthworms influence organic matter quantity and quality in tropical soils? Soil Biol. Biochem. 43 (2), 223–230. https://doi.org/10.1016/j.soilbio.2010.09.033.

Hussain, N., Abbasi, T., Abbasi, S.A., 2018. Generation of highly potent organic fertilizer from pernicious aquatic weed Salvinia molesta. Environ. Sci. Pollut. Res. 25, 4989–5002. https://doi.org/10.1007/s11356-017-0826-0.

Jayakumar, M., Sivakami, T., Ambika, D., Karmegam, N., 2011. Effect of Turkey litter (*Meleagris gallopavo* L.) vermicompost on growth and yield characteristics of paddy, oryza sativa (ADT-37). Afr. J. Biotechnol. 10 (68), 15295–15304. https://doi.org/10.5897/AJB11.2253.

Karmegam, N., Daniel, T., 2008. Effect of vermicompost and chemical fertilizer on growth and yield of hyacinth bean, *Lablab purpureus* (L.) sweet. In: Dynamic Soil, Dynamic Plant.

Karmegam, N., Daniel, T., 2009. Effect of application of vermicasts as layering media for an ornamental plant, *Codiaeum variegatum* (L.) Bl. In: Dynamic Soil, Dynamic Plant.

Khan, M.H., Meghvansi, M.K., Gupta, R., Veer, V., Singh, L., Kalita, M.C., 2014. Foliar spray with vermiwash modifies the arbuscular mycorrhizal dependency and nutrient stoichiometry of bhut jolokia (*Capsicum assamicum*). PLoS One 9 (3), e92318. https://doi.org/10.1371/journal.pone.0092318.

Khan, V.M., Manohar, R.S., Verma, H.P., 2015. Effect of vermicompost and biofertilizer on symbiotic efficiency and yield of cowpea in arid zone of Rajasthan. Asian J. Bio Sci. 10, 113–115. https://doi.org/10.15740/has/ajbs/10.1/113-115.

Kumar, A., Prakash, C.H.B., Brar, N.S., Kumar, B., 2018. Potential of vermicompost for sustainable crop production and soil health improvement in different cropping systems. Int. J. Curr. Microbiol. App. Sci. 7 (10), 1042–1055. https://doi.org/10.20546/ijcmas.2018.710.116.

Kumar, K., Gupta, S.C., Baidoo, S.K., Chander, Y., Rosen, C.J., 2005. Antibiotic uptake by plants from soil fertilized with animal manure. J. Environ. Qual. 34 (6), 2082–2085. https://doi.org/10.2134/jeq2005.0026.

Lavelle, P., 1988. Earthworm activities and the soil system. Biol. Fertil. Soils 6 (3), 237–251.

Lemtiri, A., Colinet, G., Alabi, T., Cluzeau, D., Zirbes, L., Haubruge, É., Francis, F., 2014. Impacts of Earthworms on Soil Components and Dynamics. A Review.

Liu, X.P., Zhang, W.J., Hu, C.S., Tang, X.G., 2014. Soil greenhouse gas fluxes from different tree species on Taihang Mountain, North China. Biogeosciences 11 (6), 1649.

Louis Mary, L.C., Sujatha, R., Chozhaa, A.J., Navas, P.M.A., 2015. Influence of organic manures (biofertilizers) on soil microbial population in the rhizosphere of mulberry (*Morus indica* L.). Int. J. Appl. Sci. Biotechnol. 3 (1). https://doi.org/10.3126/ijasbt.v3i1.12137.

MacLaren, C., Storkey, J., Menegat, A., Metcalfe, H., Dehnen-Schmutz, K., 2020. An ecological future for weed science to sustain crop production and the environment. A review. In: Agronomy for Sustainable Development., https://doi.org/10.1007/s13593-020-00631-6.

Oroka, F.O., 2015. Influence of municipal solid waste vermicompost on soil organic carbon stock and yield of okra (*Abelmoschus esculentus* Moench) in a tropical agroecosystem. J. Environ. Earth Science 5 (12), 61–66.

Padmavathiamma, P.K., Li, L.Y., Kumari, U.R., 2008. An experimental study of vermi-biowaste composting for agricultural soil improvement. Bioresour. Technol. 99, 1672–1681. https://doi.org/10.1016/j.biortech.2007.04.028.

Pathma, J., Sakthivel, N., 2015. Microbial diversity of vermicompost bacteria that exhibit useful agricultural traits and waste management potential. In: Biological Treatment of Solid Waste., https://doi.org/10.1201/b18872-11.

Rajpal, A., Arora, S., Bhatia, A., Kumar, T., Bhargava, R., Chopra, A.K., Kazmi, A.A., 2014. Co-treatment of organic fraction of municipal solid waste (OFMSW) and sewage by vermireactor. Ecol. Eng. 73, 154–161. https://doi.org/10.1016/j.ecoleng.2014.09.012.

Rekha, G.S., Kaleena, P.K., Elumalai, D., Srikumaran, M.P., Maheswari, V.N., 2018. Effects of vermicompost and plant growth enhancers on the exo-morphological features of *Capsicum annum* (Linn.) Hepper. Int. J. Recycl. Organ. Waste Agric. 7, 83–88. https://doi.org/10.1007/s40093-017-0191-5.

Rodríguez-Canché, L.G., Cardoso Vigueros, L., Maldonado-Montiel, T., Martínez-Sanmiguel, M., 2010. Pathogen reduction in septic tank sludge through vermicomposting using Eisenia fetida. Bioresour. Technol. 101, 3548–3553. https://doi.org/10.1016/j.biortech.2009.12.001.

Searchinger, T.D., Wirsenius, S., Beringer, T., Dumas, P., 2018. Assessing the efficiency of changes in land use for mitigating climate change. Nature 564 (7735). https://doi.org/10.1038/s41586-018-0757-z.

Seufert, V., Ramankutty, N., 2017. Many shades of gray—the context-dependent performance of organic agriculture. Sci. Adv. 3, e1602638. https://doi.org/10.1126/sciadv.1602638.

Sheikh, M.A., 2015. Impact of chemical fertilizer and organic manure on the germination and growth of soybean (*Glycine max* L.). Adv. Life Sci. Technol. 31, 73–78.

Shouche, S., Bhati, P., Jain, S., 2014. Recycling wastes into valuable organica fertilizer: vermicomposting. Int. J. Res. Biosci. Agric. Technol. 2 (8). https://doi.org/10.29369/ijrbat.2014.02.ii.0037.

Singh, J., Kaur, A., Vig, A.P., Rup, P.J., 2010a. Role of Eisenia fetida in rapid recycling of nutrients from bio sludge of beverage industry. Ecotoxicol. Environ. Saf. 73, 430–435. https://doi.org/10.1016/j.ecoenv.2009.08.019.

Singh, R., Gupta, R.K., Patil, R.T., Sharma, R.R., Asrey, R., Kumar, A., Jangra, K.K., 2010b. Sequential foliar application of vermicompost leachates improves marketable fruit yield and quality of strawberry (Fragaria × ananassa Duch.). Sci. Hortic. 124, 34–39. https://doi.org/10.1016/j.scienta.2009.12.002.

Suge, J.K., Omunyin, M.E., Omami, E.N., 2011. Effect of organic and inorganic sources of fertilizer on growth, yield and fruit quality of eggplant (*Solanum melongena* L). Arch. Appl. Sci. Res. 3, 470–479.

Truong, H.D., Wang, C.H., Kien, T.T., 2018. Effect of vermicompost in media on growth, yield and fruit quality of cherry tomato (*Lycopersicon esculentun* Mill.) under net house conditions. Compost Sci. Utiliz. 26, 52–58. https://doi.org/10.1080/1065657X.2017.1344594.

Willers, C., Jansen van Rensburg, P.J., Claassens, S., 2015. Phospholipid fatty acid profiling of microbial communities – a review of interpretations and recent applications. J. Appl. Microbiol. 119, 1207–1218. https://doi.org/10.1111/jam.12902.

Xing, M., Yang, J., Wang, Y., Liu, J., Yu, F., 2011. A comparative study of synchronous treatment of sewage and sludge by two vermifiltrations using an epigeic earthworm Eisenia fetida. J. Hazard. Mater. 185, 881–888. https://doi.org/10.1016/j.jhazmat.2010.09.103.

Yan, G., Xing, Y., Wang, J., Li, Z., Wang, L., Wang, Q., et al., 2018. Sequestration of atmospheric CO_2 in boreal forest carbon pools in northeastern China: effects of nitrogen deposition. Agric. For. Meteorol. 248, 70–81.

Zaller, J.G., 2007. Vermicompost as a substitute for peat in potting media: effects on germination, biomass allocation, yields and fruit quality of three tomato varieties. Sci. Hortic. 112, 191–199. https://doi.org/10.1016/j.scienta.2006.12.023.

Zaller, J.G., Heigl, F., Grabmaier, A., Lichtenegger, C., Piller, K., Allabashi, R., et al., 2011. Earthworm-mycorrhiza interactions can affect the diversity, structure and functioning of establishing model grassland communities. PLoS One 6 (12), e29293.

Zuo, Y., Zhang, J., Zhao, R., Dai, H., Zhang, Z., 2018. Application of vermicompost improves strawberry growth and quality through increased photosynthesis rate, free radical scavenging and soil enzymatic activity. Sci. Hortic. 233, 132–140. https://doi.org/10.1016/j.scienta.2018.01.023.

CHAPTER 7

Avenues of sustainable pollutant bioremediation using microbial biofilms

Basma A. Omran*
Petroleum Biotechnology Laboratory, Processes Design and Development Department, Egyptian Petroleum Research Institute (EPRI), Cairo, Egypt
*Corresponding author: e-mail address: dbasmaomran@yahoo.com

Abbreviations

2D	two dimensional
3D	three dimensional
AHL	acyl-homoserine lactone
BPS	periphyton-based system
BTEX	2-benzene, toluene, ethylbenzene, dichlorodiphenyltrichloroethane xylene
CFU	colony forming unit
COC	*Camellia oleifera* Cake
CV	crystal violet
DDT	dichloro-diphenyl-trichloroethane
DNA	deoxyribonucleic acid
eDNA	extracellular DNA
EIS	electrochemical impedance spectroscopy
EPA	Environmental Protection Agency
EPSs	extracellular polymeric substances
FTIR	Fourier transform infrared spectroscopy
HAAS	hormonal active agents
HRT	hydraulic retention time
MFC	microbial fuel cell
NMR	nuclear magnetic resonance
OMPs	outer membrane proteins
PAH	polycyclic aromatic hydrocarbons
PCPs	polychlorinated biphenyls
POPs	persistent organic pollutants
Qpcr	quantitative polymerase chain reaction
QS	quorum sensing
SBMMBR	sequencing batch moving bed biofilm reactor
SR-lXRF	synchrotron radiation micro X-ray fluorescence microscopy
TNT	2,4,6-trinitrotoluene
TTC	2,3,5-triphenyl-2H-tetrazolium chloride
US EPA	US Environmental Protection Agency
VBNC	viable but non-culturable
XANES	X-ray absorption near-edge structure
XTT	[(2,3-bis-(2-methoxy-4-nitro-5-sulfophenyl) 2 htetrazolium-5-carboxanilide inner salt

Relationship Between Microbes and the Environment for Sustainable Ecosystem Services, Volume 1
https://doi.org/10.1016/B978-0-323-89938-3.00007-4

1. Introduction

The remarkable wide-spread of environmental pollution during the last few decades has been attributed to the extensive pace in the anthropogenic activities, urbanization, industrialization, and the major technological advancements (Mohapatra et al., 2020). Lately, biological ecosystems have been threatened by air, water, soil, and sediment pollution with countless contaminants, such as industrial dyes and toxic heavy metals. These contaminants are associated with different industrial operations, involving mining, metallurgical, electro-osmosis, tanneries, electrolysis electroplating, surface finishing, photography, distilleries, production of steel and iron, pesticides, fertilizers, aerospace, manufacturing of electrical appliances, varnishes, paints, plastics, textiles, leather, rubber, cosmetics, paper, food, pharmaceuticals and atomic energy installations (Mohapatra et al., 2017). One of the deleterious marine and soil contaminants is polycyclic aromatic hydrocarbons (PAHs). PAHs are generated as a result of the partial combustion of organic matter and coal, automobile emissions, power plants generated electricity, etc. (Samanta et al., 2002). These carbon compounds are prevalent in nature. They are made up of two or more aromatic ring structures and cause carcinogenic and mutagenic impacts to living organisms (Seo et al., 2009). Exposure to PAHs may take place via polluted air, water, and soil. According to the US Environmental Protection Agency (US EPA), 16 PAHs have been listed as priority lethal pollutants that require rapid remediation as illustrated in Fig. 1. Lately, biological methods proved to be economically feasible and safe approaches for the efficient removal of PAHs from the contaminated environment (Singh et al., 2006). It is worth mentioning that, when it comes to biological treatment of PAHs, bacteria become of particular interest (Chauhan et al., 2008).

Pigment-based and dye manufacturing industries generate toxic effluents, which endure several organic compounds and synthetic dye stuffs (Sinha et al., 2018). Major concerns arise regarding the contamination of aquatic bodies with organic dye stuffs. As per reported by Reddy and Kotaiah (2006), nearly 10,000 different types of pigments and dyes are commercially available worldwide with annual production of more than 7×10^5 tons. Metals with a density exceeding $5\,g/cm^3$ are designated as "heavy metals". Most of the heavy metals are found in the transition elements (i.e., d-block elements) (Duruibe et al., 2007). They lack a filled d orbital; hence their cations form complex compounds. Heavy metals are considered as the most contaminating agents worldwide because they are non-degradable in nature and can be easily accumulated in the environment. Several inorganic metals are vital and needed in small quantities for several redox, metabolic, and physiological functions. They are known as trace elements including nickel (Ni), magnesium (Mg), manganese (Mn), chromium (Cr^{3+}), calcium (Ca), copper (Cu), zinc (Zn) and sodium (Na). Contrary, some heavy metals, such as aluminum (Al), cadmium (Cd), lead (Pb), mercury (Hg), silver (Ag), and gold (Au) can pose lethal effects on living organisms and do not have any vital biological importance (Siddiquee et al., 2015). When heavy metals exceed the prescribed concentration limits, they become

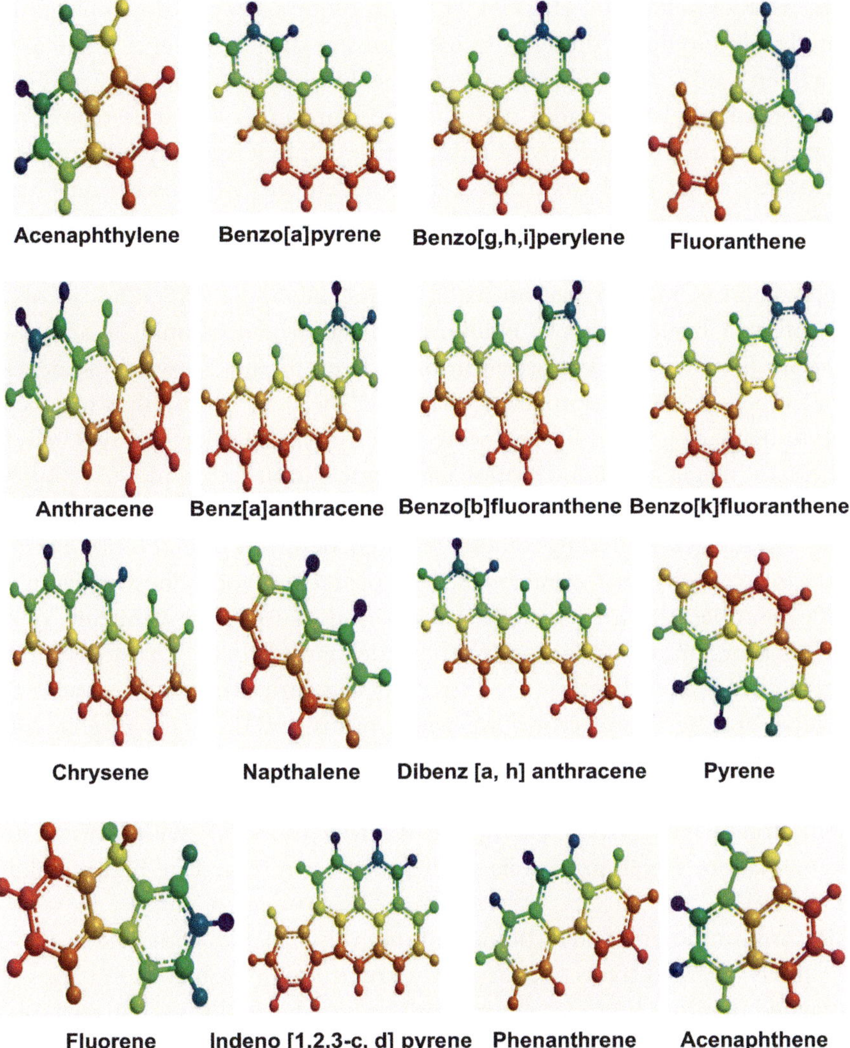

Fig. 1 Molecular structure of 16 polyaromatic hydrocarbon compounds listed as priority pollutants by the US Environmental Protection Agency (EPA).

mutagenic, carcinogenic, teratogenic, and become highly toxic (Shukla et al., 2017). They cause significant damage to humans, animals, and plants, causing cell membrane disruption, DNA denaturation, alterations in the enzymatic activities, which in turn may cause birth defects, kidney and liver damage, skin lesions, and physical and mental retardation.

Accordingly, a novel approach has arisen, which chiefly depends on the usage of microbes as remediators. This innovative approach is more environmentally compatible compared with the physical and chemical remediation techniques (Delangiz et al., 2020).

Moreover, it is cost-effective, efficient, with no requirement of the application of any complicated technologies. Different macroorganisms, such as bacteria, fungi, and microalgae have been used act as effective remediators. This approach can be achieved directly using microbes themselves or indirectly using their enzymes or their metabolic pathways, biofilms, and microbial fuel cells (MFC) (Igiri et al., 2018). Generally, different pollutants can be removed either by detoxification or remediation. Bioremediation of pollutants refers to the transformation of the more toxic form to less toxic ones by the help of living organisms (Gautam et al., 2015). Nonetheless, certain parameters might inhibit the widespread application of bioremediation technologies, including difficulties in obtaining a reliable biomass. Characteristics of effluents and biosorbents should be addressed prior to application. Hence, more investigations have been directed toward the development of sustainable bioremediation processes, including the usage of biofilms to mediate bioremediation, usage of genetically engineered microbes, and microbial fuel cells to overcome the disadvantages of the conventional bioremediation techniques. According to Das et al. (2012), biofilm-mediated bioremediation is a more promising technique because it has several advantages, including validity of high densities of microbial biomass with immobilization capacity that confer excellent protective and adhesive features. Singh et al. (2006) ascribed the improved bioremediation of toxic contaminants by biofilm-mediated microorganisms to the proficient gene transfer between the biofilm species and bacterial chemotaxes. Biofilms are extremely organized and a stable matrix that contains bacteria and few eukaryotic microorganisms, such as fungi and microalgae (Costerton, 1995). Industries can benefit from the bioremediation potential of biofilms to remediate the contaminated sites to clean and safe sites (Karn et al., 2017). In 1989, the oil spill around the shorelines of Prince William Sound in Alaska was treated using biofilm remediation, which proved its super potential to remediate highly polluted sites (Karn et al., 2017).

Biofilms are composed of single or multiple microbial species with a viscous, sticky, and slimy negatively-charged layers of particularly structured microbial assemblages. Biofilms can be attached to either biotic or abiotic surfaces via the enclosing within a microbially synthesized matrix composed of extracellular polymeric substances (EPSs) (Gupta and Diwan, 2017). EPSs contain high molecular weight biopolymers such as proteins, polysaccharides, lipids, nucleic acid, uronic acids, and humic constituents (Shukla et al., 2017). The EPS matrix secreted by the encased microorganisms do not only aid in cell growth but also provide water channels to permit the transport of oxygen and nutrients (Shukla et al., 2014). It has been observed that the anionic charges of biofilms facilitate the sequestration of positively charged and persistent organic contaminants, such as heavy metals, dye molecules, and polycyclic aromatic hydrocarbons (PAHs) (Mohapatra et al., 2020). Interestingly, EPS originated from biofilms contain microbial surface active metabolites, which are referred to as "biosurfactants." Biosurfactants are known to be metal complexing agents, which have a significant role during the bioremediation processes (Mitra and Mukhopadhyay, 2016). Furthermore, biofilm-established communities introduce a beneficial physicochemical interaction for the provision of nutrients as well as genetic exchange among the embedded microbial species. Besides, it provides protection from the surrounding harsh environmental conditions, including starvation, extreme pH and temperature, chemical stress, and predatory protozoa (Dzionek et al., 2016). Several investigations have emphasized the proficiency of microbial biofilms than the free floating planktonic species during pollutant bioremediation. This has been mainly ascribed to the high microbial biomass of biofilms and their superior capacity to immobilize pollutants within the EPS matrix (Shukla et al., 2017). Biofilms are more tolerable to pollutants and possess diverse catabolic pathways, which can be helpful during bioremediation processes. By taking the advantages of biofilms into consideration, they can be exploited beneficially in industrial treatment plants for bioremediation/bioimmobilization of industrial and municipal wastewater treatment that contain a wide range of contaminants. In this context, a growing interest has been directed toward the biotechnological applications of microbial biofilms for the bioremediation of lethal industrial contaminants. Henceforth, this chapter focuses on the beneficial side of using biofilm-producing microorganisms for the bioremediation of several environmental pollutants, such as heavy metals, dyes, and PAHs. It also spots the light on the different characteristic features of microbial biofilms, developmental stages, composition, factors affecting the development of microbial biofilms, and the interactions between bacterial biofilms and the toxic industrial effluents.

2. Environmental pollution

Environmental pollution is the undesirable alteration of the surrounding environments, which is largely or wholly caused by the activities of human kind. It refers to the "buildup, release, and accumulation of unfavorable matters into the environment beyond the allowed limits, where lethal effects start to manifest" (Pandey and Singh, 2019). It is a global phenomenon, which has emerged because of the increase of human population worldwide, rapid industrialization, anthropogenic activities, the corrupted usage of synthetic chemicals, and changes in radiation levels and energy patterns. In 1979, Holdgate defined environmental pollution as "the introduction of substances by man into the environment that cause interference with legitimate uses of the surrounding environment" (Holdgate, 1979). In 1991, Singh defined pollution in a simple way as "the disequilibrium manner from equilibrium conditions in any system" (Singh et al., 1991). This definition can be applied to all kinds of pollution starting from the physical to social, political, economic, and religious pollution types. Whereas, pollutants refer to "any substances causing pollution". Pollutants have been found in the environment since thousands of years ago since the pre-industrial era. Pollutants can be originated from chemical substances (e.g., toxic heavy metals, radionuclides, gases, and organophosphorus compounds) or geochemical substances (e.g., sediment and dust), biologically or physically originated (e.g., radiation, heat, and sound waves). These pollutants can be released intentionally or unintentionally by man into the environment, causing deleterious, adverse, unpleasant, and troublesome effects. Tables 1 and 2 show a list of different categories of pollutants, sources, characteristics and their toxic effects.

Table 1 List of contaminants causing environmental pollution.

Type of pollutants

Organic compounds

Types	Sources	Composition and characteristics	Examples	Toxicity
Polycyclic aromatic hydrocarbons (PAHs)	Incomplete combustion of fossil fuels, manufacturing of dyes, pesticides, and plastics	Two or more fused aromatic rings Highly hydrophobic in nature, tend to adsorb onto the surface of soil or sediments in marine environments	Pyrene, Napthalene, Phenanthrene, Benzo(*a*)pyrene	Lethal effects on liver and kidney, carcinogenic
Nitro-aromatic compounds	Agrochemicals, textile, pesticides, explosives and chemicals in pharmaceutical industries	Minimum of one nitro group ($-NO_2$) attached to an aromatic ring	2,4,6-Trinitrotoluene (TNT), hexanitrobenzene	Toxic to aquatic systems, animals, and humans (cause skin irritation, immuno-toxicity, and methaemoglobinemia)
Organo-chlorine compounds	Manufacture of polymers, solvents in textile industry and dry cleaning, pesticides, and insecticides	Organic compounds containing chlorine (Cl^-) atoms bounded by covalent attachment to the carbon structure, they have both aliphatic chains and aromatic rings. They are designated in the list of persistent organic pollutants (POPs) due to their toxicity and environmental persistence	Aliphatic organo-chlorides (e.g., vinyl chloride, polychlorinated biphenyls and chloromethane), Aromatic organo-chlorines such as dichlorodiphenyltrichloroethane (DDT), endosulfan, chlordane, aldrin, dieldrin, and endrin and mirex)	Xenoestrogenic and endocrine disruptors

Phthalates	Manufacture of plastics, insect repellents, synthetic fibers, lubricants, cosmetics, blood transfusion bags and intravenous fluid bags and tubes. Enteric coatings of pills and nutritional supplements as viscosity control and gelling agents in pharmaceutical industries, paper and plastic industries	Esters of phthalic acid	Butyl benzyl phthalate and di (2-ethylhexyl) phthalate	Endocrine disruptors or hormonal active agents (HAAS), eczema and rhinitis in children and carcinogenic
Azo dyes	Textile industries	Azo dyes have the functional group $R-N=N-R'$ (R and R' can be either aryl or alkyl)	Azobenzene	Toxic to aquatic fauna and flora, carcinogenic

Table 2 Examples of toxic heavy metals.

Metal	Toxicity	Source
Arsenic (As)	Carcinogenic	Arsenical pesticides, or inappropriate disposal of arsenical chemicals
Lead (Pb)	Affect the development of brain and nervous system, increased risk of high blood pressure and kidney damage	Industrial processes, gasoline, house paint, lead bullets, plumbing pipes, pewter pitchers, storage batteries, toys, faucets, vehicle exhausts
Mercury (Hg)	Neurological disorders, malfunctioning of nerves, kidneys and muscles. Disruption to membrane potential and interrupt with intracellular calcium homeostasis.	Agriculture, municipal wastewater discharges, mining, incineration, discharges of industrial wastewater, pulp and paper industries and dental preparations
Cadmium	Tracheo-bronchitis, pneumonitis, and pulmonary edema.	By-product of zinc production, rechargeable Batteries, alloys production and tobacco smoke
Chromium	Dermatitis, allergic and eczematous skin reactions, mucous membrane and skin ulcerations, perforation of the nasal septum, allergic asthmatic reactions, gastro-enteritis and bronchial carcinomas.	Burning of oil and coal, petroleum materials, pigment oxidants, catalyst, chromium steel, fertilizers, oil well drilling, sewage, fertilizers, and metal plating tanneries

3. Biofilms (remarkable biological communities)

Since the early beginnings of biological science, microorganisms have been mostly described as free-floating microbial cells. Nonetheless, in most natural biota, microorganisms are usually existed within cellular aggregates in close association with interfaces and surfaces affixed by the secreted slippery matrix which is designated as "biofilms" (Nazir et al., 2019). The term biofilms was first coined by Costerton et al. (1978). Costerton and colleagues described biofilms as "microcosm of bacterial consortium enclosed with abiotic constituents within exopolymeric matrix at interfacial regions". Microbial biofilms can be also defined as "sessile microbial consortia embedded in a three-dimensional structure and contains multicellular communities made up of prokaryotic and/or eukaryotic cells embedded within a matrix composed of exopolymeric materials secreted by the embedded cells" (Azeredo et al., 2017). Biofilms contain microbial cells embedded within a sticky sheath matrix, which is made up of EPSs. The secreted EPSs have a significant role in conferring high resistance to biofilms by acting as a molecular sieve, diffusion barrier, and as an adsorbent. Biofilms are formed in several environments such as natural waters, sediments, soil, and industrial systems or power plant water distribution systems (Shukla et al., 2014). Additionally, they can be formed on a number of surfaces such as living tissues, human prostheses, and within interior medical devices. It has been noted that biofilm bacterial cells are greatly different from their planktonic bacterial cell types at the physiological and metabolic levels (Costerton and Stewart, 2000). The produced exopolysaccharide matrix acts as a protective layer for the microbial cells embedded within. When the biofilm starts to grow and thicken, it begins to create a heterogeneous matrix with multiple channels to allow the penetration of oxygen and nutrients into the deepest depths of the biofilm. Bacteria in particular seem to develop biofilms as a survival strategy to protect themselves from the surrounding environment. Cells embedded within the biofilm display a high resistance to antimicrobials and biocides when compared with planktonic cells. A number of explanations have been suggested to explain this phenomenon. The rate of antimicrobials uptake is affected by the different physiological states present within the biofilm community. Moreover, the biofilm exopolysaccharide matrix acts as a physical barrier to prevent the penetration of antimicrobials (Stewart, 1996).

3.1 Diversity of biofilms

Biofilms have been found to harbor significant microbial diversity including; bacteria, archaea, fungi, viruses, algae, and protozoa. Regarding bacterial biofilms, Proteobacteria, Bacteroidetes, Cyanobacteria, and Beta-proteobacteria are the most prominent bacterial groups dominating freshwater biofilms such as rivers, streams, and lakes (Newton et al., 2011). In marine environments, alpha-proteobacteria are typically the most dominant bacterial groups (Anderson-Glenna et al., 2008). Other taxonomic groups can also be present in biofilms such as Acidobacteria, Actinobacteria, Firmicutes, Gamma- and Delta-proteobacteria, Gemmatimonadetes, Verrucomicrobia, and Planctomycetes. Archaea may also be present within rivers and stream (Wilhelm et al., 2014). Microbial eukaryotes are also involved in the construction of biofilms, involving microalgae such as Chlorophyta and Bacillariophyta, which release lysed products and exudates that act as carbon sources for other heterotrophic biofilm microbes (Romaní et al., 2014). Fungi, particularly Ascomycota, are prominent microbes within biofilms because they aid in the decomposition of existed organic matter. Protists such as flagellates, ciliates, and amebae affect the growth of biofilms and alter their architecture, diversity, and metabolic functions.

3.2 Developmental stages of biofilm formation

Generally, biofilms can be formed either naturally (*in vivo*) or can be developed inside laboratories on an artificial surface (*in vitro*). As mentioned earlier, biofilms represent communities of microorganisms in which microbial cells are embedded within a polymeric matrix. A typical biofilm is very complex and contains different microbial species with high cell densities ranging between 10^8 and 10^{11} cells/g wet weight (Yadav and Chandra, 2020). Biofilms are formed on various surfaces such as medical devices, wastewater treatment systems, and industrial or potable water pipes (Shukla et al., 2017). Recent investigations have discovered the stepwise biofilm formation. Biofilms are formed via specific sequential steps involving;

- Attachment of bacterial cells either reversibly or irreversibly onto a solid support.
- Irreversible attachment of bacterial cells to surfaces and formation of a matrix (expansion phase).
- Quorum sensing (QS) starts to regulate microbial cells in layered assemblages (maturation phase).
- Colonies start to take a 3D structure and reach the highest thickness.
- Planktonic bacteria start to detach from the matrix and move toward another surface (dispersion phase).

Bacterial cells differ physiologically at every phase of biofilm development. Initial development of biofilms includes the interaction between planktonic bacterial cells and the target surface. This stage lasts for few minutes only and is referred to as "reversible adsorption". Fuqua et al. (1994) ascribed this description because some bacterial cells attach to the substrate surface for only a brief period and then start to detach from it. The second phase is the irreversible attachment, which lasts for nearly 2 h and it is the phase in which the bacterial cells adhere strictly to the substrate surface and lose their mobility. Afterwards, bacterial cells start to attach to each other and to the substrate surface, leading to the formation of bacterial microcolonies. The third phase is the biofilm maturation (I) in which the EPS matrix is produced (Fuqua and Greenberg, 1998). This phase continues for three successive days and microcolonies form multi-layered structure with a thickness up to 10 µm. The next phase is biofilm maturation (II) and it lasts for 6 days. In this phase, bacterial microcolonies start to increase in size till reaching the maximum thickness of approximately 100 µm. The last phase is the dispersion phase, and it represents the last phase in biofilm maturation. Dispersion phase lasts between 9 and 12 days. In this phase, the structure of microcolonies changes as the bacterial cells existed in the central part regain their mobility and detach from the previously established biofilm. Accordingly, at this stage the biofilm becomes loose. At this point, biofilms lose their mushroom or 3D-shaped structure and adopt a shell-like structure (Wagner et al., 2006). The process of dispersion takes place to allow the bacterial cells to find another favorable substrate, most probably to obtain better access to nutrients. Fig. 2 represents an illustration for the different stages of biofilm formation.

It is noteworthy to mention that, during the attachment step, biofilm-forming bacteria attach to any suitable surface by the help of anchoring appendages that facilitate the movement and attachment of bacteria to an appropriate solid surface or interface. An ideal environment for the bacterial attachment usually takes place at the solid–liquid interface (Costerton et al., 1999). Different microbial structures play a significant role

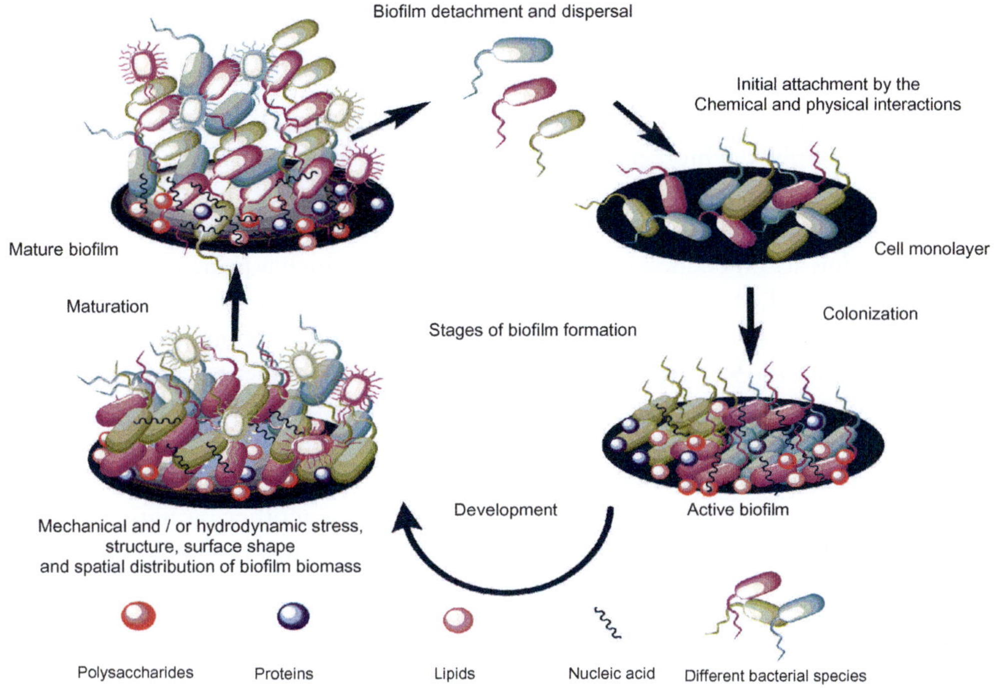

Fig. 2 Diagrammatic representation describes the cyclic stages for the formation of active biofilm. Cells initially attach to a surface via physical and chemical interactions to form cell monolayer. Afterwards, cells begin to proliferate in the monolayer to form an active biofilm and release the EPSs. At this point, the surface becomes preconditioned for further development of biofilms which is affected by several environmental conditions like mechanical and hydrodynamic stresses. Then, biofilms reach maturity and regain mobility and undergo chemotaxis. Biofilms detach and start spreading to another surface.

in biofilm formation such as flagella, fimbriae, curli (a proteinaceous and non-flagellar structure), EPSs, and outer membrane proteins (OMPs). Pili or fimbriae and curli are adhesion molecules, which enable the proficient attachment of biofilm-forming bacteria to the desired surface. The functions of these structures differ according to the surrounding environmental conditions. Pili facilitate the attachments of cell-to-cell and cell-to-surface. Curli and OMPs contribute in the transfer of genetic materials upon attachment of bacterial cells to the solid surfaces. Most fimbriae possess high quantities of hydrophobic amino acid residues. Fimbriae decrease electrostatic repulsion forces between bacterial cells and surfaces, hence facilitating the bacterial attachment to surfaces. Attachment of biofilm-forming bacterial cells was found to be prominent on hydrophobic materials rather than hydrophilic ones. Prigent-Combaret et al. (2000) managed to isolate the OmpR gene, which is mainly involved in curli production. Upon the successful attachment of bacterial cells to a particular surface, bacteria start to grow in aggregates and form colonies, which are designated as "microcolonies" (Nazir et al., 2019). Under favorable environmental conditions, number of bacterial cells increase rapidly and produce a two-dimensional (2D) structure on the solid surface. After cell growth and proliferation, these microcolonies mature into a large defined three-dimensional (3D) architecture. Interestingly, mature biofilms exhibit different shapes such as flat monolayers, 3D structures, tulip-like, and mushroom-shaped structures. All of these biofilm structures contain intervening water channels that facilitate the exchange of the required nutrients and wastes. Bacterial biofilms are embedded within a biofilm matrix, which is principally made up of a gelatinous material of EPSs. EPSs are composed of polysaccharides, proteins, lipids, nucleic acids, non-viable bacterial cells, and other polymeric components. It is hydrated with water up to 85%–95%. EPSs play a vital role in the attachment of bacterial cells to either biotic or abiotic surfaces, providing resistance against harsh conditions, supporting the access of nutrients from the surroundings, and inhibiting desiccation. The final step of biofilm lifecycle is the dispersion of cells to the surrounding environment. The detached cells turn into their planktonic state and they start to form another biofilm on another surface when the conditions become preferable. Detachment of bacterial cells during the dispersal phase of biofilm may occur because of several environmental changes, including nutrient availability, movement of surrounding liquid, and other external forces. Detachment of bacterial cells enables the biofilm spread to start another new cycle.

Biofilm formation is affected by several factors such as nutrient concentration. Microbial biofilms obtain nutrients via different means via using waste products of secondary colonizers and consuming the trace organic matters on the surface. The pH of the media can disturb various processes of microorganisms because microbial biofilms are influenced by changes in pH. Optimum pH varies from species to species, but pH around 7 is more favorable for the majority of microbes. Furthermore, microbial performance is significantly affected by temperature. For optimum growth of microbial populations, optimum temperature is crucial, whereas minor changes in optimum temperature may result in a decrease in microbial growth, which could be ascribed to the reduction of the rate of enzyme activities (Nazir et al., 2019).

Quorum sensing (QS) is also referred to as "density sensing", which controls several cellular processes (Zhao et al., 2020). Microbial communities depend on QS as a mechanism of cell-to-cell communication. QS not only modulates gene expression but also assists in controlling biocide/antimicrobial/disinfectant resistance, production of virulence factors, regulation of bacterial luminescence, toxin production, drug resistance, spore formation, and biofilm establishment and maintenance (Turan et al., 2017; Bäuerle et al., 2018). Moreover, QS in some microorganisms helps to induce pathogenicity and biofilm formation as reported by Zhao et al. (2020). QS mechanisms occur in both Gram-negative and Gram-positive bacteria but the signal molecules differ in both. Based on the different self-inducible signaling molecules, bacterial QS mechanistic systems are categorized into three types. The first one is the QS system with acyl-homoserine lactone (AHL) (i.e., self-inducible molecule) and is found in Gram-negative bacteria. A variety of AHLs are found in bacterial communities and they differ from each other in the length of the acyl side chain (Worthington et al., 2012). N-butanoylhomoserine lactone is the most existed type of AHL in *Pseudomonas aeruginosa*. While, in Gram-positive bacteria, the oligopeptides represent the QS systems and act as self-inducing molecules. However, QS systems based on furan borate diesters as self-inducing molecules are found in both Gram-negative and Gram-positive bacteria (Jagadeesh and Karabasappa, 2020).

3.3 Composition of biofilm matrix

Biofilm matrix contains a variety of extracellular polymeric substances secreted by the microbes embedded within. These extracellular polymeric substances include proteins, enzymes, polysaccharides, DNA, RNA, and water as revealed in Fig. 3. Water constitutes the biggest part of biofilms, which reaches up to 97% and assists the flow of nutrients inside the biofilm matrix. The biofilm architecture contains two distinctive components involving; the water channels that are responsible for nutrient transport and the cell-packed region that lacks any notable pores (Lear and Lewis, 2012). The water channels resemble a primitive circulatory system that help in distributing the nutrients to the microcolonies and receive the harmful metabolites (Stoodley et al., 1998). Microcolonies represent the basic structural units of a biofilm. Costerton (1995) reported that the

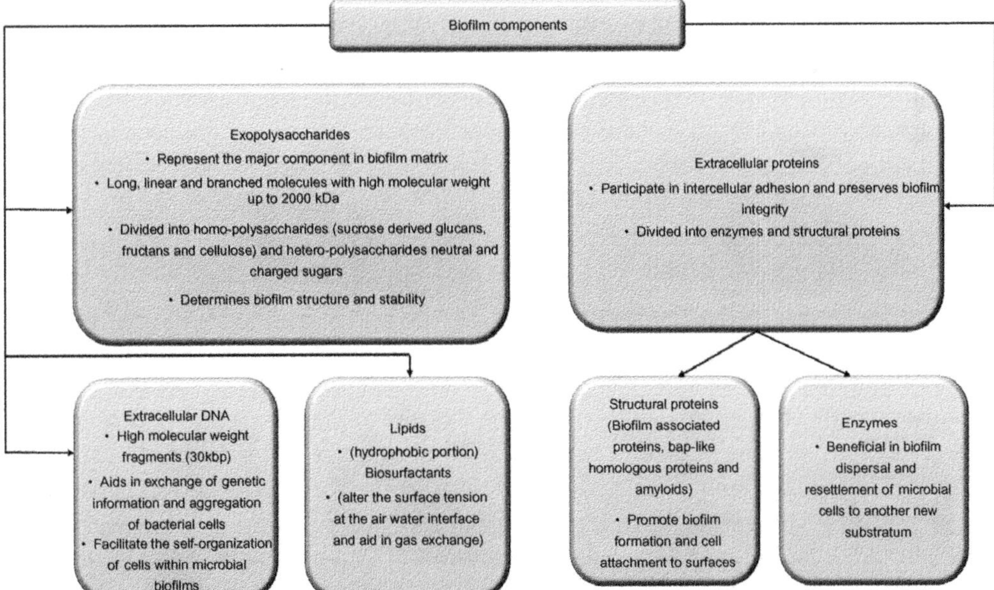

Fig. 3 Representation of biofilm main components.

percentage of biofilm matrix within biofilm-forming microcolonies differ according to the bacterial type. Generally, it is composed of 10%–25% of cells and 79%–90% of EPS matrix. EPS matrix provide microcolonies with protection against harsh environmental conditions such as UV radiations, draining, and pH abrupt changes (Nazir et al., 2019).

3.4 Methods used for detection, characterization, and assaying of microbial biofilms

A wide variety of methods, techniques, and assays have been investigated for studying biofilms. Listed in Table 3, different devices and assays used for biofilm cultivation, their main applications, aims, and pros and cons. Amongst, the microtiter plate, Calgary biofilm device, biofilm ring test, Robbins device, modified Robbins device, and drip flow biofilm reactor. Furthermore, viability of biofilm biomass can be assessed using different methodologies, which can be categorized into microbiological, molecular, physical, and chemical methods. Microscopic techniques are extremely important tools to assess biofilm distinctive features as they provide details regarding biofilm spatial organization and biofilm heterogeneity.

3.4.1 Microbiological and molecular methodologies

Total colony forming unit (CFU) is the most commonly used technique to study biofilm cell viability on agar media. It is a widespread technique in all microbiological laboratories and is mainly based upon the dilution series approach for cell quantification. Yet, this technique poses certain disadvantages and limitations (Li et al., 2014); including (i) the obtained detached sample may not represent the initial biofilm population and (ii) some biofilm cells can be viable but non-culturable (VBNC), which cannot be detected by the CFU methodology. On the other hand, flow cytometry coupled with few fluorophores enabled the accurate and rapid determination of biofilm cell viability (Oliveira et al., 2015). However, it is somehow expensive but it has resolved the drawbacks of CFU counting because it allows the differentiation between total, dead, and VBNC biofilm cells. Quantitative polymerase chain reaction (qPCR) has been used to enumerate living biofilm cells as an alternative technique for culturing biofilm bacterial cells. Nonetheless, this approach may lead to overestimation of the number of viable cells because of the presence of free extracellular DNA (eDNA) (Klein et al., 2012) in addition to the DNA resulting from the dead bacterial cells. Treatment of samples using propidium monoazide (PMA) aided in preventing the quantification of DNA derived from dead cells. Samples can be treated prior to DNA extraction (Krüger et al., 2014) and are then exposed to robust visible light that leads to the formation of a stable covalent bond between PMA and DNA. Hence, during DNA extraction, this modified DNA is lost and will not be amplified during the reaction (Nocker et al., 2009). Still, this technique endorses some disadvantages, involving (i) membrane integrity is the sole factor that discriminates between viable and dead cells and accordingly the action of

Table 3 Main application, description, advantages, and disadvantages of devices and tests used for biofilm cultivation.

Device	Aim and description	Merits	Demerits
Devices and tests used for biofilm cultivation			
Microtiter plate	➢ Most commonly used method for biofilm cultivation and quantification. Helpful in evaluating the efficacy of anti-biofilm compounds ➢ Madilyn Fletcher developed microtiter plates to investigate bacterial attachment (Fletcher, 1977) ➢ In this assay, bacterial cells are allowed to grow inside the wells of a polystyrene microtiter plate, then wells are emptied and washed to get rid of planktonic cells. Afterwards, the biomass attached to the surface of wells are stained to enable biofilm quantification	➢ High-Throughput ➢ Cost-effective ➢ No advanced equipment are needed except for a plate reader	➢ Inaccurate measurement of the loosely attached biofilm (this might occur during washing steps) ➢ Sensitive to cell sedimentation at the bottom of the wells ➢ Difficulty in direct inspection difficult ➢ Possible interference with liquid–air interface ➢ Poor reproducibility ➢ Results are dependent upon the working personnel ➢ Assessment is valid only with the presence of high cell density ➢ Investigation of early stages of biofilm formation is somehow difficult
Calgary biofilm device	➢ CBD was first developed by Ceri et al. (1999) ➢ It is used for the rapid and reproducible screening of biofilm susceptibility to antimicrobials ➢ Biofilm formation is assessed at the coverlid which is made	➢ High throughput ➢ Not sensitive to cell sedimentation ➢ No advanced equipment are needed except for a plate reader	➢ Not all sessile cells can be suspended by sonication ➢ Loosely attached biofilm may not be measured accurately during washing steps ➢ Sonication may not be capable of removing remove

Continued

Table 3 Main application, description, advantages, and disadvantages of devices and tests used for biofilm cultivation—cont'd

Device	Aim and description	Merits	Demerits
	up of pegs which fit into the wells of the microtiter plate. The biofilm formed on the pegs does not result from cell sedimentation but only from sessile development		firmly adhered bacterial cells ➢ Not favorable for studying early stages of biofilm formation
Biofilm ring test	➢ Biofilm Ring test has been developed by Chavant et al. (2007) ➢ Helpful in studying early stages of biofilm formation ➢ In this test, a bacterial suspension is mixed along with paramagnetic microbeads prior loading into the wells of a microtiter plate. Then, the microtiter plate is incubated for direct measurement. The main principle of this assay is to block the beads by the effect of the developed biofilm matrix	➢ Data are provided faster than the two other described above methods (i.e., only a couple of hours) ➢ No fixation, staining or washing steps are needed	➢ Does not provide information regarding the structure or thickness of mature biofilms ➢ Quite expensive ➢ Requires a magnetic device and a scanner ➢ Sensitivity to sedimentation owing to gravity ➢ Not helpful in investigating late stages of biofilm formation ➢ Not suitable for studying biofilms at the air–liquid interface
Robbins device and modified Robbins device	➢ The design of Robbins device belongs to Jim Robbins and Bill McCoy. Then, later on it was patented in a revised version to the Shell Oil Company	➢ Modified Robbins device has a variety of applications ranging from the biomedical applications to several industrial disciplines	➢ Artifacts in samples may take place during movement of coupons as the device is not designed to permit the direct assessment of biofilm development

Table 3 Main application, description, advantages, and disadvantages of devices and tests used for biofilm cultivation—cont'd

Device	Aim and description	Merits	Demerits
	➢ It is composed of a pipe with a number of threaded holes in which coupons are mounted on the end of screws existed in the liquid stream. Coupons are allocated parallel to the fluid flow and can be removed individually ➢ Original Robbins device monitors biofilm formation under different fluid velocities in a simulated drinking water facility (McCoy et al., 1981) ➢ The modified Robbins device consists of a square channel pipe with sampling ports at equal spaces attached to sampling plugs and aligned with the inner surface, without disturbing the flow characteristics. This device is appropriate for studying biofilms under different hydrodynamic conditions		➢ Expensive ➢ Not applicable for *in situ* investigations of biofilms ➢ Requires flow systems ➢ Heterogenic biofilms may develop on coupons ➢ Prior knowledge of the device flow operations is required

Continued

Table 3 Main application, description, advantages, and disadvantages of devices and tests used for biofilm cultivation—cont'd

Device	Aim and description	Merits	Demerits
Drip flow biofilm reactor	➢ It was first developed by Darla and co-colleagues from the Center for Biofilm Engineering, Montana State University (Goeres et al., 2009) ➢ It is composed of four chambers and vented lids. Each chamber has a coupon in which the biofilm can grow. The microbial culture medium and cell inoculum enter each chamber using a gauge needle which is inserted through the lid septum	➢ Small space is only needed ➢ Easy operation ➢ Simultaneous usage of different surfaces, materials ➢ Non-invasive sample analysis	➢ Biofilm heterogeneity on coupons ➢ Low shear stress ➢ Low similarity with industrial conditions ➢ Limited number of samples
Rotary biofilm reactors	➢ Rotary biofilm reactors have three different types involving the rotary annular reactor, the rotary disk reactor and the concentric cylinder reactor (Azeredo et al., 2017) ➢ The development in the design of rotary annular reactor was credited to Kornegay and Andrews (1968). The reactor contains a stationary external cylinder	➢ Several materials can be tested at the same nutritional and hydrodynamic conditions ➢ Shear stress and feed flow rate can be adopted individually ➢ Tolerable to high shear stress	➢ Analysis can be performed to a low number of microbial strains ➢ The geometry of the coupon cannot be altered as it is determined by the reactor design (i.e., coupon holder) ➢ Costly

Table 3 Main application, description, advantages, and disadvantages of devices and tests used for biofilm cultivation—cont'd

Device	Aim and description	Merits	Demerits
	and an inner rotating cylinder. The rotation speed is controlled via a variable speed at the internal cylinder to facilitate efficient mixing of liquid phase. Extremely helpful in investigating biofilm growth in drinking water plants and ship hulls ➢ The rotary disk reactor is composed of a disk, which holds several coupons. The disk is attached to a magnet which provides the needed rotational speed when the reactor is located on top of a magnetic stirrer. Rotation initiates a liquid surface shear through the coupons. This reactor is useful in studying biofilm resistance, to determine biofilm control strategic techniques and to study interactions between the different species in multispecies biofilms. Additionally, it is used in the ASTM		

Continued

Table 3 Main application, description, advantages, and disadvantages of devices and tests used for biofilm cultivation—cont'd

Device	Aim and description	Merits	Demerits
	standard methods E2196-12 and E2562-12 for quantification of *Pseudomonas aeruginosa* biofilms (Azeredo et al., 2017) ➢ The concentric cylinder reactor contains four cylindrical sections which can be rotated at different speeds within four concentric chambers. This allows testing under shear stresses		

antimicrobials (Tavernier and Coenye, 2015), (ii) viable cells with a slight damage in cell membrane cannot be counted (Sträuber and Müller, 2010), and (iii) presence of an increased number of dead cells can affect the process of viable cell quantification (Fittipaldi et al., 2012).

3.4.2 Physical methodologies

Dry or wet weight measurements enable the determination of a total biofilm biomass. In 1982, Trulear and Characklis determined biofilm biomass by calculating the weight difference before and after cleaning a slide with biofilm (Trulear and Characklis, 1982). In addition, authors calculated the volumetric biofilm density as a unit of dry biofilm mass per unit of wet volume. Another approach has been followed to assess the weight of biofilm biomass. Samples of attached cells can be taken from the test surfaces, vortexed, and then the released biofilm components are filtered (Jackson et al., 2014). In this approach, biofilm biomass is calculated based on the weight of a dried filter containing biofilm cells against the weight of a sterile control filter. However, this approach may result in the underestimation of biofilm biomass as it does not detach the whole biofilm from the test surface, and small molecules can penetrate through the filter. Additionally, it is time-consuming and not sensitive enough when assessing minor changes in biofilm production. Another important technique is the electrochemical impedance spectroscopy (EIS), which has been widely used to study the interactions between the microbial and electrochemical systems and it can be applied for the assessment of biofilm biomass (Dominguez-Benetton et al., 2012). It's basic principle lies in measuring an electrochemical interaction on the electrode through tracking the changes in the diffusion coefficient of a redox solute. EIS can be used to detect the developed biofilm using a digital camera (Cachet et al., 2001). Ultrasonic time–domain reflectometry is another physical technique that is extensively used for measuring biofilm thickness (Sim et al., 2013). Biofilm thickness can be estimated by measuring the difference between acoustic impedance on each side of the biofilm interface. Other important physical techniques include nuclear magnetic resonance (NMR) imaging, X-ray computed tomography, and small-angle X-ray/neutron scattering. They can be used to study biofilm structures and provide information regarding shape, size, and orientation of biofilm components rather than determining biofilm thickness (Chen et al., 2004).

3.4.3 Chemical methodologies

Chemical methodologies mainly depend on using dyes or fluorochromes, which have the potential to either bind to or adsorb onto biofilm components. They are indirect methods used for the detection and assessment of specific biofilm components such as EPS constituents. According to Christensen et al. (1985), crystal violet (CV) stain remains the most widely used stain for biofilm quantification, particularly in microtiter plate assays. However, it is only beneficial in quantifying the total biomass of bacterial biofilms because it stains both of live and dead cells as well as some constituents within the biofilm matrix. Moreover, the washing step is required to remove the free unattached cells and the unbound dye. However, dye rigidity may result in detachment of some sessile bacterial cells. Cell detachment during the passage of an air-bubble is greatly dependent on the microbial surface, conditioning film, and the speed of air-bubble passage. Moreover, rinsing and dipping lead to the detachment of an unknown number of sessile microorganisms (Gómez-Suárez et al., 2001). Henceforth, automatic pipetting is preferred than using hands to overcome such problem. Safranin is another stain that is helpful in staining bacterial biomass (Christensen et al., 1982). Concentrated ethanol has been applied to release the unbound dye, however, acetic acid solution (33%) demonstrated better efficiency and a better and quick action. Furthermore, to limit the detachment of a high number of sessile bacterial cells, a fixation step has been recommended using absolute ethanol, methanol, or via heat fixation at 60 °C for 1 h. This step is applied before using dyes for staining (Stepanović et al., 2007). The fixation step can also enhance the reproducibility of the assay. The microtiter plate dye-staining method has three main advantages, including (i) versatility, as it can be applied for a broad range of different bacterial species, as well as eukaryotic cells such as yeasts or fungi; (ii) no detachment of the tested microorganisms from the support as it is required in plate counting method; and

(iii) high-throughput because it permits testing of several conditions altogether. Contrary, some limitations take place, including (i) loose biofilms cause bias estimation of sessile cells due to the washing step; (ii) estimation of any attached bacterial biomass either sessile bacteria at the surface or from the sedimentation/adhesion of planktonic bacterial cells due to gravitation; and (iii) lack of reproducibility.

Colorimetric methodologies have also been applied to investigate cellular physiology in biofilms. The main idea relies on the conversion of specific cellular metabolic substrates into colored products, which can be detected and measured using a spectrophotometer. Koban and co-colleagues showed the ability of XTT [(2,3-bis-(2-methoxy-4-nitro-5-sulfophenyl) 2 Htetrazolium-5-carboxanilide inner salt)], a tetrazolium salt to cleave dehydrogenase enzymes into strongly colored formazan (Koban et al., 2012). Sabaeifard et al. (2014) conducted a study, in which another tetrazolium salt, TTC (2,3,5-triphenyl-2H-tetrazolium chloride) had the capacity to quantify the metabolic activities in biofilms. Resazurin is designated as Alamar Blue and it is a stable redox indicator, which is reduced to resorufin by metabolically active cells. This dye has been reported to be extensively applied for studying microbial biofilms (Van den Driessche et al., 2014). Resazurin has several advantages when compared with tetrazolium salt assays, involving (i) the alteration of the blue non-fluorescent resazurin to the highly fluorescent pink colored resorufin can be observed visually, spectrophotometrically or spectrofluorometrically; (ii) less time-consuming; (iii) resazurin is a harmless indicator to eukaryotic and prokaryotic cells; and (iv) cost-effective.

4. Biofilms and bioremediation

Along with biofilm formation, microorganisms start to attach to a surface and secrete EPSs. Accordingly, microbial phenotypes become modified in both gene transcription and growth rate. In a biofilm, aggregates of microbial cells become strongly and irreversibly attached to surfaces. They grow in moisty and sticky conditions where sufficient nutrients are valid to support their attachment to surfaces. Biofilms are initiated by the attachment of free-floating bacteria to a suitable living or non-living surface. EPSs act as efficient adhesive materials, which facilitate the first step of cellular attachment to surfaces. EPSs provide microorganisms with the required support to adhere closely to each other, which in turn leads to the development of a complex rounded-shaped 3D structure. Multiple environmental parameters determine the extent to which a biofilm can grow (thin or thick layered cells) (Rathinam et al., 2019). It has been found that microorganisms that are capable of releasing high amounts of EPSs can develop and form thick layers of biofilms even if the present nutrients are in small quantities. Among the important environmental factors is shear stress, which also affects biofilm formation. Usually, when there is a water flow, the biofilm becomes fairly thin and in contrast the biofilm becomes very thick when the water flow is slow. The last stage of biofilm growth cycle is the dispersal phase where the biofilm cells can detach from the surface as cell clusters or as individual cells. Then, the detached cells adhere to a new surface and start to create a new biofilm. It has been observed that free-living planktonic bacterial cells possess less capacity for the remediation of pollutants and toxins when compared with their biofilm forming bacterial cells (Heffernan et al., 2009). Though, the reduced resistance of free-floating planktonic bacterial cells in stress conditions has been ascribed to the decreased protection (i.e., absence of EPSs protective sheath) and the decreased metabolic activities (Lerch et al., 2017). Contrary, the presence of the complex polymeric matrix provides bacterial cells growing in a biofilm mode with enough protection against stress conditions. Other biofilm survival strategies include genetic diversity, presence of aerobic and anaerobic microorganisms, and different metabolic pathways.

The diversity in microbes and in the metabolization processes make biofilms trigger more capacity for pollutant bioremediation. In soils, indigenous biofilm-forming microorganisms bioremediate pollutants, which is considered as a self-purification system and a part of the nutrient cycle. Biofilms are extremely beneficial in treating wastewater from heavy metals, hydrocarbons, radioactive substances and persistent organic pollutants (POPs) (Das et al., 2012). This is achieved by converting these lethal substances into less harmful constituents and hence become less hazardous to humans and the surrounding environment. The tremendous environmental pollution around the globe resulted in the contamination of soil, air, and water. The presence of POPs and other pollutants caused successive environmental disasters. As a result, several processes and technologies have been advocated for environmental clean up. Bioremediation is an eco-friendly process, which is can be employed for the detoxification of harmful contaminants using microorganisms. In bioremediation, the destructive influence of complex and hazardous substances on the environment can be minimized. Bioremediation is more advantageous over the other remediation techniques,because it causes nearly no or tiniest disruptions to ecosystems surrounding the treated zone (Chaudhary and Kim, 2019). In addition, the usage of harsh chemicals is avoided. Bioremediation is more economically acceptable for the detoxification and degradation of complex pollutants at a large scale when compared with the other conventional treatment methodologies. Fig. 4 shows the different strategies by which bacteria can remediate toxic heavy metal ions in contaminated environments. These strategies include biosorption, biotransformation, bioaccumlation, bioleaching, bioassilmilation, and bioprecipitation.

Romans were the first to start bioremediation (i.e., around 600 BC) for wastewater treatment. In 1960s, bioremediation was publically performed by George Robinson (Yadav and Chandra, 2020). In this context, bioremediation using biofilm-forming microorganisms proved to be more efficient and a safer alternative than using planktonic microorganisms. This might be attributed to the high survival capacity, resistance, and adaptation against abiotic stress and antimicrobials assisted by the polysaccharide matrix. Two types of bioremediation take place either *in situ* or *ex situ*, relying on the location of the pollutant (Azubuike et al., 2016). *In situ* bioremediation describes the onsite treatment, whereas in *ex*

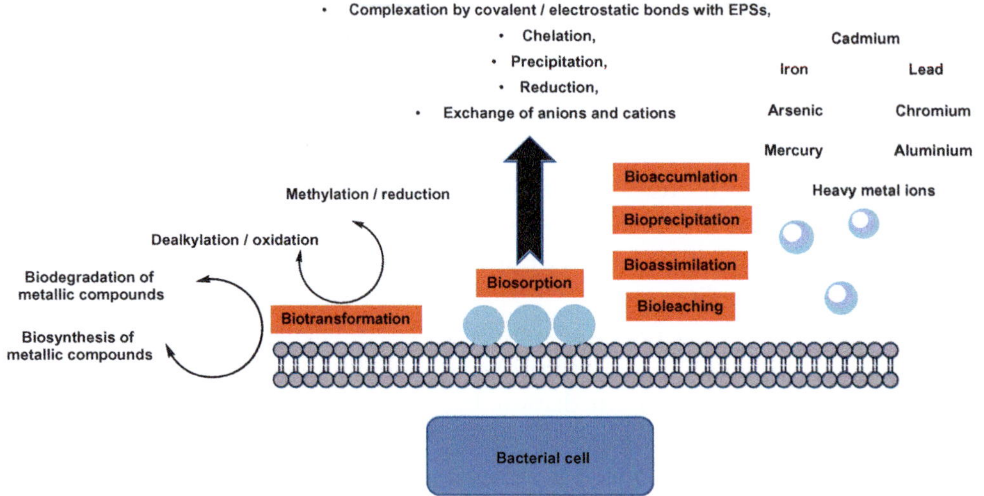

Fig. 4 Different types of interaction between bacterial cells and heavy metal ions in polluted environments.

situ bioremediation, the polluted samples are treated offsite. *In situ* bioremediation is a very interesting treatment strategy than *ex situ* bioremediation because it minimizes the transportation costs and sample disruption (Azubuike et al., 2016). *In situ* bioremediation can be boosted by the optimization of chemical and physical parameters such as pH, moisture, aeration, and nutrients. Lately, industrial waste bioremediation using biofilms in reactors has evolved as a core area of interest for scientists and industrial personnel. Bioremediation depends on the beneficial usage of various microbes to degrade and detoxify xenobiotic compounds such as PAHs, POPs, herbicides, pesticides, heavy metals, dye molecules, volatile organic compounds, crude oil, and petroleum by-products (Maximillian et al., 2019). Among the most persistent complex pollutants in the environment; pentachlorophenols, PAH, dichlorodiphenyltrichloroethane (DDT), polychlorinated biphenyls (PCBs) and 2-benzene, toluene, ethylbenzene, xylene (BTEX), 2-methyl-4-keto-2-trimethylsiloxypentane, 4-ethyl-2-methoxyphenol, ethyl-2-octynoate, 2-methyl-4-keto-2-pentan 2-ol, 3,7-dioxa-2,8-disilanonane-2,2,8,8-tetramethyl5[(trimethylsilyl)oxy], hexadecanoic acid, *cis*-9-hexadecenoic acid, octadecenoic acid, and trinitrotoluene (TNT) (Yadav and Chandra, 2020). Microbes, particularly bacteria, can change the nature of the complex toxic pollutants into less harmful ones. Moreover, fungi possess the potential to remove various pollutants (Singh et al., 2020). Heterogeneity and microbial interactions within biofilms are extremely beneficial to enhance the degradation of complex pollutants via the diverse catabolic pathways. Diverse degradation catabolic pathways are dependent on the microbial enzymatic activities for the transformation and degradation of severe environmental pollutants into less harmless components such as water (H_2O) and carbon dioxide (CO_2). Such metabolic pathways require electron transfer from electron donors to electron acceptors. Electron donors serve as nutrients for the microbes. Microorganisms have the capacity to degrade pollutants in the presence and absence of oxygen. In this context, aerobic and anaerobic degradation can take place. During aerobic degradation, microorganisms use oxygen as the final electron acceptor to degrade organic and inorganic pollutants to less toxic products, usually H_2O and CO_2. Contrary, microbes can break down organic contaminants into CO_2 and methane (CH_4) during the anaerobic degradation where iron, manganese, nitrate, and sulfate act as electron acceptors other than O_2. Normally, aerobic macroorganisms degrade pollutants more rapidly than anaerobic ones. Pollutants that can donate electrons can be degraded in presence of oxygen by aerobic microbes, while those contaminants that lack electron donors can be degraded under anaerobic conditions. Many redox reactions contribute to the immobilization of trace elements in contaminated sites. Additionally, biosorption (microbial- or plant cell-mediated) and biotransformation (enzyme- or metabolite-mediated) are the most extensively explored biological strategies used for the removal of metal and organic hazardous pollutants by biofilm-forming bacteria. The biological remediation processes proved to be more superior rather than the physical and chemical ones in terms of proficiency and economic feasibility.

5. Applications of biofilms in bioremediation of different pollutants

Use of biofilms for *in situ* remediation processes can be applied in several routes. Natural attenuation processes are based upon the natural existed strains without the addition of particular strains for bioremediation. Microbial biofilm communities present in the soil can biotransform certain contaminants into less hazardous components. Natural attenuation is based on the assumption that some pollutants can be degraded, transformed, immobilized, and detoxified under favorable conditions without any interference by humans (Sayler et al., 1995). This type of remediation process involves the use of the existed microbes, which can be present in biofilm mode (Vogt and Richnow, 2013). Natural attenuation technique is usually used when contaminant rates are fairly low. This technique has been used extensively in the remediation of petroleum hydrocarbon sites (Speight, 2018). However, extra nutrients such as sources for carbon and phosphorus and agitation can enhance the microbial activity. Additives or stimuli may be supplied to increase microbial growth rate for rapid degradation of pollutants. This process is referred to as "biostimulation" (Raimondo et al., 2020). In comparison, bioaugmentation or bioenhancement relies on inoculation of different competent microbes or microbial consortia for pollutant degradation (Tribedi et al., 2018). Bioaugmentation is much preferable technique for the treatment of contaminated sites where indigenous microbes can display reduced efficient degradation potential. Bioaugmentation may be enhanced by the introduction of genetically manipulated microbes (Wang and Tam, 2018). A study performed by Liu et al. (2011) showed that bioaugmentation and biostimulation were more effective in remediation for petroleum hydrocarbon oil. Another study conducted by Cui et al. (2014), revealed that both of bioaugmentation and biostimulation improved nitrification efficiency.

Biofilm reactors are characterized by distinctive advantages over the different conventional remediation processes. Among which, retention of highly concentrated biomass for long periods of time, improved metabolic activates, increased flow rates, high tolerance to lethal pollutants, large mass transfer area, improved biodegradation capacity, availability of anoxic and oxic metabolic activities, and decreased interruption within the bioreactor (Asri et al., 2018). In a typical biofilm reactor, a support medium is required to facilitate the adhesion of microbes and biofilm development. Biofilm bioreactors include several types such as batch, continuous stirred tank, trickle bed, air-lift reactors, upflow anaerobic sludge blanket, fluidized bed, expanded granular sludge blanket, and biofilm airlift suspension batch reactors (Mitra and Mukhopadhyay, 2016).

Long et al. (2015) reported the use of X-ray fluorescence and absorption spectra to assess the accumulation, spatial distribution, and reduction of chromium in the biofilms of *Pseudochrobactrum saccharolyticum* LY10. The resultant data indicated the selective accumulation of Cr along with other elements, including Fe, Mn, Ca, K, and Zn in strain LY10 biofilms. After 24h of Cr exposure, it was found that Cr was mainly found in 10–45 μm layer near the bottom membrane as revealed by synchrotron radiation micro-X-ray fluorescence microscopy (SR-lXRF) analysis. X-ray absorption near-edge structure (XANES) spectroscopy analysis further confirmed the bioreduction of Cr(VI) to Cr(III). The precipitate formed in the biofilms was most similar to $CrPO_4$. This study provided a systematic understanding of the accumulation, distribution, and reduction of Cr in LY10 biofilms, which can elaborate the bioremediation efficiencies for Cr detoxification.

An efficient technique was established for the removal of different heavy metal-containing effluents (Wu et al., 2017). In this study, *Camellia oleifera* cake (COC) was used as a culture medium or sorbent for the tested bacteria. Sulfate reducing bacteria (SRB) and *B. cereus* were isolated from heavy metal contaminated wastewater and soil, respectively. COC degradation assisted by *B. cereus* enabled the presence of an anoxic and a reductive environment to SRB. SRB were able to reduce sulfate to hydrogen sulfide, which in turn can precipitate onto the dissolved toxic metals. In natural ecosystems, polymeric organic compounds such as cellulose, starch, fats, proteins, and nucleic acids (DNA and RNA) cannot be used as direct substrates for SRB. Accordingly, SRB depends on other microorganisms to degrade such polymeric substrates and ferment them to other simple products to support their living. In this context, Wu and co-authors used *B. cereus* to degrade COCs into small molecules, which were then indirectly utilized by SRB. Owing to oxygen consumption by the aerobic biofilms of *B. cereus*, a suitable niche for the growth of anaerobic SRB was valid. The products fermented by *B. cereus* such as lactate acted as electron donors for SRB. A series of electrochemical reactions took place during this process. The treatment process was time-dependent and the removal potential increased with the increase in contact time. Almost 97% of heavy metals (e.g., cadmium, copper, and zinc) were removed by the proposed approach. A flotation column was also used as a subsequent separation technique for the collection of metal–laden biomass. The biomass particulates could be reused for another treatment after flotation. This study is considered an intriguing treatment approach via the synergy of biomass and microbes for the removal of heavy metals. A biofilm-based bioreactor was used for the bio-removal of zinc and manganese as reported by Pani et al. (2017). This was achieved using indigenous bacteria isolated from tannery sludge obtained from Ranipet, Tamil Nadu. The isolate was identified as *Pseudomonas beteli* using 16S rRNA gene sequencing. The isolated strain was able to tolerate up to 2000 mg/L of zinc and manganese. Authors further used *P. beteli* to develop a biofilm on the peels of *Cucumis sativus*. Atomic force microscope estimated the uptake efficiency of the formed biofilm to be 69.9% and 78.4% for zinc and manganese, respectively. Images taken by scanning electron microscopye revealed the formation of biofilm on the substrate along with the presence of adsorbed heavy metal ions on the biofilm. The adsorption kinetics followed Freundlich isotherm kinetics.

Zhu et al. (2018) managed to detoxify arsenic via the application of periphytic biofilm. Periphytic biofilm contains both heterotrophic and phototrophic microorganisms, embedded within self-produced slimy matrix of EPSs. A biochar and periphyton-based system (BPS) composed of periphyton bioreactor and a biochar column was established to avoid the toxicity concerns associated with As(III) removal from wastewater. The obtained data revealed that the periphyton was able to grow in environment with less than 5.0 mg/L of As (III). A high As(III) removal percentage was attained by the BPS and the removal rate ranged between 90.2% and 95.4% at a flow rate of 1.0 mL/ min with initial concentration of As(III) (2.0 mg/ L). Approximately 60% of As(III) was pretreated (adsorbed) within the biochar column and the remaining As(III) was removed by the aid of the periphyton bioreactor. The As(III) removal process fitted a pseudo-second-kinetic model. Authors attributed the removal of As(III) to the calcite within the periphytic biofilm surfaces and to the AOH and AC@O functional groups. Based on the resultant data, BPS can be feasible for the detoxification of As(III) practically.

The interactive cooperation between microbial cells and the secreted EPSs in biofilms is crucial to increase the resistance of biofilms toward heavy metals. Nonetheless, the exact resistance mechanisms of biofilm components have not been fully understood. Lin et al. (2020) conducted a study in which *Pseudomonas putida* CZ1 was isolated from the rhizosphere of *Elsholtzia splendens* (a copper-tolarable plant). In this study, the spatial distribution of copper in colloidal and capsular EPS, cell walls, and membranes of *P. putida* CZ1 biofilms were significantly assessed at the subcellular level. Results showed that 60%–67% of copper was positioned in the extracellular portion of biofilms, with 44.7%–42.3% within the capsular EPS. Besides, fractions ranging between

(15.5%–20.1%) and (17.2%–21.2%) of copper were found in the cell walls and membranes, respectively. X-ray absorption spectral analysis revealed that copper was mainly bound by carboxyl-, phosphate and hydrosulfide-like bondages inside the extracellular polymeric matrix, cell walls, and membranes. Fourier transform infrared spectroscopy spectra (FTIR) and sulfur K-edge X-ray absorption near edge structure analysis further confirmed the presence of carboxyl-rich acidic polysaccharides in EPS, phospholipids in the cell walls and membranes, with thiol rich intracellular proteins. Hence, this study introduced an opportunity for understanding the interaction mechanisms between biofilms and heavy metals for the effective use of biofilms in the remediation of heavy metal ions.

Raghavan and co-authors introduced a novel approach for heavy metal bioremediation using axenic cultures of the photoautotrophic cyanobacterium, *Nostoc muscorum* (Raghavan et al., 2020). *N. muscorum* was immobilized onto a glass surface to form a biofilm mode of growth. As mentioned earlier, biofilm growth modes exhibit a high cellular integrity upon exposure to high concentrations of harsh industrial effluents rather than free cell suspension cultures. Because *N. muscorum* is a photosynthetic organism, biofilm was attained with minimal nutritional requirements The strong adherence of *N. muscorum* biofilm onto the glass surface was beneficial at many levels, including absence of cell mass diffusion even with the use of top-end stirrer, no costly separation techniques were needed for the separation of biomass from the treated solution, and better interaction was attained between the biofilm surface and the metal-containing solution. The *N. muscorum* biofilm had the potential to sequester Cd(II) at pH range 5–9 with a concentration range $<0.05\,mg/L$ to $100\,mg/L$ from different waste waters. The mode of Cd(II) sequestration was chemisorption through the $C{=}O$ and $C{=}N$ functional groups. Cd(II) sequestration followed the Langmuir adsorption isotherm..

The most common types of inorganic nitrogen ions are ammonia (NH^{4+}), nitrate (NO^{3-}) and nitrite (NO^{2-}). Inorganic nitrogen pollution of water ecosystems caused several environmental threats like water acidification, eutrophication, incidence of toxic algal blooms, and direct toxicity toward aquatic fauna. Additionally, such pollution can have negative effects on human health and economy (Camargo and Alonso, 2006). Henceforth, an urgent need is necessary to apply an effective approach to eliminate inorganic nitrogen pollution. In this context, Hong et al. (2020) conducted a study in which a strain with high biofilm-formation and aerobic denitrification capabilities was isolated and was identified as *Pseudomonas mendocina* IHB602. In pure culture, the strain IHB602 was capable to remove almost all NO_3^--N, NO_2^--N, and $NH^{4+}-N$ (with an initial concentration $50\,mg/L$) within $24\,h$. The strain secreted large amounts of EPSs (maximum $430.33\,mg/g$ cell dry weight). The produced EPSs was rich in protein with almost no humic acid. The high concentrations of EPSs, strong auto-aggregation and hydrophobicity imparted the strain IHB602 with biofilm-forming potential. Additionally, it was observed that a batch biofilm reactor bioaugmented with strain IHB602 (SBBR1) managed to form biofilms stronger than the control without inoculation of strain IHB602 inoculation (SBBR2). Accordingly, the removal efficiency of nitrogen-containing ions was significantly better in SBBR1 than in SBBR2.

The menace of dye-containing wastewater has reached extreme high levels and as result, considerable attention has been directed toward the remediation of discharged effluents containing synthetic dyes. In this theme, Ong et al. (2020) introduced a study in which wastewater containing mono azo dye, reactive orange 16, was treated via a novel integrated anoxic-aerobic REACT sequencing batch moving bed biofilm reactor (SBMBBR). Two schemes were designed, involving (i) dye concentration under shock load approach, and (ii) hydraulic retention time (HRT) coupled with increasing bio carrier filling ratio. Results showed a complete biodecolorization , with more than 97% of chemical oxygen demand (COD) removal during scheme (i). A three-fold enhancement of microbial strength was observed during the removal of RO16 molecules (i.e., from 0.005 to 0.015 mg RO16/mg biomass). This was congruent with the improvement in microbial strength in terms of mg COD/mg biomass (i.e., from 0.0677 to 0.0711). In scheme (ii), a discernible decrease in COD and RO16 removal efficiencies were observed (i.e., both COD and RO16 removals were decreased to approximately 40%). The decrease in COD removal occurred in accordance with the decline in microbial strength from 0.071 to 0.0452. The presence of the EPS in the formed biofilm constructed a shield to protect the microbial biomass system against the toxic shocks of RO16. Accordingly, the attached-growth biomass system (i.e., biofilm mode) provided more advantages more than the suspended-growth biomass systems.

6. Conclusions

Microorganisms have the potential to build up flocs of microbial communities in the form of biofilms. Biofilms are stable and organized communities containing single- or multi-bacterial species in addition to some eukaryotes. The rapid development in several industrial disciplines has triggered significant threats to the environment over the last few decades. Among which, heavy metals, dyestuffs, pesticides, chemical fertilizers, lubricants, disinfectants, polyaromatic hydrocarbons, antibiotics, paints, detergents, polychlorinated biphenyls, etc. These lethal contaminants need to be treated and monitored to achieve sustainable environmental developments. As a result, numerous studies have investigated the use of biofilms as promising tools for the bioremediation of these potentially hazardous contaminants. The use of microorganisms in biofilm mode of growth as well as the secreted EPSs to degrade and clean up a broad variety of pollutants is beneficial. No disruption takes place in the polluted site and no release of hazardous byproducts. In this chapter, formation of biofilm, biofilm components, and diversity of biofilms have been overviewed. The latest investigations dealing with the possible applications of biofilms in bioremediation processes have been discussed.

References

Anderson-Glenna, M.J., Bakkestuen, V., Clipson, N.J.W., 2008. Spatial and temporal variability in epilithic biofilm bacterial communities along an upland river gradient. FEMS Microbiol. Ecol. 64 (3), 407–418.

Asri, M., et al., 2018. Biofilm-based systems for industrial wastewater treatment. In: Handbook of Environmental Materials Management. Springer, Cham, pp. 1–21.

Azeredo, J., et al., 2017. Critical review on biofilm methods. Crit. Rev. Microbiol. 43 (3), 313–351.

Azubuike, C.C., Chikere, C.B., Okpokwasili, G.C., 2016. Bioremediation techniques—classification based on site of application: principles, advantages, limitations and prospects. World J. Microbiol. Biotechnol. 32 (11), 1–18.

Bäuerle, T., et al., 2018. Self-organization of active particles by quorum sensing rules. Nat. Commun. 9 (1), 1–8.

Cachet, H., et al., 2001. Characterization of deposits by direct observation and by electrochemical methods on a conductive transparent electrode. Application to biofilm and scale deposit under cathodic protection. Electrochim. Acta 46 (24–25), 3851–3857.

Camargo, J.A., Alonso, Á., 2006. Ecological and toxicological effects of inorganic nitrogen pollution in aquatic ecosystems: a global assessment. Environ. Int. 32 (6), 831–849.

Ceri, H., et al., 1999. The calgary biofilm device: new technology for rapid determination of antibiotic susceptibilities of bacterial biofilms. J. Clin. Microbiol. 37 (6), 1771–1776.

Chaudhary, D.K., Kim, J., 2019. New insights into bioremediation strategies for oil-contaminated soil in cold environments. Int. Biodeterior. Biodegrad. 142 (December 2018), 58–72.

Chauhan, A., Oakeshott, J.G., Jain, R.K., 2008. Bacterial metabolism of polycyclic aromatic hydrocarbons: strategies for bioremediation. Indian J. Microbiol. 48 (1), 95–113.

Chavant, P., et al., 2007. A new device for rapid evaluation of biofilm formation potential by bacteria. J. Microbiol. Methods 68 (3), 605–612.

Chen, V., Li, H., Fane, A.G., 2004. Non-invasive observation of synthetic membrane processes—a review of methods. J. Membr. Sci. 241 (1), 23–44.

Christensen, G.D., et al., 1982. Adherence of slime-producing strains of *Staphylococcus epidermidis* to smooth surfaces. Infect. Immun. 37 (1), 318–326.

Christensen, G.D., et al., 1985. Adherence of coagulase-negative *Staphylococci* to plastic tissue culture plates: a quantitative model for the adherence of staphylococci to medical devices. J. Clin. Microbiol. 22 (6), 996–1006.

Costerton, J.W., 1995. Overview of microbial biofilms. J. Ind. Microbiol. 15 (3), 137–140.

Costerton, J.W., Stewart, P.S., 2000. Biofilms and device-related infections. In: Persistent Bacterial Infections. American Society of Microbiology, pp. 423–439.

Costerton, J.W., Geesey, G.G., Cheng, K.-J., 1978. How bacteria stick. Sci. Am. 238 (1), 86–95.

Costerton, J.W., Stewart, P.S., Greenberg, E.P., 1999. Bacterial biofilms: a common cause of persistent infections. Science 284 (5418), 1318–1322.

Cui, D., et al., 2014. Improvement of nitrification efficiency by bioaugmentation in sequencing batch reactors at low temperature. Front. Environ. Sci. Eng. 8 (6), 937–944.

Das, N., et al., 2012. Application of biofilms on remediation of pollutants—an overview. J. Microbiol. Biotechnol. Res. 2, 783–790.

Delangiz, N., et al., 2020. Beneficial microorganisms in the remediation of heavy metals. In: Molecular Aspects of Plant Beneficial Microbes in Agriculture. Elsevier, pp. 417–423.

Dominguez-Benetton, X., et al., 2012. The accurate use of impedance analysis for the study of microbial electrochemical systems. Chem. Soc. Rev. 41 (21), 7228–7246.

Duruibe, J.O., Ogwuegbu, M.O.C., Egwurugwu, J.N., 2007. Heavy metal pollution and human biotoxic effects. Int. J. Phys. Sci. 2 (5), 112–118.

Dzionek, A., Wojcieszyńska, D., Guzik, U., 2016. Natural carriers in bioremediation: a review. Electron. J. Biotechnol. 19 (5), 28–36.

Fittipaldi, M., Nocker, A., Codony, F., 2012. Progress in understanding preferential detection of live cells using viability dyes in combination with DNA amplification. J. Microbiol. Methods 91 (2), 276–289.

Fletcher, M., 1977. The effects of culture concentration and age, time, and temperature on bacterial attachment to polystyrene. Can. J. Microbiol. 23 (1), 1–6.

Fuqua, C., Greenberg, E.P., 1998. Self perception in bacteria: quorum sensing with acylated homoserine lactones. Curr. Opin. Microbiol. 1 (2), 183–189.

Fuqua, W.C., Winans, S.C., Greenberg, E.P., 1994. Quorum sensing in bacteria: the LuxR-LuxI family of cell density-responsive transcriptional regulators. J. Bacteriol. 176 (2), 269275.

Gautam, R.K., Soni, S., Chattopadhyaya, M.C., 2015. Functionalized magnetic nanoparticles for environmental remediation. In: Handbook of Research on Diverse Applications of Nanotechnology in Biomedicine, Chemistry, and Engineering. IGI Global, pp. 518–551.

Goeres, D.M., et al., 2009. A method for growing a biofilm under low shear at the air–liquid interface using the drip flow biofilm reactor. Nat. Protoc. 4 (5), 783–789.

Gómez-Suárez, C., Busscher, H.J., van der Mei, H.C., 2001. Analysis of bacterial detachment from substratum surfaces by the passage of air-liquid interfaces. Appl. Environ. Microbiol. 67 (6), 2531–2537.

Gupta, P., Diwan, B., 2017. Bacterial exopolysaccharide mediated heavy metal removal: a review on biosynthesis, mechanism and remediation strategies. Biotechnol. Rep. 13, 58–71.

Heffernan, B., Murphy, C.D., Casey, E., 2009. Comparison of planktonic and biofilm cultures of *Pseudomonas fluorescens* DSM 8341 cells grown on fluoroacetate. Appl. Environ. Microbiol. 75 (9), 2899–2907.

Holdgate, M.W., 1979. A Perspective of Environmental Pollution. Cambridge University Press, p. 169.

Hong, P., et al., 2020. Bioaugmentation treatment of nitrogen-rich wastewater with a denitrifier with biofilm-formation and nitrogen-removal capacities in a sequencing batch biofilm reactor. Bioresour. Technol. 303 (7), 122905–122914.

Igiri, B.E., et al., 2018. Toxicity and bioremediation of heavy metals contaminated ecosystem from tannery wastewater: a review. J. Toxicol. 2018, 2568038.

Jackson, S., et al., 2014. Biofilm development by blastospores and hyphae of *Candida albicans* on abraded denture acrylic resin surfaces. J. Prosthet. Dent. 112 (4), 988–993.

Jagadeesh, N.M., Karabasappa, M., 2020. Control of microbial biofilms: application of natural and synthetic compounds. In: New and Future Developments in Microbial Biotechnology and Bioengineering: Microbial Biofilms. Elsevier, pp. 101–115.

Karn, S.K., Duan, J., Jenkinson, I.R., 2017. Book review: role of biofilms in bioremediation. Front. Environ. Sci. 5, 22.

Klein, M.I., et al., 2012. Molecular approaches for viable bacterial population and transcriptional analyses in a rodent model of dental caries. Mol Oral Microbiol 27 (5), 350–361.

Koban, I., et al., 2012. XTT assay of ex vivo saliva biofilms to test antimicrobial influences. GMS Krankenhaushyg. Interdiszip. 7 (1), 1–10.

Kornegay, B.H., Andrews, J.F., 1968. Kinetics of fixed-film biological reactors. J. Water Pollut. Control Fed. 40, R460–R468.

Krüger, N.-J., et al., 2014. "Limits of control"-crucial parameters for a reliable quantification of viable Campylobacter by real-time PCR. PLoS One 9 (2), e88108.

Lear, G., Lewis, G.D., 2012. Microbial Biofilms: Current Research and Applications. Horizon Scientific Press.

Lerch, T.Z., et al., 2017. Biofilm vs. planktonic lifestyle: consequences for pesticide 2,4-D metabolism by Cupriavidus necator JMP134. Front. Microbiol. 8, 1–11.

Li, L., et al., 2014. The importance of the viable but non-culturable state in human bacterial pathogens. Front. Microbiol. 5 (June), 1.

Lin, H., et al., 2020. A subcellular level study of copper speciation reveals the synergistic mechanism of microbial cells and EPS involved in copper binding in bacterial biofilms. Environ. Pollut. 263, 114485–114495.

Liu, P.-W.G., et al., 2011. Bioremediation of petroleum hydrocarbon contaminated soil: effects of strategies and microbial community shift. Int. Biodeterior. Biodegradation 65 (8), 1119–1127.

Long, D., et al., 2015. Determination of the accumulation, spatial distribution and reduction of Cr in unsaturated *Pseudochrobactrum saccharolyticum* LY10 biofilms by X-ray fluorescence and absorption methods. Chem. Eng. J. 280, 763–770.

Maximillian, J., et al., 2019. Pollution and environmental perturbations in the global system. In: Environmental and Pollution Science, third ed. Elsevier Inc.

McCoy, W.F., et al., 1981. Observations of fouling biofilm formation. Can. J. Microbiol. 27 (9), 910–917.

Mitra, A., Mukhopadhyay, S., 2016. Biofilm mediated decontamination of pollutants from the environment. AIMS Bioeng. 3 (1), 44–59.

Mohapatra, R.K., et al., 2017. Biodetoxification of toxic heavy metals by marine metal resistant bacteria-a novel approach for bioremediation of the polluted saline environment. In: Microbial Biotechnology. Springer, pp. 343–376.

Mohapatra, R.K., et al., 2020. Potential application of bacterial biofilm for bioremediation of toxic heavy metals and dye-contaminated environments. In: New and Future Developments in Microbial Biotechnology and Bioengineering: Microbial Biofilms. Elsevier B.V, pp. 267–281.

Nazir, R., Zaffar, M.R., Amin, I., 2019. Bacterial biofilms: the remarkable heterogeneous biological communities and nitrogen fixing microorganisms in lakes. In: Freshwater Microbiology: Perspectives of Bacterial Dynamics in Lake Ecosystems. Elsevier Inc, pp. 3017–3340.

Newton, R.J., et al., 2011. A guide to the natural history of freshwater lake bacteria. Microbiol. Mol. Biol. Rev. 75 (1), 14–49.

Nocker, A., et al., 2009. Selective detection of live bacteria combining propidium monoazide sample treatment with microarray technology. J. Microbiol. Methods 76 (3), 253–261.

Oliveira, F., et al., 2015. Evidence for inter-and intraspecies biofilm formation variability among a small group of coagulase-negative staphylococci. FEMS Microbiol. Lett. 362 (20), fnv175.

Ong, C., Lee, K., Chang, Y., 2020. Biodegradation of mono azo dye-reactive Orange 16 by acclimatizing biomass systems under an integrated anoxic-aerobic REACT sequencing batch moving bed biofilm reactor. J. Water Process Eng. 36, 101268–101282.

Pandey, V.C., Singh, V., 2019. Exploring the Potential and Opportunities of Current Tools for Removal of Hazardous Materials from Environments, Phytomanagement of Polluted Sites. Elsevier Inc., pp. 501–516.

Pani, T., Das, A., Osborne, J.W., 2017. Bioremoval of zinc and manganese by bacterial biofilm: a bioreactor-based approach. J. Photochem. Photobiol. B Biol. 175 (August), 211–218.

Prigent-Combaret, C., et al., 2000. Developmental pathway for biofilm formation in curli-producing *Escherichia coli* strains: role of flagella, curli and colanic acid. Environ. Microbiol. 2 (4), 450–464.

Raghavan, P.S., et al., 2020. Axenic cyanobacterial (*Nostoc muscorum*) biofilm as a platform for Cd(II) sequestration from aqueous solutions. Algal Res. 46, 101778–101790.

Raimondo, E.E., et al., 2020. Coupling of bioaugmentation and biostimulation to improve lindane removal from different soil types. Chemosphere 238, 124512–124523.

Rathinam, N.K., et al., 2019. Role of exopolysaccharides in biofilm formation. ACS Symp. Ser. 1323, 17–57.

Reddy, S.S., Kotaiah, B., 2006. Comparative evaluation of commercial and sewage sludge based activated carbons for the removal of textile dyes from aqueous solutions. Iran. J. Environ. Health Sci. Eng. 3 (4), 239–246.

Romaní, A.M., et al., 2014. Shifts in microbial community structure and function in light-and dark-grown biofilms driven by warming. Environ. Microbiol. 16 (8), 2550–2567.

Sabaeifard, P., et al., 2014. Optimization of tetrazolium salt assay for *Pseudomonas aeruginosa* biofilm using microtiter plate method. J. Microbiol. Methods 105, 134–140.

Samanta, S.K., Singh, O.V., Jain, R.K., 2002. Polycyclic aromatic hydrocarbons: environmental pollution and bioremediation. Trends Biotechnol. 20 (6), 243–248.

Sayler, G.S., et al., 1995. Molecular site assessment and process monitoring in bioremediation and natural attenuation. Appl. Biochem. Biotechnol. 54 (1–3), 277–290.

Seo, J.-S., Keum, Y.-S., Li, Q.X., 2009. Bacterial degradation of aromatic compounds. Int. J. Environ. Res. Public Health 6 (1), 278–309.

Shukla, S.K., et al., 2014. Biofilm-mediated bioremediation of polycyclic aromatic hydrocarbons. In: Microbial Biodegradation and Bioremediation. Elsevier, pp. 203–232.

Shukla, S.K., et al., 2017. Bacterial biofilms and genetic regulation for metal detoxification. In: Handbook of Metal-Microbe Interactions and Bioremediation. CRC Press, p. 317 (Chapter 19).

Siddiquee, S., et al., 2015. Heavy metal contaminants removal from wastewater using the potential filamentous fungi biomass: a review. J. Microb. Biochem. Technol. 7 (6), 384–395.

Sim, S.T.V., et al., 2013. Monitoring membrane biofouling via ultrasonic time-domain reflectometry enhanced by silica dosing. J. Membr. Sci. 428, 24–37.

Singh, S.K., et al., 1991. Air pollution tolerance index of plants. J. Environ. Manag. 32 (1), 45–55.

Singh, R., Paul, D., Jain, R.K., 2006. Biofilms: implications in bioremediation. Trends Microbiol. 14 (9), 389–397.

Singh, R.K., et al., 2020. Fungi as potential candidates for bioremediation. In: Abatement of Environmental Pollutants. Elsevier Inc, pp. 177–191.

Sinha, S., et al., 2018. Removal of Congo red dye from aqueous solution using Amberlite IRA-400 in batch and fixed bed reactors. Chem. Eng. Commun. 205 (4), 432–444.

Speight, J.G., 2018. Biological transformations. In: Reaction Mechanisms in Environmental Engineering. Elsevier.

Stepanović, S., et al., 2007. Quantification of biofilm in microtiter plates: overview of testing conditions and practical recommendations for assessment of biofilm production by *Staphylococci*. APMIS 115 (8), 891–899.

Stewart, P.S., 1996. Theoretical aspects of antibiotic diffusion into microbial biofilms. Antimicrob. Agents Chemother. 40 (11), 2517–2522.

Stoodley, P., et al., 1998. Influence of hydrodynamics and nutrients on biofilm structure. J. Appl. Microbiol. 85 (S1), 19S–28S.

Sträuber, H., Müller, S., 2010. Viability states of bacteria—specific mechanisms of selected probes. Cytometry A 77 (7), 623–634.

Tavernier, S., Coenye, T., 2015. Quantification of *Pseudomonas aeruginosa* in multispecies biofilms using PMA-qPCR. PeerJ 3, e787.

Tribedi, P., et al., 2018. Bioaugmentation and biostimulation: a potential strategy for environmental remediation. J. Microbiol. Exp. 6 (5), 223–231.

Trulear, M.G., Characklis, W.G., 1982. Dynamics of biofilm processes. J. Water Pollut. Control Fed. 54, 1288–1301.

Turan, N.B., et al., 2017. Quorum sensing: little talks for an effective bacterial coordination. Trends Anal. Chem. 91, 1–11.

Van den Driessche, F., et al., 2014. Optimization of resazurin-based viability staining for quantification of microbial biofilms. J. Microbiol. Methods 98, 31–34.

Vogt, C., Richnow, H.H., 2013. Bioremediation via in situ microbial degradation of organic pollutants. In: Geobiotechnology II. Springer, pp. 123–146.

Wagner, V.E., et al., 2006. Quorum sensing: dynamic response of *Pseudomonas aeruginosa* to external signals. Trends Microbiol. 14 (2), 55–58.

Wang, Y., Tam, N.F.Y., 2018. Microbial remediation of organic pollutants. In: World Seas: An Environmental Evaluation Volume III: Ecological Issues and Environmental Impacts, second ed. Elsevier Ltd.

Wilhelm, L., et al., 2014. Rare but active taxa contribute to community dynamics of benthic biofilms in glacier-fed streams. Environ. Microbiol. 16 (8), 2514–2524.

Worthington, R.J., Richards, J.J., Melander, C., 2012. Small molecule control of bacterial biofilms. Org. Biomol. Chem. 10 (37), 7457–7474.

Wu, M., et al., 2017. Decontamination of multiple heavy metals-containing effluents through microbial biotechnology. J. Hazard. Mater. 337, 189–197.

Yadav, S., Chandra, R., 2020. Biofilm-mediated bioremediation of pollutants from the environment for sustainable development. In: New and Future Developments in Microbial Biotechnology and Bioengineering: Microbial Biofilms. Elsevier B.V, pp. 177–203.

Zhao, X., Yu, Z., Ding, T., 2020. Quorum-sensing regulation of antimicrobial resistance in bacteria. Microorganisms, 1–21.

Zhu, N., et al., 2018. Arsenic removal by periphytic biofilm and its application combined with biochar. Bioresour. Technol. 248, 49–55.

CHAPTER 8

Endophytic bacteria in a biocontrol perspective

Riddha Dey[a] and Richa Raghuwanshi[b,*]
[a]Department of Botany, Institute of Science, Banaras Hindu University, Varanasi, UP, India
[b]Department of Botany, Mahila Mahavidyalaya, Banaras Hindu University, Varanasi, UP, India
*Corresponding author: e-mail address:

1. Introduction

In the present scenario, pest management is a serious global concern. Loss of 30% of the total yield is perceived in the major crops due to phytoinfections (Savary et al., 2019). To keep up with the increasing global population, food security will be one of the severe challenges in upcoming years. Therefore, further increase in the efficiency of plant disease control and reduction of yield loss can significantly contribute to the increasing demand for food worldwide (Valin et al., 2014). Biological control drew the attention of the researchers over the past few years driven by the need for an alternative form of chemical pesticides as the prolonged use of chemical pesticides harms the resistant microbial population immensely as well as damages the soil health. The concept of biological control is both system favorable and environmentally sustainable.

Bacteria colonizing the internal tissues of healthy plants can enhance plant growth and health making it appropriate biocontrol agents (Frommel et al., 1991; Kloepper et al., 1992; Turner et al., 1993; Sturz et al., 1999). The life cycle of bacterial endophytes is much shorter than the host plant due to which they can evolve faster; consequently, they have a potential selection of antagonistic forms against the phytopathogens (Carroll, 1988). There is substantial literature describing the prospects of endophytic bacteria managing plant health. In spite of occupying different ecological niches, the bacterial endophytes manifest some common mechanisms like production of inhibitory allelochemicals, competition, detoxification, induced systemic resistance to control the broad-spectrum phytopathogens (Haas et al., 2002; Ryu et al., 2004). The chapter focuses on the biocontrol activity of bacterial endophytes and provides information about the mechanisms used by the endophytes to impart defense to the host plants as well as improve phytoimmunity. It also emphasizes the strategies that can be employed to improve the productivity of crops through manipulating plant microbial ecosystems. A comprehensive study about the bacterial endophytes helps to understand its ecological role as well as the host-microbe interaction. This will aid to identify useful compounds produced or the mechanisms imparted through the biotechnological applications.

Relationship Between Microbes and the Environment for Sustainable Ecosystem Services, Volume 1
https://doi.org/10.1016/B978-0-323-89938-3.00008-6
     155

2. Biocontrol activity of endophytes

Several researchers reported that inoculation of bacterial endophytes inside the plant species can inhibit diseases caused by the wide variety of phytopathogens (Ping and Boland, 2004; Berg and Hallmann, 2006). The intimate relationship between host and endophytic bacteria makes them promising and potential biocontrol agents (Chen et al., 1995; van Buren et al., 1993). These bacteria occupy the same niche as phytopathogens and compete with the pathogens acting as a biocontrol agent (Hallmann et al., 1997; Berg et al., 2005). Ryan et al. (2008) anticipated that bacterial endophytes can be sustainably transmitted to the next generation making it more suitable as a biocontrol agent compared to the rhizospheric bacteria. On the other hand, rhizospheric bacteria attribute poor rhizosphere competence which makes it an unstable biocontrol agent (Schroth et al., 1984; Weller, 1988). In study of 61 out of 192 isolated endophytic bacteria from potato, stem tissue exhibited significant biocontrol activity against *Clavibacter michiganensis* subsp. *Sepedonicus* (van Buren et al., 1993). Bacterial endophytes were reported to implicate symptom inhibition in the case of bacterial blight of rice (Poon et al., 1977), as well as they, were also biologically active against oak wilt pathogen (Brooks et al., 1994). Microbial endophytes also colonize both external and internal tissues to enhance the phytoimmunity. Chen et al. (1995) isolated 170 endophytes from cotton tissues, among which 40 showed biocontrol activity and 25 of them induced systemic resistance against *Colletotrichum orbiculare* in cucumber. Various bacterial endophytes reported as biocontrol agents are enlisted in Table 1. The use of beneficial bacteria as a plant protection tool in integrated pest management depends on various factors like isolation and identification of the BCAs, mode of incorporation, mechanism of biocontrol activity, prolonged functional activity, commercial implementation, and cost-efficiency. Management of diseases through biological control should have a holistic approach rather than focusing on single crop disease.

3. Mechanisms of biocontrol

The bacterial endophytes modulate the morphology and physiology of the plants and undergo specific mechanisms to check the growth of phytopathogens and the damage done by them. The various mechanisms adopted by the biocontrol agents are discussed below and illustrated in Fig. 1.

3.1 Production of siderophores

Growth of plants highly depends on the bioavailability of iron. It is an essential growth element for living organisms. The scarcity of iron leads to high competition in the soil niche. In such conditions, bacterial endophytes produce siderophores which is a low molecular weight compound and acquire ferric ions competitively (Whipps, 2001).

Table 1 Possible mechanisms of disease control by bacterial endophytes.

Sl. No.	Disease	Causal organism	Symptoms	Biocontrol agent	Host plant	Mechanism	References
1.	Root rots	*Fusarium solani*	Roots turn brown or black, desiccation of root tips	*Pseudomonas stutzeri* YPL–1	Ginseng rhizosphere	Production of lytic extracellular enzymes	Lim et al. (1991)
2.	Oak wilt	*Ceratocystis fagacearum*	Yellow or brown coloration of leaves, curling around midrib, immature shedding	*Pseudomonas denitrificans*	*Quercus fusiformis*	Competition	Brooks et al. (1994)
3.	Chestnut blight	*Cryphonectria parasitica*	Reddish-brown lesions and cracked cankers	*Bacillus subtilis*	Chestnut trees	Growth suppression	Wilhelm et al. (1997)
4.	Red rot disease of sugarcane	*Colletotrichum falcatum*	Sudden discoloration from orange to yellow, leaves gradually dry, red blotches, and white patches in stem	*Pseudomonas fluorescens* EP1	Sugarcane	Induced systemic resistance	Viswanathan and Samiyappan (1999)
5.	Fusarium Wilt	*Fusarium oxysporum*	Stunted growth, discoloration of plant parts, dark streaks in the xylem vascular tissue	*Streptomyces* sp. strain NRRL 30562	*Kennedia nigriscans*	Antibiosis	Castillo et al. (2002)

Continued

Table 1 Possible mechanisms of disease control by bacterial endophytes—cont'd

Sl. No.	Disease	Causal organism	Symptoms	Biocontrol agent	Host plant	Mechanism	References
6.	Blackleg of potatoes	*Erwinia carotovora* subsp. *atroseptica*	Inky black and slimy soft rot of stems	*Clavibacter michiganensis*	Potato	Quorum sensing	Reiter et al. (2002)
7.	Gray mold	*Botrytis cinerea*	Circular gray-brown spots on leaves and stems, underdeveloped flowers and buds, premature desiccation	*Serratia plymuthica*	Melon	Lysis by production of chitinolytic enzyme	Kamensky et al. (2003)
8.	Verticillium wilt	*Verticillium dahliae*	Leaves turn yellow, leaf margins turn brown and appear scorched, wilting of leaf, a decline in new twig formation	*Paenibacillus* K165	Eggplant, potato	Induced systemic resistance	Tjamos et al. (2004)
9.	White root rot of Avocado	*Rosellinia necatrix*	Small, pale green leaves, leaf wilting	*Pseudomonas pseudoalcaligenes* AVO110	Avocado	Competition for nutrients and niches	Pliego et al. (2008)
10.	Black pepper root rot	*Phytophthora capsici*	Stunted growth, root appear black or brown, decay of root tips	*Pseudomonas aeruginosa, Pseudomonas putida*	*Piper nigrum*	Secreation of lytic enzymes	Aravind et al. (2012)
11.	Necrosis	*Botrytis cinerea* B05–10	Foliar infection, wilting, decolorization, and shedding of premature leaves	*Trichoderma hamatum T382*	*Arabidopsis thaliana*	Induced systemic resistance	Mathys et al. (2012)

12.	Anthracnose in strawberry	*Colletotrichum gloeosporioides*	Whitish or light brown lesions, discoloration of the stem, wilting	*Bacillus amyloliquefaciens* strain S13–3	Strawberry	Induced systemic resistance	Yamamoto et al. (2015)
13.	Charcoal rot in soybean	*Macrophomina phaseolina*	Stunted or wilted plants in patches, discoloration	*Bacillus* sp. PGPBacCA1	Soybean	Antibiosis	Torres et al. (2016)
14.	Grapevine crown gall	*Agrobacterium vitis*	Wilting and yellowing of the canopy, stunted shoot growth, and cluster desiccation	*Curtobacterium flaccumfaciens*	*Vitis vinifera*	Antibiosis	Ferrigo et al. (2017)
15.	Bacterial canker in kiwifruit	*Pseudomonas syringae* pv. *actinidiae*	Necrotic spots on leaves, canker, and dieback on canes and trunks, twig wilting	*Pseudomonas* sp. T1R21 *Pseudomonas* sp. T4MS32AP *Pseudomonas* sp. T4MS33	*Leptospermum scoparium*	Antibiosis	Wicaksono et al. (2018)

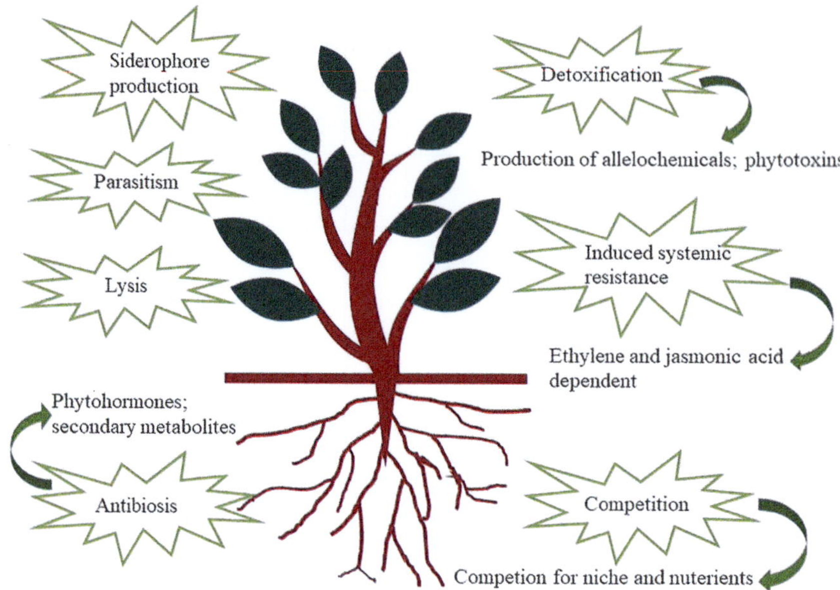

Fig. 1 Mechanisms adopted by biocontrol agents.

Some of the beneficial bacteria withdraw iron ions from heterologous siderophores, produced by the surrounding microbes (Loper and Henkels, 1999). Siderophore production is modulated by several factors like iron availability, iron ion production, availability of carbon, phosphate, and nitrogen, pH (Duffy and Defago, 1999). The regulation of siderophore synthesis is done by the global regulators (GacA and GacS), Fur proteins which are iron sensitive, quorum-sensing autoinducers (N-acyl homoserine lactone), sigma factors (RpoS, PvdS, and FpvI), and site-specific recombinases (Cornelis and Matthijs, 2002; Ravel and Cornelis, 2003).

3.2 Antibiosis

Antibiosis is a process of biocontrol mechanism which is achieved by the production of secondary metabolites like volatile compounds and antibiotics by the beneficial microbes which help to check the growth of pathogens in the host plants (Fravel, 1988). Antibiotics like D-gluconic acid, 2-hexyl-5-propyl resorcinol, DMDS, and the volatiles 2,3-butanediol, 6-pentyl-α-pyrone, volatile HCN, which are produced by bacterial endophytes are responsible for antibiosis (Haas and Defago, 2005; Dandurishvili et al., 2011). Pseudomonads produce amphisin, 2,4-diacetylphloroglucinol (DAPG), hydrogen cyanide, oomycin A, phenazine, pyoluteorin, pyrrolnitrin, tensin, tropolone, and cyclic lipopeptides which facilitate antibiosis (de Souza et al., 2003; Nielsen et al., 2002; Nielsen and Sørensen,

2003; Raaijmakers et al., 2002). *Bacillus*, *Streptomyces*, and *Stenotrophomonas* sp. produce antibiotics such as oligomycin A, kanosamine, zwittermicin A, and xanthobaccin causing faster antibiosis (Hashidoko et al., 1999; Kim et al., 1999; Nakayama et al., 1999). *Bacillus* widely secretes peptide antibiotics, low molecular weight volatile compounds, and lipopeptides which are antagonistic to phytopathogenic agents (Tabbene et al., 2009; Zhang et al., 2013). The antibiosis mechanism of iturins is reported to suppress *Pectobacterium carotovorum* and *Xanthomonas campestris* (Zeriouh et al., 2011). Toure et al. (2004) studied that the production of fungicin can check gray mold disease caused by *Botrycis cinerea* in apple plants. Furthermore, the *Bacillus* PGPBacCA1 produces antibiotic consortia of surfactin, iturin, and fengycinin which suppress charcoal rot in soybean (Torres et al., 2016). GacA/GacSor GrrA/GrrS, RpoD, and RpoS, N-acyl homoserine lactone derivatives are involved in positive autoregulation of the antibiotics (Schnider-Keel et al., 2000; Bloemberg and Lugtenberg, 2001; Brodhagen et al., 2004). The synthesis of antibiotics depends on the metabolic status of the bacteria, nutrient availability, temperature, pH, salinity of the soil, and biotic conditions (Bender et al., 1999; Duffy and Defago, 1999; Ownley et al., 2003). The production of DAPG by *Pseudomonas fluorescens* CHA0 can be affected by bacterial metabolites like salicylates and pyoluteorin (Schnider-Keel et al., 2000). Smith and Goodman (1999) reported that plant host genotype significantly alters the disease suppressive capacity of the endophyte–associated host plants.

3.3 Competition

The most beneficial mechanism the bacterial endophytes follow to control phytopathogens is competition for the microbial habitat and nutrient availability. It provides a wide challenge to pathogenic growth. These bacteria can easily register into the soil environment. They show a highly efficient regulatory system that enhances the phytoimmunity of the plants. Malfanova et al. (2013) reported that these bacterial endophytes often merge the mechanism of antibiosis which provides a more efficient result.

3.4 Parasitism and lysis

Another widely followed mechanism is a hyperparasitic activity for biocontrol activity. The bacterial endophytes secrete hydrolytic enzymes which kills the phytopathogen (Chernin and Chet, 2002). Frankowski et al. (2001) reported that *Serratia plymuthica* C48 is responsible for inhibition of spore germination and germ–tube elongation in *Botrytis cinerea*. Production of extracellular chitinases by *Serratia marcescens* and *Paenibacillus* sp. strain 300 makes it antagonistic against *Sclerotium rolfsii* and *Fusarium oxysporum* f. sp. *Cucumerinum*, respectively (Ordentlich et al., 1988). Lim et al. (1991) reported that *Pseudomonas stutzeri* secretes extracellular chitinase and laminarinase to lyse and degrade the mycelia of *Fusarium solani*. Moreover, protease facilitates the growth suppression of

Sclerotinia sclerotiorum and *Botrytis cinerea* (Kamensky et al., 2003). The enzyme 1,3-glucanase synthesized by *Paenibacillus* sp. strain 300 and *Streptomyces* sp. strain 385 can lyse the cell walls of *Fusarium oxysporum* f. sp. *cucumerinum* (Singh et al., 1999). The regulatory systems involved in modulating the production of lytic enzymes like proteases and chitinases is GacA/GacS, GrrA/GrrS, and the colony phase variation (Corbell and Loper, 1995; Lugtenberg et al., 2001; Ovadis et al., 2004).

3.5 Detoxification

Beneficial endophytes show biocontrol activity through detoxification of virulence factor of the phytopathogens. There are two types of mechanisms: reversible detoxification and irreversible detoxification. In reversible detoxification, the protein synthesized is bound reversibly to the toxin in bacteria like *Klebsiella oxytoca* (Walker et al., 1988) and *Alcaligenes denitrificans* (Basnayake and Birch, 1995). Later mechanism was observed in *Pantoea dispersa* where irreversible detoxification of albicidin was mediated by esterase (Zhang and Birch, 1997). It has been studied that *Bacillus cepacia* and *Ralstonia solanacearum* hydrolyses phytotoxin like fusaric acid which is synthesized by *Fusarium* sp. (Toyoda et al., 1988; Toyoda and Utsumi, 1991). A broad-spectrum activity is exhibited by the toxins secreted by the pathogens as a self-defense mechanism which includes growth suppression of microbial competitors, detoxification of antibiotics produced by the biocontrol agents (Duffy et al., 2003; Schouten et al., 2004). Studies reported that some beneficial bacteria quench the quorum sensing capacity of the pathogens by deactivating the autoinducer signals, which results in suppression of virulence genes (Dong et al., 2000, 2004; Molina et al., 2003; Morello et al., 2004; Newton and Fray, 2004; Uroz et al., 2003). This approach is very effective in alleviating plant diseases as most of the phytopathogen uses autoinducer mediated quorum sensing to activate the major virulence gene like toxin-producing genes and cell degrading genes (von Bodman et al., 2003). Microbial endophytes synthesize allelochemicals as a biocontrol mechanism in response to phytopathogens. It has been extensively studied in free-living beneficial bacteria (Lodewyckx et al., 2002). *Streptomyces* sp. NRRL 30562 which was isolated from *Kennedia nigriscans*, produces munumbicins which inhibit the growth of pathogenic fungi *Pythium ultimum* and *Fusarium oxysporum* (Castillo et al., 2002). Detoxification process is exhibited by potato-associated endophytes along with the production of siderophore and antibiotic compounds which can diminish the growth of *Streptomyces scabies* and *X. campestris in vitro* (Sessitsch et al., 2004). *P. fluorescens* strain FPT 9601 synthesizes 2,4-diacetylphloroglucinol (DAPG) and deposits the crystals of DAPG along the roots of the tomato plant which lead to detoxification of phytopathogens (Aino et al., 1997). However, it is also studied that the biocontrol activity of the endophytes is affected by the bacterial mass colonized in the interior tissue of the host plants (Sturz et al., 1999).

3.6 Induced resistance

Plants exhibit an induced resistance which is a state of enhanced defensive capacity when exposed to biotic stress (van Loon et al., 1998). The defensive capability effectively resists a wide range of pathogens including bacteria, viruses, nematodes, and even parasitic plants and insects (Kessler and Baldwin, 2002). Induced resistance has two forms of action which are systemic acquired resistance (SAR) and induced systemic resistance (ISR) depending on the elicitor nature and regulatory pathways (Schenk et al., 2000; van Wees et al., 2000; Yan et al., 2002). Resistance can be stimulated in plants through applying chemicals or necrosis-producing pathogens, a process termed systemic acquired resistance (SAR). Working on the plant immune system, Pieterse et al. (1998) proposed that ISR and SAR pathways can be differentiated not only based on the elicitor but also by the signal transduction pathways they elicit within a plant. Induced systemic resistance is elicited by rhizobacteria and/or nonpathogenic microorganisms, while SAR is elicited by pathogens or chemical compounds. Further, the signal transduction pathway of ISR is dependent on jasmonate and ethylene and independent of salicylic acid while the pathway of SAR is dependent on salicylic acid and variably depends on jasmonate and ethylene. However, recent studies report that few rhizobacteria, including some of the endophytic strains, elicit systemic protection that is dependent on salicylic acid and independent of ethylene or jasmonate (Kloepper and Ryu, 2006). Therefore, ISR cannot be marked separate from SAR based only on signal transduction pathways. Molecular tools can be helpful in understanding the critical turnover points between the two pathways and the differential responses of bacterial strains in eliciting ISR in plants.

3.6.1 Systemic acquired resistance

Plants develop a defense mechanism as a preliminary response toward the infection caused by phytopathogens. The phytopathogen induces a hypersensitive reaction which causes visible symptoms in the host plants which elicit systemic acquired resistance (SAR) (van Loon et al., 1998). SAR can be triggered chemically using salicylic acid, 2,6-dichloro-isonicotinic acid (INA), or benzo (1,2,3) thiadiazole-7-carbothioic acid S-methyl ester (BTH) (Sticher et al., 1997). A specific time period is essential for the establishment of SAR depending on the accumulation of salicylic acid and pathogenesis-related protein throughout the host plant (Cameron et al., 1994; Uknes et al., 1992). Disruption in salicylic acid accumulation causes attenuation in SAR response (Lawton et al., 1995). The induction of SAR broadly depends on the nature of the eliciting agent and the site of elicitor action on the plant. SAR is effective in a wide range of plant species.

3.6.2 Induced systemic resistance

The beneficial bacteria associated with the host plants trigger an immunity response mechanism called induced systemic resistance (ISR). The induction of ISR elicited by beneficial bacteria was first observed in *Dianthus caryophillus* and *Cucumis sativus* which was less susceptible toward *Fusarium* wilt and *Colletotrichum orbiculare*, respectively (van Peer et al., 1991; Wei et al., 1991). A cucumber plant surviving in a field heavily infested with cucurbit wilt disease, caused by *Erwinia tracheiphila* was found to be colonized by *Bacillus pumilus* strain INR7 (Wei et al., 1996). This process has been reported by van Loon and Bakker (2003), Kamilova et al. (2005), Van Wees et al. (2008) in *P. fluorescens* strains WCS417R and WCS365. ISR has been reported to be dependent on jasmonate and ethylene-regulated pathways (Yan et al., 2002; Van Loon and Bakker, 2003). Endophytes trigger ISR by producing siderophore, salicylic acid, c-LPs, pyocyanins (Audenaert et al., 2002; Ryu et al., 2003; Schuhegger et al., 2006; Pérez-García et al., 2011). The induction of ISR largely depends on the biochemical responses incited within the host plant, different bacterial surface molecules, secreted metabolites, and volatiles (Lugtenberg et al., 2013). PGPR strains show specificity to trigger ISR on a particular genotype (Yan et al., 2002). The ISR immunized plant can protect itself from a wide range of phytopathogens of different origins. However, the inducing beneficial bacterial strain does not cause any visible symptoms (van Loon et al., 1998). The bacterial endophytes responsible for eliciting ISR are *Bacillus amyloliquefaciens*, *B. pumilus*, *Bacillus subtilis*, *P. fluorescens*, *Pseudomonas syringae*, and *S. marcescens* (Kloepper and Ryu, 2006). The ISR mechanism is analogous to inherent immunity and uses toll-like receptors (De Weert et al., 2007). Beneficial bacteria produce salicylic acid which induces the production of pathogenesis-related proteins via jasmonic acid and ethylene production which in turn facilitate ISR (Hoffland et al., 1995; Pieterse et al., 1998; Romeiro, 2000). Phytoalexins enhance the production of lipoxygenase which inhibits the incidence of diseases (Li et al., 1991). The increase in the secretion of enzymes induces pathogenesis through ISR, although it depends on the nature of the host and disease-inflicting pathogens (Nakkeeran et al., 2006; Saikia et al., 2006). Wei et al. (1991) exhibited the mechanism of ISR for suppression of anthracnose caused by *Colletotrichum orbiculare* in cucumber. Chen et al. (2000a,b) and Saikia et al. (2004) reported that endophytes produce enzymes like peroxidases, lipoxygenases, chitinases, and glucanases which cause pathogenesis.

4. Strategies to enhance biocontrol efficiency

Considering the wide range of mechanisms used by the beneficial bacteria to provide a defense to the host plant against the phytopathogens, it was determined that modulation of plant microbiota can provide efficient and consistent biocontrol agents, improving agricultural production. This led to an increase in the exploration of endophytes antagonistic to pathogens and discovering ways to improve its proficiency. In the following section, some of the strategies are discussed to enhance biocontrol efficiency.

4.1 Optimized formulation and inoculation techniques

Studies report major advancements in formulations of biocontrol agents (Burghes, 1998). During the processing and field application, gram-negative bacteria become more vulnerable and susceptible toward deleterious factors due to lack of resting spores (O'Callaghan, 2016). Therefore, suitable and protective formulations are needed to escalate the efficiency at the target site. There are two types of formulations: dry products and liquid suspensions. Dry formulations are used for gram-positive endophytes. It is applied as seed coatings or soil amendments (Bashan et al., 2014). Standard carriers like soil-derived carriers (charcoal, turfs), organic carriers (vermicompost, sawdust, animal manure, sewage sludge), inert materials (silicates, polymers, bentonite, perlite) are used for dry formulations. Pure lyophilized cultures added with lyoprotectant can be used either directly or by combining with the solid carrier (Malusa et al., 2012). Liquid formulation is used in the case of gram-negative bacteria. It can be directly coated onto the seed prior to sowing (Bashan et al., 2014). Liquified suspensions can be used as a foliar spray for above-ground parts of the plants (Jambhulkar et al., 2016). Optimized formulation should ensure the survival of bacteria as well as harbor the efficacy of bacterial activity.

Successful biocontrol can be achieved through the effective delivery of biocontrol agents. Musson et al. (1995) reported eight methods of delivering BCAs in *Gossypium* sp. that include soaking seed in bacterial suspensions, stab-inoculation of bacteria into stems, soil drench, methylcellulose seed coating, foliar spray, vacuum infiltration, the pruned-root dip, and bacteria-impregnated granules applied in-furrow. Fahey et al. (1991) reported a unique seed inoculation technique in which the bacterial suspension is imbibed into the seeds under pressure followed by redrying of seeds. Implementation of these techniques may lead to suppression of phytopathogens as well as the emergence of promising biocontrol agents.

4.2 Integrated biocontrol strategies

Integrated biocontrol technique is another strategy for escalating biocontrol activity. Bacterial endophytes are combined with other strains of beneficial bacteria or biochemicals. Microbial consortium can enhance the plant defense mechanism by activating antioxidant enzymes and phenylpropanoid pathway followed by accumulation of total phenolics, proline, and PR proteins. It was reported that a consortium of *Pseudomonas aeruginosa* PJHU15, *Trichoderma harzianum* TNHU27, and *Bacillus subtilis* BHHU100 can suppress soft rot caused by *Sclerotinia sclerotiorum* (Jain et al., 2012). Integrated biological control strategies suggest using a combination of antagonistic endophytes with complementary modes-of-action, colonization sites, with synthetic control agents which can be explored further. Antagonistic potential can be enhanced by altering the biochemical elements of the plants. Hallmann et al. (1999) showed that the application of chitin can increase the antagonistic activity of *Burkholderia cepacian*. Furthermore, a study proved that root rot disease of pepper can be controlled by bacterial endophytes *B. subtilis* and *Bacillus licheniformis* along with the application of chitin (Ahmed et al., 2003). Chitin derivate chitosan when used in combination with *Bacillus pumilus*, successfully enhanced plant resistance toward *Fusarium* wilt of tomato (Benhamou et al., 1998). It can be assumed that both chitin and chitosan establish chitinolytic microflora which is responsible for the decomposition of the fungal cell wall.

4.3 Management of the indigenous endophytic microflora

Indigenous bacterial endophytes play an important role in enhancing phytoimmunity of the host. They enter and evade plant defense mechanisms. The bacterial strains present in the organic system of the plant have the potential to enhance suppressive activity. The activity of disease suppression is regulated by the quorum-sensing process. Suppression of *Verticillium* in oilseed rape is done by the process of quorum sensing of *Serratia plymuthica* HRO-C48 (Muller et al., 2009). Soil is stated as 'microbial seed bank' as it possesses the largest microbial biomass (Philippot et al., 2013). It contributes toward soil health and crop management. Many studies reported that altering soil properties affects the composition of the plant microbiome (Reeve et al., 2016; Estendorfer et al., 2017). Xia et al. (2015) isolated 239 indigenous bacterial endophytes from tomato, corn, melon, and potato grown in the organic management system. Breeding plant genotypes increases the biocontrol efficiency of the antagonistic endophytes. Fray (2002) reported that plant breeds which express signaling molecules like N-acylhomoserine lactones or molecules which deactivate the bacterial communication also play significant role on the biocontrol activity of the indigenous strains.

5. Genetic engineering

Transgenic means are also a promising way to promote bacterial endophytes as biocontrol agents. The genes regulating the modes of action, survival, fitness, colonization, adaptation are targeted for genetic engineering. Crop Genetics International (Hanover, MD) exhibited biocontrol mechanisms in various genetically modified agricultural crops (Fahey et al., 1991). Bioengineered endophyte *Clavibacter xylii* is used to deliver delta-endotoxin of *Bacillus thuringensis* subsp. kurtaki in corn to make it resistant against European corn borer. Shields et al. (1995) naturally inserted the pTOM plasmid from *B. cepacia* G4 into lupine endophyte *B. cepacia* BU0072 using conjugation process to transfer the toluene degrading gene. Downing and Thomson (2000) genetically modified *Pseudomonas fluorescens* which was isolated from apple plantlets to extract a gene encoding chitinase in *Serratia marcescens*. The cloning of the gene chiA was done in broad-host-range plasmid pKT240 and the integration vector pJFF350 using tac promoter. These genetically engineered *P. fluorescens* bacteria carrying tac-chiA prominently suppressed the growth of *Rhizoctonia solani* on beans. However, such technologies are still difficult and expensive to produce transgenic biocontrol agents on broad-scale for commercial purposes.

5.1 Antimicrobial peptides

Antimicrobial peptides (AMPs) are short peptide sequences constituting less than 50 amino acid residues. These are considered as first-line defense in living organisms. A broad range of antimicrobial peptides is produced by microorganisms including, bacteriocins, fungal defensins, cyclopeptides, pseudopeptides, and peptaibols. Bacteriocins and fungal defensins are produced by ribosomal synthesis and cyclopeptides, pseudopeptides, and peptaibols are produced by non-ribosomal synthesis.

The bacteriocins and fungal defensins have a compact structure of antiparallel strands with disulfide bonds. These are strongly antagonistic against closely related species. AFP peptides produced by *Aspergillus giganteus* (Lacadena et al., 1995) are reported to control the growth of *Pyricularia oryzae, Botrytis cinerea,* and *Fusarium* sp. (Vila et al., 2001; Moreno et al., 2003, 2005). Cyclopeptides are composed of amino acid residues arranged in cyclic rings with D- and L- forms and allo and diamino derivatives, lacking disulfide bonds. Lipidic cyclopeptides are cytotoxic in nature having antimicrobial properties. It is produced by beneficial bacteria. *Rhizoctonia solani* and *Rhodococcus fascians* are checked by tolaasins (Bassarello et al., 2004). *P. fluorescens* BRG100 producing Pseudophomins is antagonistic against *Sclerotinia sclerotiorum* and *Leptosphaeria maculans* (Pedras et al., 2003). Pseudopeptides constituted of a few peptide bonds and modified amino acids. *Pantoea agglomerans* synthesize pantocines, a derivative of alanine, which helps to control *Erwinia amylovora,* by inhibiting the synthesis of transaminase catalyzed aminoacid (Brady et al., 1999; Jin et al., 2003). Polyoxins are used as commercial antifungal agents who inhibit the synthesis of chitin and hence act against *Alternaria* sp., *Botrytis cinerea,* and *Rhizoctonia solani*. Blasticidin checks the growth of *Pyricularia oryzae* by preventing the protein synthesis process. *B. subtilis* produces many pseudopeptides like bacilysin and rhizocticin which show antimicrobial property (Stein, 2005). Peptaibols are straight-chain peptides with an amino alcohol C terminus and with an acyl group in N terminus. These are highly rich in dialkylated amino acids like α-diaminobutyric acid (Degenkolb et al., 2003). These disrupt the membrane of pathogenic fungi and bacteria leading to the death of the organism. Peptaibols like trichokonins show active antimicrobial activity against bacteria *Clavibacter michiganensis* and fungus *Bipolaris sorokiniana, Rhizoctonia solani, Colletotrichum* sp. (Xiao-Yan et al., 2006). Direct and natural application of AMPs has limitations and is not adequately significant as biocontrol agents. These are synthesized in ribosome which makes it accessible for developing it synthetically as well as can be genetically engineered using transgenic tools. These artificially produced AMPs confer complete resistance against phytopathogens.

Chemically analogs of AMPs are designed using 6–47 amino acids. These were produced by the solid phase method and combinatorial chemistry (Monroc et al., 2006b). Combinatorial chemistry is broadly used for molecule mimicking, designing new improved molecules. Chemical analogs of AMPs focus on specificity, it acts against selected phytopathogen. Moreover, it is susceptible toward protease digestion diminishing toxicity toward host plants (Oh et al., 1999; Monroc et al., 2006a). A synthetically produced cecropin-melittin hybrid, Pep3, is antagonistic to *Phytophthora infestans* and *Thielaviopsis basicola* (Andreu et al., 1992; Cavallarin et al., 1998). D4E1, an analog of cecropin B checks the growth of *Verticillium dahliae, Fusarium moniliforme, Thielaviopsis basicola* as well as pathogenic bacteria *Xanthomonas campestris* pv. *malvacearum* and *Pseudomonas syringae* pv. *tabaci* (DeLucca and Walsh, 1999). Further another analogous of cecropin B, MB39 is actively against *Erwinia carotovora* sp. *betavasculorum, C. michiganensis, P. infestans,* and *Rhizoctonia solani* (Owens and Heutte, 1997).

Genes coding for AMPs are constructed and expressed in model plants through genetic engineering which provides resistance against phytopathogens. Plant defensins are also expressed in plants and reported to be of great success (Montesinos, 2007). The radish defensin Rs-AFP2, expressed in Solanaceae members like tobacco and potato which acted against *Alternaria longipes* (Terras et al., 1995). *Verticillium dahliae* was controlled by expressing Alf-AFP, alfalfa defensin in potato (Gao et al., 2000). Expression of DRR206, pea defensin in canola, and tobacco demonstrated activity against *Leptosphaeria maculans* (Wang et al., 1999). *Phytophthora parasitica* was checked through the expression of BSD1, cabbage defensin in tobacco (Park et al., 2002). Dm-AMP1 dahlia defensin when expressed in eggplant showed defense against *Verticillium alboatrum* and *Botrytis cinerea* (Turrini et al., 2004). Barley hordothionin acted against *P. syringae* pv. *Tabaci* and *C. michiganensis* when expressed in tobacco (Carmona et al., 1993). Mj-AMP1 jalapa defensin when expressed in tomato conferred defense against *Alternaria solani* (Schaefer et al., 2005). Microorganisms secreting the AMPs can be used as successful biocontrol agents. Advanced combinatorial chemical methods can provide more specific and improved AMPs with low cytotoxicity and higher protease stability.

6. Future research prospects

The agri-food sector is emphasizing toward increasing productivity in a sustainable and eco-friendly method. Employing new biotechnological techniques as well as managing crop production strategies will put forward greater productivity. Bacterial endophytes demonstrated their biocontrol potential extensively. Further research venturing the association of ISR and the plant parasitic model will enhance biological disease control capacity. Moreover, researchers should focus on determining efficient techniques to isolate and identify the beneficial bacteria easily. Detailed study of genetic level regulation of mechanisms involved in disease resisting capacity of bacteria can enhance the understanding of the microbial system which will help its better exploration. The host specificity and receptivity toward beneficial endophytes can be determined through germplasm selection. Sufficient study of inoculant formulation including microbial viability, effective ingredients, and reasonable production cost is much needed. Commercialization of bacteria-based biocontrol agents should be enhanced and appreciated and practical implications should be encouraged. However, assured implementation is required for successful sustainable agricultural practice.

References

Ahmed, A.S., Ezziyyani, M., Pérez Sánchez, C., Candela, M.E., 2003. Effect of chitin on biological control activity of *Bacillus* spp. and *Trichoderma harzianum* against root rot disease in pepper (*Capsicum annuum*) plants. Eur. J. Plant Pathol. 109, 633–637.

Aino, M., Maekawa, Y., Mayama, S., Kato, H., 1997. Biocontrol of bacterial wilt of tomato by producing seedlings colonized with endophytic antagonistic pseudomonads. In: Ogoshi, A., Kobayashi, K., Homma, Y., Kodama, F., Kondo, N., Akino, S. (Eds.), Plant Growth Promoting Rhizobacteria: Present Status and Future Prospects. Nakanishi Printing, Sapporo, Japan, pp. 120–123.

Andreu, D., Ubach, J., Boman, A., Wahlin, B., Wade, D., Merrifield, R.B., Boman, H.G., 1992. Shortened cecropin A-melittin hybrids. Significant size reduction retains potent antibiotic activity. FEBS Lett. 296, 190–194.

Aravind, R., Kumar, A., Eapen, S.J., 2012. Pre-plant bacterisation: a strategy for delivery of beneficial endophytic bacteria and production of disease-free plantlets of black pepper (*Piper nigrum* L.). Arch. Phytopathol. Pflanzenschutz. 45 (9), 1115–1126.

Audenaert, K., Pattery, T., Cornelis, P., Höfte, M., 2002. Induction of systemic resistance to *Botrytis cinerea* in tomato by *Pseudomonas aeruginosa* 7NSK2: role of salicylic acid, pyochelin and pyocyanin. Mol. Plant-Microb. Interact. 15, 1147–1156.

Bashan, Y., de Bashan, L.E., Prabhu, S.R., Hernandez, J.P., 2014. Advances in plant growth-promoting bacterial inoculant technology: formulations and practical perspectives (1998–2013). Plant Soil 378, 1–33.

Basnayake, W.S., Birch, R.G., 1995. A gene from *Alcaligenes denitrificans* that confers albicidin resistance by reversible antibiotic binding. Microbiology 141 (3), 551–560.

Bassarello, C., Lazzaroni, S., Bifulco, G., Lo Cantore, P., Iacobellis, N.S., Riccio, R., Gomez-Paloma, L., Evidente, A., 2004. Tolaasins A–E, five new lipodepsipeptides produced by *Pseudomonas tolaasii*. J. Nat. Prod. 67, 811–816.

Bender, C.L., Rangaswamy, V., Loper, J., 1999. Polyketide production by plant-associated pseudomonads. Annu. Rev. Phytopathol. 37, 175–196.

Benhamou, N., Kloepper, J.W., Tuzun, S., 1998. Induction of resistance against *Fusarium* wilt of tomato by combination of chitosan with an endophytic bacterial strain. Planta 204, 153–168.

Berg, G., Hallmann, J., 2006. Control of plant pathogenic fungi with bacterial endophytes. In: Microbial Root Endophytes. Springer, Berlin, Heidelberg, pp. 53–69.

Berg, G., Krechel, A., Ditz, M., Sikora, R.A., Ulrich, A., Hallmann, J., 2005. Endophytic and ectophytic potato-associated bacterial communities differ in structure and antagonistic function against plant pathogenic fungi. FEMS Microbiol. Ecol. 51 (2), 215–229.

Bloemberg, G.V., Lugtenberg, B.J., 2001. Molecular basis of plant growth promotion and biocontrol by rhizobacteria. Curr. Opin. Plant Biol. 4 (4), 343–350.

Brady, S.F., Wright, S.A., Lee, J.C., Sutton, A.E., Zumoff, C.H., Wodzinski, R.S., Beer, S.V., Clardy, J., 1999. Pantocin B, an antibiotic from *Erwinia herbicola* discovered by heterologous expression of cloned genes. J. Am. Chem. Soc. 121, 11912–11913.

Brodhagen, M., Henkels, M.D., Loper, J.E., 2004. Positive autoregulation of the antibiotic pyoluteorin in the biological control organism *Pseudomonas fluorescens* Pf-5. Appl. Environ. Microbiol. 70, 1758–1766.

Brooks, D.S., Gonzalez, C.F., Appel, D.N., Filer, T.H., 1994. Evaluation of endophytic bacteria as potential biological control agents for oak wilt. Biol. Control 4, 373–381.

Burghes, H.D., 1998. Formulation of Biopesticides. Kluwer, Dordrecht.

Cameron, R.K., Dixon, R.A., Lamb, C.J., 1994. Biologically induced systemic acquired resistance in *Arabidopsis thaliana*. Plant J. 5, 715–725.

Carmona, M.J., Molina, A., Fernández, J.A., López-Fando, J.J., García-Olmedo, F., 1993. Expression of the α-thionin gene from barley in tobacco confers enhanced resistance to bacterial pathogens. Plant J. 3 (3), 457–462.

Carroll, G., 1988. Fungal endophytes in stems and leaves: from latent pathogen to mutualistic symbiont. Ecology 69 (1), 2–9.

Castillo, U.F., Strobel, G.A., Ford, E.J., Hess, W.M., Porter, H., Jensen, J.B., Albert, H., Robison, R., Condron, M.A., Teplow, D.B., Stevens, D., 2002. Munumbicins, wide-spectrum antibiotics produced by *Streptomyces* NRRL 30562, endophytic on *Kennedia nigriscansaa* the GenBank accession number for the sequence determined in this work is AY127079. Microbiology 148 (9), 2675–2685.

Cavallarin, L., Andreu, D., San Segundo, B., 1998. Cecropin A derived peptides are potent inhibitors of fungal plant pathogens. Mol. Plant-Microbe Interact. 11, 218–227.

Chen, C., Bauske, E.M., Musson, G., Rodriguez-Kabana, R., Kloepper, J.W., 1995. Biological control of *Fusarium* wilt on cotton by use of endophytic bacteria. Biol. Control 5 (1), 83–91.

Chen, C., Belanger, R.R., Benhamou, N., Paulitz, T.C., 2000a. Defense enzymes induced in cucumber roots by treatment with plant-growth promoting rhizobacteria (PGPR). Physiol. Mol. Plant Pathol. 56, 13–23.

Chen, J., Abawi, G.S., Zucherman, B.M., 2000b. Efficacy of *Bacillus thuringiensis*, *Paecilomyces marquandii* and *Streptomyces costaricanus* with organic amendment against *Meloidogyne hapla* infecting lettuce. J. Nematol. 32, 70–77.

Chernin, L., Chet, I., 2002. Microbial enzymes in biocontrol of plant pathogens and pests. In: Burns, R.G., Dick, R.P. (Eds.), Enzymes in the Environment: Activity, Ecology, and Applications. Marcel Dekker, New York, pp. 171–225.

Corbell, N., Loper, J.E., 1995. A global regulator of secondary metabolite production in *Pseudomonas fluorescens* Pf-5. J. Bacteriol. 177 (21), 6230–6236.

Cornelis, P., Matthijs, S., 2002. Diversity of siderophore-mediated iron uptake systems in fluorescent pseudomonads: not only pyoverdines. Environ. Microbiol. 4 (12), 787–798.

Dandurishvili, N., Toklikishvili, N., Ovadis, M., Eliashvili, P., Giorgobiani, N., Keshelava, R., Tediashvili, M., Vainstein, A., Khmel, I., Szegedi, E., Chernin, L., 2011. Broad-range antagonistic rhizobacteria *Pseudomonas fluorescens* and *Serratia plymuthica* suppress *Agrobacterium* crown gall tumours on tomato plants. J. Appl. Microbiol. 110 (1), 341–352.

de Souza, J.T., de Boer, M., de Waard, P., van Beek, T.A., Raaijmakers, J.M., 2003. Biochemical, genetic, and zoosporicidal properties of cyclic lipopeptide surfactants produced by *Pseudomonas fluorescens*. Appl. Environ. Microbiol. 69 (12), 7161–7172.

De Weert, S., Kuiper, I., Kamilova, F., Mulders, I.H.M., Bloemberg, G.V., Kravchenko, L., Azarova, T., Eijkemans, K., Preston, G.M., Rainey, P., Tikhonovich, I., 2007. The role of competitive root tip colonization in the biological control of tomato foot and root rot. In: Chincolkar, S.B., Mukerji, K.G. (Eds.), Biological Control of Plant Diseases. The Haworth Press, New York, London, Oxford, pp. 103–122.

Degenkolb, T., Berg, A., Gams, W., Schlegel, B., Griafe, U., 2003. The occurrence of peptaibols and structurally related peptaibiotics in fungi and their mass spectrophotometric identification via diagnostic fragment ions. J. Pept. Sci. 9, 666–678.

DeLucca, A.J., Walsh, T.J., 1999. Antifungal peptides: novel therapeutic compounds against emerging pathogens. Antimicrob. Agents Chemother. 43, 1–11.

Dong, Y.H., Xu, J.L., Li, X.Z., Zhang, L.H., 2000. AiiA, an enzyme that inactivates the acylhomoserine lactone quorum-sensing signal and attenuates the virulence of *Erwinia carotovora*. Proc. Natl. Acad. Sci. U. S. A. 97, 3526–3531.

Dong, Y.H., Zhang, X.F., Xu, J.L., Zhang, L.H., 2004. Insecticidal *Bacillus thuringiensis* silences *Erwinia carotovora* virulence by a new form of microbial antagonism, signal interference. Appl. Environ. Microbiol. 70, 954–960.

Downing, K.J., Thomson, J.A., 2000. Introduction of the *Serratia marcescens* chiA gene into an endophytic *Pseudomonas fluorescens* for the biocontrol of phytopathogenic fungi. Can. J. Microbiol. 46 (4), 363–369.

Duffy, B., Schouten, A., Raaijmakers, J.M., 2003. Pathogen self-defense: mechanisms to counteract microbial antagonism. Annu. Rev. Phytopathol. 41, 501–538.

Duffy, B.K., Defago, G., 1999. Environmental factors modulating antibiotic and siderophore biosynthesis by *Pseudomonas fluorescens* biocontrol strains. Appl. Environ. Microbiol. 65 (6), 2429–2438.

Estendorfer, J., Stempfhuber, B., Haury, P., Vestergaard, G., Rillig, M.C., Joshi, J., Schröder, P., Schloter, M., 2017. The influence of land use intensity on the plant-associated microbiome of *Dactylis glomerata* L. Front. Plant Sci. 8, 930.

Fahey, J.W., Dimock, M.B., Tomasino, S.F., Taylor, J.M., Carlson, P.S., 1991. Genetically engineered endophytes as biocontrol agents: a case study from industry. In: Andrews, J.H., Hirano, S.S. (Eds.), Microbial Ecology of Leaves. Springer, Berlin, Heidelberg, New York, pp. 401–411.

Ferrigo, D., Causin, R., Raiola, A., 2017. Effect of potential biocontrol agents selected among grapevine endophytes and commercial products on crown gall disease. BioControl 62 (6), 821–833.

Frankowski, J., Lorito, M., Scala, F., Schmid, R., Berg, G., Bahl, H., 2001. Purification and properties of two chitinolytic enzymes of *Serratia plymuthica* HRO-C48. Arch. Microbiol. 176 (6), 421–426.

Fravel, D.R., 1988. Role of antibiosis in the biocontrol of plant diseases. Annu. Rev. Phytopathol. 26 (1), 75–91.

Fray, R.G., 2002. Altering plant-microbe interaction through artificially manipulating bacterial quorum sensing. Ann. Bot. 89, 245–253.

Frommel, M.I., Nowak, J., Lazarovits, G., 1991. Growth enhancement and developmental modifications of in vitro grown potato (*Solanum tuberosum* spp. *tuberosum*) as affected by a nonfluorescent *Pseudomonas* sp. Plant Physiol. 96 (3), 928–936.

Gao, A.G., Hakimi, S.M., Mittanck, C.A., Wu, Y., Woerner, B.M., Stark, D.M., Shah, D.M., Liang, J., Rommens, C.M., 2000. Fungal pathogen protection in potato by expression of a plant defensin peptide. Nat. Biotechnol. 18 (12), 1307–1310.

Haas, D., Defago, G., 2005. Biological control of soil-borne pathogens by fluorescent pseudomonads. Nat. Rev. Microbiol. 3 (4), 307–319.

Haas, D., Keel, C., Reimmann, C., 2002. Signal transduction in plant-beneficial rhizobacteria with biocontrol properties. Antonie. Leeuwenhoek. 81 (1–4), 385–395.

Hallmann, J., Quadt-Hallmann, A., Mahaffee, W.F., Kloepper, J.W., 1997. Bacterial endophytes in agricultural crops. Can. J. Microbiol. 43 (10), 895–914.

Hallmann, J., Rodríguez-Kábana, R., Kloepper, J.W., 1999. Chitin-mediated changes in bacterial communities of the soil, rhizosphere and within roots of cotton in relation to nematode control. Soil Biol. Biochem. 31, 551–560.

Hashidoko, Y., Nakayama, T., Homma, Y., Tahara, S., 1999. Structure elucidation of xanthobaccin A, a new antibiotic produced from *Stenotrophomonas* sp. strain SB-K88. Tetrahedron Lett. 40 (15), 2957–2960.

Hoffland, E., Hakulinen, J., van Pelt, J.A., 1995. Comparison of systemic resistance induced by avirulent and nonpathogenic *Pseudomonas* species. Phytopathology 86, 757–762.

Jain, A., Singh, S., Kumar Sarma, B., Bahadur Singh, H., 2012. Microbial consortium–mediated reprogramming of defence network in pea to enhance tolerance against *sclerotinia sclerotiorum*. J. Appl. Microbiol. 112 (3), 537–550.

Jambhulkar, P.P., Sharma, P., Yadav, R., 2016. Delivery systems for introduction of microbial inoculants in the field. In: Singh, D.P., Singh, H.B., Prabha, R. (Eds.), Microbial Inoculants in Sustainable Agricultural Productivity. Functional Applications, Vol. 2. Springer, New Delhi, India, pp. 199–218.

Jin, M., Wright, S., Beer, S., Clardy, J., 2003. The biosynthetic gene cluster of pantocin a provides insights into biosynthesis and a tool for screening. Angew. Chem. Int. Ed. 42, 2902–2905.

Kamensky, M., Ovadis, M., Chet, I., Chernin, L., 2003. Soil-borne strain IC14 of *Serratia plymuthica* with multiple mechanisms of antifungal activity provides biocontrol of *Botrytis cinerea* and *Sclerotinia sclerotiorum* diseases. Soil Biol. Biochem. 35 (2), 323–331.

Kamilova, F., Validov, S., Azarova, T., Mulders, I., Lugtenberg, B., 2005. Enrichment for enhanced competitive plant root tip colonizers selects for a new class of biocontrol bacteria. Environ. Microbiol. 7, 1809–1817.

Kessler, A., Baldwin, I.T., 2002. Plant responses to insect herbivory: the emerging molecular analysis. Annu. Rev. Plant Biol. 53, 299–328.

Kim, B.S., Moon, S.S., Hwang, B.K., 1999. Isolation, identification and antifungal activity of a macrolide antibiotic, oligomycin A, produced by *Streptomyces libani*. Can. J. Bot. 77, 850–858.

Kloepper, J.W., Ryu, C.M., 2006. Bacterial endophytes as elicitors of induced systemic resistance. In: Schulz, B., Boyle, C., Sieber, T. (Eds.), Microbial Root Endophytes. Springer, Berlin, Heidelberg, pp. 33–52.

Kloepper, J.W., Schippers, B., Bakker, P.A.H.M., 1992. Proposed elimination of the term endorhizosphere. Phytopathology 82 (7), 726–727.

Lacadena, J., Martinez del Pozo, A., Gasset, M., Campos-Olivas, R., Vazquez, C., Martinez-Ruiz, A., 1995. Characterization of the antifungal protein secreted by the mould *Aspergillus giganteus*. Arch. Biochem. Biophys. 324, 273–281.

Lawton, K., Weymann, K., Friedrich, L., Vernooij, B., Uknes, S., Ryals, J., 1995. Systemic acquired resistance in Arabidopsis requires salicylic acid but no ethylene. Mol. Plant-Microbe Interact. 6, 863–870.

Li, W.X., Kodama, O., Akatsuka, T., 1991. Role of oxygenated fatty acids in rice phytoalexin production. Agric. Biol. Chem. 55, 1041–1147.

Lim, H.S., Kim, Y.S., Kim, S.D., 1991. *Pseudomonas stutzeri* YPL-1 genetic transformation and antifungal mechanism against *Fusarium solani*, an agent of plant root rot. Appl. Environ. Microbiol. 57, 510–516.

Lodewyckx, C., Vangronsveld, J., Porteous, F., Moore, E.R.B., Taghavi, S., Mezgeay, M., van der Lelie, D., 2002. Endophytic bacteria and their potential applications. Crit. Rev. Plant Sci. 21, 583–606.

Loper, J.E., Henkels, M.D., 1999. Utilization of heterologous siderophores enhances levels of iron available to *Pseudomonas putida* in the rhizosphere. Appl. Environ. Microbiol. 65 (12), 5357–5363.

Lugtenberg, B., Malfanova, N., Kamilova, F., Berg, G., 2013. Chapter 53: Plant growth promotion by microbes. In: de Bruijn, F.J. (Ed.), Molecular Microbial Ecology of the Rhizosphere. Hoboken, Wiley, Blackwell, pp. 561–573.

Lugtenberg, B.J., Dekkers, L., Bloemberg, G.V., 2001. Molecular determinants of rhizosphere colonization by *Pseudomonas*. Annu. Rev. Phytopathol. 39 (1), 461–490.

Malfanova, N., Kamilova, F., Validov, S., Chebotar, V., Lugtenberg, B., 2013. Is L-arabinose important for the endophytic lifestyle of *Pseudomonas* spp.? Arch. Microbiol. 195 (1), 9–17.

Malusa, E., Sas-Paszt, L., Ciesielska, J., 2012. Technologies for beneficial microorganisms inocula used as biofertilizers. Sci. World J. 12, 1–12.

Mathys, J., De Cremer, K., Timmermans, P., Van Kerkhove, S., Lievens, B., Vanhaecke, M., Cammue, B., De Coninck, B., 2012. Genome-wide characterization of ISR induced in *Arabidopsis thaliana* by *Trichoderma hamatum* T382 against *Botrytis cinerea* infection. Front. Plant Sci. 3, 108.

Molina, L., Constantinescu, F., Reimmann, C., Duffy, B., Defago, G., 2003. Degradation of pathogen quorum-sensing molecules by soil bacteria: a preventive and curative biological control mechanism. FEMS Microbiol. Ecol. 45, 71–81.

Monroc, S., Badosa, E., Besalu, E., Planas, M., Bardaji, E., Montesinos, E., Feliu, L., 2006a. Improvement of cyclic decapeptides against plant pathogenic bacteria using a combinatorial chemistry approach. Peptides 27, 2575–2584.

Monroc, S., Badosa, E., Feliu, L., Planas, M., Montesinos, E., Bardaji, E., 2006b. De novo designed cyclic cationic peptides as inhibitors of plant pathogenic bacteria. Peptides 27, 2567–2574.

Montesinos, E., 2007. Antimicrobial peptides and plant disease control. FEMS Microbiol. Lett. 270 (1), 1–11.

Morello, J.E., Pierson, E.A., Pierson, L.S., 2004. Negative cross communication among wheat rhizosphere bacteria: effect on antibiotic production by the biological control bacterium *Pseudomonas aureofaciens* 30–84. Appl. Environ. Microbiol. 70, 3103–3109.

Moreno, A.B., Martinez, A., Borja, M., SanSegundo, B., 2003. Activity of the antifungal protein from *Aspergillus giganteus* against *Botrytis cinerea*. Mol. Plant-Microbe Interact. 93, 1344–1353.

Moreno, A.B., Penas, G., Rufat, M., Bravo, J.M., Estopa, M., Messeguer, J., SanSegundo, B., 2005. Pathogen-induced production of the antifungal AFP protein from *Aspergillus giganteus* confers resistance to the blast fungus *Magnaporthe grisea* in transgenic rice. Mol. Plant-Microbe Interact. 18, 960–972.

Muller, H., Westendorf, C., Leitner, E., Chernin, L., Riedel, K., Schmidt, S., Eberl, L., Berg, G., 2009. Quorum-sensing effects in the antagonistic rhizosphere bacterium *Serratia plymuthica* HRO-C48. FEMS Microbiol. Ecol. 67 (3), 468–478.

Musson, G., Mc Inroy, J.A., Kloepper, J.W., 1995. Development of delivery systems for introducing endophytic bacteria into cotton. Biocontrol Sci. Tech. 5, 407–416.

Nakayama, T., Homma, Y., Hashidoko, Y., Mizutani, J., Tahara, S., 1999. Possible role of xanthobaccins produced by *Stenotrophomonas* sp. strain SB-K88 in suppression of sugar beet damping-off disease. Appl. Environ. Microbiol. 65, 4334–4339.

Nakkeeran, S., Kavitha, K., Chandrasekar, G., Renukadevi, P., Fernando, W.G.D., 2006. Induction of plant defence compounds by *Pseudomonas chlororaphis* PA23 and *Bacillus subtilis* BSCBE 4 in controlling damping-off of hot pepper caused by *Pythium aphanidermatum*. Biocontrol Sci. Tech. 16, 403–416.

Newton, J.A., Fray, R.G., 2004. Integration of environmental and host derived signals with quorum sensing during plant–microbe interactions. Cell. Microbiol. 6 (3), 213–224.

Nielsen, T.H., Sørensen, D., Tobiasen, C., Andersen, J.B., Christeophersen, C., Givskov, M., Sørensen, J., 2002. Antibiotic and biosurfactant properties of cyclic lipopeptides produced by fluorescent *Pseudomonas* spp. from the sugar beet rhizosphere. Appl. Environ. Microbiol. 68, 3416–3423.

Nielsen, T.H., Sørensen, J., 2003. Production of cyclic lipopeptides by *Pseudomonas fluorescens* strains in bulk soil and in the sugar beet rhizosphere. Appl. Environ. Microbiol. 69, 861–868.

O'Callaghan, M., 2016. Microbial inoculation of seed for improved crop performance: issues and opportunities. Appl. Microbiol. Biotechnol. 100 (13), 5729–5746.

Oh, J.E., Hong, S.Y., Lee, K.H., 1999. Structure-activity relationship study: short antimicrobial peptides. J. Pept. Res. 53, 41–46.

Ordentlich, A., Elad, Y., Chet, I., 1988. The role of chitinase of *Serratia marcescens* in biocontrol of *sclerotium rolfsii*. Phytopathology 78 (1), 84–88.

Ovadis, M., Liu, X., Gavriel, S., Ismailov, Z., Chet, I., Chernin, L., 2004. The global regulator genes from biocontrol strain *Serratia plymuthica* IC1270: cloning, sequencing, and functional studies. J. Bacteriol. 186 (15), 4986–4993.

Owens, L.D., Heutte, T.M., 1997. A single amino acid substitution in the antimicrobial defense protein cecropin B is associated with diminished degradation by leaf intercellular fluid. Mol. Plant-Microbe Interact. 10, 525–528.

Ownley, B.H., Duffy, B.K., Weller, D.M., 2003. Identification and manipulation of soil properties to improve the biological control performance of phenazine-producing Pseudomonas fluorescens. Appl. Environ. Microbiol. 69 (6), 3333–3343.

Park, H.C., Kang, Y.H., Chun, H.J., Koo, J.C., Cheong, Y.H., Kim, C.Y., Kim, M.C., Chung, W.S., Kim, J.C., Yoo, J.H., Koo, Y.D., 2002. Characterization of a stamen-specific cDNA encoding a novel plant defensin in Chinese cabbage. Plant Mol. Biol. 50 (1), 57–68.

Pedras, M.S., Ismail, N., Quail, J.W., Boyetchko, S.M., 2003. Structure, chemistry, and biological activity of pseudophomins A and B, new cyclic lipodepsipeptides isolated from the biocontrol bacterium *Pseudomonas fluorescens*. Phytochemistry 62, 1105–1114.

Pérez-García, A., Romero, D., De Vicente, A., 2011. Plant protection and growth stimulation by microorganisms: biotechnological applications of *Bacilli* in agriculture. Curr. Opin. Biotechnol. 22 (2), 187–193.

Philippot, L., Raaijmakers, J.M., Lemanceau, P., Van Der Putten, W.H., 2013. Going back to the roots: the microbial ecology of the rhizosphere. Nat. Rev. Microbiol. 11 (11), 789.

Pieterse, C.M.J., van Pelt, J.A., Knoester, M., Laan, R., Gerrits, H., Weisbeek, P.J., van Loon, L.C.A., 1998. Novel signaling pathway controlling induced systemic resistance in *Arabidopsis*. Plant Cell 10, 1571–1580.

Ping, L., Boland, W., 2004. Signals from the underground: bacterial volatiles promote growth in *Arabidopsis*. Trends Plant Sci. 9 (6), 263–266.

Pliego, C., De Weert, S., Lamers, G., De Vicente, A., Bloemberg, G., Cazorla, F.M., Ramos, C., 2008. Two similar enhanced root-colonizing *Pseudomonas* strains differ largely in their colonization strategies of avocado roots and *Rosellinia necatrix* hyphae. Environ. Microbiol. 10 (12), 3295–3304.

Poon, E.S., Huang, T.C., Kuo, T.T., 1977. Possible mechanism of symptom inhibition of bacterial blight of rice by an endophytic bacterium isolated from rice. Bot. Bull. Acad. Sin. 18, 61–70.

Raaijmakers, J.M., Vlami, M., De Souza, J.T., 2002. Antibiotic production by bacterial biocontrol agents. Antonie Leeuwenhoek 81, 537–547.

Ravel, J., Cornelis, P., 2003. Genomics of pyoverdine-mediated iron uptake in pseudomonads. Trends Microbiol. 11, 195–200.

Reeve, J.R., Hoagland, L.A., Villalba, J.J., Carr, P.M., Atucha, A., Cambardella, C., Davis, D.R., Delate, K., 2016. Organic farming, soil health, and food quality: considering possible links. In: Advances in Agronomy. Vol. 137. Academic Press, pp. 319–367.

Reiter, B., Pfeifer, U., Schwab, H., Sessitsch, A., 2002. Response of endophytic bacterial communities in potato plants to infection with *Erwinia carotovora* subsp. *atroseptica*. Appl. Environ. Microbiol. 68 (5), 2261–2268.

Romeiro, R.S., 2000. PGPR e indução de resistência sistêmica em plantas a patógenos. Summa Phytopathol. 26, 177–184.

Ryan, R.P., Germaine, K., Franks, A., Ryan, D.J., Dowling, D.N., 2008. Bacterial endophytes: recent developments and applications. FEMS Microbiol. Lett. 278 (1), 1–9.

Ryu, C.M., Farag, M.A., Hu, C.H., Reddy, M.S., Wei, H.X., Paré, P.W., Kloepper, J.W., 2003. Bacterial volatiles promote growth in *Arabidopsis*. Proc. Natl. Acad. Sci. 100 (8), 4927–4932.

Ryu, C.M., Murphy, J.F., Mysore, K.S., Kloepper, J.W., 2004. Plant growth promoting rhizobacteria systemically protect *Arabidopsis thaliana* against cucumber mosaic virus by a salicylic acid and NPR1 independent and jasmonic acid dependent signaling pathway. Plant J. 39 (3), 381–392.

Saikia, R., Kumar, R., Arora, D.K., Gogoi, D.K., Azad, P., 2006. *Pseudomonas aeruginosa* inducing rice resistance against *rhizoctonia solani*: production of salicylic acid and peroxidases. Folia Microbiol. 51, 375–380.

Saikia, R., Kumar, R., Singh, T., Srivastava, A.K., Arora, D.K., Gogoi, D.K., Lee, M.W., 2004. Induction of defense related enzymes and pathogenesis related proteins in *Pseudomonas fluorescens* treated chickpea in response to infection by *Fusarium oxysporum* f. sp. *ciceri*. Mycobiology 32, 47–52.

Savary, S., Willocquet, L., Pethybridge, S.J., Esker, P., McRoberts, N., Nelson, A., 2019. The global burden of pathogens and pests on major food crops. Nat. Ecol. Evol. 3 (3), 430–439.

Schaefer, S.C., Gasic, K., Cammue, B., Broekaert, W., Van Damme, E.J., Peumans, W.J., Korban, S.S., 2005. Enhanced resistance to early blight in transgenic tomato lines expressing heterologous plant defense genes. Planta 222 (5), 858.

Schenk, P.M., Kazan, K., Wilson, I., Anderson, J.P., Richmond, T., Somerville, S.C., Manners, J.M., 2000. Coordinated plant defense responses in *Arabidopsis* revealed by microarray analysis. Proc. Natl. Acad. Sci. U. S. A. 97, 11655–11660.

Schnider-Keel, U., Seematter, A., Maurhofer, M., Blumer, C., Duffy, B., Gigot-Bonnefoy, C., Reimmann, C., Notz, R., Défago, G., Haas, D., Keel, C., 2000. Autoinduction of 2, 4-diacetylphloroglucinol biosynthesis in the biocontrol agent *Pseudomonas fluorescens* CHA0 and repression by the bacterial metabolites salicylate and pyoluteorin. J. Bacteriol. 182 (5), 1215–1225.

Schouten, A., van den Berg, G., Edel-Hermann, V., Steinberg, C., Gautheron, N., Alabouvette, C., De Vos, C.H., Lemanceau, P., Raaijmakers, J.M., 2004. Defense responses of *fusarium oxysporum* to 2, 4-diacetylphloroglucinol, a broad-spectrum antibiotic produced by *Pseudomonas fluorescens*. Mol. Plant-Microbe Interact. 17 (11), 1201–1211.

Schroth, M.N., Loper, J.E., Hilderbrand, D.C., 1984. Bacteria as bio-control agents of plant disease. In: Klug, M.J., Reddy, C.A. (Eds.), Current Perspectives in Microbial Ecology. American Society of Microbiology, Washington, DC, pp. 362–369.

Schuhegger, R., Ihring, A., Gantner, S., Bahnweg, G., Knappe, C., Vogg, G., Hutzler, P., Schmid, M., Van Breusegem, F., Eberl, L.E.O., Hartmann, A., 2006. Induction of systemic resistance in tomato by N-acyl-L-homoserine lactone producing rhizosphere bacteria. Plant Cell Environ. 29 (5), 909–918.

Sessitsch, A., Reiter, B., Berg, G., 2004. Endophytic bacterial communities of field-grown potato plants and their plant growth-promoting and antagonistic abilities. Can. J. Microbiol. 50, 239–249.

Shields, M.S., Reagin, M.J., Gerger, R.R., Campbell, R., Somerville, C., 1995. TOM, a new aromatic degradative plasmid from *Burkholderia* (*Pseudomonas*) *cepacia* G4. Appl. Environ. Microbiol. 61 (4), 1352–1356.

Singh, P.P., Shin, Y.C., Park, C.S., Chung, Y.R., 1999. Biological control of fusarium wilt of cucumber by chitinolytic bacteria. Phytopathology 89 (1), 92–99.

Smith, K.P., Goodman, R.M., 1999. Host variation for interactions with beneficial plant-associated microbes. Annu. Rev. Phytopathol. 37 (1), 473–491.

Stein, T., 2005. *Bacillus subtilis* antibiotics: structures, syntheses and specific functions. Mol. Microbiol. 56, 845–857.

Sticher, L., Mauch-Mani, B., Me Traux, J.P., 1997. Systemic acquired resistance. Annu. Rev. Phytopathol. 35, 235–270.

Sturz, A., Christie, B., Matheson, B., Arsenault, W., Buchanan, N., 1999. Endophytic bacterial communities in the periderm of potato tubers and their potential to improve resistance to soil-borne plant pathogens. Plant Pathol. 48 (3), 360–369.

Tabbene, O., Slimene, I.B., Bouabdallah, F., Mangoni, M.L., Urdaci, M.C., Limam, F., 2009. Production of anti-methicillin-resistant *staphylococcus* activity from *Bacillus subtilis* sp. strain B38 newly isolated from soil. Appl. Biochem. Biotechnol. 157 (3), 407–419.

Terras, F.R., Eggermont, K., Kovaleva, V., Raikhel, N.V., Osborn, R.W., Kester, A., Rees, S.B., Torrekens, S., Van Leuven, F., Vanderleyden, J., 1995. Small cysteine-rich antifungal proteins from radish: their role in host defense. Plant Cell 7 (5), 573–588.

Tjamos, E.C., Tsitsigiannis, D.I., Tjamos, S.E., Antoniou, P.P., Katinakis, P., 2004. Selection and screening of endorhizosphere bacteria from solarized soils as biocontrol agents against *Verticillium dahliae* of solanaceous hosts. Eur. J. Plant Pathol. 110 (1), 35–44.

Torres, M.J., Brandan, C.P., Petroselli, G., Erra-Balsells, R., Audisio, M.C., 2016. Antagonistic effects of *Bacillus subtilis* subsp. *subtilis* and *B. amyloliquefaciens* against *Macrophomina phaseolina*: SEM study of fungal changes and UV-MALDI-TOF MS analysis of their bioactive compounds. Microbiol. Res. 182, 31–39.

Toure, Y., Ongena, M.A.R.C., Jacques, P., Guiro, A., Thonart, P., 2004. Role of lipopeptides produced by *Bacillus subtilis* GA1 in the reduction of grey mould disease caused by *Botrytis cinerea* on apple. J. Appl. Microbiol. 96 (5), 1151–1160.

Toyoda, H., Hashimoto, H., Utsumi, R., Kobayashi, H., Ouchi, S., 1988. Detoxification of fusaric acid by a fusaric acid-resistant mutant of *pseudomonas solanacearum* and its application to biological control of fusarium wilt of tomato. Phytopathology 78, 1307–1311.

Toyoda, H., Utsumi, R., 1991. Method for the prevention of fusarium diseases and microorganisms used for the same. U.S. patent 4,988,586.

Turner, J.T., Kelly, J.L., Carlson, P.S., 1993. Endophytes: an alternative genome for crop improvement. Int. Crop Sci. I, 555–560.

Turrini, A., Sbrana, C., Pitto, L., Ruffini Castiglione, M., Giorgetti, L., Briganti, R., Bracci, T., Evangelista, M., Nuti, M.P., Giovannetti, M., 2004. The antifungal Dm-AMP1 protein from Dahlia merckii expressed in *Solanum melongena* is released in root exudates and differentially affects pathogenic fungi and mycorrhizal symbiosis. New Phytol. 163 (2), 393–403.

Uknes, S., Mauch-Mani, B., Moyer, M., Potter, S., Williams, S., Dincher, S., Chandler, D., Slusarenko, A., Ward, E., Ryals, J., 1992. Acquired resistance in *Arabidopsis*. Plant Cell 4, 645–656.

Uroz, S., D'Angelo-Picard, C., Carlier, A., Elasri, M., Sicot, C., Petit, A., Oger, P., Faure, D., Dessaux, Y., 2003. Novel bacteria degrading N-acylhomoserine lactones and their use as quenchers of quorum-sensing-regulated functions of plant-pathogenic bacteria. Microbiology 149 (8), 1981–1989.

Valin, H., Sands, R.D., Van der Mensbrugghe, D., Nelson, G.C., Ahammad, H., Blanc, E., Bodirsky, B., Fujimori, S., Hasegawa, T., Havlik, P., Heyhoe, E., 2014. The future of food demand: understanding differences in global economic models. Agric. Econ. 45 (1), 51–67.

van Buren, A.M., Andre, C., Ishimaru, C.A., 1993. Biological control of the bacterial ring rot pathogen by endophytic bacteria isolated from potato. Phytopathology 83, 1406.

van Loon, L.C., Bakker, P.A.H.M., 2003. In: De Kroon, H., Visser, W.J.W. (Eds.), Root ecology. Springer, Berlin, pp. 297–330.

van Loon, L.C., Bakker, P.A.H.M., Pieterse, C.M.J., 1998. Systemic resistance induced by rhizosphere bacteria. Annu. Rev. Phytopathol. 36 (1), 453–483.

van Peer, R., Niemann, G.J., Schippers, B., 1991. Induced resistance and phytoalexin accumulation in biological control of *fusarium* wilt of carnation by *Pseudomonas* sp. strain WCS417r. Phytopathology 81, 728–734.

van Wees, S.C.M., de Swart, E.A.M., van Pelt, J.A., van Loon, L.C., Pieterse, C.M.J., 2000. Enhancement of induced disease resistance by simultaneous activation of salicylate and jasmonate dependent defense pathways in *Arabidopsis thaliana*. Proc. Natl. Acad. Sci. U. S. A. 97, 8711–8716.

van Wees, S.C.M., Van der Ent, S., Pieterse, C.M.J., 2008. Plant immune responses triggered by beneficial microbes. Curr. Opin. Plant Biol. 11, 443–448.

Vila, L., Lacadena, V., Fontanet, P., Martinez, A., SanSegundo, B., 2001. A protein from the mold *Aspergillus giganteus* is a potent inhibitor of fungal plant pathogens. Mol. Plant-Microbe Interact. 14, 1327–1331.

Viswanathan, R., Samiyappan, R., 1999. Induction of systemic resistance by plant growth promoting rhizobacteria against red rot disease in sugarcane. Sugar Tech 1 (3), 67–76.

von Bodman, S.B., Bauer, W.D., Coplin, D.L., 2003. Quorum sensing in plant-pathogenic bacteria. Annu. Rev. Phytopathol. 41 (1), 455–482.

Walker, M.J., Birch, R.G., Pemberton, J.M., 1988. Cloning and characterization of an albicidin resistance gene from *klebsiella oxytoca*. Mol. Microbiol. 2 (4), 443–454.

Wang, Y., Nowak, G., Culley, D., Hadwiger, L.A., Fristensky, B., 1999. Constitutive expression of pea defense gene DRR206 confers resistance to blackleg (*Leptosphaeria maculans*) disease in transgenic canola (*Brassica napus*). Mol. Plant-Microbe Interact. 12 (5), 410–418.

Wei, G., Kloepper, J.W., Tuzun, S., 1991. Induction of systemic resistance of cucumber to *Colletotrichum orbiculare* by select strains of plant growth-promoting rhizobacteria. Phytopathology 81, 1508–1512.

Wei, G., Kloepper, J.W., Tuzun, S., 1996. Induced systemic resistance to cucumber diseases and increased plant growth by plant growth-promoting rhizobacteria under field conditions. Phytopathology 86, 221–224.

Weller, D.M., 1988. Biological control of soilborne plant pathogens in the rhizosphere with bacteria. Annu. Rev. Phytopathol. 26, 379–407.

Whipps, J.M., 2001. Microbial interactions and biocontrol in the rhizosphere. J. Exp. Bot. 52, 487–511.

Wicaksono, W.A., Jones, E.E., Casonato, S., Monk, J., Ridgway, H.J., 2018. Biological control of *Pseudomonas syringae* pv. *actinidiae* (Psa), the causal agent of bacterial canker of kiwifruit, using endophytic bacteria recovered from a medicinal plant. Biol. Control 116, 103–112.

Wilhelm, E., Arthofer, W., Schafleitner, R., Krebs, B., 1997. *Bacillus subtilis* an endophyte of chestnut (*Castanea sativa*) as antagonist against chestnut blight (*Cryphonectria parasitica*). In: Pathogen and Microbial Contamination Management in Micropropagation. Springer, Dordrecht, pp. 331–337.

Xia, Y., DeBolt, S., Dreyer, J., Scott, D., Williams, M.A., 2015. Characterization of culturable bacterial endophytes and their capacity to promote plant growth from plants grown using organic or conventional practices. Front. Plant Sci. 6, 490.

Xiao-Yan, S., Qing-Tao, S., Shu-Tao, X., Xiu-Lan, C., Cai-Yun, S., Yu-Zhong, Z., 2006. Broad-spectrum antimicrobial activity and high stability of trichokonins from *Trichoderma koningii* SMF2 against plant pathogens. FEMS Microbiol. Lett. 260, 119–125.

Yamamoto, S., Shiraishi, S., Suzuki, S., 2015. Are cyclic lipopeptides produced by *Bacillus amyloliquefaciens* S13-3 responsible for the plant defence response in strawberry against *Colletotrichum gloeosporioides*? Lett. Appl. Microbiol. 60 (4), 379–386.

Yan, Z., Reddy, M.S., Yyu, C.M., Mc Inroy, J.A., Wilson, M., Kloepper, J.W., 2002. Induced systemic protection against tomato late blight by plant growth-promoting rhizobacteria. Phytopathology 92, 1329–1333.

Zeriouh, H., Romero, D., García-Gutiérrez, L., Cazorla, F.M., de Vicente, A., Pérez-García, A., 2011. The iturin-like lipopeptides are essential components in the biological control arsenal of *Bacillus subtilis* against bacterial diseases of cucurbits. Mol. Plant-Microbe Interact. 24 (12), 1540–1552.

Zhang, L., Birch, R.G., 1997. The gene for albicidin detoxification from *Pantoea dispersa* encodes an esterase and attenuates pathogenicity of *Xanthomonas albilineans* to sugarcane. Proc. Natl. Acad. Sci. 94 (18), 9984–9989.

Zhang, X., Li, B., Wang, Y., Guo, Q., Lu, X., Li, S., Ma, P., 2013. Lipopeptides, a novel protein, and volatile compounds contribute to the antifungal activity of the biocontrol agent *Bacillus atrophaeus* CAB-1. Appl. Microbiol. Biotechnol. 97 (21), 9525–9534.

CHAPTER 9

Microbiome stimulants and their applications in crop plants

Shristi Bhandari[a], Sarvjeet Kukreja[b], Vijay Kumar[a], Abhijit Dey[c], and Umesh Goutam[a],*

[a]School of Bioengineering and Biosciences, Lovely Professional University, Phagwara, Punjab, India
[b]School of Agriculture, Lovely Professional University, Phagwara, Punjab, India
[c]Department of Life Sciences, Presidency University, Kolkata, India
*Corresponding author: e-mail address: umeshbiotech@gmail.com

1. Introduction

The connection between plants and microorganisms is believed to be old, and the mutualism of arbuscular mycorrhizal is likely to have played a crucial role in the growth and diversity of plants in terrestrialization (Atamna-Ismaeel et al., 2012). Plant microbiome research has improved considerably in recent years. The study of the plant microbiome involves binding the biology and operation of the plant host on the microbial ecology and seeing microbes as a storage place for additional genes and host activities (Bulgarelli et al., 2013). There is widespread recognition of the role of cooperating microbial symbionts in their hosts' lives and fitness. A plant may therefore be regarded as a holobiont comprising both the plant host and its microbiota (Zilber-Rosenberg and Rosenberg, 2008). The term holobiont is an overview of the activities and interactions between a macroorganism host and its surrounding molecules (i.e., a single dynamic entity). The holobiome is described as the genetic representativeness of the complex symbiotic network connecting a person from a taxon with their microbiome. The holobiome is the genetically modified representation (Guerrero et al., 2013). The rhizosphere is the initial habitat of soil microorganisms impacted by plants located on the perimeter of the root. Plant metabolism impacts a thin layer of soil around the roots through oxygen release and the separation of a whole range of exudates that contain carbon-rich compounds that microbes may use as energy sources and antibacterial substances. In general, it is a dynamic environment for the rhizospheres between soil and roots and, therefore, the variety of the microbial population (Schreiter et al., 2014). Plant-profitable microorganisms are well known as feasible alternatives to traditional fertilizers and insecticides in agriculture. The influence of plant-associated and soil microbiome management methods was examined on a range of crop species. It is essential to have a deeper ecological awareness of the complex interactions in a changing environment between the beneficial strains imported and the local microbiota. The formulation of beneficial microbial strains and the characteristics of

Relationship Between Microbes and the Environment for Sustainable Ecosystem Services, Volume 1
https://doi.org/10.1016/B978-0-323-89938-3.00009-8

substratum growth and commercial agriculture practices influence these complex interactions considerably (Vassilev et al., 2021).

2. Plant microbiome

For the survival of Earth life, microbial is necessary. The plant microbiome is a major contributor to plant health and productivity and has lately received considerable attention. Isolated microorganisms have been carefully studied using culturally based techniques, while molecular techniques such as metagenomics allow more and more the identification of bacteria *in situ* (Turner et al., 2013). New depths are characterized by microbial Community Profiling, but functional insights are obtained largely from investigations utilizing individual bacteria. The fitness of the plant is damaged by whole microbial, which might promote and defeat growth (Schlaeppi and Bulgarelli, 2015). Due to the variety of microbes colonized above-ground plant areas, such as bacteria, yeasts, and filamentous fungi, protists, the full variety of these herbal microbes, as well as the factors forming complex plant microbial communities from host colonization to plant senescence, have been lacking in knowledge. Plants and behavior are profoundly impacted under natural settings by associated microbial communities, known as the microbiome (Agler et al., 2016). The ground–plant interface is the well-studied rhizosphere, whereas the phyllosphere represents the air–plant contact. This microhabitat is of particular importance because of its vast and exposed area, as well as the connection with air microbiomas, especially airborne illnesses. Plants can support plants to eradicate disease, stimulate growth, occupy areas normally occupied by a pathogen, enhance stress tolerance, and affect crop yield and quality by mobilizing and transporting nutrients. Different types of microorganisms that are creating microbiome around plants have been shown in Fig. 1 (Vassilev et al., 2021). Many bacteria colonize the microbiome of the plant as colonizers of the phyllosphere or rhizosphere, which produce plant exudates. The development and fitness of plants are usually not impacted by these symptoms. With molecular methods, the plant microbiome can now be rapidly characterized. Other microorganisms, such as arbuscular, nitrhoeal, endophytes, and disease-suppressive, influences the development of plants and the acquisition of nutrients (Berg et al., 2014). The constant advent of a high throughput sequence in conjunction with a range of "omics" techniques now allows researchers to recognize both the structure and dynamics of the microbiome and host interactions at an unprecedented level. The identification and relative abundance of microbiological companions of a plant may be very detailed by modern sequencing techniques since environmental samples create sequences immediately (Lucaciu et al., 2019). According to reductionist approaches, plant immunity and the plant microbiome are bidirectional, and plants, bacteria, and the environment together produce a complex chemical conversation that orchestrates the plant microbiome collectively. The next stage in plant microbiome research will

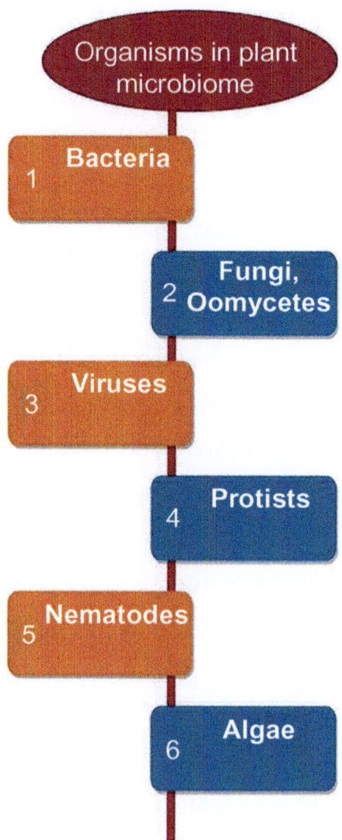

Fig. 1 Different types of organisms present in plant microbiome.

entail blending ecology and reductionist techniques to build a wide understanding of the assembly and function in natural and controlled contexts (Mahmud et al., 2021). A better understanding of the mechanisms behind the plant–microbe interactions will aid in the creation of plant probiotics that will improve agricultural production and provide plant resistance to biotic and abiotic stresses while reducing chemical inputs. Advances in molecular biology and high-throughput -omics are paving the door for a better knowledge of key microbiome gene activity that aids plant performance (Verma et al., 2021).

The promotion of plant growth by using soil microbiomes has enormous potential to provide the world's fast-growing population with an environmentally friendly solution to the growing demand for nutrition while helping to ease the associated environmental and social problems related to large-scale food production. Many studies on rhizosphere microbiome inhibiting disease and promoting plant development have been recently reported (Zhang et al., 2021). Weeds and plants battle for the light, food, and water, but interact differently with soil microorganisms. A further investigation of the interactions between "unwanted" plants and cultures under various management plans is feasible due to the emergence of new sequencing technologies for studying soil microbiome. These findings will help us to understand the functions of microorganisms in agricultural yields and plant health, weed production, and management (Trognitz et al., 2016).

3. Microbiome: A stimulant in crop plants

Agricultural products will be demanded globally with the growing world population projected to increase. As a result, there is an urgent need for sustainable improvement in agricultural production while dealing with small areas of cultivation and difficult climatic circumstances. The microbiome of plants may be a long-term approach to improve farm output, food and soil quality, and major crops having microbiome are shown in Fig. 2 (Verma, 2018). Plant microbiomes are just as important for plant health as human microbiomes are for human health. While rhizospheric bacteria have been extensively studied for decades, the more intimate interactions between plants and endophytes, or microorganisms that reside wholly within plants, have recently been studied. However, it is now clear that the plant microbiome substantially impacts plant growth and health (Doty, 2017). The term farming bio-energizers is ordinarily used to depict equations utilized for plants or soils to expand the strength, yields, quality, and stress resistance of synthetics, substances, or microorganisms. Plants normally remember microorganisms for seeds and in seeds that can advance seedlings' development and early endurance. Biostimulants remember humic materials for plants that seem, by all accounts, to be the sign atoms in plants and cause improved disguise of soil microorganisms to root cells and tissues (Vassilev et al., 2021). Basic microbiome contemplates, which assume a significant part in the upkeep and advantageous abuse of the equilibrium of the inside and outside conditions in plants, will forestall the prompt food lack fiasco and other financial and ecological factors issues (Song et al., 2021). Microbiomes with crop plant affiliations can invigorate development or control illnesses in plants. Microbiome reproducing, relocating, and focusing on microbiome designing might be new farming strategies. Microbiome designing control of the plant holobiont is a creating biotechnology

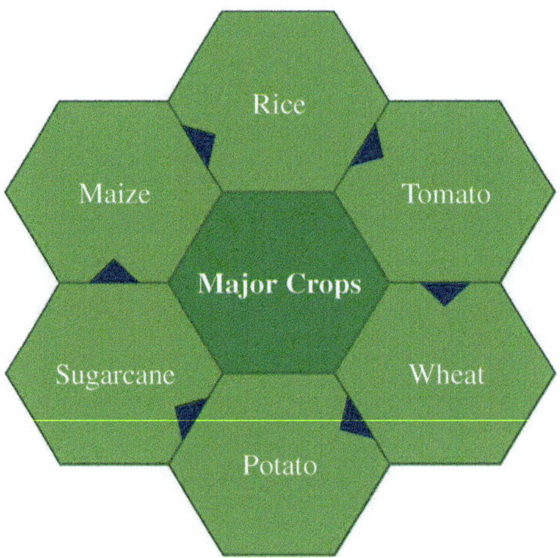

Fig. 2 Major crops having microbiome stimulants.

procedure to improve rural yield and versatility (Arif et al., 2020). Numerous microbial soothing frameworks, including microbial items, compounds, physiological, atomic cycles, pathway disturbance, signals, and plant administrative systems, are being utilized to fight pressure and increment crop development and creation (Sankaranarayanan et al., 2021). Rice is the most broadly developed food crop on the planet, adding to human food security and monetary flourishing. Rice plants use root exudates to choose and shape their microbiomes, outstandingly in their rhizospheres. Simultaneously, rice microbiome flagging influences have plant development and improvement, just as sustenance (Sugavanam et al., 2021).

Microbial energizers (MS, for example, arbuscular mycorrhizal parasites (AMFs) and plant development advancing rhizobacterias (PGPRs)), which abide in the rhizosphere, have as of late been among the issues that have gotten a ton of consideration as far as manageable horticulture in vegetable creation. The utilization of AMFs and PGPRs builds the plant's supplement assimilation from the dirt, which assists with plant advancement, yield, and natural product quality (Seymen et al., 2021). Sierra Mixe maize is a far-off landrace type developed in Oaxaca's nitrogen-insufficient districts. Mexico's nourishing necessities are met without the utilization of engineered manures by shaping partnerships with free-living diazotrophs found in the microbiota of its airborne root adhesive. Marker qualities for a few direct plant development upgrade techniques were found in diazotrophic groupings (PGP). These organic entities might empower potato and conventional maize development by utilizing an assortment of advancement techniques (Higdon et al., 2020). Broomrapes (*Phelipanche/Orobanche* spp.) are holoparasitic plants that feed on the underlying foundations of numerous horticultural yields, framing direct associations with the host vascular framework. The microbiome of the Parasitic Weed *Phelipanchea egyptiaca* and Tomato-*Solanum lycopersicum* (Mill.) as a host contains microscopic organisms that assume significant valuable parts in numerous parts of their host plant's science, for example, upgraded have development rate, speeding up seed germination, expanded pressure resilience, and the arrangement of basic supplements to the host (IasurKruh et al., 2017).

4. Microbiome stimulants in major crops

4.1 Tomato

Early plant growth was shown to be driven by the spatial development of available nutrients through root growth promotion, primarily through improved inoculants and changes in rhizosphere microbes. Soil microorganisms interact with roots which impact plants' growth and nutrient acquisition (Eltlbany et al., 2019). Colonizing plant rhizosphere microorganisms can help plant health, growth, and productivity. In conjunction with one crop in seven different soil and growing soil sources, variation, assembly, and composition of rhizobacterial communities in 11 tomato cultivars have been carefully investigated. Microbiota Rhizosphere Tomato has dominated Phyla Proteobacteria, Acidobacteria, and Bacteroidetes bacteria (Cheng et al., 2020). The *Ralstonia solanacearum*, a soilborne pathogen, remains resistant to the Hawaiian 7996 tomato variety but is vulnerable to the Moneymaker type. Bacteria of RRSP have been found to decrease disease symptoms from resistant plants as compared to sensitive plants. A rhizosphere metagenomes from the resistant and susceptible plants were analyzed to allow for the detection and assembly of a *Flavobacterium* that was considerably more common in resistant plant rhizosphere than in susceptible plant rhizosphere microbiome (Kwak et al., 2018). Endophytes were widely investigated in order to improve plant growth and biocontrol agents like biostimulants. Nine endophytic bacterial species were discovered as members of the phyla firmicutes and proteobacteria. The results revealed that the most beneficial plant properties identified in tomato root endophytes were bacteria of three genera: *Pseudomonas*, *Rhizobium*, and *Bacillus*, respectively (Tian et al., 2017). Bacillus representatives are progressively used in agricultural plant development and plant disease protection. The influences of Bacillus biostimulants on indigenous microbiota plants were investigated exclusively by species *Bacillus amyloliquefaciens*, respectively. Tomato plants and the elimination of soil transmission conditions were encouraged by *Bacillus subtilis* root colonizer PTS-394. Roche 454 pyrosequencing showed that the PTS-394 only had a transient effect on the microbial community of the tomato rhizosphere (Qiao et al., 2017). The construction of the microbiome is a spatially and dynamically root-driven process regulated by the soil type, plant development, and genotype. The effect of plant genetic species on microbiome assembly has been documented for farmed crop species and their wild families, for genotypes of the same plants and plants with mutations in specific genes and pathways (Cordovez et al., 2021).

4.2 Rice

Rice plants choose microbes that aid their growth, nutrient uptake, induced resistance, or pathogen antagonism. The usage of cyanobacteria as bioinoculants has been demonstrated to benefit rice plants. Cyanobacteria are important microbiological members in rice fields, and they are employed as bioinoculants to boost fertility, soil structure, and crop yields (Priya et al., 2015). Endophytes are microorganisms that live between plant cells and can be toxic or beneficial. The connection they make with the plant and their potential role in plant health are mostly unknown, and it is believed that there is considerable potential for uncovering unique and fascinating endophytes among the varied plant kingdom. A total of 1318 putative endophytes were isolated from rice roots, leaves, and stems growing in submerged and dry conditions, yielding a working collection of 229 isolates. Endophytes were discovered in many isolates, and a few of them exhibited the potential to support plant growth (Bertani et al., 2016). Plants rely on the interaction of roots and microbes for nutrition availability, growth stimulation, and disease control. Rice cultivation is a major source of global methane emissions, and potential microbial consortia engaged in certain methanol cycle activities was discovered in every geographical area of field-grown rice and study. Dynamic changes in microbiome acquisition, as well as consistent compositions of spatial compartments, support a multistage model for root microbiome soils in which the rhizoplane plays a particular gateway function (Edwards et al., 2015). Plants are exposed to a variety of biological and abiotic stimuli throughout their lives. Phytobiomes are the complete system that includes plants, their environment, and all animals in these settings. Abiotic and biotic components of the phytobiome interact dynamically, affecting natural ecosystems and agroecosystems.

The plant-related microbiome is used among biotic components to describe the plant microbiome or microbiota (Kim and Lee, 2020). Rice infected with endophytic bacteria improves plant growth and grain production significantly. The inoculation of rice seeds with endophytic bacteria stimulates robust seedling growth, critical for plant establishment and grain production (Hardoim, 2015). The use of low (LN), standard (SN), and high (HN) N fertilizers in paddy rice ecosystems has been explored as one of the field management goals for sustainable agricultural operations and changes in bacterial populations. The relative abundance of Burkholderia, Bradyrhizobium, and Methylosinus have significantly increased in the root microbiome of the field LN compared to the field SN (Ikeda et al., 2014). Interactions between the host, bacteria, and environment form the structure and function of microbial communities. Interactions have been established. Microbial hubs have been identified to regulate microbiome and rice genomic regions of network structure that monitor the quantity of these hubs, enriched for activities linked to stress responses and glucose metabolism (Roman-Reyna et al., 2019).

4.3 Wheat

Wheat has been one of the oldest domesticated plants between 7000 and 9000 BCE and has spread to include worldwide farming ever since. Broadwheat, *Triticum aestivum* L., with more than 20,000 variations, is the most commonly cultivated species. It represents 17% of all cultivated land in the globe and supplies food to 35% of the world's population. It is one of the most significant crops in the world (Bell and Lupton, 1987). A viable agricultural intensification process, resulting in increased crop tolerances for biotic and abiotic stresses, improved efficiency in nutrient use, and the development of new biofertilizers, should be implemented to achieve considerable crop improvement (Misra et al., 2020). Various factors were examined alone or in connection with the niche to detect their influence on the wheat microbiome. This includes human (anthropogenic), soil (edaphic), environment variables linked to natural conditions and herbal reliability, and the microorganisms that help to survive in the abiotic stress condition have been given in Table 1 (Kavamura et al., 2021). Previous studies on wheat microbiomes

Table 1 The wheat microbiome that induces plant growth in different abiotic stress conditions.

Abiotic stress condition	Microorganisms	References
Low temperature	*Mycobacterium phlei* MbP18, *Mycobacterium* sp. 44, *Mycoplana bullata* MpB46, *Pantoeaagglomerans 050309* and *Pseudomonas fluorescens* PsIA12 *Pseudomonas* sp. NARs9, *P. fluorescens* PPRs4, *Pseudomonas jessani* PGRs1, *Pseudomonas koreensis* PBRs7, *Pseudomonas lurida* NPRs3 and *Pseudomonas putida* PGRs4	Mishra et al. (2011), Abbaspoor et al. (2009), and Ashraf et al. (2004)
Salinity	*P. fluorescens* 153 *P. putida* 108 *Aeromonas hydrophila* MAS-765 *Bacillus insolitus* MAS17, *Bacillus* sp. MAS617/620/820 *Achromobacter xylosoxidans* 249 *Enterobacter* sp. 12 *Pseudomonas* sp. 33 *Serratia marcescens* 73	Ali et al. (2011) and Alvarez et al. (1996)
Drought and heat stress	*P. putida* AKMP7 *Azospirillumbrasilense* sp245 *Azospirillumlipoferum* AZ1 *Bacillus safensis* W10 *Ochrobactrum pseudogregnonense* IP8 *Pantoea theicola* NBRC 110557T	Arzanesh et al. (2011), Bender et al. (2016), and Pokharel (2011)

have largely centered on the identification of microorganisms' roots or rhizosphere, with much less focus on overgrown organisms. The complete wheat microbiota, both in overgrowth and below ground, has been examined using high-performance sequencing techniques by no published studies. This category is divided into three bacterial, fungal microbiome systems (leaves and roots) grown by four land management approaches (no-till, conventional tillage, low input, and organic) (Gdanetz and Trail, 2017). Plant microorganisms that stimulate plant growth are directly or indirectly connected to plant roots and encourage plant development. The microbial use of PGP is a promising agricultural technology that plays an essential role in the preservation of wheat, growth, and managing biological illnesses and soil fertility. The efficient application of either rhizobial or PGP microbial inoculants in wheat needs the transfer of sustainable bacteria to the root zone by means of coated seeds or bulk inoculations, most commonly through the use of inoculations for dormant bacterial cell preparedness (Yadav, 2019).

4.4 Potato

Genome studies have been carried out in recent years to investigate cells, organisms, and populations with high performance and to detect many processes and interactions between nematodes and bacteria (Buchholz et al., 2021). Two main vegetable-parasite nematodes in the San Luis Valley of Colorado potatoes are Columbia nematode root-knot (*Meloidogyne chitwoodi*) and root-lesion nematode (*Pratylenchus neglectus*). In commercial potatoes in the Northwest USA, *Meloidogyne chitwoodi* is a major pest and contains other hosts, such as tomatoes, oat, garlic, rye, and wheat (Malviya et al., 2021). In the past, potato plants have been carefully investigated in order to enhance growth and fitness in microbiota and rhizosphere. However, the microbiota and its role in the storage stability of post-harvest potato tuber are not well-defined. The stability of the pumpkin stock is based on the genotype and storage conditions. However, tubers can also be grown on the soil (Fischer et al., 2012).

4.5 Sugarcane

Production of sugar cane has to be sustainably enhanced, and techniques must be developed to deal with disease, abiotic stress, and pests. The employment of beneficial microorganisms as biofertilizers is increasingly important in sustainably cultivating sugar cane crops. In this regard, a sugar cane microbiome is an effective approach for sustainable production (Figueiredo et al., 2010). Sugar cane is also a source of sugar, renewables, and biomaterials and is one of the world's leading crops. Sugar cane is one of the world's most important crops. Sugar cane is cultivated in 110 tropical and subtropical countries and produced worldwide in Brazil and India by roughly 50% (Yoneyama et al., 1997). Rhizospheric bacteria, including diazotrophic bacteria, with cultivar-specific interactions, can be used for special controls for particular diazotrophic attractions. Variable responses of crops and growing areas often restrict the efficiency of rhizobacterial biofertilizers that cultivate diazotrophic and other plant cultivations (PGPR) (Reganold et al., 1990). *Azaspirillum*, *Gluconacetobacter*, *Burkholderia*, and *Herbaspirillum* diazotrophic bacteria have been discovered from sugar cane intercellular spaces, roots, and rhizospheres it remains difficult to evaluate BNF participation in the crop N budget. Diazotrophic bacteria vary greatly in sugar cane crops (Busby et al., 2017).

5. Microbiome stimulants and their application process

The major applications of the microbiome of the crop can be seen in Fig. 3.

5.1 Sustainable agriculture

Sustainable agriculture is not a reversal of preindustrial practices but a mixture of traditional methods to agricultural conservation with contemporary technologies. To

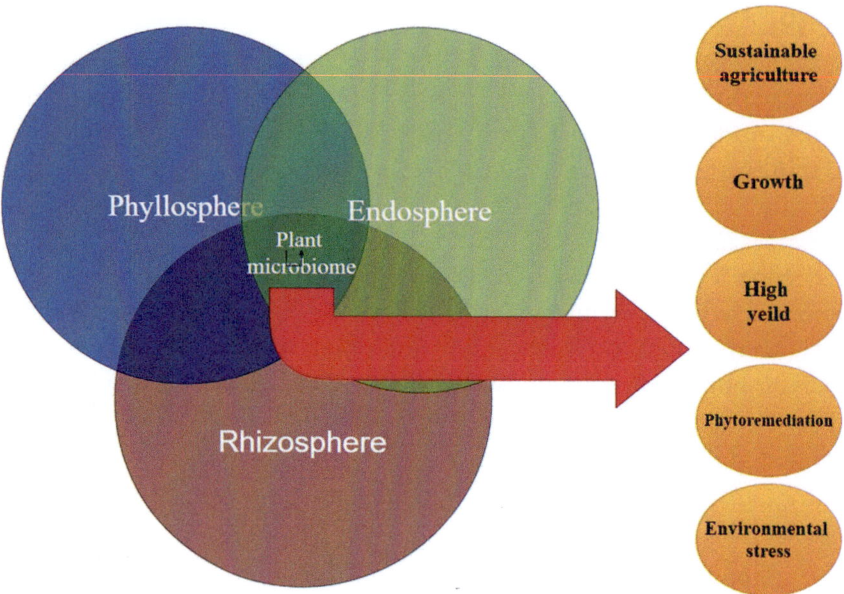

Fig. 3 Plant microbiome and its application in crop development.

understand soil significance for sustainable agriculture, the logic must first be understood (Igiehon and Babalola, 2018). Food production must be optimized in terms of dependability, resources, and the environmental effect to feed a growing world population while addressing climate change. The integration into sustainable agricultural production of useful plant microbiomes, which enhance plant growth, nutrient efficiencies, abiotic stress tolerance, and disease resistance, is one approach for achieving these aims (Barea, 2015).

Therefore international research of the structure and function of the crop microbiome is urgently needed to create effective and rationally designed microbiome technologies for sustainable agriculture. This effort will produce fresh knowledge on the crucial environmental and evolutionary links that may be exploited to promote farming productivity between plants and their microbiomes (Ke et al., 2021). All major issues relating to human life and the long-term health of global ecosystems include effective ways to recycle and manage pests and illnesses and decrease the negative impacts of abiotic stress factors. These are common microbial services that may control beneficial bacteria and their functioning properly (Hunter, 2016). Mutualistic plant-associated microorganisms provide enormous potential for economic and sustainable agriculture. Plant microbiome engineering may take two approaches: a lower up, involving isolation, engineering, and reintroduction of particular bacteria, or the top–down, involving synthetic ecology, involving horizontal transmission of genes to a wide range of hosts in the field and then a microbiome phenotyping (Parasuraman et al., 2019).

Plant microbiome investigation precedes the animal and human study, as microbial, notably nitrogen and phosphorus, perform a significant function in plant nutrient fixation. More than 80% of terrestrial plants appear to be the closest and earliest mycorrhizal fungus. Legumes have developed a tripartite combination of rhizobia and mycorrhiza to collect nutrients as effectively as possible (Castillo et al., 2017). In-depth analyses and studies on the dynamics and consequences of plant-associated phyllosphere microbiome can give new dimensions to sustainable agriculture strategies such as microbial inoculants as plant growth and development biofertilizers and improved crop production improvement of nutrient mobilization, quality management, and biocontrol agents (Jat et al., 2021). *Meloidogyne chitwoodi* and *Pratylenco neglectus* are the major plant-parasite nematodes in the San Luis Valley, Colorado potato crop. There have been five different bacterial microbiomes of potato farms (16S rRNA copies per gram of soil) and nematode groups (nematodes per 200 g of soil). *M. chitwoodi* was favorably connected to the Rhodoplanes, Phenylobacterium, and Kaistobacter proteobacteria, whereas *P. neglectus* was positive for Bacteroidia and Proteobacteria (Yadav et al., 2017). Along with potato, maize microbiome was investigated, and *B. safensis*, *Paenibacillus alvei*, *B. pumilus*, and *Brevundimonas vesicularis* were shown to improve yield by 24%–34%. The microbiome in maize systems is mostly studied in the endosphere, rhizosphere, and non-rhizosphere, with a little research on the phyllosphere as well. The root of native Mexican maize was discovered to contain nitrogen-fixing bacteria, which may be used to develop nitrogen-fixing maize varieties (Rascovan et al., 2016). Microbiome abundance in all environments as a PGP (biofertilizing) is known to work in sustainable agriculture, for micronutrient biologies like Fe and Zn and as a probiotic for novel functional nutrients. Extreme pH, temperature, the salinity of water, and stress include (Parnell et al., 2016).

5.2 Plant growth in crops

Beneficial, neutral, and harmful microbes make up the plant microbiome. Plant growth-promoting bacteria (PGPB) can stimulate plant development in a number of ways, both directly and indirectly. Auxin, cytokinin, and gibberellin are phytohormones produced by certain PGPB that impact plant development by regulating endogenous hormone levels in connection with a plant. Furthermore, certain PGPB can secrete an enzyme called 1-aminocyclopropane-1-carboxylate (ACC) deaminase, which lowers ethylene levels in the plant (Sarkar et al., 2018). Microbial consortia are a new way to get data from the lab to the field (Rolli et al., 2015). Traditionally, agricultural applications of helpful microorganisms have focused on a few well-studied microbes, such as mycorrhizal fungi or rhizobia bacteria. The processes behind plant growth promotion are well recognized. Furthermore, the majority of this research was primarily focused on the capacity of the applied microorganisms to promote certain plant growth-promoting characteristics such as phosphate solubilization, nitrogen fixation, and ACC deaminase synthesis (Vrieze et al., 2018).

Furthermore, metagenomics-based methods have revealed previously unknown populations of microorganisms that may have new or improved characteristics that might be useful in agriculture, bioremediation, and human health. Comparative studies of rhizosphere metagenomes from resistant and susceptible tomato plants, for example, resulted in the discovery and assembly of a flavobacteria genome that was significantly more common in the resistant plant rhizosphere microbiome than in the susceptible plant microbiome (Berg and Koskella, 2018). Microbial consortia can also include strains with similar modes of action that tolerate various environmental conditions or plant genotypes. Microbial combinations have the potential to improve plant growth-promoting (PGP) effects when compared to single inoculants, according to research on the grapevine, potato, tomato, Arabidopsis, and maize (Molina-Romero et al., 2017; Ahmad et al., 2019; Wang et al., 2020; Sarpong et al., 2020). Soil microbe benefits include biological nitrogen fixation, phytohormone generation, mobilization of and solubilization of nutrients, biocontrol activity, hydrolytic enzyme synthesis, and pressure tolerance induction. These characteristics of helpful microorganisms may be exploited by improving soil health, plant growth and production, and enhancing crop stresses. Improvement in beneficial microbial populations by designing or using rhizosphere microbial inoculants and/ or their metabolites can assist in altering the microbiome of the soil, leading to enhanced agroecosystem production (Ahmad et al., 2019). Biochar soil amendment enhanced the beneficial bacteria Bacillus and Lysobacter while suppressing pathogens Fusarium and Ilyonectria, according to a study of soil microbial composition. Biochar addition enhanced soil pH, available potassium, electrical conductivity, organic matter, available phosphorus, and C/N ratio substantially (Hussain et al., 2018). Microbial-based agricultural supplements are boosting nutrient-use efficiency and crop production across the board, and they may soon be a viable alternative to certain standard plant nutrient management techniques (Ansari, 2018).

5.3 Stress management

The advantages of the green revolution are over since the present agricultural production has affected efficiency limitations in growing crop yields. Shrinking agricultural areas, costing work, and biotic and abiotic stressors aggravate this dilemma. Indeed, using and using microorganisms of agricultural significance as an Alternative Strategy for greater crop yield would be crucial to global agriculture and increased output. The combined management of biotic and abiotic stressors as well as the management of nutrients play a major part in efficient microorganisms to decrease chemical usage and enhance crop performance (Chen et al., 2019). Stress is the main factor in reducing farm output, and the major sources of plant stress are shown in Fig. 4. The production of reactive

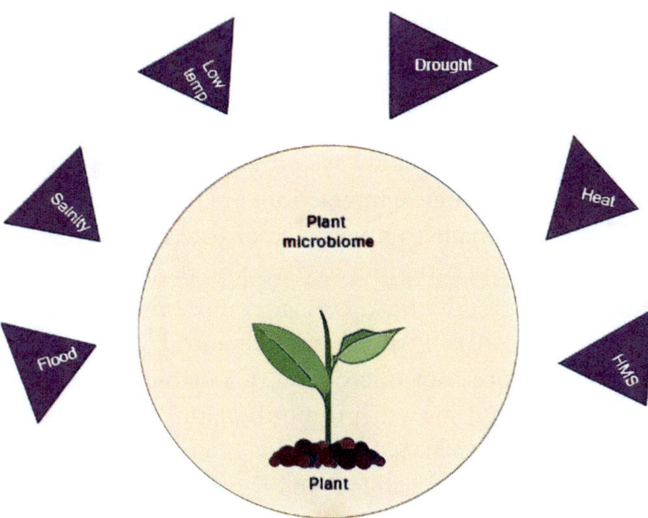

Fig. 4 Plant microbiome in abiotic stress management.

oxygen species in different cell compartments is responsible for prolonged stress situations. ROS assaults and disrupts the cell's normal mechanism that ultimately leading to cell death. Crops needed to acclimate to external stress caused by environmental circumstances overcome by their natural biological systems, which sustain both growth and production. Microbes have developed an elaborate physiological and metabolic system for managing hazardous oxygen species created by stress, the supreme natural inhabitants in different habitats (Yadav, 2017). In the biosphere, antibiotic resistance is rising, but, in comparison to its prevalence in soil and water habitats, the antibiotic-resistant in the plant microbiome is neglected. The plant microbiome can serve as the interface between the natural and human microbiome and is a critical route for human exposure to environmental resistance to antibiotics. The spread of antibiotic resistance via the food chain can be compounded by plant-based microorganisms, direct contact, and globalization (Thijs et al., 2016). The extreme natural biosources are plant microbiomes and microbiomes of extreme conditions. These are significant bio-resources for the justified use of chemical fertilizers in agriculture because of the diversity of activities and the number of microorganisms in various environments worldwide. In order to substitute fertilizer chemical products and improve abiotic stress in normal or stressed circumstances, microorganisms with diverse PGP characteristics can be used as biofertilizers (Thijs et al., 2017).

5.4 Phytoremediation

Phytoremediation is a potential method for purifying polluted soils based on plants and microorganisms' synergistic activity. But a greater knowledge of plant-microbe interactions is necessary in order to be a generally accepted and predictable remediation solution.

Some studies relate the achievement of phytoremediation with the microbiome functioning linked with plants (Niu et al., 2021). Phytoremediation is being used as a technique for soil remediation that is more sustainable. However, major improvements in efficiency are still needed in order to achieve greater environmental and economic sustainability. Current treatments may not always provide desired results in field environments as the multicomponent biological interactions are not fully grasped. New progress in omics is progressively being made to investigate on-site contaminated land microbial populations. This creates new opportunities to detect biodegradable strains and provides us with new strategies to interfere with microbial communities to increase the pace of bioremediation (Lampis et al., 2015). The metal-tolerant *H. cannibunus* plant was probably chosen for a particular core root microbiota. Cannabinus, which has the greatest number of metals toller and plant growth-promoting families (PGPB), are *Pseudomonadaceae*, *Enterobacteriaceae*, and *Comamonadaceae*. The microbial community of the root tended to develop niche constructed patterns and forms a smaller bacterial pool predominating in metal-polluted circumstances, including proteobacteria, actinobacteria, and chloroflexi (Jong et al., 2015). Table 2 has been given that provides the list of plants used in phytoremediation with the help of their microbiome.

6. Reduce pathogen infection

In the order of tens of thousands of species, the variety of microorganisms associated with root plants is tremendous. This complex plant-related microbial population is important to plant health, sometimes referred to as the second genome of the plant. Recent developments in interacting plant/microbe interactions have shown that plants are able, as is shown by the fact that plants have unique microbial communities when grown on a single ground. This study examines evidence that plants can recruit defensive microorganisms after infections or insect attacks and increase microbial activity to eliminate rhizospheric

Table 2 Plant species having microbiome with phytoremediation traits.

Plants	References
P. vittata	Senthilkumar et al. (2005)
Stanleya pinnata	Khan et al. (2015)
Astragalus bisulcatus	Khan et al. (2015)
Physaria bellis	Khan et al. (2015)
Medicago sativa	Khan et al. (2015)
Prosopis juliflora	Ma et al. (2015)
Lolium multiflorum L.	Ma et al. (2015)
Ricinus communis	Ternes et al. (2007)
Brassica juncea	Ternes et al. (2007)
Phragmites australis	Ternes et al. (2007)

diseases (Berendsen et al., 2012). In natural complex microbial communities, *Pseudomonas* species consortiums assessed how significant *Pseudomonas* Community Diversity is to their survival and the inhibition of *Ralstonia solanacearum* bacterial plant disease in the microbiome of tomato rhizosphere. With growing variety, the survival of the imported Pseudomonas consortium has risen.

Moreover, the great diversity of Pseudomonas reduces rhizosphere pathogen density and reduces the frequency of illness due to both increased competition of resources and pathogen interference (Hu et al., 2016). A fast-developing new topic in plant biology is the description of plant microbiomes by sequencing DNA, comprehensive bioinformatics analyses, and the creation of large-scale culture component collection. This is viewed as an application for the reduction of plant disease in novel methods (Ellis, 2017). The theory is that healthy plants' microbiome protects them against pathogenic diseases, which leads to improved yielding and wealth-generating disease control technology by establishing and promoting a "healthy microbiome" by using a complex consortium of seeds, cuttings, and tubers or directly on soils and crops. This idea is partially based on many years of study of "disease suppressive soils," where soil microbiota, even in pathogens, is considered to eliminate crop illnesses (Sessitsch and Mitter, 2015; Zhalnina et al., 2021; Qiu et al., 2020; da Silva et al., 2014).

7. Conclusion

Food insecurity has been worse over the last few decades as a result of population growth and climate change. The introduction of agrochemicals provided some relief, but it came with its own set of drawbacks. Microbes are able to inhibit the growth of phytopathogens and promote plant growth as a result of their interactions. Microbial interactions, metabolic accumulation of waste products, and competition for limited resources all restrict the best use of consortia for improving agricultural yield. Recent advances in proteomics, metabolomics, and computational modeling have proven beneficial in enhancing the utility of targeted microorganisms for consortia usage.

References

Abbaspoor, A., et al., 2009. The efficiency of plant growth promoting rhizobacteria (PGPR) on yield and yield components of two varieties of wheat in salinity condition. Am Eurasian J Sustain Agric 3 (4), 824–828.

Agler, M.T., et al., 2016. Microbial hub taxa link host and abiotic factors to plant microbiome variation. PLoS Biol. 14 (1), e1002352.

Ahmad, M., Nadeem, S.M., Zahir, Z.A., 2019. Plant-microbiome interactions in agroecosystem: an application. In: Microbiome in Plant Health and Disease. Springer, Singapore, pp. 251–291.

Ali, S.Z., et al., 2011. Effect of inoculation with a thermotolerant plant growth promoting Pseudomonas putida strain AKMP7 on growth of wheat (Triticum spp.) under heat stress. J. Plant Interact. 6 (4), 239–246.

Alvarez, M.I., Sueldo, R.J., Barassi, C.A., 1996. Effect of Azospirillum on coleoptile growth in wheat seedlings under water stress. Cereal Res. Commun., 101–107.

Ansari, M.I., 2018. Plant microbiome and its functional mechanism in response to environmental stress. Int. J. Green Pharm. 12 (01).

Arif, I., Batool, M., Schenk, P.M., 2020. Plant microbiome engineering: expected benefits for improved crop growth and resilience. Trends Biotechnol. 38 (12), 1385–1396.

Arzanesh, M.H., et al., 2011. Wheat (Triticum aestivum L.) growth enhancement by Azospirillum sp. under drought stress. World J. Microbiol. Biotechnol. 27 (2), 197–205.

Ashraf, M., et al., 2004. Inoculating wheat seedlings with exopolysaccharide-producing bacteria restricts sodium uptake and stimulates plant growth under salt stress. Biol. Fertil. Soils 40 (3), 157–162.

Atamna-Ismaeel, N., et al., 2012. Microbial rhodopsins on leaf surfaces of terrestrial plants. Environ. Microbiol. 14 (1), 140–146.

Barea, J.M., 2015. Future challenges and perspectives for applying microbial biotechnology in sustainable agriculture based on a better understanding of plant-microbiome interactions. J. Soil Sci. Plant Nutr. 15 (2), 261–282.

Bell, G.D.H., Lupton, F.G.H., 1987. Wheat breeding. pp. 31–50.

Bender, S.F., Wagg, C., van der Heijden, M.G.A., 2016. An underground revolution: biodiversity and soil ecological engineering for agricultural sustainability. Trends Ecol. Evol. 31 (6), 440–452.

Berendsen, R.L., Pieterse, C.M.J., Bakker, P.A.H.M., 2012. The rhizosphere microbiome and plant health. Trends Plant Sci. 17 (8), 478–486.

Berg, M., Koskella, B., 2018. Nutrient-and dose-dependent microbiome-mediated protection against a plant pathogen. Curr. Biol. 28 (15), 2487–2492.

Berg, G., et al., 2014. Unraveling the plant microbiome: looking back and future perspectives. Front. Microbiol. 5, 148.

Bertani, I., et al., 2016. Rice bacterial endophytes: isolation of a collection, identification of beneficial strains and microbiome analysis. Environ. Microbiol. Rep. 8 (3), 388–398.

Buchholz, F., et al., 2021. 16S rRNA gene-based microbiome analysis identifies candidate bacterial strains that increase the storage time of potato tubers. Sci. Rep. 11 (1), 1–12.

Bulgarelli, D., et al., 2013. Structure and functions of the bacterial microbiota of plants. Annu. Rev. Plant Biol. 64, 807–838.

Busby, P.E., et al., 2017. Research priorities for harnessing plant microbiomes in sustainable agriculture. PLoS Biol. 15 (3), e2001793.

Castillo, J.D., Vivanco, J.M., Manter, D.K., 2017. Bacterial microbiome and nematode occurrence in different potato agricultural soils. Microb. Ecol. 74 (4), 888–900.

Chen, Q.-L., et al., 2019. Antibiotic resistomes in plant microbiomes. Trends Plant Sci. 24 (6), 530–541.

Cheng, Z., et al., 2020. Revealing the variation and stability of bacterial communities in tomato rhizosphere microbiota. Microorganisms 8 (2), 170.

Cordovez, V., et al., 2021. Successive plant growth amplifies genotype-specific assembly of the tomato rhizosphere microbiome. Sci. Total Environ. 772, 144825.

da Silva, D.P., et al., 2014. Bacterial multispecies studies and microbiome analysis of a plant disease. Microbiology 160 (3), 556–566.

Doty, S.L., 2017. Functional importance of the plant endophytic microbiome: implications for agriculture, forestry, and bioenergy. In: Functional Importance of the Plant Microbiome. Springer, Cham, pp. 1–5.

Edwards, J., et al., 2015. Structure, variation, and assembly of the root-associated microbiomes of rice. Proc. Natl. Acad. Sci. 112 (8), E911–E920.

Ellis, J.G., 2017. Can plant microbiome studies lead to effective biocontrol of plant diseases? Mol. Plant Microbe Interact. 30 (3), 190–193.

Eltlbany, N., et al., 2019. Enhanced tomato plant growth in soil under reduced P supply through microbial inoculants and microbiome shifts. FEMS Microbiol. Ecol. 95 (9), fiz124.

Figueiredo, M.V.B., et al., 2010. Plant growth and health promoting bacteria. Microbiol. Monogr.

Fischer, D., et al., 2012. Molecular characterisation of the diazotrophic bacterial community in uninoculated and inoculated field-grown sugarcane (Saccharum sp.). Plant and Soil 356 (1), 83–99.

Gdanetz, K., Trail, F., 2017. The wheat microbiome under four management strategies, and potential for endophytes in disease protection. Phytobiomes 1 (3), 158–168.

Guerrero, R., Margulis, L., Berlanga, M., 2013. Symbiogenesis: the holobiont as a unit of evolution. Int. Microbiol. 16 (3), 133–143.

Hardoim, P.R., 2015. Heading to the origins-rice microbiome as functional extension of the host. J. Rice Res. 3 (2), 1–3.

Higdon, S.M., et al., 2020. Diazotrophic bacteria from maize exhibit multifaceted plant growth promotion traits in multiple hosts. PLoS One 15 (9), e0239081.

Hu, J., et al., 2016. Probiotic diversity enhances rhizosphere microbiome function and plant disease suppression. MBio 7 (6), e01790-16.

Hunter, P., 2016. Plant microbiomes and sustainable agriculture: deciphering the plant microbiome and its role in nutrient supply and plant immunity has great potential to reduce the use of fertilizers and biocides in agriculture. EMBO Rep. 17 (12), 1696–1699.

Hussain, S.S., Mehnaz, S., Siddique, K.H.M., 2018. Harnessing the plant microbiome for improved abiotic stress tolerance. In: Plant Microbiome: Stress Response. Springer, Singapore, pp. 21–43.

IasurKruh, L., et al., 2017. Host-parasite-bacteria triangle: the microbiome of the parasitic weed Phelipancheaegyptiaca and tomato-*Solanum lycopersicum* (Mill.) as a host. Front. Plant Sci. 8, 269.

Igiehon, N.O., Babalola, O.O., 2018. Rhizosphere microbiome modulators: contributions of nitrogen fixing bacteria towards sustainable agriculture. Int. J. Environ. Res. Public Health 15 (4), 574.

Ikeda, S., et al., 2014. Low nitrogen fertilization adapts rice root microbiome to low nutrient environment by changing biogeochemical functions. Microbes Environ., ME13110.

Jat, S.L., et al., 2021. Microbiome for sustainable agriculture: a review with special reference to the corn production system. Arch. Microbiol., 1–23.

Sura-de Jong, M., et al., 2015. Selenium hyperaccumulators harbor a diverse endophytic bacterial community characterized by high selenium resistance and plant growth promoting properties. Front. Plant Sci. 6, 113.

Kavamura, V.N., et al., 2021. Defining the wheat microbiome: towards microbiome-facilitated crop production. Comput. Struct. Biotechnol. J. 19, 1200.

Ke, J., Wang, B., Yoshikuni, Y., 2021. Microbiome engineering: synthetic biology of plant-associated microbiomes in sustainable agriculture. Trends Biotechnol. 39 (3), 244–261.

Khan, M.U., et al., 2015. Cr-resistant rhizo- and endophytic bacteria associated with Prosopis juliflora and their potential as phytoremediation enhancing agents in metal-degraded soils. Front. Plant Sci. 5, 755.

Kim, H., Lee, Y.-H., 2020. The rice microbiome: a model platform for crop holobiome. Phytobiomes J. 4 (1), 5–18.

Kwak, M.-J., et al., 2018. Rhizosphere microbiome structure alters to enable wilt resistance in tomato. Nat. Biotechnol. 36 (11), 1100–1109.

Lampis, S., et al., 2015. Promotion of arsenic phytoextraction efficiency in the fern *Pteris vittata* by the inoculation of As-resistant bacteria: a soil bioremediation perspective. Front. Plant Sci. 6, 80.

Lucaciu, R., et al., 2019. A bioinformatics guide to plant microbiome analysis. Front. Plant Sci. 10, 1313.

Ma, Y., et al., 2015. Serpentine bacteria influence metal translocation and bioconcentration of Brassica juncea and Ricinus communis grown in multi-metal polluted soils. Front. Plant Sci. 5, 757.

Mahmud, K., et al., 2021. Rhizosphere microbiome manipulation for sustainable crop production. Curr. Plant Biol., 100210.

Malviya, M.K., et al., 2021. Sugarcane microbiome: role in sustainable production. In: Microbiomes and Plant Health. Academic Press, pp. 225–242.

Mishra, P.K., et al., 2011. Alleviation of cold stress in inoculated wheat (Triticum aestivum L.) seedlings with psychrotolerant pseudomonads from NW Himalayas. Arch. Microbiol. 193 (7), 497–513.

Misra, M., Sachan, A., Sachan, S.G., 2020. Current aspects and applications of biofertilizers for sustainable agriculture. Plant Microbiomes Sustain. Agricult. 25, 445.

Molina-Romero, D., et al., 2017. Compatible bacterial mixture, tolerant to desiccation, improves maize plant growth. PLoS One 12 (11), e0187913.

Niu, H., et al., 2021. Behaviors of cadmium in rhizosphere soils and its interaction with microbiome communities in phytoremediation. Chemosphere 269, 128765.

Parasuraman, P., Pattnaik, S., Busi, S., 2019. Phyllosphere microbiome: functional importance in sustainable agriculture. In: New and Future Developments in Microbial Biotechnology and Bioengineering. Elsevier, pp. 135–148.

Parnell, J.J., et al., 2016. From the lab to the farm: an industrial perspective of plant beneficial microorganisms. Front. Plant Sci. 7, 1110.

Pokharel, R., 2011. Importance of Plant Parasitic Nematodes in Colorado Crops. Diss. Colorado State University Libraries.

Priya, H., et al., 2015. Influence of cyanobacterial inoculation on the culturable microbiome and growth of rice. Microbiol. Res. 171, 78–89.

Qiao, J., et al., 2017. Addition of plant-growth-promoting Bacillus subtilis PTS-394 on tomato rhizosphere has no durable impact on composition of root microbiome. BMC Microbiol. 17 (1), 1–12.

Qiu, Z., et al., 2020. Plant microbiomes: do different preservation approaches and primer sets alter our capacity to assess microbial diversity and community composition? Front. Plant Sci. 11, 993.

Rascovan, N., et al., 2016. Integrated analysis of root microbiomes of soybean and wheat from agricultural fields. Sci. Rep. 6 (1), 1–12.

Reganold, J.P., Papendick, R.I., Parr, J.F., 1990. Sustainable agriculture. Sci. Am. 262 (6), 112–121.

Rolli, E., et al., 2015. Improved plant resistance to drought is promoted by the root-associated microbiome as a water stress-dependent trait. Environ. Microbiol. 17 (2), 316–331.

Roman-Reyna, V., et al., 2019. The Rice Leaf Microbiome Has a Conserved Community Structure Controlled by Complex Host-Microbe Interactions. Available at SSRN 3382544,.

Sankaranarayanan, A., et al., 2021. Soil microbiome to maximize the benefits to crop plants—a special reference to rhizosphere microbiome. In: Microbiome Stimulants for Crops. Woodhead Publishing, pp. 125–140.

Sarkar, A., et al., 2018. Enhancement of growth and salt tolerance of rice seedlings by ACC deaminase-producing Burkholderia sp. MTCC 12259. J. Plant Physiol. 231, 434–442.

Sarpong, C.K., et al., 2020. Improvement of plant microbiome using inoculants for agricultural production: a sustainable approach for reducing fertilizer application. Can. J. Soil Sci. 101 (1), 1–11.

Schlaeppi, K., Bulgarelli, D., 2015. The plant microbiome at work. Mol. Plant Microbe Interact. 28 (3), 212–217.

Schreiter, S., et al., 2014. Effect of the soil type on the microbiome in the rhizosphere of field-grown lettuce. Front. Microbiol. 5, 144.

Senthilkumar, P., et al., 2005. Prosopis juliflora—a green solution to decontaminate heavy metal (Cu and Cd) contaminated soils. Chemosphere 60 (10), 1493–1496.

Sessitsch, A., Mitter, B., 2015. 21st century agriculture: integration of plant microbiomes for improved crop production and food security. J. Microbial. Biotechnol. 8 (1), 32.

Seymen, M., et al., 2021. Potential effect of microbial biostimulants in sustainable vegetable production. In: Microbiome Stimulants for Crops. Woodhead Publishing, pp. 193–237.

Song, S., et al., 2021. Mechanisms in plant–microbiome interactions: lessons from model systems. Curr. Opin. Plant Biol. 62, 102003.

Sugavanam, R., et al., 2021. Microbiome establishment, adaptation, and contributions to anaerobic stress tolerance and nutrient acquisition in rice. In: Microbiome Stimulants for Crops. Woodhead Publishing, pp. 369–379.

Ternes, T.A., et al., 2007. Irrigation of treated wastewater in Braunschweig, Germany: an option to remove pharmaceuticals and musk fragrances. Chemosphere 66 (5), 894–904.

Thijs, S., et al., 2016. Towards an enhanced understanding of plant–microbiome interactions to improve phytoremediation: engineering the metaorganism. Front. Microbiol. 7, 341.

Thijs, S., et al., 2017. Phytoremediation: state-of-the-art and a key role for the plant microbiome in future trends and research prospects. Int. J. Phytoremediation 19 (1), 23–38.

Tian, B., et al., 2017. Beneficial traits of bacterial endophytes belonging to the core communities of the tomato root microbiome. Agric. Ecosyst. Environ. 247, 149–156.

Trognitz, F., et al., 2016. The role of plant–microbiome interactions in weed establishment and control. FEMS Microbiol. Ecol. 92, 10.

Turner, T.R., James, E.K., Poole, P.S., 2013. The plant microbiome. Genome Biol. 14 (6), 1–10.

Vassilev, N., et al., 2021. Plant root interaction with associated microbiomes to improve plant resiliency and crop biodiversity. Front. Plant Sci. 12, 1269.

Verma, J.P., 2018. Functional importance of the plant microbiome: implications for agriculture, forestry and bioenergy: a book review. J. Clean. Prod. 178, 877–879.

Verma, S.K., et al., 2021. The roles of endophytes in modulating crop plant development. In: Microbiome Stimulants for Crops. Woodhead Publishing, pp. 33–39.

De Vrieze, M., et al., 2018. Combining different potato-associated Pseudomonas strains for improved biocontrol of Phytophthora infestans. Front. Microbiol. 9, 2573.

Wang, W., et al., 2020. Biochar application alleviated negative plant-soil feedback by modifying soil microbiome. Front. Microbiol. 11, 799.

Yadav, A.N., 2017. Agriculturally important microbiomes: biodiversity and multifarious PGP attributes for amelioration of diverse abiotic stresses in crops for sustainable agriculture. Biomed. J. Sci. Technol. Res. 1 (4), 861–864.

Yadav, A.N., 2019. Microbiomes of wheat (Triticum aestivum L.) endowed with multifunctional plant growth promoting attributes. EC Microbiol. 15 (4), 1–6.

Yadav, A.N., et al., 2017. Beneficial microbiomes: biodiversity and potential biotechnological applications for sustainable agriculture and human health. J. Appl. Biol. Biotechnol. 5 (6), 45–57.

Yoneyama, T., et al., 1997. The natural 15N abundance of sugarcane and neighbouring plants in Brazil, the Philippines and Miyako (Japan). Plant and Soil 189 (2), 239–244.

Zhalnina, K., et al., 2021. Managing plant microbiomes for sustainable biofuel production. Phytobiomes J., 3–13.

Zhang, J., et al., 2021. Harnessing the plant microbiome to promote the growth of agricultural crops. Microbiol. Res., 126690.

Zilber-Rosenberg, I., Rosenberg, E., 2008. Role of microorganisms in the evolution of animals and plants: the hologenome theory of evolution. FEMS Microbiol. Rev. 32 (5), 723–735.

CHAPTER 10

Microbes: A sustainable tool for healthy and climate smart agriculture

Surojit Bera[a], Richa Arora[b], Collins Njie Ateba[c], and Ajay Kumar[d],*

[a]Department of Microbiology, School of Bioengineering and Biosciences, Lovely Professional University, Jalandhar, Punjab, India
[b]Department of Microbiology, Punjab Agricultural University, Ludhiana, Punjab, India
[c]Food Security and Safety Niche Area, Faculty of Agriculture, Science and Technology, North West University, Mmabatho, Mafikeng, South Africa
[d]Department of Postharvest Science, Agriculture Research Organization, Volcani center, Rishon Leziyon, Israel
*Corresponding author.

1. Introduction

Soil is an important mediator for the growth of various vegetations on this earth including crops. The balance of various factors viz., physical, chemical, and biological reflecting the soil health status which is important from the point of view of a sustainable ecosystem especially the terrestrial one (Doran and Zeiss, 2000; Tahat et al., 2020). The good quality soil has the efficiency of generating the soil organic matter and helps in the decomposition of it, sequestering of carbon dioxide which is in excess amount helps in reduction of the level of carbon dioxide in the atmosphere. There are various factors that regulate/influence the carbon amount in soils like the diversity of microorganisms in the soil, type of soil, vegetation age and type, climate, land topography (Jenny, 1941). The population/diversity of microbes and biochemical reactions performed by them are influenced by the soil environment. At microscale level, the diverse microorganisms present in the soil assist in various catabolic and anabolic reactions which directly or indirectly influence the pH, the cycling of the nutrients, also influences the carbon dioxide and oxygen level in the soil (Brady and Weil, 2010).

A healthy soil has organic matter in sufficient amount and has diverse microbial inhabitants which help the soil to be resistant to various disturbances like drought, floods, and tillage (Doran and Zeiss, 2000; Tahat et al., 2020), and these soil microbial populations are in continuous interaction with various plant diversity (Berendsen et al., 2012; Kumar and Dubey, 2020). The diverse and efficient microbial population represents one of the good characteristics of healthy soil which also influences the physic-chemical parameters of healthy soil. The important microbial population includes bacteria, fungi, and archaea (Spence and Bais, 2013) which plays a crucial role in ecosystem processes. These microbes play an important role in the biochemical cycling of elements and make these elements in usable forms to the plants. These microbes have synergistic as well as antagonistic interactions with plants.

Relationship Between Microbes and the Environment for Sustainable Ecosystem Services, Volume 1
https://doi.org/10.1016/B978-0-323-89938-3.00010-4

For global food security, it is necessary to have intensive agricultural production systems but these systems utilize large amounts of chemical fertilizers, pesticides, and consumption of a huge amount of water. Excessive, continuous, and indiscriminate use of pesticides and chemical fertilizers leads to the destruction of natural microbial flora diversity which leads to the degradation of the soil quality (Reganold et al., 1987).

Therefore, today it is a big issue that using intensive agricultural production systems is not only unsustainable but also degrades the natural ecosystem (Horrigan et al., 2002).

Besides intensive agricultural systems, the changing climate is another challenge for global food security (Scheben et al., 2016). The climate change leads to the generation of various abiotic and biotic stresses which have adverse effects on crops (Kissoudis et al., 2015). The increased carbon dioxide in the atmosphere is one of the drivers for climate change. The increased CO_2 directly involved in the increase of temperature and precipitation decreases (Peters et al., 2013; Dai, 2013) on the other hand increased CO_2 indirectly degrades the agricultural land due to increased salinity in soil and due to drought period time ext. extension (Munns and Tester, 2008; Zhao and Running, 2010). The increased concentration of CO_2 alters the geographic distribution of pests directly by affecting the fertilization of the pests and indirectly by changing the interaction of pests with environmental variables (Kashyap et al., 2017). So, under climate change to increase crop production is another challenge to feed the increasing population. To overcome this issue many strategies are used like genetically engineered crops and improved agronomy practices but these strategies have limited scope for improved crop production (Huang and Han, 2014; Abberton et al., 2016; Scheben et al., 2016).

In the agriculture ecosystem, microbes have an important role without which food production and agriculture could not exist. Microbes are involved in various biochemical cycling especially nitrogen and carbon cycles and are associated with the consumption and production of greenhouse gases like nitrous oxide, carbon dioxide, and methane. Therefore, these microbes play an important and a crucial role in climate-smart agriculture by improving crop production and reducing the adverse effect of climate change (Mohanty and Swain, 2018).

2. Important role of microbes in agriculture

A plant is growing in the soil is not alone but it is in association with complex interaction with microbes which are well-regulated community and well structured (Lundberg et al., 2012; Chaparro et al., 2014; Bulgarelli et al., 2015; Kumar and Dubey, 2020). The microbes associated with plant forms phytomicrobiome whereas plant plus phytomicrobiome forms holobiont (Theis et al., 2016; Smith et al., 2017). The various factors like soil environment, exchange of carbon between organic matter in the soil, exchange of energy, exchange of nutrients, atmosphere, and aquatic ecosystem are there which influence the agricultural productivity, climate change, and quality of water (Lehmann and

Kleber, 2015). In an estimate of FAO, 38.47% of earth land is used for agriculture purposes and out of this 28.43% land is suitable for growing crops and only 3.13% land is permanently used for cultivation (Gouda et al., 2018). Every year 20%–25% of land is degraded worldwide (Abhilash et al., 2016). The excessive and indiscriminate use of chemical fertilizers (especially phosphorus and nitrogen) leads to soil pollution by reduction exchangeable bases and reduction in pH which led to the scarcity of nutrients to the crop plant which ultimately results into decrease crop production (Gupta et al., 2015). To combat abiotic stresses the co-evolution of plant-associated microbes is an essential part that results into a sustainable environment and increases soil productivity (Khan et al., 2016; Compant et al., 2016). To increase crop production and maintain soil in a good health, plant growth-promoting rhizobacteria (PGPR) are a sustainable tool (Bhat et al., 2019; Disi et al., 2019; Guo et al., 2019). These beneficial PGPR are associated with plants as endophytes or epiphytes. This PGPR affect plant growth by direct and indirect mechanisms (Fig. 1). The direct mechanism involved the availability of nutrients or compounds which directly influence plant growth (Goswami et al., 2016). Also, PGPR helps the plant to overcome abiotic (temperature, light, etc.) and biotic (disease, insect, etc.) stresses (Gabriela et al., 2015). On the other hand, indirect mechanism involved the suppression of plant pathogenic microorganisms by PGPR using various strategies like exclusion by competition, induction of systemic resistance, and antibiosis (Tripathi et al., 2012).

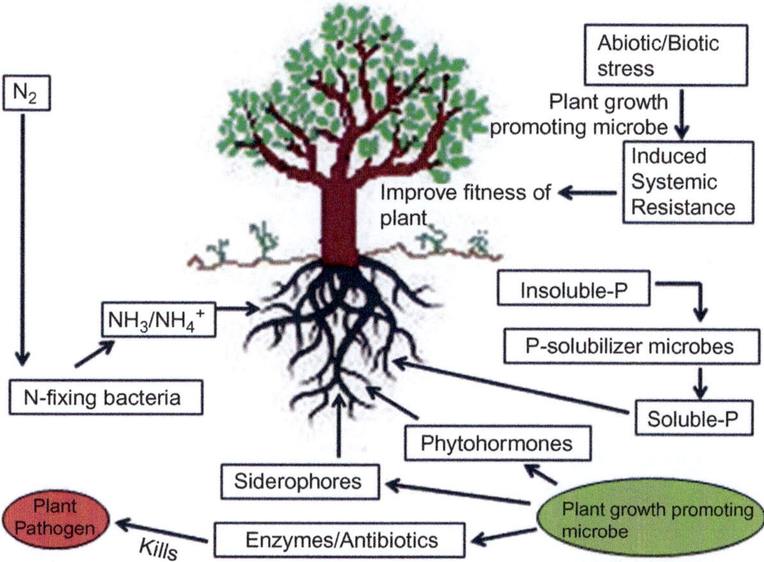

Fig. 1 Various mechanisms utilized by plant growth-promoting microbes for improving plant growth.

2.1 Biological nitrogen fixation

Nitrogen is one of the most important nutrients for the growth of any living organism including microbes and plants. The atmosphere contains 78% of nitrogen but it is not utilized by the plants until unless it is reduced to ammonia and plants are unable to perform this task (Aeron et al., 2020). Under such circumstances, some PGPR bacteria play an important role in fixing this atmospheric nitrogen and reducing it into ammonia which is absorbed by plant roots (Soumare et al., 2020) and metabolized to become a part of the integral protein (Fig. 1). These nitrogen-fixing bacteria are classified into two categories: symbiotic and asymbiotic (free-living). The symbiotic nitrogen-fixing bacteria are highly specific for their legume plant host and produce nodules specifically in roots. The symbiotic bacteria penetrate deep inside the plant root cortex, activate nodule formation, multiply inside the nodule, and finally producing nitrogen-fixing bacteroids. These bacteroids have the capability to produce nitrogenase enzymes that have a catalytic activity to convert atmospheric nitrogen into ammonia. The most common and well-studied symbiotic bacteria are *Rhizobium*. Besides this other symbiotic bacteria include *Mesorhizobium, Bradyrhizobium sinorhizobium*, etc. (Sulieman et al., 2015; Aeron et al., 2020). On the other hand, a symbiotic or free-living nitrogen-fixing living is free in the soil and they do not have specificity for the host plant (Oberson et al., 2013). The well-known free-living nitrogen-fixing bacteria include, *Azotobacter, Azospirillum, Paenibacillus, Herbaspirillum*, etc. (Goswami et al., 2015, 2016). These free-living bacteria do not penetrate into plant roots and live in the close vicinity of plant roots. Here free-living nitrogen-fixing bacteria convert atmospheric nitrogen into ammonia using nitrogenase enzyme. The reduced nitrogen is utilized by bacteria for its metabolic activities and left excess ammonia is released into the soil which is absorbed by the plant roots (Goswami et al., 2016). The nitrogenase enzyme is a complex enzyme that is produced by nif genes present in the nitrogen-fixing bacteria. The nif genes in blend with other structural genes produce a molybdenum iron cofactor. This cofactor in combination with other regulatory genes directly influences the activity of nitrogenase enzyme (Reed et al., 2011). In one report 20–30 kg nitrogen/hectare/year is fixed by nitrogen-fixing bacteria (Stacey et al., 1992). So, overall, these nitrogen-fixing bacteria play an important role in increasing crop production in a sustainable manner maintaining the soil health in good condition.

2.2 Phosphate solubilization

After nitrogen, phosphorus is the second most important nutrient essential for plant growth. Phosphorus is required by the plants for the proper functioning of various metabolic activities like photosynthesis, signal transduction, biosynthesis of macromolecules, energy transfer, and respiration (Anand et al., 2016). Although, plenty of phosphorus is available in the soil worldwide most of it (95%–99%) is in unusable or fixed form. The most common forms of phosphorus which are easily absorbed by plant roots are mono-basic ($H_2PO_4^-$) or dibasic (HPO_4^{-2}) phosphate ions (Aeron et al., 2020). Generally in the soil, soluble phosphorus is precipitated by calcium in alkaline soil whereas in acidic soil aluminum and iron are responsible for the precipitation (Kalayu, 2019). To overcome the deficiency of phosphate to plants, chemical phosphatic fertilizers are used to increase crop production but the use of chemical fertilizers leads to degradation of the microbiological quality of the soil. Under such circumstances, PGPR bacteria play an important role in providing a usable form of phosphorus to plants and improving soil quality. The various form of organic phosphorus (aluminum phosphate, rock phosphate, tricalcium phosphate, etc.) is solubilized to inorganic form which is plant usable form (Fig. 1). Different mechanisms are employed by the phosphate solubilizing bacteria (PSB) to solubilize insoluble phosphorus. The most common mechanism reported for the solubilization of phosphorus is the production of organic acids which is the product of the metabolism of sugars and secreted by the PSB in the soil. In the rhizospheric soil, PSB have the access to plenty of sugars from root exudates (Goswami et al., 2016). The produced organic acids chelate the calcium, aluminum, and iron ion and release free phosphate in the soil which is absorbed by the plant roots (Kalayu, 2019). The production of various acids and protons also decreases the pH of the soil which results in the increased solubility of calcium and the release of phosphate in soil which is soluble (Goswami et al., 2016; Kalayu, 2019). The organic form of phosphorus (remains of animals and plants) includes polyphosphate, nucleic acid, phytic acids, phospholipids, etc. The mineralization of organic phosphorus releases the free inorganic phosphorus in the soil which is absorbed by the plant roots. The PSB produces acid and alkaline phosphatases like phytase which hydrolyze organic phosphate and convert it into inorganic phosphate which is uptake by plant roots (Satyaprakash et al., 2017; Kumar et al., 2018). The various efficient PSB (*Bacillus, Pseudomonas, Erwinia, Burkholderia*, etc.) are reported in the literature (Khan et al., 2009; Kumar et al., 2018). Some non-phosphate solubilizing bacteria (NPSB) indirectly provide phosphorus to the plants. The NPSB acquires less soluble phosphorus from the soil through its transporter proteins and after their death by mineralization process, this soluble phosphorus is made available to the plant roots (Aeron et al., 2020).

2.3 Phytohormones production

Phytohormones are organic compounds that at low concentration is able to influence plant growth. These phytohormones are responsible for the growth of plant shoot as well as increasing the rate of cell division of root parts (Khatoon et al., 2020). The production of lateral roots increases the surface area for more absorption of water and nutrients from the soil which directly influences the plant growth (Sureshbabu et al., 2016). Many PGPR is reported to produce these phytohormones and help in plant growth (Fig. 1).

The most common phytohormones include auxins, cytokinins, gibberellins, ethylene, etc. (Patten and Glick, 1996). Indole acetic acid (IAA) is an auxin that is produced by many PGPR. The IAA control cell division, cell elongation, helps in apical dominance and differentiation of the tissue. Since, IAA-producing PGPR is present in the rhizospheric soil, therefore, roots are the primary target. The IAA increases the rate of cell root division which ultimately increases the surface area for the absorption of water and nutrients. So, all these events improved plant growth (Ramos-Solano et al., 2008). PGPR are able to produce IAA by two pathways: tryptophan dependent and tryptophan independent. In tryptophan-dependent pathway the tryptophan is made available to PGPR which is present in the plant root exudates (Goswami et al., 2016). Another important phytohormone is gibberellins (GA) which help in fruit and flower setting, elongation of the stem, and germination of seed (Hedden and Phillips, 2000). GA shows more effective results in plants as this phytohormone is absorbed by plant roots and translocated to aerial parts of the plant. Many PGPR in the rhizospheric region are reported to produce GA. Cytokinins (CK) is another phytohormone that is produced by many PGPR. This phytohormone influences the plant's developmental and physiological processes. CK is reported to boost root development, help in shoot imitation, increase cell division and inhibit the elongation of roots (Amara et al., 2015; Jha and Saraf, 2015; Goswami et al., 2016). Besides this CK is also involved in the expansion of leaf, senescence delay, production of chlorophyll, nutritional signaling, germination of seed, root growth, formation of the vascular system in embryo, etc. (Wong et al., 2015).

2.4 Production of antibiotics

Some of the PGPR isolates were reported to produce an antimicrobial compound (Fig. 1) which is an environmentally friendly approach to control plant pathogens as compared to chemical pesticides (Khatoon et al., 2020). The two important bacterial genera, i.e., *Pseudomonas* and *Bacillus* are reported to produce antibiotic compounds. These antibiotic compounds are secreted by bacteria extracellular which are highly potent at low concentrations. *Pseudomonas* are reported to produce antibacterial antibiotics (andazomycin and pseudomonic acid), antifungal antibiotics (pyocyanin, 2,4-diacetylphloroglucinol, phenazines, rhamnolipids, etc.), antiviral antibiotics (Karalicine), and antitumor antibiotics (cepafungins, FR901463) (Hammer et al., 1997; Ramadan et al., 2016). Another important bacterial genera of *Bacillus* is also reported to be a produced variety of antimicrobial compounds which have a ribosomal or non-ribosomal origin. The compounds like sublancin, subtilosin A, subtilin, and TasA are ribosomal originating. On the other hand, compounds like mycobacillin, chlorotetain, bacillaene, rhizocticins, difficidin, and bacilysin are non-ribosomal originating. Lipopolypeptide antibiotics are also produced by *Bacillus* species like bacillomycin, iturins, and surfactin, etc. (Leclere et al., 2005; Wang et al., 2015).

2.5 Siderophore production

Siderophores are iron–chelating agent which is produced by many PGPR bacteria (Guo et al., 2020; Khatoon et al., 2020). Iron is a very essential component for plant growth as it requires for various important plant metabolic activities like nitrogen fixation, photosynthesis, and respiration. Abundant iron is present in the soil but it is in the unavailable form. The soluble or available form of iron in the soil is Fe^{3+} which reacts to form insoluble hydroxides and oxides which makes this iron inaccessible to the microbes as well as to the plants (Saha et al., 2016). Some bacterial strains evolved a method to overcome the scarcity of iron by secreting siderophores (Fig. 1). Functional groups are generally present in the siderophores which bind in a reversible manner with iron ions and make this iron soluble so that microbes or plant roots can absorb it easily. Catechols and hydroxamates are two common functional groups present in siderophores (Ghosh et al., 2020). *Pseudomonas* is the most common bacteria which is reported to produce siderophores. Siderohore producing PGPR also helps in the suppression of plant pathogens like fungi by limiting the access of iron to the pathogen as complex of iron-siderophore is not absorbed by the fungi (Shen et al., 2013).

2.6 Enzyme production

Many PGPR bacteria are reported to produce enzymes (especially cell wall degrading) that have a detrimental effect on the plant fungal pathogens (Fig. 1). The most common enzymes produced include chitinase, protease, cellulase, and β-1,3-glucanase (Goswami et al., 2016). Commonly cell wall of fungus is constituted of chitin and β-1,4-N-acetylglucosamine which is easily degraded by the enzymes chitinase and β-1,3-glucanase. The strains of *Pseudomonas, Bacillus, Enterobacter*, etc., are reported to produce enzymes that have a direct effect on the structural integrity of the fungal cell wall (Sadfi et al., 2001; Ramadan et al., 2016).

2.7 Induced systemic resistance

Induced systemic resistance (ISR) is defined as improvement in the immunity/defense system of the plant by some inducer. Many PGPR is reported to elicit plant resistance (Fig. 1) to various stresses like pathogen, environment, etc. (Prathap and Ranjitha, 2015; Meena et al., 2020). During ISR activation many protective enzymes are activated or secreted like peroxidase, catalase, chitinase, superoxide dismutase, polyphenol oxidase, glucanase, ascorbate peroxidase, etc. During the invasion of the pathogen in the plant, signals are produced which leads to the production of plant protective enzymes by the vascular system (Gouda et al., 2018). ISR is not a pathogen–specific response but is a general response by the plant which is effective to control various pathogens. The different components of PGPR are able to elicit the ISR in plants like volatile components (acetoin, etc.), siderophores, homoserine lactones, lipopolysaccharides (Berendsen et al.,

2015). Generally, to evade plant pathogens the plant induced ISR involves ethylene hormone signaling. The PGPR in the rhizosphere induces ISR signaling in the roots which systemically spread to the other distant parts of the plant which prevents the further spreading of infection (Goswami et al., 2016).

2.8 Management of plant stress

Plants in natural conditions are exposed to various stress conditions which could be biotic or abiotic in nature. Under various abiotic stress conditions, plants produce reactive oxygen species (ROS) which are lethal to the metabolism of plant-like hydroxyl radical (OH^-), superoxide ions (O^{2-}), singlet oxygen ($_1O^2$), and hydrogen peroxide (H_2O_2). The highly reactive ROS damage important plant biomolecules like proteins, nucleic acids, lipids, photosynthetic pigments (Ahmad and Prasad, 2011). The plant has developed an anti-ROS system especially the scavenging enzymes which are reported to be present in mitochondria and chloroplast. The common anti-ROS enzymes include glutathione reductase, catalase, superoxide dismutase, peroxidase, and ascorbate peroxidase (Koyro et al., 2012; Agami et al., 2016). Many researchers reported that the application of PGPR to plants under salinity stress modulates these ROS scavenging enzymes (Chunjuan et al., 2012; Naveed et al., 2014; Bhat et al., 2020; Goswami and Deka, 2020). L-proline an amino acid is another stress indicated marker produced by the plant under abiotic stress. This amino acid helps in cell membrane stabilization, maintain proper folding of proteins, act as a scavenger of hydroxyl radical and it serves as nitrogen and energy source, The application of PGPR to the stressed plant modulate the expression level of L-proline in plants (Razi and Sen, 1996; Ngumbi and Kloepper, 2016; El Moukhtari et al., 2020).

Besides abiotic stresses, plants are exposed to biotic stress like a plant can be attacked by various pathogens viz., bacteria, viruses, fungi, nematodes, etc., which induce the Induced systemic resistance (ISR) in plants (a defense mechanism). The ISR is already discussed in Section 2.7.

3. Combating climate change by microbes

The abundance and diversity of microbes in the soil directly influence the productivity of the soil, make availability of nutrients to the plants, controls biogeochemical cycles, and have influence on greenhouse gases (GHG) (Fig. 2) emissions (Wagg et al., 2014). Climate change has a direct influence on the microbial diversity present in the soil and this alters the role of microbes in the soil (ecosystem) which ultimately leads to the reduction in crop production. The climate-smart agriculture (CSA) is an integrated approach to overcome the issues of low soil productivity, emission of GHG, etc. (Lipper et al., 2014; Paustian et al., 2016).

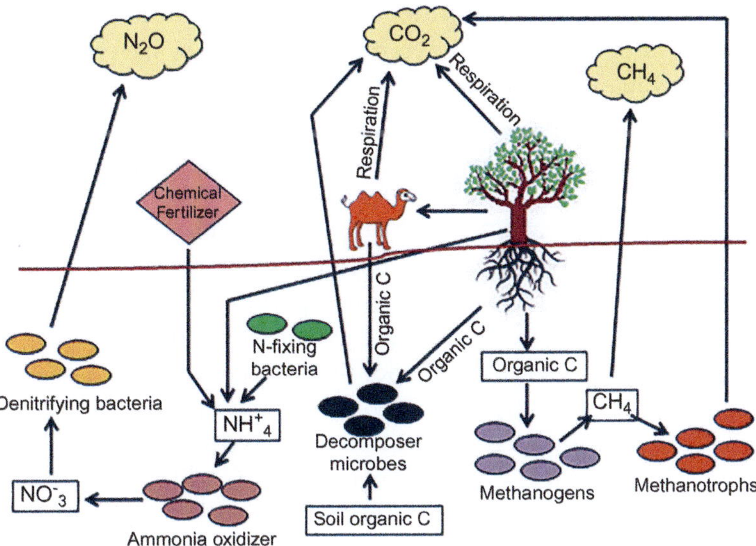

Fig. 2 Generation of greenhouse gases in the soil ecosystem.

In the current scenario, high temperature and increasing carbon dioxide concentration are the big issues that have adverse effects on soil microbes and crop production (Mueller et al., 2016). Yu et al. (2018) in their study in semiarid grassland ecosystems reported that under the warmer condition the process of nitrogen cycle decreases whereas under elevated CO$_2$ concentration the nitrogen (N) and carbon (C) cycling processes increase. Under elevated CO$_2$ concentration, the genes involved in N mineralization, denitrification, nitrogen fixation, methane metabolism, C fixation, and degradation are stimulated whereas, under the warmer condition the genes involved in ammonification and denitrification are inhibited. In another study by Waghmode et al., (2018) they studied the effect of warmer environments on microbial diversity. Their study results showed that under the warmer condition with regular irrigation increased concentration of nitrate and rate of nitrification and soil moisture decreased. Under warming conditions, profuse ammonia-oxidizing bacteria dominate whereas the abundance of denitrifying bacteria and ammonia-oxidizing archaea decreased. On the other hand, under the warmer condition with high irrigation, the abundance of denitrifying bacteria and ammonia-oxidizing archaea increased. So, the diversity and richness of these microbes have a direct influence on the turnover of N in agro-ecosystem.

Meisner and de Boer (2018) in their perspective article discuss the effect of drought and rainfall on biocontrol microbes which kill the plant pathogens. If biocontrol microbes are unable to survive under extreme weather conditions then plant pathogens will proliferate which finally leads to decreased crop production. Under extreme conditions such as drought and rainfall (anoxic), the biocontrol soil microbes have to use various strategies to survive under these conditions like production of extracellular peptides, production of osmolytes, moves into dormancy or have thick cell wall, etc. (Or et al., 2007; Schimel et al., 2007; Warren, 2014; Shoemaker and Lennon, 2018).

Another important focus area of CSA is the mitigation of greenhouse gases emission and sequestering the carbon in the soil (Lipper et al., 2014). The potential of nitrous oxide (N_2O) and methane (CH_4) is very high as compared to carbon dioxide (CO_2). Climate change directly or indirectly affects the microbial growth in the soil. The direct effect includes temperature, moisture, extreme climatic conditions, etc., whereas the indirect effect includes changes in the soil physiochemical properties by climate change. The soil microbes play an important role in the soil decomposition process and increase carbon dioxide (CO_2) efflux in the atmosphere. So, the high concentration of carbon dioxide in the atmosphere will shape the soil microbial community with more heterotrophic microbes. On the other hand increased carbon dioxide concentration enhances the rate of photosynthesis by the plants which ultimately transfers the fixed carbon dioxide to roots and soil microbes living in the rhizosphere.

The soil microbe plays an important role in ecosystem processes like decomposition of organic matter, nutrient cycling, and regulate climate by controlling the emissions of GHG (Lupatini et al., 2014; Bhattacharya et al., 2016; Delgado-Baquerizo et al., 2017). Soil respiration which involved plant respiration and microbial respiration contributes to the emission of GHG. The soil organic matter decomposition not only increases the carbon content in the soil but also this process leads to the emission of GHG. Besides this, intensive agriculture practices reduce soil carbon. The stock of carbon in the soil is the equilibrium between carbon sequestered by the plant through photosynthesis and carbon released by microbial respiration. This equilibrium is very important as this will shape the microbial diversity in the soil, influence the soil organic matter recycling and energy flow pathways in soil (Johnson et al., 2014). In agriculture tillage practice at a low level or no-tillage reduce the loss of soil organic carbon and it favors the growth of communities in the soil especially fungi (Castro et al., 2010). Nitrogen fertilizer used in agriculture is the major source of nitrous oxide emission. So, to overcome this issue use of nitrogen fertilizer which is less available to microbes is beneficial. It is better to use nitrogen fertilizer in less quantity, slow-release fertilizer, and use it at the time when crops require it. Also, improved drainage reduces the anaerobic conditions in the soil which slow down the denitrification process, and finally, emission of nitrous oxide is reduced. The application of nitrification inhibitor is also recommended to reduce the emission of nitrous oxide (Smith, 2008).

In CSA biochar application proved to be beneficial in sequestration of carbon in the soil and mitigating GHG. Biochar is produced from various organic feedstocks and rich sources of carbon. The addition of biochar in the soil protects the soil organic matter from microbial degradation by the formation of micro-aggregates between biochar and minerals, organic carbon (dissolved), and soil organic matter (Jiang et al., 2020) thus sequestering the C in the soil. The application of biochar in soil inhibits the methanogenic bacterial activity and increases methanotrophic bacterial activity which reduces the emission of methane gas in the environment (Qin et al., 2016; Jiang et al., 2020). By the application of biochar, the soil aeration is improved which suppresses the process of denitrification which reduces the emission of nitrous oxide (Ramlow and Cotrufo, 2018).

The paddy field is reported to produce methane (GHG) but the paddy field which is incorporated with straw as a nutrient source shows 3–11 times more emissions of methane gas. The addition of biochar reduces the production of methane and nitrous oxide production because there is a competition for electrons between biochar and methanogens which finally helps in the proliferation of nitrifiers, methanotrophs, and denitrifiers (Wang et al., 2018a). Slag is also tested in the agricultural soil where its application mitigates the methane emission. In anoxic soil, iron present in the slag act as an electron acceptor, and iron-reducing bacteria are stimulated over methanogens (Das et al., 2019). The mineral carbon dioxide sequestration spontaneously occurs by silicate mineral dissolution and carbonate mineral precipitation (Oelkers et al., 2008). Wang et al. (2018b) reported the application of slag increased the C sequestration in the paddy field but only in the early crop. Nitrogen is an important component for the profuse growth of plants. But in agricultural soils, ammonium is rapidly converted into nitrate by nitrification process (by *Nitrosomonas* sp. *Nitrobacter* sp.) that leads to the increased mobility of nitrate in the soil which leaches out in the soil and access of this nitrate is away from the plant roots (Norton and Ouyang, 2019). Scientists are working out how to minimize this nitrification process to decrease the loss of nitrogen in the soil. The main approaches which are used to minimize this nitrogen loss include the control mechanism of availability of ammonium for nitrification and, inhibition of nitrifier bacteria (Norton and Ouyang, 2019). A study by Mohanty et al., (2019) found that biogenic nitrate (in soil aliphatic hydrocarbon reacts with soil nitrate to form biogenic nitrate) in the soil minimize denitrification process which maintains the soil nitrogen stock. On the other hand, microbial organic volatile compounds (mVOC) enhance the rate of nitrification and enhance nitrifying bacteria abundance. So, overall still we need exhaustive and detailed knowledge of plant microbes interaction and the ecology of the microbes in the soil under climate change scenarios so that we can utilize these microbes for mitigation and adaptation to climate change.

4. Conclusions

Microbes play an important role in the ecosystem as well as in agriculture crop production. The microbial diversity present in the soil especially in the rhizosphere sense the chemical messengers secreted by the plants and in response, these microbes also produce chemical components which play an important role plant–microbe interactions. A healthy plant–microbe interaction improves plant growth by providing a usable form of N, P, plant growth-promoting components (phytohormones, siderophores, etc.), improves the plant defense system by ISR, help in nutrients recycling, decomposition of organic matter, and these microbes help in the survival of plants under biotic and abiotic stress conditions. The deeper knowledge of various mechanisms utilized by these plant growth-promoting microbes has opened a new door that help in better utilization of these microbes. The climate change has a great impact on the agriculture system and this also generates adaptation processes in microorganisms and plants. Under climate-smart agriculture these microbes have a crucial role in the survival of plants and maintain the productivity of the soil. Due to greenhouse gases (GHG) emissions (CO_2, CH_4, N_2O, etc.) the global temperature is increasing. The various strategies can be used to decrease the emission of this GHG like the application of biochar for the sequestration of carbon dioxide, an abundance of methanotrophic bacteria will decrease methane emission and biochar also help in the decrease emission of methane and nitrous oxide. For the implementation of microbial technology under climate change mitigation and adaptation, a deeper knowledge of the ecology of microbes and soil–plant-microbe interactions is essential.

References

Abberton, M., Batley, J., Bentley, A., Bryant, J., Cai, H., Cockram, J., de Oliveira, A.C., Cseke, L.J., Dempewolf, H., De Pace, C., Edwards, D., Gepts, P., Greenland, A., Hall, A.E., Henry, R., Hori, K., Howe, G.T., Hughes, S., Humphreys, M., Lightfoot, D., Marshall, A., Mayes, S., Nguyen, H.-T., Ogbonnaya, F.C., Ortiz, R., Paterson, A.H., Tuberosa, R., Valliyodan, B., Varshney, R.K., Yano, M., 2016. Global agricultural intensification during climate change: a role for genomics. Plant Biotechnol. J. 14, 1095–1098.

Abhilash, P.C., Tripathi, V., Edrisi, S.A., Dubey, R.K., Bakshi, M., Dubey, P., Singh, H.B., Ebbs, S., 2016. Sustainability of crop production from polluted lands. Energy Ecol. Environ. 1, 54–65.

Aeron, A., Khare, E., Jha, C.K., Meena, V.S., Aziz, S.M.A., Islam, M.T., Kim, K., Meena, S.K., Pattanayak, A., Rajashekara, H., Dubey, R.C., Maurya, B.R., Maheshwari, D.K., Saraf, M., Choudhary, M., Verma, R., Meena, H.N., Subbanna, A.R.N.S., Parihar, M., Shukla, S., Muthusamy, G., Bana, R.S., Bajpai, V.K., Han, Y.K., Rahman, M., Kumar, D., Singh, N.P., Meena, R.-K., 2020. Revisiting the plant growth-promoting rhizobacteria: lessons from the past and objectives for the future. Arch. Microbiol. 202, 665–676.

Agami, R.A., Medani, R.A., Abd El-Mola, I.A., Taha, R.S., 2016. Exogenous application with plant growth promoting rhizobacteria (PGPR) or proline induces stress tolerance in basil plants (*Ocimum basilicum* L.) exposed to water stress. Int. J. Environ. Agricult. Res. 2, 78–92.

Ahmad, P., Prasad, M.N.V., 2011. Environmental Adaptations and Stress Tolerance of Plants in the Era of Climate Change. Springer Science & Business Media, Berlin.

Amara, U., Khalid, R., Hayat, R., 2015. Soil bacteria and phytohormones for sustainable crop production. In: Maheshwari, D.K. (Ed.), Bacterial Metabolites in Sustainable Agroecosystem. Springer International, pp. 87–103.

Anand, K., Kumari, B., Mallick, M.A., 2016. Phosphate solubilizing microbes: an effective and alternative approach as bio-fertilizers. Int. J. Pharm. Pharm. Sci. 8, 37–40.

Berendsen, R.L., Pieterse, C.M., Bakker, P.A., 2012. The rhizosphere microbiome and plant health. Trends Plant Sci. 17, 478–486.

Berendsen, R.L., Verk, M.C.V., Stringlis, I.A., Zamioudis, C., Tommassen, J., Pieterse, C.M.J., Bakker, P.-A.H.M., 2015. Unearthing the genomes of plant-beneficial *Pseudomonas* model strains WCS358, WCS374 and WCS417. BMC Genomics 16, 539.

Bhat, M.A., Rasool, R., Ramzan, S., 2019. Plant growth promoting rhizobacteria (PGPR) for sustainable and eco-friendly agriculture. Acta Sci. Agricult. 3, 23–25.

Bhat, M.A., Kumar, V., Bhat, M.A., Wani, I.A., Dar, F.L., Farooq, I., Bhatti, F., Koser, R., Rahman, S., Jan, A.T., 2020. Mechanistic insights of the interaction of plant growth-promoting rhizobacteria (PGPR) with plant roots toward enhancing plant productivity by alleviating salinity stress. Front. Microbiol. 11, 1952.

Bhattacharya, S.S., Kim, K.H., Das, S., Uchimiya, M., Jeon, B.H., Kwon, E., Szulejko, J.E., 2016. A review on the role of organic inputs in maintaining the soil carbon pool of the terrestrial ecosystem. J. Environ. Manage. 167, 214–827.

Brady, N.C., Weil, R.R., 2010. Elements of the Nature and Properties of Soils. Upper Saddle River, Pearson Prentice Hall, NJ.

Bulgarelli, D., Garrido-Oter, R., Munch, P.C., Weiman, A., Droge, J., Pan, Y., McHardy, A.C., Schulze-Lefert, P., 2015. Structure and function of the bacterial root microbiota in wild and domesticated barley. Cell Host Microbe 17, 392–403.

Castro, H.F., Classen, A.T., Austin, E.E., Norby, R.J., Schadt, C.W., 2010. Soil microbial community responses to multiple experimental climate change rivers. Appl. Environ. Microbiol. 76, 999–1007.

Chaparro, J.M., Badri, D.V., Vivanco, J.M., 2014. Rhizosphere microbiome assemblage is affected by plant development. ISME J. 8, 790.

Chunjuan, W., YaHui, G., Chao, W., HongXia, L., DongDong, N., YunPeng, W., JianHua, G., 2012. Enhancement of tomato (*Lycopersicon esculentum*) tolerance to drought stress by plant-growth-promoting rhizobacterium (PGPR) *Bacillus cereus* AR156. J. Agricult. Biotechnol. 20, 1097–1105.

Compant, S., Saikkonen, K., Mitter, B., Campisano, A., Blanco, J.M., 2016. Soil, plants and endophytes. Plant and Soil 405, 1–11.

Dai, A., 2013. Increasing drought under global warming in observations and models. Nat. Clim. Change 3, 52–58.

Das, S., Kim, G.W., Hwang, H.Y., Verma, P.P., Kim, P.J., 2019. Cropping with slag to address soil, environment, and food security. Front. Microbiol. 10, 1320.

Delgado-Baquerizo, M., Trivedi, P., Trivedi, C., Aldridge, D.J., Jeffries, T.C., Reich, P., Singh, B.K., 2017. Microbial richness and composition independently drive soil multifunctionality. Funct. Ecol. 31, 2330–2343.

Disi, J.O., Mohammad, H.K., Lawrence, K., Kloepper, J., Fadamiro, H.A., 2019. Soil bacterium can shape belowground interactions between maize, herbivores and entomopathogenic nematodes. Plant and Soil 437, 83–92.

Doran, J.W., Zeiss, M.R., 2000. Soil health and sustainability: managing the biotic component of soil quality. Appl. Soil Ecol. 15, 3–11.

El Moukhtari, A., Cabassa-Hourton, C., Farissi, M., Savoure, A., 2020. How does proline treatment promote salt stress tolerance during crop plant development? Front. Plant Sci. 11, 1127.

Gabriela, F., Casanovas, E.M., Quillehauquy, V., Yommi, A.K., Goni, M.G., Roura, S.I., Barassi, C.A., 2015. *Azospirillum* inoculation effects on growth, product quality and storage life of lettuce plants grown under salt stress. Sci. Hortic. 195, 154–162.

Ghosh, S.K., Bera, T., Chakrabarty, A.M., 2020. Microbial siderophore—a boon to agricultural sciences. Biol. Control 144, 104214.

Goswami, M., Deka, S., 2020. Plant growth-promoting rhizobacteria—alleviators of abiotic stresses in soil: a review. Pedosphere 30, 40–61.

Goswami, D., Parmar, S., Vaghela, H., Dhandhukia, P., Thakker, J., 2015. Describing *Paenibacillus mucilaginosus* strain N3 as an efficient plant growth promoting rhizobacteria (PGPR). Cogent Food Agricult. 1, 1000714.

Goswami, D., Thakker, J.N., Dhandhukia, P.C., 2016. Portraying mechanics of plant growth promoting rhizobacteria (PGPR): a review. Cogent Food Agricult. 2, 1–19.

Gouda, S., Kerry, R.G., Das, G., Paramithiotis, S., Shin, H.S., Patra, J.K., 2018. Revitalization of plant growth promoting rhizobacteria for sustainable development in agriculture. Microbiol. Res. 206, 131–140.

Guo, Q., Li, Y., Lou, Y., Shi, M., Jiang, Y., Zhou, J., Sun, Y., Xue, Q., Lai, H., 2019. *Bacillus amyloliquefaciens* Ba13 induces plant systemic resistance and improves rhizosphere micro ecology against tomato yellow leaf curl virus disease. Appl. Soil Ecol. 137, 154–166.

Guo, Q., Shi, M., Chen, L., Zhou, J., Zhang, L., Li, Y., Xue, Q., Lai, H., 2020. The biocontrol agent *Streptomyces pactum* increases *Pseudomonas koreensis* populations in the rhizosphere by enhancing chemotaxis and biofilm formation. Soil Biol. Biochem. 144, 107755.

Gupta, G., Parihar, S.S., Ahirwar, N.K., Snehi, S.K., Singh, V., 2015. Plant growth promoting Rhizobacteria (PGPR): Current and future prospects for development of sustainable agriculture. J. Microbiol. Biochem. 7, 96–102.

Hammer, P.E., Hill, D.S., Lam, S.T., Van Pee, K.H., Ligon, J.M., 1997. Four genes from *Pseudomonas fluorescens* that encode the biosynthesis of pyrrolnitrin. Appl. Environ. Microbiol. 63, 2147–2154.

Hedden, P., Phillips, A.L., 2000. Gibberellin metabolism: new insights revealed by the genes. Trends Plant Sci. 5, 523–530.

Horrigan, L., Lawrence, R.S., Walker, P., 2002. How sustainable agriculture can address the environmental and human health harms of industrial agriculture. Environ. Health Perspect. 110, 445–456.

Huang, X., Han, B., 2014. Natural variations and genome-wide association studies in crop plants. Annu. Rev. Plant Biol. 65, 531–551.

Jenny, H., 1941. Factors of soil formation. Soil Sci. 52, 415.

Jha, C.K., Saraf, M., 2015. Plant growth promoting rhizobacteria (PGPR): a review. E3 J. Agricult. Res. Dev. 5, 108–119.

Jiang, Z., Lian, F., Wang, Z., Xing, B., 2020. The role of biochars in sustainable crop production and soil resiliency. J. Exp. Bot. 71, 520–542.

Johnson, J.A., Runge, C.F., Senauer, B., Foley, J., Polasky, S., 2014. Global agriculture and carbon tradeoffs. Proc. Natl. Acad. Sci. U. S. A. 111, 12342–12347.

Kalayu, G., 2019. Phosphate solubilizing microorganisms: promising approach as biofertilizers. Int. J. Agron. 2019, 4917256.

Kashyap, P.L., Rai, P., Srivastava, A.K., Kumar, S., 2017. *Trichoderma* for climate resilient agriculture. World J. Microbiol. Biotechnol. 33, 155.

Khan, A., Jilani, V., Akhtar, M.S., Naqvi, S.M.S., Rasheed, M., 2009. Phosphorus solubilizing bacteria: occurrence, mechanisms and their role in crop production. J. Agricult. Biol. Sci., 148–158.

Khan, Z., Rho, H., Firrincieli, A., Hung, H., Luna, V., Masciarelli, O., Kim, S.H., Doty, S.L., 2016. Growth enhancement and drought tolerance of hybrid poplar upon inoculation with endophyte consortia. Curr. Plant Biol. 6, 38–47.

Khatoon, Z., Huang, S., Rafique, M., Fakhar, A., Kamran, M.A., Santoyo, G., 2020. Unlocking the potential of plant growth-promoting rhizobacteria on soil health and the sustainability of agricultural systems. J. Environ. Manage. 273, 111118.

Kissoudis, C., Chowdhury, R., van Heusden, S., Clemens van de Wiel, C., Finkers, R., Visser, R.G.F., Bai, Y., van der Linden, G., 2015. Combined biotic and abiotic stress resistance in tomato. Euphytica 202, 317–332.

Koyro, H.W., Ahmad, P., Geissler, N., 2012. Abiotic stress responses in plants: an overview. In: Ahmad, P., P., Prasad, M.N.V. (Eds.), Environmental Adaptations and Stress Tolerance of Plants in the Era of Climate Change. Springer, New York, NY, pp. 1–28.

Kumar, A., Dubey, A., 2020. Rhizosphere microbiome: engineering bacterial competitiveness for enhancing crop production. J. Adv. Res. 29, 337–352.

Kumar, A., Kumar, A., Patel, H., 2018. Role of microbes in phosphorus availability and acquisition by plants. Int. J. Curr. Microbiol. App. Sci. 7, 1344–1347.

Leclere, V., Bechet, M., Adam, A., Guez, J.S., Wathelet, B., Ongena, M., Thonart, P., 2005. Mycosubtilin overproduction by *Bacillus subtilis* BBG100 enhances the organism's antagonistic and biocontrol activities. Appl. Environ. Microbiol. 71, 4577–4584.

Lehmann, J., Kleber, M., 2015. The contentious nature of soil organic matter. Nature 528, 60–68.

Lipper, L., Thornton, P., Campbell, B.M., Baedeker, T., Braimoh, A., Bwalya, M., Caron, P., Cattaneo, A., Garrity, D., Henry, K., Hottle, R., Jackson, L., Jarvis, A., Kossam, F., Mann, W., McCarthy, N.,

Meybeck, A., Neufeldt, H., Remington, T., Sen, P.T., Sessa, R., Shula, R., Tibu, A., Torquebiau, E.F., 2014. Climate-smart agriculture for food security. Nat. Clim. Change 4, 1068–1072.

Lundberg, D.S., Lebeis, S.L., Paredes, S.H., Yourstone, S., Gehring, J., Malfatti, S., Tremblay, J., Engelbrektson, A., Kunin, V., del Rio, T.G., Edgar, R.C., Eickhorst, T., Ley, R.E., Hugenholtz, P., Tringe, S.G., Dangl, J.L., 2012. Defining the core *Arabidopsis thaliana* root microbiome. Nature 488, 86–90.

Lupatini, M., Suleiman, A.K.A., Jacques, R.J.S., Antoniolli, Z.I., de Siqueira Ferreira, A., Kuramae, E.E., Roesch, L.F.W., 2014. Network topology reveals high connectance levels and few key microbial genera within soils. Front. Environ. Sci. 2, 10.

Meena, M., Swapnil, P., Divyanshu, K., Kumar, S., Harish, Tripathi, Y.N., Zehra, A., Marwal, A., Upadhyay, R.S., 2020. PGPR-mediated induction of systemic resistance and physiochemical alterations in plants against the pathogens: current perspectives. J. Basic Microbiol. 60, 828–861.

Meisner, A., de Boer, W., 2018. Strategies to maintain natural biocontrol of soil-borne crop diseases during severe drought and rainfall events. Front. Microbiol. 9, 2279.

Mohanty, S.R., Nagarjuna, M., Parmar, R., Ahirwar, U., Patra, A., Dubey, G., Kollah, B., 2019. Nitrification rates are affected by biogenic nitrate and volatile organic compounds in agricultural soils. Front. Microbiol. 10, 772.

Mohanty, S., Swain, C., 2018. Role of microbes in climate smart agriculture. In: Panpatte, D.G., Jhala, Y.K., Shelat, H.N., Vyas, R.V. (Eds.), Microorganisms for Green Revolution. vol. 2. Springer, Singapore, pp. 129–140.

Mueller, K.E., Blumenthal, D.M., Pendall, E., Carrillo, Y., Dijkstra, F.A., Williams, D.G., Follett, R.F., Morgan, J.A., 2016. Impacts of warming and elevated CO_2 on a semi-arid grassland are non-additive, shift with precipitation, and reverse over time. Ecol. Lett. 19, 956–966.

Munns, R., Tester, M., 2008. Mechanisms of salinity tolerance. Annu. Rev. Plant Biol. 59, 651–681.

Naveed, M., Hussain, M.B., Zahir, Z.A., Mitter, B., Sessitsch, A., 2014. Drought stress amelioration in wheat through inoculation with *Burkholderia phytofirmans* strain PsJN. Plant Growth Regul. 73, 121–131.

Ngumbi, E., Kloepper, J., 2016. Bacterial-mediated drought tolerance: current and future prospects. Appl. Soil Ecol. 105, 109–125.

Norton, J., Ouyang, Y., 2019. Controls and adaptive management of nitrification in agricultural soils. Front. Microbiol. 10, 1931.

Oberson, A., Frossard, E., Buhlmann, C., Mayer, J., Mader, P., Luscher, A., 2013. Nitrogen fixation and transfer in grass-clover leys under organic and conventional cropping systems. Plant Soil 371, 237–255.

Oelkers, E.H., Gislason, S.R., Matter, J., 2008. Mineral carbonation of CO_2. Elements 4, 333–337.

Or, D., Smets, B.F., Wraith, J.M., Dechesne, A., Friedman, S.P., 2007. Physical constraints affecting bacterial habitats and activity in unsaturated porous media—a review. Adv. Water Resour. 30, 1505–1527.

Patten, C.L., Glick, B.R., 1996. Bacterial biosynthesis of indole-3-acetic acid. Can. J. Microbiol. 42, 207–220.

Paustian, K., Lehmann, J., Ogle, S., Reay, D., Robertson, G.P., Smith, P., 2016. Climate-smart soils. Nature 532, 49–57.

Peters, G.P., Andrew, R.M., Boden, T., Canadell, J.G., Ciais, P., Le Quere, C., Marland, G., Raupach, M.-R., Wilson, C., 2013. The challenge to keep global warming below 2 °C. Nat. Clim. Change 3, 4–6.

Prathap, M., Ranjitha, K.B.D., 2015. A Critical review on plant growth promoting rhizobacteria. J. Plant Pathol. Microbiol. 6, 1–4.

Qin, X., Li, Y., Wang, H., Liu, C., Li, J., Wan, Y., Gao, Q., Fan, F., Liao, Y., 2016. Long-term effect of biochar application on yield-scaled greenhouse gas emissions in a rice paddy cropping system: a four-year case study in south China. Sci. Total Environ. 569, 1390–1401.

Ramadan, E.M., AbdelHafez, A.A., Hassan, E.A., Saber, F.M., 2016. Plant growth promoting rhizobacteria and their potential for biocontrol of phytopathogens. Afr. J. Microbiol. Res. 10, 486–504.

Ramlow, M., Cotrufo, M.F., 2018. Woody biochar's greenhouse gas mitigation potential across fertilized and unfertilized agricultural soils and soil moisture regimes. Glob. Change Biol. Bioenergy. 10, 108–122.

Ramos-Solano, B., Barriuso, J., Gutierrez-Manero, F.J., 2008. Physiological and molecular mechanisms of plant growth promoting rhizobacteria (PGPR). In: Ahmad, I., Pichtel, J., Hayat, S. (Eds.), Plant–Bacteria Interactions: Strategies and Techniques to Promote Plant Growth. Wiley VCH, Weinheim, pp. 41–54.

Razi, S.S., Sen, S.P., 1996. Amelioration of water stress effects on wetland rice by urea-N, plant growth regulators, and foliar spray of a diazotrophic bacterium *Klebsiella* sp. Biol. Fertil. Soils 23, 454–458.

Reed, S.C., Cleveland, C.C., Townsend, A.R., 2011. Functional ecology of free-living nitrogen fixation: a contemporary perspective. Annu. Rev. Ecol. Evol. Syst. 42, 489–512.

Reganold, J., Elliott, L., Unger, Y., 1987. Long-term effects of organic and conventional farming on soil erosion. Nature 330, 370–372.

Sadfi, N., Cherif, M., Fliss, I., Boudabbous, A., Antoun, H., 2001. Evaluation of bacterial isolates from salty soils and *Bacillus thuringiensis* strains for the biocontrol of *Fusarium* dry rot of potato tubers. J. Plant Pathol. 83, 101–117.

Saha, M., Sarkar, S., Sarkar, B., Sharma, B.K., Bhattacharjee, S., Tribedi, P., 2016. Microbial siderophores and their potential applications: a review. Environ. Sci. Pollut. Res. 23, 3984–3999.

Satyaprakash, M., Nikitha, T., Reddi, E.U.B., Sadhana, B., Vani, S.S., 2017. A review on phosphorous and phosphate solubilising bacteria and their role in plant nutrition. Int. J. Curr. Microbiol. App. Sci. 6, 2133–2144.

Scheben, A., Yuan, Y., Edwards, D., 2016. Advances in genomics for adapting crops to climate change. Curr. Plant Biol. 6, 2–10.

Schimel, J., Balser, T.C., Wallenstein, M., 2007. Microbial stress-response physiology and its implications for ecosystem function. Ecology 88, 1386–1394.

Shen, X., Hu, H., Peng, H., Wang, W., Zhang, X., 2013. Comparative genomic analysis of four representative plant growth-promoting rhizobacteria in Pseudomonas. BMC Genomics 14, 271.

Shoemaker, W.R., Lennon, J.T., 2018. Evolution with a seed bank: the population genetic consequences of microbial dormancy. Evol. Appl. 11, 60–75.

Smith, P., 2008. Land use change and soil organic carbon dynamics. Nutr. Cycl. Agroecosyst. 81, 169–178.

Smith, D.L., Gravel, V., Yergeau, E., 2017. Editorial: signaling in the phytomicrobiome. Front. Plant Sci. 8, 611.

Soumare, A., Diedhiou, A.G., Thuita, M., Hafidi, M., Ouhdouch, Y., Gopalakrishnan, S., Kouisni, L., 2020. Exploiting biological nitrogen fixation: a route towards a sustainable agriculture. Plan. Theory 9, 1011.

Spence, C., Bais, H., 2013. Probiotics for plants: rhizospheric microbiome and plant fitness. In: de Bruijn, F.-J., F.J. (Eds.), Molecular Microbial Ecology of the Rhizosphere. vol. 2. Wiley Blackwell, Hoboken, NJ, pp. 713–721.

Stacey, G., Burris, R.H., Evans, H.J., 1992. Biological Nitrogen Fixation. Springer Science & Business Media, Berlin.

Sulieman, S., Chien, V., Esfahani, M., Yasuko, W., Rie, N., Chung, T., Dong, V., Tran, L., 2015. DT2008: a promising new genetic resource for improved drought tolerance in soybean when solely dependent on symbiotic N_2 fixation. Biomed. Res. Int. 2015, 687213.

Sureshbabu, K., Amaresan, N., Kumar, K., 2016. Amazing multiple function properties of plant growth promoting rhizobacteria in the rhizosphere soil. Int. J. Curr. Microbiol. App. Sci. 5, 661–683.

Tahat, M.M., Alananbeh, K.M., Othman, Y.A., Leskovar, D.I., 2020. Soil health and sustainable agriculture. Sustainability 12, 4859.

Theis, K.R., Dheilly, N.M., Klassen, J.L., Brucker, R.M., Baines, J.F., Bosch, T.C.G., Cryan, J.F., Gilbert, S.F., Goodnight, C.J., Lloyd, E.A., Sapp, J., Vandenkoornhuyse, P., Zilber-Rosenberg, I., Rosenberg, E., Bordenstein, S.R., 2016. Getting the hologenome concept right: an ecoevolutionary framework for hosts and their microbiomes. msystems 1, e00028-16.

Tripathi, D.K., Singh, V.P., Kumar, D., Chauhan, D.K., 2012. Impact of exogenous silicon addition on chromium uptake, growth, mineral elements, oxidative stress, antioxidant capacity, and leaf and root structures in rice seedlings exposed to hexavalent chromium. Acta *Physiologiae Plantarum* 34, 279–289.

Wagg, C., Bender, S.F., Widmer, F., van der Heijden, M.G., 2014. Soil biodiversity and soil community composition determine ecosystem multifunctionality. Proc. Natl. Acad. Sci. U. S. A. 111, 5266–5270.

Wang, X., Mavrodi, D.V., Ke, L., Mavrodi, O.V., Yang, M., Thomashow, L.S., Zheng, N., Weller, D.M., Zhang, J., 2015. Biocontrol and plant growth-promoting activity of rhizobacteria from Chinese fields with contaminated soils. J. Microbial. Biotechnol. 8, 404–418.

Waghmode, T.R., Chen, S., Li, J., Sun, R., Liu, B., Hu, C., 2018. Response of nitrifier and denitrifier abundance and microbial community structure to experimental warming in an agricultural ecosystem. Front. Microbiol. 9, 474.

Wang, Y.Q., Bai, R., Di, H.J., Mo, L.Y., Han, B., Zhang, L.M., He, J.Z., 2018a. Differentiated mechanisms of biochar mitigating straw-induced greenhouse gas emissions in two contrasting paddy soils. Front. Microbiol. 9, 2566.

Wang, W., Lai, D., Abid, A., Neogi, S., Xu, X., Wang, C., 2018b. Effects of steel slag and biochar incorporation on active soil organic carbon pools in a subtropical paddy field. Agronomy 8, 1–17.

Warren, C.R., 2014. Response of osmolytes in soil to drying and rewetting. Soil Biol. Biochem. 70, 22–32.

Wong, W.S., Tan, S.N., Ge, L., Chen, X., Yong, J.W.H., 2015. The importance of phytohormones and microbes in biofertilizers. In: Maheshwari, D.K. (Ed.), Bacterial Metabolites in Sustainable Agroecosystem. Springer International, pp. 105–158.

Yu, H., Deng, Y., He, Z., Van Nostrand, J.D., Wang, S., Jin, D., Wang, A., Wu, L., Wang, D., Tai, X., Zhou, J., 2018. Elevated CO_2 and warming altered grassland microbial communities in soil top-layers. Front. Microbiol. 9, 1790.

Zhao, M.S., Running, S.W., 2010. Drought induced reduction in global terrestrial net primary production from 2000 through 2009. Science 329, 940–943.

CHAPTER 11

Microbes as biocontrol agent: From crop protection till food security

C.R. Vanshree[a], Muskan Singhal[b], Mansi Sexena[b], Mahipal Singh Sankhla[b,*], Kapil Parihar[c], Ekta B. Jadhav[d], Kumud Kant Awasthi[e], and Chandra Shekhar Yadav[f]

[a]Department of Forensic Science, Institute of Sciences, SAGE University, Indore, Madhya Pradesh, India
[b]Department of Forensic Science, Vivekananda Global University, Jaipur, Rajasthan, India
[c]State Forensic Science Laboratory, Jaipur, Rajasthan, India
[d]Government Institute of Forensic Science, Aurangabad, Maharashtra, India
[e]Department of Life Sciences, Vivekananda Global University, Jaipur, Rajasthan, India
[f]School of Forensic Science, NFSU, Gandhinagar, India
*Corresponding author.

1. Introduction

Agriculture has suffered a significant economic decline in recent years as a result of low output or a collapse caused by a variety of reasons, the most prevalent of which are diseases and plagues (Galindo et al., 2015). The latter, often known as "natural enemies," is divided into three categories: parasites, predators, and pathogens (Badii and Abreu, 2006). The procedures or substances utilized to resist and safeguard the harvests of this disadvantage are primarily synthetic or chemistry-based goods (Brimner and Boland, 2003). However, in recent years, the usage of pesticides and herbicides has sparked debate since it has negative impacts on consumers' health and the environment. Even though these appear to be inexpensive, in the long run, they pollute the environment and have negative effects on food and human health due to the high concentrations of such compounds, the prolonged periods of degradation required, carcinogenic potential and teratogenic effect, and the fact that they can promote the generation of resistant organisms due to the persistence of such compounds. As a result, it is very desired to reduce or eliminate the use of synthetic pesticides in agriculture. This has prompted researchers to look into additional control methods, and as a result, biological control agents have received a lot of attention (Vos et al., 2014). The employment of an organism to oppose another organism is known as biological control (Mmbaga et al., 2008). This method can take one of two approaches: (1) using native organisms in the system or (2) including nonnative organisms that can combat harmful diseases in the crop in issue (Sharma et al., 2009). The use of natural chemicals is the premise of this option, which works for hand in hand with sustainable development and protection of natural resources. New species with a definite potential to be part of a select group known as "biological control agents," as well as their action mechanisms, have been the subject of research. This has already led

Relationship Between Microbes and the Environment for Sustainable Ecosystem Services, Volume 1
https://doi.org/10.1016/B978-0-323-89938-3.00011-6

to the commercialization of certain biological control organisms, and the so-called "biopesticides market" is likely to grow in the future years (Wyckhuys et al., 2013). This study will cover the fundamentals of biological control (CB), including its origins, kinds of agents, trends, and future prospects.

2. Microbes as biocontrol agents

One of the primary issues in achieving the objective of food security for the world's rising population is plant diseases. The loss of agricultural yields owing to widespread diseases in cereals and fruits has a significant impact on crop yields. To manage plant diseases, a variety of approaches have been used, including crop rotation, the use of resistant plant types, and so on. Chemical pesticides, on the other hand, are still widely used to control numerous plant diseases. Chemical pesticides are huge health and environmental threat, thus more environmentally acceptable options must be investigated to mitigate the impact. Biocontrol with natural antagonistic microorganisms, also known as biocontrol agents (BCA), is an environmentally friendly technique of disease prevention that is often the only alternative available. Furthermore, in addition to avoiding plant diseases, BCA promotes plant development. They also help plants cope with stress, acquire nutrients, and develop disease resistance. BCA's products can be utilized as biofertilizers, plant strengtheners, and biopesticides based on their processes and effects. The employment of non-pathogenic hostile microorganisms by humans can be described as the use of biological control to eliminate the disease-conscious pathogen (Cook, 1993). The US National Research Council says it should describe the use of biocontrol agents as a way of reducing adverse effects of pathogens by the use of natural or genetically modified organisms and their genes or gene products.

Two fundamental biological control principles exist:

1. In order to suppress the population of harmful plant patents, the application of biological controls depends on the phenomena of "natural control."
2. There is no removal of pests by biological control. It simply reduces the amount of pests, in order to retain both the pest as well as the natural adversary at reduced agro-ecosystem densities (Babbal et al., 2017).
3. BCA is used to reduce disease and/or suppress pathogens in different legumes, grains, flowers, fruits, and ornamental plants in greenhouses and field crops. By harnessing the actions of natural enemies' biological control can also be described as the elimination of dangerous diseases (Cook, 1993). The employment of a single antagonist against a single disease inside a single-crop system is the most restricted, biological control (Cook, 1993). Biocontrol agents, which are beneficial in the management of plant diseases and pests, are:
   - The ladybird, the bird with red and black patterns, and the dragonflies.

- Bacteria *Bacillus thuringiensis* (Bt) is used to remove butterfly caterpillars combined with water and scattered to susceptible plant species like brassicas and fruits, which are consumed by the insect larvae and larvae in the gut, releasing the toxin and killing the larvae.
- Trichoderma species are free-lived fungi that are present in the root ecosystem as biocontrol agents of several diseases of plants.
- The pathogens employed as biological control agents are the baculoviruses which target insects and other arthropods and most of them are in the genus Nucleopolyhedrovirus.

3. Microbial biocontrol for crop protection and food security

Biocontrol refers to the use of biological methods to combat plant pests and diseases. In today's society, these problems are largely due to the use of chemicals—through insecticides and pesticides. These chemicals are toxic and very dangerous to humans and animals alike and have been polluting our environment (soil, groundwater), fruits, vegetables, and crops. Our soil is also contaminated by the use of weedicides to kill weeds. These are conventional farming practices that often use chemical methods to indiscriminately kill both useful and harmful microorganisms. There is a method of pest control in agriculture that relies on natural predation rather than introduced chemicals. This helps to develop appropriate means of biocontrol to protect the crops from various diseases and enhances food security. Apart from this Plants are exposed to a wide variety of environmental Stresses that limit agricultural productivity. Environmental stress on plants can be divided into abiotic and biotic stress (Fig. 1).

3.1 Microbes as biofertilizers

The use of chemical fertilizers to meet the growing demand for agricultural products has been a major contributor to environmental pollution. The excessive use of chemical

Fig. 1 Illustration of types of stress on plants.

fertilizers is associated with many problems, which is why a switch to organic farming is necessary. Organic farming is the cultivation of useful plants through the use of Biofertilizers that provide good nutrition to the plants. The term biofertilizer refers to a formulation that contains live microbes that help improve soil fertility by fixing atmospheric nitrogen, dissolving phosphorus and other nutrients, and increasing plant growth through the production of plant hormones (Barman et al., 2017). The main sources of organic fertilizers are bacteria, fungi, and cyanobacteria.

3.1.1 Nitrogen fixing microbes

Assessing nitrogen fixation by free-living bacteria is difficult, but for some plants, such as *Medicago sativa*, it is estimated to vary between 3 and $10 \, kg \, N \, ha^{-1}$ (Roper et al., 1995). The free-living aqueous extract of Frankia (Actinomycetes) forms a symbiotic association and fixes atmospheric nitrogen in the rhizosphere of its host plant (Smolander and Sarsa, 1990). Free-living cyanobacteria (blue-green algae) were used in rice cultivation in India, which under ideal conditions can produce up to $20–30 \, kg \, N \, ha^{-1}$ (Kannaiyan, 2002). The leaves of some plants such as Ardisia develop special internal cavities that house symbiotic nitrogen-fixing bacteria such as Xanthomonas and Mycobacterium, and these are a source of nitrogen fertilizer for the soil (Miller, 1990). Another ecologically crucial group of microorganisms is Blue-green Algae (Cyanobacteria), some of which Trichodesmium, Nostoc, and Anabaena contribute about 36% of the world's nitrogen fixation and which have been reported to be useful in improving the fertility of rice fields for growing rice in the World (Kundu and Ladha, 1995; Gallon, 2001; Irisarri et al., 2001). Several studies have proven that due to nitrogen fixation and the production of growth-promoting substances, Azospirillum stimulates the growth and yield of rice, wheat, cotton, sugar beet, sunflower, etc. (Okon, 1985; Bashan et al., 1989). Thus, the ability to fix nitrogen in a significant amount of these microorganisms makes them attractive candidates for their use as biofertilizers (Giri et al., 2019) (Fig. 2).

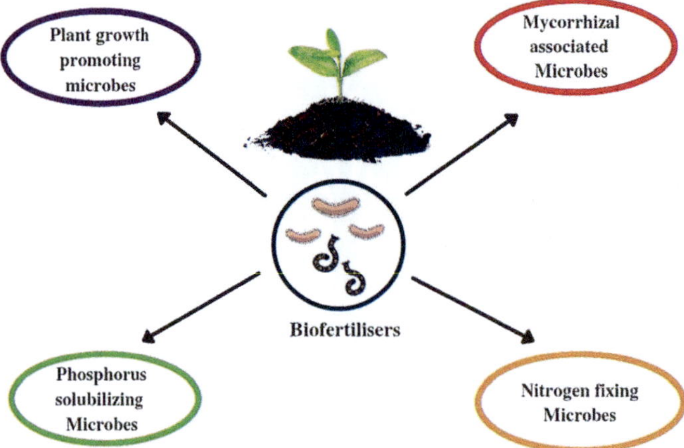

Fig. 2 Illustration of microbes as Biofertilizers.

3.1.2 Phosphorus solubilizing microbes

Soil has high levels of phosphorus, but most of it is in unavailable form making it the second most limiting plant nutrient after nitrogen (Schachtman et al., 1998). Phosphorus-solubilizing bacteria such as Bacillus and Pseudomonas can increase the availability of phosphorus to plants by mobilizing it from forms that are not available in the soil (Richardson, 2001). An active group of microbes, such as Pseudomonas, Bacillus, Aspergillus, Penicillium, Micrococcus, flavobacterium, Fusarium, and Sclerotium has been reported to be effective in the solubilizing process (Krishnaprabu, 2020). A phosphate-solubilizing bacterial strain NII 0909 from Micrococcus exhibits polyvalent properties, including the solubilization of phosphate and the production of siderophores (Dastager et al., 2010). Similarly, two fungi, namely, *Aspergillus fumigatus* and *Aspergillus niger*, were isolated from rotting cassava pods that were formed to convert cassava waste into phosphate organic fertilizer using the semi-solid fermentation technique (Ogbo, 2010). The stress-tolerant bacteria, *Burkholderia vietnamiensis* produces gluconic and 2-ketogluconic acids, which are involved in the solubilization of phosphate (Park et al., 2010a). Enterobacter and Burkholdenia, isolated from the rhizosphere of sunflowers, have been found to produce siderophores and indole compounds that can dissolve phosphate (Ambrosini et al., 2012).

3.1.3 Plant growth-promoting microbes

In addition to nitrogen-fixing and phosphorus-solubilizing microbes, there are a few other microbes that are appropriate for use as organic fertilizers, as they promote plant growth through the synthesis of growth-promoting chemicals (Bashan, 1998). *Bacillus licheniformis* and Rhizosphere *Bacillus pumilus* have been found to produce significant amounts of the physiologically active plant hormone gibberellin (Gutiérrez-Mañero et al., 2001). Similarly, *Paenibacillus polymyxa* exhibited a variety of useful properties including production of antibiotics, production of plant hormones such as, cytokinins phosphorus solubility, nitrogen fixation, and improved soil porosity (Timmusk et al., 1999). Some species of Azospirillum have also been reported to produce a variety of plant hormones (Bashan and Holguin, 1997). The inoculated microbes produce various hormones that promote plant growth, such as indole acetic acid, cytokinins, Gibberellic acid, and ethylene (Hayat et al., 2010). Streptomyces sp. and Actinomycetes microbispora sp., both produce the stress buster ACC deaminase and the plant growth hormone IAA (Glick, 2014). Plant growth microbes enhance the agronomy performance by decreasing the production costs reducing pollution and the usage of chemical fertilizers (Souza et al., 2015). The hybrid culture of microbes enhances and maintains soil physical–chemical structures, thereby altering balance and increasing plant growth, yield, and health (Bhattacharyya and Jha, 2012). Enterobacter sp. UPMR 18 helped the okra plant to grow by improving salt tolerance growth parameters, germination rate (Habib et al., 2016). A strain of *Bacillus flexus* was also found to improve salt tolerance in crops (Wang et al., 2017).

3.1.4 Mycorrhiza fungi association

The fungus also forms obligate or facultative symbiotic interactions with more than 80% of roots of higher plants, known as Mycorrhiza. The symbiont takes phosphorus from the soil and transfers it to plants. Plants with such associations show resistance to root-borne pathogens, tolerance to salinity and drought, and a general increase in plant growth and development. In return, the fungi receive shelter and food from this association (Smith and Read, 2010; Thakur and Singh, 2018). The important function of these fungi is to absorb water and minerals by increasing the surface area of the roots, dissolving the organic matter of the soil humus and absorbing nutrients, and secreting antimicrobial substances that protect the plants from various diseases (Giri et al., 2019). Mycorrhizal fungi are known to improve soil quality, soil aeration, water, and heavy metal dynamics, drought tolerance of plants, and to make plants less susceptible to root-borne pathogens (Thakur and Singh, 2018; Rillig et al., 2002). Researchers have found that fungal mycorrhizal inoculant can be used as a biological fertilizer and biological control agent for orchid root diseases (Bhattacharjee and Dey, 2014).

3.2 Microbes as biopesticides

Biopesticides are living organisms or their products that have the ability to kill/repel certain insects or pests. Biopesticides from bacteria, viruses, fungi, algae, nematodes, and protozoa, as well as some other compounds made directly from these microbes such as metabolites are the main microbial pest control agents (Van Lenteren, 2012). The overuse of chemical/synthetic pesticides has resulted in the emergence of new strains of pests that are resistant to insecticides and synthetic pesticides (Nawaz et al., 2016) thus, there is a need to switch on organic farming based on biopesticides. More than 100 bacteria have been identified/diagnosed as pathogenic insects, among which *Bacillus thuringiensis* (Bt) has given paramount importance as a microbial agent. Similarly, more than 1000 viruses have been isolated from insect species, such as baculoviruses of the genus Nucleopolyhedrovirus (NPV), which infect insects and other arthropods, have species-specific narrow-spectrum insecticidal applications (Nawaz et al., 2016).

3.2.1 Bacteria as biopesticides

The most commonly used bacteria are *B. thuringiensis* and *Bacillus sphaericus*. B thuringiensis is a specific, environmentally friendly, and effective pest control tool (Roh et al., 2007). The insecticidal property of *B. thuringiensis* lies in the cry protein, which is responsible for the ingestion and death of insects (Khachatourians, 1986). Similarly, *B. sphaericus* is another Gram-positive Aerobic bacteria that produces an intracellular toxin at the time of sporulation. *B. sphaericus*-based products are widely used to control mosquitoes (Park et al., 2010a; De Barjac et al., 1985) (Fig. 3).

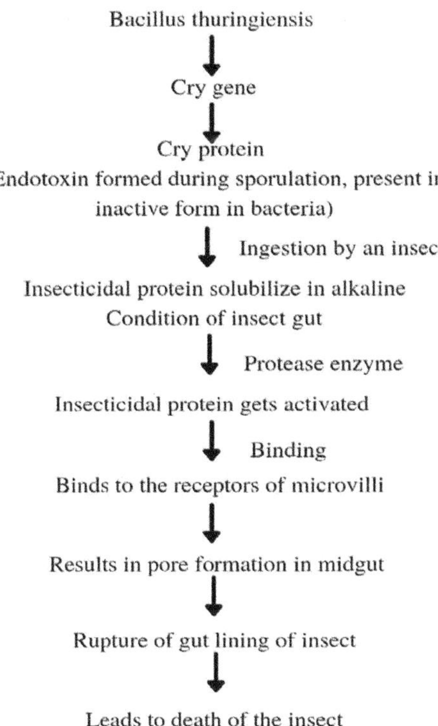

Fig. 3 Illustration of *Bacillus thuringiensis* crystal toxin mechanism.

3.2.2 Viruses as biopesticides

Viruses used to control insects or pests are the DNA containing such as Baculovirus (BV) belonging to the genus nucleopolyhedrovirus (NPV), granulovirus (GV), and RNA-containing nodavirus, reovirus, etc., these bacteria are widely used to control vegetables and crops around the world, and are effective in combating insect chewing on plants (Koul, 2011). Codling moth is controlled by Cydia pomonella GVs on fruit trees (Lacey et al., 2008), and potato tuber worms by *Phthorimaea operculella* GVs in tubers (Arthurs et al., 2008). Viruses-based products are also useful in combating different worms such as cabbage moths, cotton leaf worms, bollworms, tobacco budworms, etc. (Koul, 2011).

3.2.3 Fungi as biopesticides

Another group of microbial pest control is pathogenic fungi, which act as a powerful microbial biopesticide (Khachatourians, 1986). It grows in both terrestrial habitats and aquatic habitats, especially when associated with insects known as entomopathogenic fungi (Koul, 2011). The pathogenic effect is contact-dependent and infects and kills sucking harmful insects, such as aphids, mosquitoes, whiteflies, bugs, etc. (Barbara and

Clewes, 2003; Pineda et al., 2007). The Trichoderma fungus is a biological control agent that is being developed to treat plant diseases. Some fungi, mainly streptomycetes, also produce toxins that are effective against insects (Dowd, 2001). Similarly, actinomycetes *Saccharopolyspora spinosa* produce active compound Spinosyns which is commercially available as biopesticidal compound effective against Diptera and Hymenoptera (Hall and Menn, 1999). It has been reported that about 50 such compounds are active against various insect species belonging to Coleoptera, Lepidoptera, and Homoptera (Cole and Rolinson, 1972).

3.2.4 Nematodes as biopesticides

Entomopathogenic nematode species of the genera Heterorhabditis and Steinernema act as obligate parasites and, due to their mutual symbiosis with insect pathogenic bacteria of the genera Photorhabdus and Xenorhabdus, have a considerable insecticidal potential that shows toxicity against insects by ingestion (Lewis and Clarke, 2012; Pioner). Nematodes, which are often used in pest control, belong to the genera Steinernema and Heterorhabditis, which attack the host as an infectious juvenile, that enter the host through the mouth, cuticles, spiracles, and anus where they release their bacterial symbionts in the hemocele of the host and kills the host in 24–48 h (Koul, 2011; Ruiu, 2018). A variety of products based on different species of nematodes are marketed worldwide to control certain types of pests (Ruiu, 2018).

3.3 Microbes as bioherbicides

The use of microorganisms as bioherbicides has the unique ability to reduce the invasive weed population through highly specific, self-sufficient effects and thus to contribute to the protection of the natural ecosystem (Van Driesche et al., 2010). Bioherbicidal microbial agents may include obligate fungal parasites, soilborne fungal pathogens, non-phytopathogenic fungi, pathogenic and non-pathogenic bacteria, and nematodes (Kremer, 2005). The potential of bioherbicides to control weeds offers advantages over herbicides because the multiple mechanisms involved mean that there is less likelihood that the target weeds will develop resistance to bioherbicides (Crump et al., 1999). Microorganisms that are largely overlooked as biological control agents for weeds are the Rhizobacteria (Schippers et al., 1987). It was reported that the live culture of syringal strain 3366 produced by Pseudomonas has sometimes slowed the growth of weed roots in controlled environments (Johnston and Booth, 1983). Similarly, ethyl acetate extracts from Pseudomonas syringal strain 3366 have been found to be effective to reduce weed roots and shoots growth in field conditions in the Pacific Northwest (Gealy et al., 1996). Mechanically transmitted viruses such as Tobacco Green Mosaic Virus, which is produced, harvested, and freeze-dried in surrogate host plants, are being studied as invasive weed herbicides, including tropical soda apple (*Solanum viarum*) (Charudattan, 2001). Soilborne fungi have been considered an important bioherbicide supplement as these fungi are used directly in the soil to reduce weed growth by seed rot before emergence or to kill seedlings soon after emergence (Jones and Hancock, 1990). According to research there are only five proven effective effects of microorganisms as bioherbicides, they are *Colletotrichum gloeosporioides aeschynomene* for control of northern joint vetch, *Chondrostereum purpureum pouzar* for control of weedy tree species, *Phytophthora palmivora* for control of Strangler vine, Acremonium sp. For control of scrambled egg bush and *Xanthomonas campestris* for control of annual bluegrass (Kremer, 2005; Bailey and Falk, 2011).

3.4 Microbes as biofungicides

Biofungicide is a generic name given to microbes and natural substances that have the ability to control plant diseases (Abbey et al., 2019). Due to their origin and the low concentration of the active ingredients, most of the preparation in this group is considered pure in nature (Khakimov et al., 2020). In several studies, the various pathogens were fought both in the field and in storage with the help of biofungicides (Abbey et al., 2019). Biofungicides are produced in wettable powders, emulsion concentrate, suspension concentrate, tablets, and other forms (Khakimov et al., 2020). Biofungicides are applied by treating the planting material, adding it to the soil, and spraying the growing plant (Cheremesin and Yakimova, 2011). *Trichoderma pseudokoningii* and *T. viride* have been reported to inhibit Botrytis cinerea in strawberry fruit by producing some secondary metabolites (Tronsmo and Dennis, 1977). Fusarium crown and root rot of tomato caused by *Fusarium oxysporum*, is a recent harmful disease of greenhouse crops. Tomato plants that have been treated with biofungicides produce more and have better fruit than those treated with the synthetic fungicide hymexazol (Hibar et al., 2006). Water Hyacinth in Florida USA is controlled by the fungal spores of *Cercospora rodmanii*. Milkweed vines in citrus orchards are controlled by spores from the *Phytophthora palmivora* fungus known as devine. Some other examples are *Trichoderma harzianum* species with biological control potential against Botrytis cinerea, Fusarium, Pythium, causing a powdery mildew hyperparasite, *Gliocladium virens* is an effective bio-control against soilborne pathogens and *Chaetomium globosum* and *C. cupreum*, which have biocontrol activity against root rot (Sharma, 2015).

4. Microbiol biocontrol for postharvest management

Plant diseases are the main barriers affecting plant production and productivity in terms of both quality and quantity (Babychan et al., 2017). It is estimated that even in industrialized countries around 20%–25% of the fruits and vegetables harvested are spoiled by disease-causing organisms (Singh and Sharma, 2018). In developing countries, the percentage loss is as high as 50% (Eckert and Ogawa, 1985). The use of synthetic fungicides is the primary means of combating postharvest diseases, but many factors for several reasons, such as the need to reduce excessive use of chemicals, concern about pollution, inability to control fungal infections due to the emergence of fungal tolerant strains in pathogens, all these have encouraged rapid development in alternative methods (Ippolito and Nigro, 2000). The use of microbial biocontrol for postharvest diseases in agriculture receives special attention and has been extensively studied (Droby, 2006). Infections with pathogenic organisms can occur during the growing season, at harvest time, during handling, storage, transportation, and marketing, or even after purchase by the consumer (Spadaro, 2012). There are two basic approaches of using microbial biocontrol in combating postharvest fruit and vegetable disease first, using preexisting microorganisms that can be promoted and managed, and second, those that are artificially introduced against postharvest pathogens (Sharma et al., 2009).

4.1 Postharvest applications

Postharvest microbial biocontrol applications are better and useful to control postharvest diseases of fruits and vegetables (Babychan et al., 2017). In this process, microbial cultures are applied as postharvest sprays or as immersion in an antagonistic solution (Irtwange, 2006). The acceptable rate of decomposition in the postharvest harvesting system is usually less than 5% the rate normally obtained with the single use of a biocontrol agent (Janisiewicz and Korsten, 2002). Wounds are created during harvesting and fruit handling can be prevented from invading pathogens with a single application of the antagonist directly to the postharvest wound using existing techniques such as sprays, dips, etc., the harvested fruits are then stored in cold storage for different periods of time, depending on the product, ranging from a few days to months. As soon as the fruit is stored in cold storage, the host's metabolic rates and the associated microflora decrease depending upon the temperature selected (Spadaro, 2012; Janisiewicz and Korsten, 2002). Mixed cultures of the microbial biocontrols are used, which tend to provide better control of postharvest diseases than individual cultures or strains (Sharma et al., 2009). Similarly, the effectiveness of the microbial antagonists can be enhanced when used with low doses of fungicides, saline additives, manipulating the environment, combining field and postharvest applications, Manipulating the formulations, physiological and genetic Manipulation of an antagonist, and last, integrating with different nutrients and methods (Janisiewicz and Korsten, 2002; Sharma et al., 2009). A high level of postharvest gray mold control in strawberries and Alternaria rot in lemons has been achieved through postharvest applications of biological pesticides, such as *Trichoderma viride*, *Trichoderma harzianum*, *Paecilomyces variotii*, and *Gliocladium roseum* (Dukare et al., 2019). A significant reduction in storage degradation was achieved by bringing various yeast species into direct control with harvested fruits, for example, the direct contact of the microbial antagonist with the infected fruit has proven to be very useful in suppressing the growth of pathogens such as *Penicillium digitatum*, *Penicillium italicum*, *Botrytis cinerea* apples (Chalutz and Wilson, 1990; Gullino et al., 1992).

5. Mechanism of microbial control activity

Organisms that are responsible for the destruction of plant growth are called pathogens. Pathogens reduce the plant growth and even result in its death (Gupta and Sharma, 2008). Therefore, biocontrol activities are needed. Biocontrol activities are the activities that involve microorganisms usually their metabolites and products to minimize the harmful effect caused by pathogens. So microbial microorganism like yeast, bacteria, fungi, actinomycetes works as a shield for plants that are affected by different pathogens and hence, help to control several plants of vegetables, and fruits from different toxic infections (Mathre et al., 1999). Some non-pathogenic microorganisms can work as a biocontrol agents. For example, Pyrrolnitrin works as an antibiotic. As the fungicidal derivative named fungicide flupidol is derived from it and is commonly used as a foliar spray, soil drenches soil treatment (Sturz, 2006). Wound pathogens are known to cause postharvest diseases, and to compete with these pathogens biocontrol agents (BCA) should be grown hastily. Biocontrol agents will colonize on plants that consume the nutrient; therefore, pathogen will not get sufficient nutrients for its functioning. As a result, growth and toxic effect of pathogens will diminish (Nunes, 2012). Some biocontrol agents can protect plants indirectly by inducing resistance in the plant, without interacting with the pathogens. The main role of biocontrol activity is to improve plant health by minimizing the toxic effect on the crop (Dwivedi, 2014). It is a disease suppression phenomenon involving internal actions in plants, biocontrol agents, pathogens, and supporting physical, chemical, and biological environment (Handelsman and Stabb, 1996). The most complex biological phenomenon is for soil chronic diseases, as it is a disease that occurs at the interference of soil and root in one of the most challenging and dynamic environments (Rovira, 1965; Hawes, 1990).

5.1 Antagonism

According to the researchers (Montesinos, 2003), activities like the growth of pathogenic microorganism was inhibited/stopped by the bacterial and fungal strain that was isolated from the analysis of biocontrol activities (Ulloa-Ogaz et al., 2015). Bacterial strains are mainly isolated from genera Agrobacterium, Bacillus, Pseudomonas and, fungal strains are isolated from the genera Candida, Trichoderma (Slininger et al., 2003). And, it is also proven that microbial strain has at least 2%–10% capabilities to work as "ANTAGONIST." Therefore, these strains are now commercially marketed and are labeled as "EPA registered biopesticide." One such strain is isolated from Rhizospheric soil (Ulloa-Ogaz et al., 2015). The stain could work as an Antagonist shield only if it has (Ulloa-Ogaz et al., 2015; Larkin and Fravel, 1998).

  i. Tendency to form colonies on the surface of plants.
  ii. Tendency to absorb nutrients faster and stronger than pathogens.
  iii. Tendency to withstand unfavorable climatic conditions.

iv. Should be genetically stable.

v. Should not have requirement of extra nutrient for survival.

vi. Microorganisms should cause any dangerous/side effect on human health neither directly by producing secondary metabolites nor indirectly by any other mechanism.

vii. Should be effective (working) even at low concentrations.

viii. Should be effective on a broad range of pathogens, so that it can protect a wide range of fruits and vegetables.

For the study of antagonism, it is important to understand antibiotic production and its resistant.

5.1.1 Antibiotic production

Association of several microorganism and soil is very remarkable. There are some chemicals in the soil that have an adverse effect on other microorganism and have tendency properties to be used as the medicine are known to have antibiotic (Abawi and Widmer, 2000; Kubicek et al., 2001). There are some chemicals that are antibiotically producing and enhancing the growth of Rhizobacteria (PGPR) (Fig. 4).

All the above-listed antibiotics may have an antiviral, insecticidal, antimicrobial, antioxidant, and cytotoxic effect (Fernando et al., 2005). Different antibiotics perform different modes of action depending upon the type of microorganism, environmental conditions, species of plant, and many more (Reid et al., 2002) (Fig. 5).

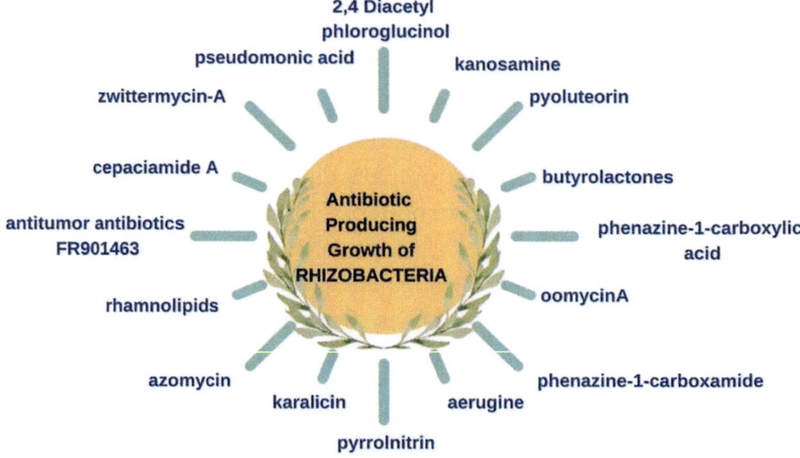

Fig. 4 Antibiotic producing growth of Rhizobacteria.

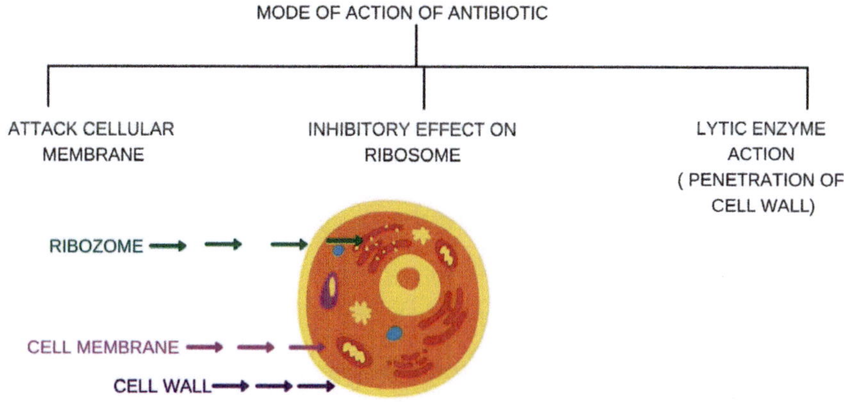

Fig. 5 Mode of action antibiotic.

5.2 Competition

The major factor that is responsible for the growth and germination of the pathogen is nutrient uptake. Pathogens compete with each other for nutrients and space therefore, competition is considered as a key antagonist factor to suppress pathogens (Nunes, 2012). Some common microbes that utilize compete as the mode of action are *Aureobasidium plants, Cryptococcus humicola, Debaryomyces hansenii, Mestchmikcoia pullnerioma* (Bencheqroun et al., 2007; Filonow et al., 1996). Some microbes by these nutrients compete and modulate the growth condition for the plant (Köhl et al., 2019).

Most of the plant pathogens especially obligate Biotrophic pathogens depend only on the nutrient of the host plant cell, i.e., infected cell, they do not take the nutrient from the external environment (Agrios, 2005). Moreover, a bacterium, fungi, oomycetes belonging to Necrotrophic plant pathogens kill the tissue of the host cell and extracts nutrient from it (Köhl et al., 2019). Necrotrophic first colonizes on the host cell and after colonization is responsible for plant death. In this competition, some non-pathogenic associations of the Saprophytic endocytic took advantage of these associations and are laterally present in infected host tissue as a result indirectly helping pathogens, for example, leaf lesions (Köhl et al., 2019). But as Necrotrophic plant pathogen depends on the external nutrient during spore formation. Therefore, Obligate Biotropic pathogens develop more vulnerability toward nutrient competition (Kohl and Fokkema, 1998). Competition of biocontrol agents and microbe depends on the level of nutrient available, timings, and distribution of Biocontrol Agents (BCA), temperature, point of interaction of microbe (Köhl et al., 2019). The competent tendency of pathogen depends on various factors like temperature, nutrient availability, pH, drought, wind, the direction of the wind, sunlight (Köhl et al., 2019). According to a survey (Saravanakumar et al., 2008). Competition of iron was the major mode of action in pathogens like *M. pulchenima* (Table 1).

Table 1 List of some common pathogen and its targeting nutrient.

Pathogen	Nutrient	Biocontrol agent	Referencing
Yeast	Carbohydrate, nitrogen	*Pichia guilliermondii*	Köhl et al. (2019) and Spadaro et al. (2010)
Rhizospheres	Siderophore	*P. cinerea, Pseudomas* spp.	Raaijmakers et al. (1995) and Bakker et al. (1993)
Fungal	Iron	*Trichoderma asperellem*	Segarra et al. (2010)
Monilina lexa	Carbon	*A. pullulans*	Bauer et al. (2010)
B. cenera	Vitamin	*T. harzianum*	Gullino (1992)
Penicillium italicum	Carbon, nitrogen	*Kloeckra apiculture*	Liu et al. (2013)
Rhizopus nigricans	Beta-carotene	*Pichia guilliermondii*	Zhao et al. (2008)

5.3 Parasitism

Parasitism is the type of direct competitive interaction in which one party gets benefited. In this mode, benefits are shared by parasites or BCA (biocontrol agents) and the pathogen gets harmed. This type of interaction is mainly seen in fungi, and rarely seen in bacteria (Köhl et al., 2019). These parasites result in a negative effect on the host t individual, populational, and ecological levels. Therefore, biological control methods are developed to minimize the harmful effect of parasites. The most common biocontrol agent used for the parasitic association are parasitoids like Hymenoptera, Diptera, Coleoptera, Lepidoptera, Trichoptera, Strepsiptera (Wajnberg and Ris, 2009). The most common type of parasite is Hypovirus, a facultative parasite, an obligate parasite, and predators (Heydari and Pessarakli, 2010). And, the most common type of parasitic interaction is hyperparasitism. It is the interaction in which the host plant is also working as a parasite, i.-e., parasites are themselves infected by the parasite (Köhl et al., 2019). The interaction in which hyperparasites depends on the host plant to gain nutrient without killing the host cell is called biotropic mycoparasitism (Jeffries, 1995). Most of the mycoparasite belongs to the genera Trichoderma and Clonostachys. Whereas, nematodic parasites gain nutrients from the dead plant cells. The parasiting pathogen is called hyperparasites and the parasites pathogen is called hypoparasitism (Barbosa, 1998). Plant-parasitic nematodes are considered the most dangerous pathogens as they cause severe plant losses all over the world (Koenning et al., 1999). Management of these parasitic nematodes is most harmful as they majorly attack underground crop production (Stirling, 1991). BCA affect these parasites in a number of ways like by producing toxins, making antibiotics, enzymes, or inducing systematic resistance (Siddiqui and Mahmood, 1999). A major source of cell wall degradation in CWDE (cell wall degrading enzyme). And it is responsible for the death of the host plant. In biotropic mycoparasitism less amount of CWDE is excreted by the pathogen. Therefore, the host cell does not die. Whereas, in the case of nematodic parasite large amount of CWDE is excreted that resulting in the breakage of the cell wall. Hence, the host cell dies. Cell wall degradation also depends on the amount of chitinases, β-1,3-glucanases, proteases, celluloses (Köhl et al., 2019) (Fig. 6).

6. Emerging biocontrol strategies

In today's era diseases are mainly caused by the plant pathogen. The range of plant-pathogen differs from the simple structure to the more complex. Pathogen like bacteria, yeast, fungi, and nematodes are responsible for major crop loss. Plants by releasing their broad types of chemicals from their leaves and roots communicate with their associated microbiome. Therefore, for making new strategies it is very much necessary to understand the "plant-microbiome language" By decoding this language it will be clear that what attracts plants to give them required or needed benefits (Ab Rahman et al., 2018; Vorholt, 2012). The microbe will result in better growth of plant's health, and development via different routes, or mechanisms of action. Their microbes can directly enhance the growth by increasing the nutrient, or by increasing plant growth hormone, or indirectly by acting as an antagonist or competent for the pathogen (Ab Rahman et al., 2018).

6.1 Exudates from the roots

Roots are the hidden part of the plant that is responsible for nutrient and water uptake. Root of plant exudates some specific chemicals or fluids that attract microbial population. Discharged exudates work as the interaction between plant and neighboring microbes (Lyon and Wilson, 1921). As roots exudation is the process of Rhizodeposition and is considered as the major source of carbon (Nguyen, 2003). Some particular root exudates some particular chemical to attract particular microbe to satisfy a particular need. Attracted microbe activates the specific part of the plant that helps the plant to fight some

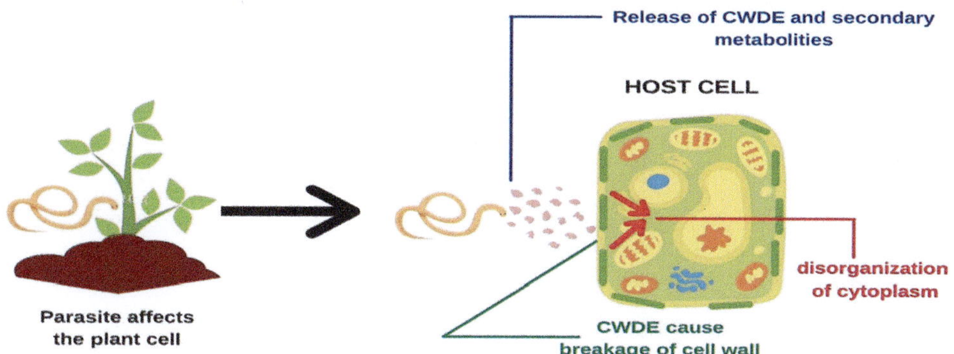

Fig. 6 Mechanism of action of the parasite.

diseases like foliar disease (Ryu et al., 2004). Especially, the microbes present in the soil get attracted toward exudates and help to prevent diseases by improving the nutrient content in the plants (Cao et al., 2011; Kavoo-Mwangi et al., 2013). Attracted microbe also helps in water supply, managing defense, and phosphate solubilization (Ab Rahman et al., 2018).

Amount of root exudates depends on several factors like temperature, sunlight, type of soil, type of plant, age of the plant, location of the plant, and many more shown in Table 2.

6.2 Use of substrate and biocidal volatiles

Substrates are the composition of different nutrient that promotes the growth of biocontrol agents. Therefore, spraying the substrates around the crops helps to promote the growth of beneficial biocontrol microbe. These substrates not only help in the growth of the plant but also help in the metabolism, and activities of the microbial cell (Bai et al., 2015). Spraying the substrate will also protect the spread of disease in the plant or around the plant. Culture of microbial agents depends on the systematic bacterial conditions. Use of substrates works as an advantage to colonize the required microbes in the crop. This not only helps to fight with the existing diseases but also prepares the crop for upcoming diseases. Moreover, the substrate is a mixture of different nutrients that helps the microbiome to survive in different habitats and climatic conditions (Ab Rahman et al., 2018).

Volatile organic compounds (VOC) have the defusing property due to which they can pass through the cellular membrane (Pichersky et al., 2006). VOC works as the communicating channel. VOCs is emitted by the bacteria or other pathogens present

Table 2 Amount COF exudates of different plant.

S. no.	Plant	Amount of exudates per plant	References
1	PEA *Pisum arvense*	2.9–4.3 mg	Meshkov (1961)
2	CORN *Zea mays*	8.4–8.2 mg	Meshkov (1961)
3	WHEAT *Triticum*	1.5–13 mg	Rovira (1969)
4	VETCH *Vicia* sp.	1.6–2.9 mg	Harmsen and Jager (1962)
5	RICE *Oryza sativa*	2–22.5 mg	Harmsen and Jager (1962)
6	OATS *Avena sativa*	12 mg	Harmsen and Jager (1962)
7	BARLEY *Hordeum vulgare*	2.5–1 mg	Rovira (1969)

in the soil. Emission took place above the ground level or deep inside the soil (Bhardwaj et al., 2020).

6.3 Use of foliar microbiome and detoxifying enzymes

Foliar diseases are mainly the fungus diseases like tan spot, stagonospora nodorum blotch, septoriatritici blotch, net blotch, scald, powdery mildew (Madden and Nutter Jr, 1995). According to the survey (Dean et al., 2012), foliar diseases are the major and serious diseases in the crop. Moreover, fungus comes under the top eighth dangerous infection in the world. Therefore, foliar biomass could be the best alternative to control fungal disease after the chemical control methods (Ab Rahman et al., 2018). Whereas, detoxifying is the enzymes that can degrade the growth of toxins produced by pathogens especially fungus by blocking the pathway of toxin at plant interference (Munkvold, 2003).

6.4 Use of iron-chelating compounds

Microbes and pathogens compete for the nutrient especially iron for its growth (Baker et al., 1997). Therefore, iron-chelating compounds can help in the transport of iron toward the microbe (Neilands, 1981; Hider and Kong, 2010). This will enhance the ability of a microbe to consume more nutrients than a pathogen and hence may make the pathogen weaker, or enable them to spread infection (Nelson, 1991). Iron-chelating method is successfully done by the growth-promoting strain of *Pseudomonas fluorescens*, such as A1, Bk1, TL3B1, and BIO against the common pathogen *Erwinia carotovora* (Kloepper et al., 1980).

7. Conclusion and future prospects

Biocontrol agents hold the potential to replace chemical pesticides. They can be employed as biopesticides, biofertilizers, and stimulators for plant development. The BCA employs various action mechanisms to control the pathogens of a plant which either involve a direct interaction of BCA with the pathogen, such as the production of hyperparasites, lytic enzymes or antibiotics, or the indirect interaction of BCA and pathogen, e.g., nutrient competition and spatial induction and plant resistance. As part of an integrated pest management (IPM) system, biocontrol agents can be created to reduce the environmental impact of chemical fungicides. Further investments are needed in biocontrol agents' research and development. Further research on dosage, formulation, impact on the action of BCA, and the effects of BCA on the native plant microflora of different environmental circumstances should be carried out. The mechanism of action of biocontrol agents is crucial to be fully understood in order to create more effective antagonists. The formulation processes must be improved so that formulations may be formulated with increased shelf-life. The encapsulation of microbial organisms is one of the potential techniques of the effective formulation. However, the most utilized encapsulation technique is still alginate-based beads. Therefore, efforts must be made to develop new techniques and enhance the present formulations and delivery methods. The interaction between the capsule matrix and physical–chemical soil or agroecosystem is also crucial to acquire understanding. Environmental factors play an important role to determine biocontrol agents' effectiveness. The environmental conditions under which the biocontrol agents function efficiently thus need to be studied. Factors leading to the rhizosphere colonization and the influence of BCA on the native land population must also be determined. The genes, gene products, and other signaling molecules that are responsible for the antagonistic activity of microorganisms are also crucial to determine. Effective techniques for mass manufacturing of BCA must also be developed and technological challenges linked to their mass production resolved and formulations improved so that gram-negative bacteria may be employed successfully.

References

Ab Rahman, S.F.S., Singh, E., Pieterse, C.M., Schenk, P.M., 2018. Emerging microbial biocontrol strategies for plant pathogens. Plant Sci. 267, 102–111.

Abawi, G.S., Widmer, T.L., 2000. Impact of soil health management practices on soilborne pathogens, nematodes and root diseases of vegetable crops. Appl. Soil Ecol. 15 (1), 37–47.

Abbey, J.A., Percival, D., Abbey, L., Asiedu, S.K., Prithiviraj, B., Schilder, A., 2019. Biofungicides as alternative to synthetic fungicide control of grey mould (Botrytis cinerea)—prospects and challenges. Biocontrol Sci. Tech. 29 (3), 207–228.

Agrios, G.N., 2005. Plant Pathology. Elsevier Academic Press.

Ambrosini, A., Beneduzi, A., Stefanski, T., Pinheiro, F.G., Vargas, L.K., Passaglia, L.M., 2012. Screening of plant growth promoting rhizobacteria isolated from sunflower (*Helianthus annuus* L.). Plant Soil 356 (1), 245–264.

Arthurs, S.P., Lacey, L.A., De La Rosa, F., 2008. Evaluation of a granulovirus (PoGV) and Bacillus thuringiensis subsp. kurstaki for control of the potato tuberworm (Lepidoptera: Gelechiidae) in stored tubers. J. Econ. Entomol. 101 (5), 1540–1546.

Babbal, Adivitiya, Khasa, Y.P., 2017. Microbes as biocontrol agents. In: Probiotics and Plant Health. Springer, pp. 507–552, https://doi.org/10.1007/978-981-10-3473-2_24.

Babychan, M., Jojy, E.T., Syriac, G.M., 2017. Bio-control agents in management of post-harvest diseases. Life Sci. Int. Res. J. 4 (1), 51–55.

Badii, M.H., Abreu, J.L., 2006. Control biológico una forma sustentable de control de plagas (biological control a sustainable way of pest control). Daena Int. J. Good Consci. 1 (1), 82–89.

Bai, Y., Müller, D.B., Srinivas, G., Garrido-Oter, R., Potthoff, E., Rott, M., Schulze-Lefert, P., 2015. Functional overlap of the Arabidopsis leaf and root microbiota. Nature 528 (7582), 364–369.

Bailey, K.L., Falk, S., 2011. Turning research on microbial bioherbicides into commercial products—a Phoma story. Pest Technol. 5 (1), 73–79.

Baker, B., Zambryski, P., Staskawicz, B., Dinesh-Kumar, S.P., 1997. Signaling in plant-microbe interactions. Science 276 (5313), 726–733.

Bakker, P.A., Raaijmakers, J.M., Schippers, B., 1993. Role of iron in the suppression of bacterial plant pathogens by fluorescent pseudomonads. In: Barton, L.L., Hemming, B.C. (Eds.), Iron Chelation and Soil Microorganisms. Academic Press, London, pp. 269–281.

Barbara, D.J., Clewes, E., 2003. Plant pathogenic Verticillium species: how many of them are there? Mol. Plant Pathol. 4 (4), 297–305.

Barbosa, P.A. (Ed.), 1998. Conservation Biological Control. Elsevier.

Barman, M., Paul, S., Choudhury, A.G., Roy, P., Sen, J., 2017. Biofertilizer as prospective input for sustainable agriculture in India. Int. J. Curr. Microbiol. App. Sci. 6 (11), 1–177.

Bashan, Y., 1998. Inoculants of plant growth-promoting bacteria for use in agriculture. Biotechnol. Adv. 16 (4), 729–770.

Bashan, Y., Holguin, G., 1997. Azospirillum–plant relationships: environmental and physiological advances (1990–1996). Can. J. Microbiol. 43 (2), 103–121.

Bashan, Y., Ream, Y., Levanony, H., Sade, A., 1989. Nonspecific responses in plant growth, yield, and root colonization of noncereal crop plants to inoculation with *Azospirillum brasilense* Cd. Can. J. Bot. 67 (5), 1317–1324.

Bauer, R., du Toit, M., Kossmann, J., 2010. Influence of environmental parameters on production of the acrolein precursor 3-hydroxypropionaldehyde by Lactobacillus reuteri DSMZ 20016 and its accumulation by wine lactobacilli. Int. J. Food Microbiol. 137 (1), 28–31.

Bencheqroun, S.K., Bajji, M., Massart, S., Labhilili, M., El Jaafari, S., Jijakli, M.H., 2007. In vitro and in situ study of postharvest apple blue mold biocontrol by *Aureobasidium pullulans*: evidence for the involvement of competition for nutrients. Postharvest Biol. Technol. 46 (2), 128–135.

Bhardwaj, K., Abraham, J., Kaur, S., 2020. Natural product as Avermectins and Milbemycins for agriculture perspectives. In: Natural Bioactive Products in Sustainable Agriculture. Springer, Singapore, pp. 259–271.

Bhattacharjee, R., Dey, U., 2014. Biofertilizer, a way towards organic agriculture: a review. Afr. J. Microbiol. Res. 8 (24), 2332–2343.

Bhattacharyya, P.N., Jha, D.K., 2012. Plant growth-promoting rhizobacteria (PGPR): emergence in agriculture. World J. Microbiol. Biotechnol. 28 (4), 1327–1350.

Brimner, T.A., Boland, G.J., 2003. A review of the non-target effects of fungi used to biologically control plant diseases. Agric. Ecosyst. Environ. 100 (1), 3–16.

Cao, Y., Zhang, Z., Ling, N., Yuan, Y., Zheng, X., Shen, B., Shen, Q., 2011. Bacillus subtilis SQR 9 can control fusarium wilt in cucumber by colonizing plant roots. Biol. Fertil. Soils 47 (5), 495–506.

Chalutz, E., Wilson, C.L., 1990. Postharvest biocontrol of green and blue mold and sour rot of citrus fruit by *Debaryomyces hansenii*. Plant Dis. 74 (2), 134–137.

Charudattan, R., 2001. Biological control of weeds by means of plant pathogens: significance for integrated weed management in modern agro-ecology. BioControl 46 (2), 229–260.

Cheremesin, A.I., Yakimova, I.A., 2011. Influence of plant growth stimulants and biofungicides on the productivity of potato microplants. Achiev. Sci. Technol. Agro-Ind. Complex 3, 26–28.

Cole, M., Rolinson, G.N., 1972. Microbial metabolites with insecticidal properties. Appl. Microbiol. 24 (4), 660–662.

Cook, R.J., 1993. Making greater use of introduced microorganisms for biological control of plant pathogens. Annu. Rev. Phytopathol. 31, 53–80.

Crump, N.S., Ash, G.J., Fagan, R.J., 1999. The development of an Australian bioherbicide. In: 12th Australian Week Conference, p. 235237.

Dastager, S.G., Deepa, C.K., Pandey, A., 2010. Isolation and characterization of novel plant growth promoting Micrococcus sp NII-0909 and its interaction with cowpea. Plant Physiol. Biochem. 48 (12), 987–992.

De Barjac, H., Larget-Thiery, I., Dumanoir, V.C., Ripouteau, H., 1985. Serological classification of Bacillus sphaericus strains on the basis of toxicity to mosquito larvae. Appl. Microbiol. Biotechnol. 21 (1), 85–90.

Dean, R., Van Kan, J.A., Pretorius, Z.A., Hammond-Kosack, K.E., Di Pietro, A., Spanu, P.D., Foster, G.D., 2012. The top 10 fungal pathogens in molecular plant pathology. Mol. Plant Pathol. 13 (4), 414–430.

Dowd, P.F., 2001. Antiinsectan compounds derived from microorganisms. In: Microbial Biopesticides. CRC Press, pp. 20–127.

Droby, S., 2006. Biological control of postharvest diseases of fruits and vegetables: difficulties and challenges. Phytopathol. Pol. 39, 105–117.

Dukare, A.S., Paul, S., Nambi, V.E., Gupta, R.K., Singh, R., Sharma, K., Vishwakarma, R.K., 2019. Exploitation of microbial antagonists for the control of postharvest diseases of fruits: a review. Crit. Rev. Food Sci. Nutr. 59 (9), 1498–1513.

Dwivedi, S.K., 2014. Role of antagonistic microbes in management of phytopathogenic fungi of some important crops. In: Microbial Diversity and Biotechnology in Food Security. Springer, New Delhi, pp. 273–292.

Eckert, J.W., Ogawa, J.M., 1985. The chemical control of postharvest diseases: subtropical and tropical fruits. Annu. Rev. Phytopathol. 23 (1), 421–454.

Fernando, W.D., Nakkeeran, S., Zhang, Y., 2005. Biosynthesis of antibiotics by PGPR and its relation in biocontrol of plant diseases. In: PGPR: Biocontrol and Biofertilization. Springer, Dordrecht, pp. 67–109.

Filonow, A.B., Vishniac, H.S., Anderson, J.A., Janisiewicz, W.J., 1996. Biological control of Botrytis cinereain apple by yeasts from various habitats and their putative mechanisms of antagonism. Biol. Control 7 (2), 212–220.

Galindo, E., Serrano-Carreón, L., Gutiérrez, C.R., Balderas-Ruíz, K.A., Muñoz-Celaya, A.L., Mezo-Villalobos, M., Arroyo-Colín, J., 2015. Desarrollo histórico y los retos tecnológicos y legales para comercializar Fungifree AB®, el primer biofungicida 100% mexicano. TIP Rev. Espec. Cien. Quím. Biol. 18 (1), 52–60.

Gallon, J.R., 2001. N2 fixation in phototrophs: adaptation to a specialized way of life. Plant Soil 230 (1), 39–48.

Gealy, D.R., Gurusiddaiah, S., Ogg, A.G., 1996. Isolation and characterization of metabolites from *Pseudomonas syringae*-strain 3366 and their phytotoxicity against certain weed and crop species. Weed Sci. 44 (2), 383–392.

Giri, B., Prasad, R., Wu, Q.S., Varma, A. (Eds.), 2019. Biofertilizers for Sustainable Agriculture and Environment. Springer International Publishing, Cham.

Glick, B.R., 2014. Bacteria with ACC deaminase can promote plant growth and help to feed the world. Microbiol. Res. 169 (1), 30–39.

Gullino, M.L., 1992. Control of botrytis rot of grapes and vegetables with Trichoderma spp. In: Biological Control of Plant Diseases. Springer, Boston, MA, pp. 125–132.

Gullino, M.L., Benzi, D.A.M.C., Testoni, A., Garibaldi, A., 1992. Biological control of Botrytisrot of apple. In: Recent Advances in Botrytis Research. Pudoc Scientific Publishers, p. 197.

Gupta, V.K., Sharma, R.C., 2008. Integrated management of soil-borne diseases. In: Manjunath, N., Rani, G.S.D. (Eds.), Advances in Soil Borne Plant Diseases. New India Publishing Agency, New Delhi, India, ISBN: 9788189422813, pp. 415–427.

Gutiérrez-Mañero, F.J., Ramos-Solano, B., Probanza, A.N., Mehouachi, J.R., Tadeo, F., Talon, M., 2001. The plant-growth-promoting rhizobacteria *Bacillus pumilus* and *Bacillus licheniformis* produce high amounts of physiologically active gibberellins. Physiol. Plant. 111 (2), 206–211.

Habib, S.H., Kausar, H., Saud, H.M., 2016. Plant growth-promoting rhizobacteria enhance salinity stress tolerance in okra through ROS-scavenging enzymes. Biomed. Res. Int. 2016, 6284547.

Hall, F.R., Menn, J.J., 1999. Biopesticides: Use and Delivery. Methods in Biotechnology. vol. 5 Humana Press, Totowa, NJ.

Handelsman, J., Stabb, E.V., 1996. Biocontrol of soilborne plant pathogens. Plant Cell 8 (10), 1855.

Harmsen, G.W., Jager, G., 1962. Determination of the quantity of carbon and nitrogen in the rhizosphere of young plants. Nature 195 (4846), 1119–1120.

Hawes, M.C., 1990. Living plant cells released from the root cap: a regulator of microbial populations in the rhizosphere? Plant Soil 129 (1), 19–27.

Hayat, R., Ali, S., Amara, U., Khalid, R., Ahmed, I., 2010. Soil beneficial bacteria and their role in plant growth promotion: a review. Ann. Microbiol. 60 (4), 579–598.

Heydari, A., Pessarakli, M., 2010. A review on biological control of fungal plant pathogens using microbial antagonists. J. Biol. Sci. 10 (4), 273–290.

Hibar, K., Daami-Remadi, M., Hamada, W., El-Mahjoub, M., 2006. Bio-fungicides as an alternative for tomato fusarium crown and root rot control. Tunis. J. Plant Prot. 1 (1), 19.

Hider, R.C., Kong, X., 2010. Chemistry and biology of siderophores. Nat. Prod. Rep. 27 (5), 637–657.

Ippolito, A., Nigro, F., 2000. Impact of preharvest application of biological control agents on postharvest diseases of fresh fruits and vegetables. Crop Prot. 19 (8–10), 715–723.

Irisarri, P., Gonnet, S., Monza, J., 2001. Cyanobacteria in Uruguayan rice fields: diversity, nitrogen fixing ability and tolerance to herbicides and combined nitrogen. J. Biotechnol. 91 (2–3), 95–103.

Irtwange, S.V., 2006. Application of biological control agents in pre-and postharvest operations. Agric. Eng. Int. CIGR J.

Janisiewicz, W.J., Korsten, L., 2002. Biological control of postharvest diseases of fruits. Annu. Rev. Phytopathol. 40 (1), 411–441.

Jeffries, P., 1995. Biology and ecology of mycoparasitism. Can. J. Bot. 73 (S1), 1284–1290.

Johnston, A., Booth, C., 1983. Plant Pathologist's Pocketbook. CABI.

Jones, R.W., Hancock, J.G., 1990. Soilborne fungi for biological control of weeds. In: Hoagland, R.E. (Ed.), Microbes and Microbial Products as Microbial Herbicides. American Chemical Society, Washington, DC, pp. 276–286.

Kannaiyan, S. (Ed.), 2002. Biotechnology of Biofertilizers. Springer Science & Business Media.

Kavoo-Mwangi, A.M., Kahangi, E.M., Ateka, E., Onguso, J., Mukhongo, R.W., Mwangi, E.K., Jefwa, J.-M., 2013. Growth effects of microorganisms based commercial products inoculated to tissue cultured banana cultivated in three different soils in Kenya. Appl. Soil Ecol. 64, 152–162.

Khachatourians, G.G., 1986. Production and use of biological pest control agents. Trends Biotechnol. 4 (5), 120–124.

Khakimov, A.A., Omonlikov, A.U., Utaganov, S.B.U., 2020. Current status and prospects of the use of biofungicides against plant diseases. GSC Biol. Pharm. Sci. 13 (3), 119–126.

Kloepper, J.W., Leong, J., Teintze, M., Schroth, M.N., 1980. Enhanced plant growth by siderophores produced by plant growth-promoting rhizobacteria. Nature 286 (5776), 885–886.

Koenning, S.R., Overstreet, C., Noling, J.W., Donald, P.A., Becker, J.O., Fortnum, B.A., 1999. Survey of crop losses in response to phytoparasitic nematodes in the United States for 1994. J. Nematol. 31 (4S), 587.

Kohl, J., Fokkema, N.J., 1998. Biological control of necrotrophic foliar fungal pathogens. In: Plant-Microbe Interactions and Biological Control. Marcel Dekker, pp. 49–88.

Köhl, J., Kolnaar, R., Ravensberg, W.J., 2019. Mode of action of microbial biological control agents against plant diseases: relevance beyond efficacy. Front. Plant Sci. 10, 845.

Koul, O., 2011. Microbial biopesticides: opportunities and challenges. CAB Rev. 6, 1–26.

Kremer, R.J., 2005. The role of bioherbicides in weed management. Biopestic. Int. 1 (3), 4.

Krishnaprabu, S., 2020. Liquid microbial consortium: a potential tool for sustainable soil health. J. Pharmacogn. Phytochem. 9 (2), 2191–2199.

Kubicek, C.P., Mach, R.L., Peterbauer, C.K., Lorito, M., 2001. Trichoderma: from genes to biocontrol. J. Plant Pathol. 83, 11–23.

Kundu, D.K., Ladha, J.K., 1995. Efficient management of soil and biologically fixed N2 in intensively-cultivated rice fields. Soil Biol. Biochem. 27 (4–5), 431–439.

Lacey, L.A., Headrick, H.L., Arthurs, S.P., 2008. Effect of temperature on long-term storage of codling moth granulovirus formulations. J. Econ. Entomol. 101 (2), 288–294.

Larkin, R.P., Fravel, D.R., 1998. Efficacy of various fungal and bacterial biocontrol organisms for control of Fusarium wilt of tomato. Plant Dis. 82 (9), 1022–1028.

Lewis, E.E., Clarke, D.J., 2012. Nematode parasites and entomopathogens. In: Insect Pathology. Academic Press, pp. 395–424.

Liu, P., Luo, L., Long, C.A., 2013. Characterization of competition for nutrients in the biocontrol of Penicillium italicum by Kloeckera apiculata. Biol. Control 67 (2), 157–162.

Lyon, T.L., Wilson, J.K., 1921. Liberation of Organic Matter by Roots of Growing Plants. vol. 40 Cornell University.

Madden, L.V., Nutter Jr., F.W., 1995. Modeling crop losses at the field scale. Can. J. Plant Pathol. 17 (2), 124–137.

Mathre, D.E., Cook, R.J., Callan, N.W., 1999. From discovery to use: traversing the world of commercializing biocontrol agents for plant disease control. Plant Dis. 83 (11), 972–983.

Meshkov M.V. (1961) Works of Symposium on Soil Microbiology. Through Krasil'nikov, "Soil Microorganisms and Higher Plants." Translated and published for the NSF and USDA by the Israel Program for Scientific Translations, 285.

Miller, I.M., 1990. Bacterial leaf nodule symbiosis. Adv. Bot. Res. 17, 163–234.

Mmbaga, M.T., Sauve, R.J., Mrema, F.A., 2008. Identification of microorganisms for biological control of powdery mildew in Cornus florida. Biol. Control 44 (1), 67–72.

Montesinos, E., 2003. Development, registration and commercialization of microbial pesticides for plant protection. Int. Microbiol. 6 (4), 245–252.

Munkvold, G.P., 2003. Cultural and genetic approaches to managing mycotoxins in maize. Annu. Rev. Phytopathol. 41 (1), 99–116.

Nawaz, M., Mabubu, J.I., Hua, H., 2016. Current status and advancement of biopesticides: microbial and botanical pesticides. J. Entomol. Zool. Stud. 4 (2), 241–246.

Neilands, J.B., 1981. Iron absorption and transport in microorganisms. Annu. Rev. Nutr. 1 (1), 27–46.

Nelson, E.B., 1991. Exudate molecules initiating fungal responses to seeds and roots. In: The Rhizosphere and Plant Growth. Springer, Dordrecht, pp. 197–209.

Nguyen, C., 2003. Rhizodeposition of organic C by plants: mechanisms and controls. Agronomie 23 (5–6), 375–396.

Nunes, C.A., 2012. Biological control of postharvest diseases of fruit. Eur. J. Plant Pathol. 133 (1), 181–196.

Ogbo, F.C., 2010. Conversion of cassava wastes for biofertilizer production using phosphate solubilizing fungi. Bioresour. Technol. 101 (11), 4120–4124.

Okon, Y., 1985. Azospirillum as a potential inoculant for agriculture. Trends Biotechnol. 3 (9), 223–228.

Park, H.W., Bideshi, D.K., Federici, B.A., 2010a. Properties and applied use of the mosquitocidal bacterium, *Bacillus sphaericus*. J. Asia Pac. Entomol. 13 (3), 159–168.

Pichersky, E., Noel, J.P., Dudareva, N., 2006. Biosynthesis of plant volatiles: nature's diversity and ingenuity. Science 311 (5762), 808–811.

Pineda, S., Alatorre, R., Schneider, M.L., Martinez, A.M., 2007. Pathogenicity of two entomopathogenic fungi on *Trialeurodes vaporariorum* and field evaluation of a Paecilomyces fumosoroseus isolate. Southwest. Entomol. 32 (1), 43–52.

Raaijmakers, J.M., Sluis, L.V.D., Bakker, P.A., Schippers, B., Koster, M., Weisbeek, P.J., 1995. Utilization of heterologous siderophores and rhizosphere competence of fluorescent Pseudomonas spp. Can. J. Microbiol. 41 (2), 126–135.

Reid, T.C., Hausbeck, M.K., Kizilkaya, K., 2002. Use of fungicides and biological controls in the suppression of Fusarium crown and root rot of asparagus under greenhouse and growth chamber conditions. Plant Dis. 86 (5), 493–498.

Richardson, A.E., 2001. Prospects for using soil microorganisms to improve the acquisition of phosphorus by plants. Funct. Plant Biol. 28 (9), 897–906.

Rillig, M.C., Wright, S.F., Eviner, V.T., 2002. The role of arbuscular mycorrhizal fungi and glomalin in soil aggregation: comparing effects of five plant species. Plant Soil 238 (2), 325–333.

Roh, J.Y., Choi, J.Y., Li, M.S., Jin, B.R., Je, Y.H., 2007. *Bacillus thuringiensis* as a specific, safe, and effective tool for insect pest control. J. Microbiol. Biotechnol. 17 (4), 547–559.

Roper, M.M., Gault, R.R., Smith, N.A., 1995. Contribution to the N status of soil by free-living N2-fixing bacteria in a lucerne stand. Soil Biol. Biochem. 27 (4–5), 67–471.

Rovira, A.D., 1965. Interactions between plant roots and soil microorganisms. Annu. Rev. Microbiol. 19 (1), 241–266.

Rovira, A.D., 1969. Plant root exudates. Bot. Rev. 35 (1), 35–57.

Ruiu, L., 2018. Microbial biopesticides in agroecosystems. Agronomy 8 (11), 235.

Ryu, C.M., Farag, M.A., Hu, C.H., Reddy, M.S., Kloepper, J.W., Paré, P.W., 2004. Bacterial volatiles induce systemic resistance in Arabidopsis. Plant Physiol. 134 (3), 1017–1026.

Saravanakumar, D., Ciavorella, A., Spadaro, D., Garibaldi, A., Gullino, M.L., 2008. Metschnikowia pulcherrima strain MACH1 outcompetes Botrytis cinerea, *Alternaria alternata* and Penicillium expansum in apples through iron depletion. Postharvest Biol. Technol. 49 (1), 121–128.

Schachtman, D.P., Reid, R.J., Ayling, S.M., 1998. Phosphorus uptake by plants: from soil to cell. Plant Physiol. 116 (2), 447–453.

Schippers, B., Bakker, A.W., Bakker, P.A., 1987. Interactions of deleterious and beneficial rhizosphere microorganisms and the effect of cropping practices. Annu. Rev. Phytopathol. 25 (1), 339–358.

Segarra, G., Casanova, E., Avilés, M., Trillas, I., 2010. Trichoderma asperellum strain T34 controls Fusarium wilt disease in tomato plants in soilless culture through competition for iron. Microb. Ecol. 59 (1), 141–149.

Sharma, P., 2015. Bio fungicides: their role in plant disease management. In: Winter School, p. 113.

Sharma, R.R., Singh, D., Singh, R., 2009. Biological control of postharvest diseases of fruits and vegetables by microbial antagonists: a review. Biol. Control 50 (3), 205–221.

Siddiqui, Z.A., Mahmood, I., 1999. Role of bacteria in the management of plant parasitic nematodes: a review. Bioresour. Technol. 69 (2), 167–179.

Singh, D., Sharma, R.R., 2018. Postharvest diseases of fruits and vegetables and their management. In: Postharvest Disinfection of Fruits and Vegetables. Academic Press, pp. 1–52.

Slininger, P.J., Behle, R.W., Jackson, M.A., Schisler, D.A., 2003. Discovery and development of biological agents to control crop pests. Neotrop. Entomol. 32, 183–195.

Smith, S.E., Read, D.J., 2010. Mycorrhizal Symbiosis. Academic Press.

Smolander, A., Sarsa, M.L., 1990. Frankia strains of soil under *Betula pendula*: behavior in soil and in pure culture. Plant Soil 122 (1), 129–136.

Souza, R.D., Ambrosini, A., Passaglia, L.M., 2015. Plant growth-promoting bacteria as inoculants in agricultural soils. Genet. Mol. Biol. 38, 401–419.

Spadaro, D., 2012. Biological control of postharvest diseases of fruits and vegetable. Agricultural sciences. In: Encyclopedia of Life Support Systems (EOLSS), Developed under the Auspices of the UNESCO. Eolss Publishers, Oxford. http://www.eolss.net. Retrieved 23.

Spadaro, D., Ciavorella, A., Dianpeng, Z., Garibaldi, A., Gullino, M.L., 2010. Effect of culture media and pH on the biomass production and biocontrol efficacy of a Metschnikowia pulcherrima strain to be used as a biofungicide for postharvest disease control. Can. J. Microbiol. 56 (2), 128–137.

Stirling, G.R., 1991. Biological Control of Plant Parasitic Nematodes: Progress, Problems and Prospects. CAB International.

Sturz, A.V., 2006. Bacterial root zone communities, beneficial allelopathies and plant disease control. In: Allelochemicals: Biological Control of Plant Pathogens and Diseases. Springer, Dordrecht, pp. 123–142.

Thakur, P., Singh, I., 2018. Biocontrol of soilborne root pathogens: an overview. Root Biol., 181–220.

Timmusk, S., Nicander, B., Granhall, U., Tillberg, E., 1999. Cytokinin production by *Paenibacillus polymyxa*. Soil Biol. Biochem. 31 (13), 1847–1852.

Tronsmo, A., Dennis, C., 1977. The use of Trichoderma species to control strawberry fruit rots. Neth. J. Plant Pathol. 83 (1), 449–455.

Ulloa-Ogaz, A.L., Muñoz-Castellanos, L.N., Nevárez-Moorillón, G.V., 2015. Biocontrol of phytopathogens: antibiotic production as mechanism of control. In: Méndez-Vilas, A. (Ed.), The Battle Against Microbial Pathogens: Basic Science, Technological Advances and Educational Programs. Formatex, pp. 305–309.

Van Driesche, R.G., Carruthers, R.I., Center, T., Hoddle, M.S., Hough-Goldstein, J., Morin, L., Van Klinken, R.D., 2010. Classical biological control for the protection of natural ecosystems. Biol. Control 54, S2–S33.

Van Lenteren, J.C., 2012. The state of commercial augmentative biological control: plenty of natural enemies, but a frustrating lack of uptake. BioControl 57 (1), 1–20.

Vorholt, J.A., 2012. Microbial life in the phyllosphere. Nat. Rev. Microbiol. 10 (12), 828–840.

Vos, C.M., Yang, Y., De Coninck, B., Cammue, B.P.A., 2014. Fungal (-like) biocontrol organisms in tomato disease control. Biol. Control 74, 65–81.

Wajnberg, E., Ris, N., 2009. Parasitism and biological control. In: Ecology and Evolution of Parasitism, pp. 107–127.

Wang, T.T., Ding, P., Chen, P., Xing, K., Bai, J.L., Wan, W., Jiang, J.H., Qin, S., 2017. Complete genome sequence of endophyte *Bacillus flexus* KLBMP 4941 reveals its plant growth promotion mechanism and genetic basis for salt tolerance. J. Biotechnol. 260, 38–41.

Wyckhuys, K.A., Lu, Y., Morales, H., Vazquez, L.L., Legaspi, J.C., Eliopoulos, P.A., Hernandez, L.M., 2013. Current status and potential of conservation biological control for agriculture in the developing world. Biol. Control 65 (1), 152–167.

Zhao, Y., Tu, K., Shao, X., Jing, W., Su, Z., 2008. Effects of the yeast Pichia guilliermondii against Rhizopus nigricans on tomato fruit. Postharvest Biol. Technol. 49 (1), 113–120.

CHAPTER 12

Composting process: Fundamental and molecular aspects

Ruchi Soni[a] and Sunita Devi[b],*
[a]Regional Centre of Organic Farming (HQ), Ghaziabad, Uttar Pradesh, India
[b]Department of Basic Sciences, Dr YS Parmar University of Horticulture and Forestry Nauni, Solan, India
*Corresponding author: e-mail address: sunitachamba@gmail.comm

1. Introduction

With the rapid development of the economy and agriculture, organic and hazardous waste, such as animal manure, sewage sludge, green waste, residues of antibiotics, municipal solid waste, and agricultural waste continue to grow rapidly and have become major sources of problems to the environment and society. Since such wastes are rich in organic matter content and nutrients, these can be applied as soil amendment or fertilizer but at the same time, these wastes also serve as a source of various types of toxic materials or elements like viruses, fungi, heavy metals, and antibiotics. Their proper and scientific disposal is a major concern nowadays. Generated waste has traditionally been disposed of either by landfilling or by in situ incineration, producing huge black smoke clouds that led to environmental pollution along with potential risks to human health. In addition, CO_2 generated during incineration is associated with global warming as greenhouse gases (Rashad et al., 2010). However, these problems associated with the increase and disposal of waste can be minimized or even cease to be a dilemma if waste was given an added value. Composting bestows a better and more fascinating way of reducing, recycling, and adding value to the amount of waste produced. It is one of the most important natural bioprocesses used to treat organic waste through microbial interventions. As the oldest and easiest method for handling organic waste, composting can simply be defined as a self-heating biological conversion in which wastes are converted into value-added products like fertilizers, substrates for biogas (methane), and mushroom cultivation (Sarkar et al., 2016).

Composting accomplishes several beneficial goals for its practitioners like (i) it reduces the bulk of waste (Zhu, 2007); (ii) it lowers the biological oxygen demand (BOD) of waste; (iii) it improves the physical characteristics of the waste and makes handling easier; (iv) it reduces a load of human, animal, and plant pathogens (Zhu, 2007); (v) it eliminates weed seeds; (vi) it reduces land use for landfilling and for surface application of waste; (vii) it transforms nitrogen from volatile ammonia into stable inorganic forms (Zhu, 2007); (viii) it improves the fertility and health of the soil that led to an increase in agricultural productivity, improvement in soil biodiversity, reduction in environmental risks, and overall development of a better environment (Zhu, 2007); (ix) compost provides nutrients to the soil, when applied to it (Lee et al., 2004); and (x) it seasonally meets fertilizer requirements. However, if composting is not carried out properly, it can also have some disadvantages like (i) lot of land is required, particularly if static windrow systems are used; (ii) it affects nearby homeowners who, as long as they are in someone else's backyard, would embrace most environmentally conscious waste management systems; (iii) composting installations are odor nuisances that is why the tendency goes to completely enclosed systems where the outlet air is treated in a biofilter before being emitted; (iv) soil pollution, if the heavy metal content of the compost is too high; and (v) groundwater pollution if composting is carried out on a surface that is not made up properly or where the runoff water is not collected.

A wide range of microbial diversity including mesophilic, thermotolerant, and thermophilic aerobic microbes like bacteria, actinomycetes, yeasts, molds, and various other fungi have an extensive role in the bioconversion of organic waste in composting. Genetic diversity among microbes recovered from compost has not been widely studied. To understand genetic diversity and microbial community composition of the compost, different molecular approaches are adopted by taxonomists to characterize and identify isolates up to species level. The novel strains identified with these techniques may further be beneficial in improving the quality of compost and play a significant role in agriculture. Keeping in view the above factors, this chapter presents the fundamentals of composting, different types of composting, and techniques to analyze diverse microbial communities involved in the composting process.

2. Composting: Fundamental aspects

Composting is a solid waste disposal technique that has been practiced since ancient times. From a practical perspective, it is a pretreatment process that transforms a variety of wastes including food and agricultural waste, biosolids, municipal solid waste, and even hazardous wastes into forms suitable for land application. Essentially, the composting process turns waste products into an organic soil amendment with the help of naturally occurring microbes in soil and optimizing their carbon cycling activities. In fact, the spatial characteristics and physical–chemical gradients found in a compost system are very similar to those found in a soil system. However, soil and compost systems are distinct in several other respects. For example, compost is primarily organic in composition and the rate of microbial activity is extremely high during compost formation and after it is added to the soil. In contrast, in soil, both organic matter content and microbial activity levels are much lower. In soils, physical factors including temperature, moisture, and bulk density are imposed externally but in composting systems, these factors are controlled internally. Finally, the soil matrix is very stable alone and undergoes change slowly. The compost matrix is an organic substrate that undergoes rapid physical change because of degradative activities.

The carbon-to-nitrogen (C:N) ratio and the moisture content are the major factors that contribute toward making the composting process more effective (Chang and Chen, 2010; Kumar et al., 2010), and bulking agents are known to be effective in maintaining the essential parameters of composting including moisture and carbon-to-nitrogen ratio. Diverse types of bulking agents viz., wheat straw, fruit and vegetable waste, wood chips, grass clippings, sawdust, corn stalks, etc. are used to achieve high quality, time-efficient, and cost-effective composting (Batham et al., 2013) as depicted in Table 1. Fundamentally, composting is an applicable technology in which organic waste is decomposed through biological means under the conditions of thermophilic temperatures, developed because of biologically produced heat, and the final product thus obtained is stable enough for its storage and application point of view without any unfavorable environmental effects too (Lim et al., 2017). Composting process is mediated by interaction amid organic matter (raw materials), moisture content, oxygen, and microorganisms. The organic matter would usually have an indigenous mixed population of air, water, and soil-derived microorganisms.

Under optimal conditions of moisture and aeration, an increase in microbial activity would be noticed. The quick assimilation of organic matter leads to the production of organic products, water, and energy along with the evolution of CO_2. The major portion of the produced energy is utilized by the microorganisms for their metabolic activities while the remaining portion is emitted off in the form of heat (Fig. 1).

The main goal of composting is, therefore, to produce a fairly stable organic material roughly like a soil that has a carbon-to-nitrogen (C:N) ratio of approximately 10:1. This low carbon content material, when applied to soil, prevents competition between growing plants and microorganisms for nitrogen. If a product with a high carbon content is applied to the soil, the density of the bacterial population increases tremendously by drawing out nitrogen from the soil that would otherwise be accessible to the plants. In composting, the biological decomposition process is mediated by microorganisms like bacteria, fungi, and protozoa. Decomposition includes two phases of biochemical transformation, mineralization, and humification in particular (Kulikowska and Gusiatin, 2015). In mineralization, readily and easily available fermentable organic constituents like sugars (carbohydrates) and amino acids (proteins) are degraded by microorganisms' metabolic activities lead to the formation of a partially stable organic residual product along with the generation of carbon dioxide, water, and heat (Vigneswaran et al., 2016). The typical pathways of biochemical reaction occurring during mineralization are:

Table 1 Different types of compostable substrates with bulking agents used in the composting process.

S. no.	Type of compost substrate	Bulking agent(s) used for composting	Salient achievements	References
1	Kitchen waste	Cornstalks, sawdust, and spent mushroom substrate	1. Mature compost was formed via reduced maturing time 2. A decline in CH_4 and N_2O emissions was observed 3. Little effect on NH_3 emissions was revealed	Yang et al. (2013)
2	Food waste	Rice bran	1. The bulking agent improved the compost's moisture content and C / N ratio to an initial value of approximately 62% and 38:1, respectively 2. The dominant phyla were Firmicutes, Proteobacteria, Bacteroidetes, and Actinobacteria 3. The microbial populations were extremely similar and relatively small in the thermophilic phase in contrast to the maturity phase wherein species diversity increased	Wang et al. (2017)
		Chopped wheat straw, chopped mature hay (80% timothy and 20% clover), and pinewood shavings	1. The only two formulations that reached thermophilic temperatures during the 10 days of successful composting were chopped wheat straw (CWS) and chopped hay (CH) when combined with FW at 8.9 and 8.6:1 wet mass ratio (FW:CWS and FW:CH), respectively 2. The least decomposed compost was provided by wood shavings (WS) at maturation with wood particles	Adhikari et al. (2009)

3	Sewage sludge generated during municipal wastewater treatment	Sawdust	1. The bulking agent improved the moisture content and given ample porosity for composting 2. The readily degradable carbon source increases CO_2 emission and decreases NH_3 emission 3. The least NH_3 emission was given by the addition of 70% sucrose/30% straw powder at 108 h	Li et al. (2013a, b)
4	Organic fraction of Municipal solid waste (OFMSW)	Wood shaving, agricultural and yard trimming waste	1. Bulking waste increased the enzymatic activity and composting pace 2. Compost quality is greatly enhanced by the inclusion of wood shaving waste as bulking waste 3. The composting cycle is shortened by inoculating the microbial consortium with the bulking agent	Awasthi et al. (2015)
5	Sludge from a slaughter-house wastewater plant	Crushed wood pallet	1. The inclusion of wet crushed wood to sludge brought a drop in the relaxation time 2. Nuclear magnetic resonance (NMR) has proven to be an effective tool for calculating the distribution of water in the mixture of sludge and bulking agents	Duval et al. (2010)
6	Household wet biodegradable waste	Garden waste	1. Adequate provisions for aeration and periodic waste turning are demonstrated to be essential to produce good quality compost 2. Addition of a suitable microbial inoculum shortened the duration of waste stabilization 3. Up to 20% of soil-blended compost showed no phytotoxicity for cress seeds	Manu et al. (2019)

Table 1 Different types of compostable substrates with bulking agents used in the composting process—cont'd

S. no.	Type of compost substrate	Bulking agent(s) used for composting	Salient achievements	References
7	Dewatered anaerobically stabilized primary sewage sludge (DASPSS) and sawdust	Clinoptilolite	1. Clinoptilolite absorbed a substantial quantity of heavy metals 2. When the sawdust content in the initial mixture increased, so did the humic substances in the final product 3. The natural zeolite amended initial mixture absorbed a substantial quantity of heavy metals 4. The final compost product contained 3.95%, 7.50%, and 11.73% (SSC1, SSC2, and SSC3, respectively) of humic substances	Zorpas and Loizidou (2008)
8	Waste activated sludge (WAS)	Vermicompost	1. Use of fresh vermicompost as bulking agent abolished the necessity of outside material 2. The composting stage resulted in ample augmentation of bulking material with the organic matter after 20 cycles of recycling and mixing with WAS and produced materials adequate for vermicomposting 3. Significant decline in pH, volatile solids (VS), basic oxygen absorption rate (SOUR), total carbon (TC), total organic carbon (TOC), C/N ratio, and pathogens while, a significant rise in electrical conductivity (EC), total nitrogen (TN), and total phosphorous (TP) were recorded during vermicomposting of the composted content	Hait and Tare (2011)

| 9 | Poultry litter | Biochar (carbonized organic materials) | 1. A substantial upsurge in CO_2 respiration rates from biochar-blended poultry litter was noticed
2. The first respiration maxima increased by 44% and total respiration increased by 28% | Steiner et al. (2014) |
| 10 | Flowers waste | Dry Leaves | 1. The reduction in total organic carbon was 21%, which was further increased by 36.48% with the inclusion of dry leaves and cow dung during the composting of flower waste
2. On the 120th day of composting, the pH was 7.33, electrical conductivity 2.77 mS/cm, total organic carbon 26.9%, total nitrogen 2.2%, and C:N ratio was 12 | Sharma and Yadav (2017) |

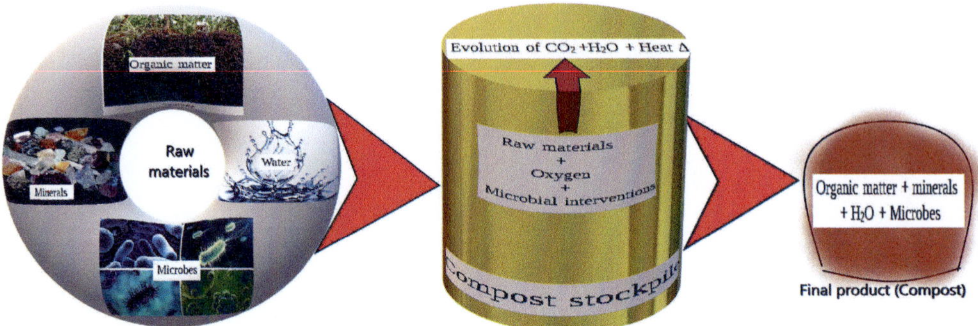

Fig. 1 Scientific process of composting.

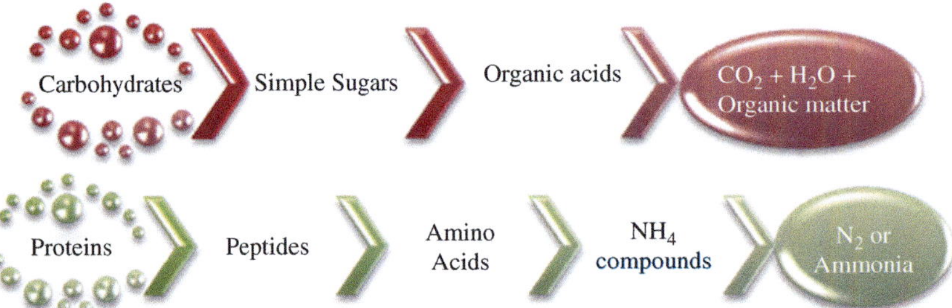

During humification, the dense and partially stabilized organic matter produced as a result of mineralization is given the characteristic humic (loose and fluffy). Humification process is mediated by the fungal and actinomycetes species. Although composting is seen as a safer and greener technique because it reflects a biological transformation of the organic waste into a fertilizer rich in nutrients, recent studies have revealed that composting practices have a detrimental effect on the environment too in terms of producing a significant amount of greenhouse gases (GHG) that has a high potential for global warming (GWP). Decomposition of organic matter, consumption of fossil fuel for machinery and transportation, turning, etc. are some of the contributory factors associated with the release of GHG (Bong et al., 2017). As far as a biological phase of the composting process is concerned, the type of GHG released and its associated GWP rely on several factors like temperature, pH, C:N ratio, mode and rate of aeration, organic waste, turning frequency, etc. (Lim et al., 2017).

3. Types of the composting process (aerobic and anaerobic)

Composting, a natural biological process, can be categorized into two types: aerobic or anaerobic based on the difference in the nature of the decomposition process, but it is much faster, more energy-efficient, and less odoriferous if done aerobically (Misra et al., 2003; Devi et al., 2012). But some of the bottlenecks in aerobic composting are comparatively more loss of nutrients from materials and associated risk of phytotoxicity though little, with the final end product. However, generally, aerobic composting comes first in mind when we talk about compost (Tanugur, 2009).

3.1 Aerobic process of composting

Oxygen-mediated decomposition of organic matter is referred to as aerobic composting that commences with proper mixing of suitable raw (organic) ingredients. To start the process, the mixing of raw materials is accompanied by the supply of a sufficient amount of air to ensure proper aeration. Breakdown of organic raw materials is accomplished by the aerobic microorganisms lead to the formation of a relatively stable final product in the form of humus along with the formation of ammonia, water, carbon dioxide, and heat but the process halts if the oxygen supply gets interrupted (AbMuttalib et al., 2016). While aerobic composting may produce organic acids like intermediate compounds, they are further decomposed by aerobic microorganisms. Relatively unstable final compost thus obtained exhibits risk of phytotoxicity though little. The prevailing aerobic conditions in the composting hasten the decomposition of complex carbohydrates (cellulose and hemicellulose), proteins, and fats due to increased microbial activities, resulting in an increase in temperature higher than desired to destroy the pathogens (human or plant) and weed seeds besides lessening processing time and nuisance odors (Bayer, 2008). Aerobic is generally more efficient and useful for agricultural production when compared to anaerobic composting.

3.2 Anaerobic process of composting

Decomposition of organic matter that takes place either in the absence or limited supply of oxygen is referred to as anaerobic composting. The onset of the anaerobic process involves the collection of organic materials in the pits accompanied by layering with a thick layer of soil. The mixture in the pits left undisturbed for about 6–8 months to ensure complete degradation. The prevailing anaerobic conditions promote the growth of anaerobic microorganisms that decompose the organic matter with the formation of some intermediates like hydrogen sulfide, methane, organic acids, and other substances. These compounds are not further metabolized but continue to accumulate. Most of these compounds possess a pungent smell while some do display phytotoxicity too. Because of incomplete conversion, the final end product thus produced may include aggregated masses, hence is called mud rather than compost as in aerobic composting (Ozturk, 1999). Being low-temperature processes, anaerobic composting fails to kill weed seeds and pathogens. Additionally, this process is slower than aerobic composting. But it also has some benefits as it is a less laborious process and nutrient loss is also limited. A comparative account of both types of composting processes is depicted in Fig. 2.

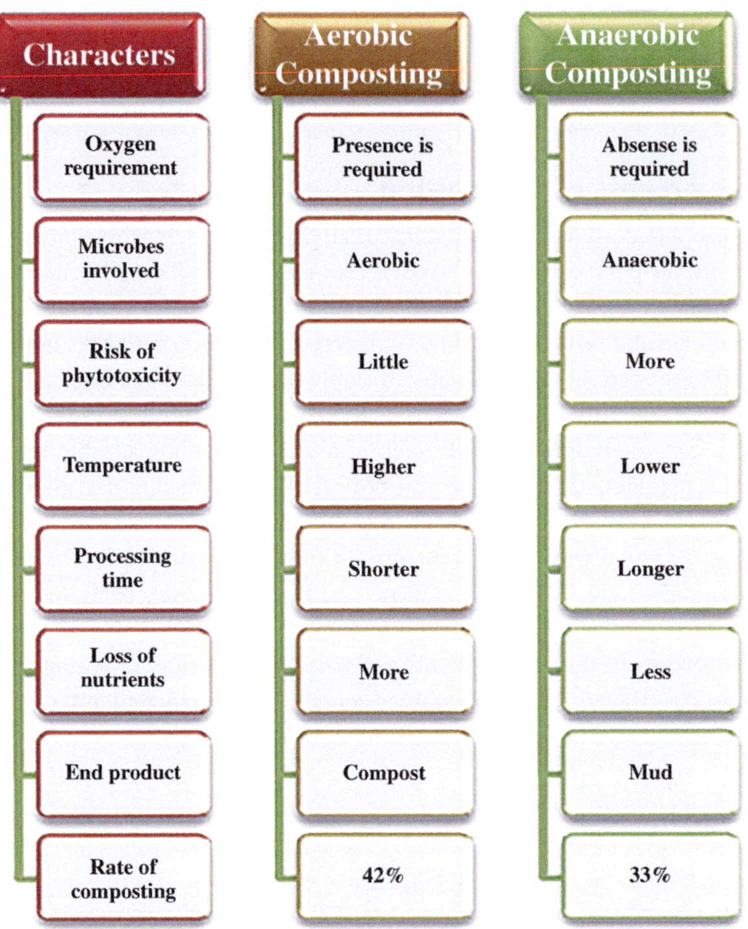

Fig. 2 Difference between aerobic and anaerobic composting.

4. Phases of composting

The organic matter when composted in a batch has four different phases viz. (i) mesophilic phase, (ii) thermophilic phase, (iii) cooling phase, and (iv) curing phase of the composting process are evident (Fig. 3). Though the same phases ensue in the continuous process of composting, they are less evident in batch.

4.1 Mesophilic phase.

4.2 Thermophilic phase.

4.3 Cooling phase.

4.4 Curing phase.

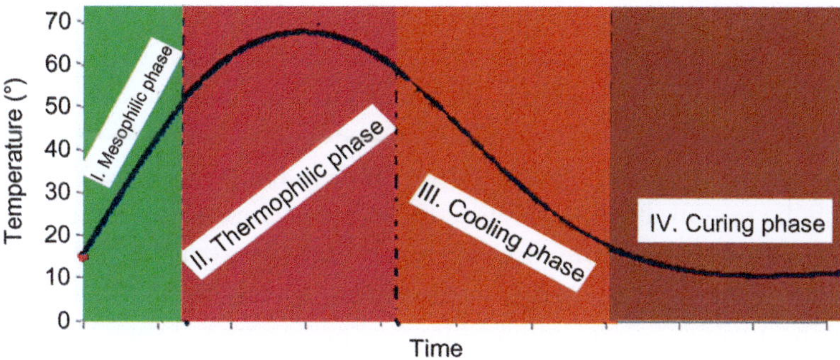

Fig. 3 Diverse phases of composting process.

4.1 Mesophilic phase

Mesophilic phase (25–40 °C) also referred to as starting the process, is the very first step of composting process wherein easily available and degradable energy-rich compounds like sugars (carbohydrates) and amino acids (proteins) are acted upon by bacteria, fungi, and actinobacteria (primary decomposers) to produce carbon dioxide and energy. Some of the energy is utilized by microorganisms for their own growth and reproduction while, the remainder is given off as heat (Mehta and Sirari, 2018). Once the composting process starts in the compost pile, mesophilic bacteria proliferate, raising the composting mass's temperature to 44 °C (111 °F) only if mechanical impacts such as turning are feeble and the compost mesofauna including worms, millipedes, mites, etc. develop that played the role of catalysts in the process. The contribution of these animals, as in vermicomposting in particular, is either negligible or substantial, depending on the composting process. Consequent upon primary decomposers' metabolic activities, a rise in the temperature is apparent, however, the counts of mesophiles in the original substrate are three times higher than those of thermophiles. These mesophilic bacteria may include *E. coli* and other human intestinal tract bacteria but their population soon gets suppressed by rising temperature as they are replaced by thermophilic bacteria in the transition range of 44–52 °C (111–125.6 °F). These onsets the thermophilic (second stage) phase of the composting process (Insam and de Bertoldi, 2007).

4.2 Thermophilic phase (35–65 °C)

Thermophilic microorganisms (high-temperature loving) become very active at this stage of the process and they produce a lot of heat. This offers the thermophilic microflora with a competitive advantage over mesophilic microorganisms where the latter eventually get completely eliminated at the end. Formerly thriving mesophiles die off and are gradually demolished along with the remaining, easily degradable substrate by the subsequent thermophiles. The process of decomposition quickly accelerated and rushed to reach a temperature of approximately 62 °C. Generally, the optimum growth rate of thermophilic fungi is reached between 35 and 55 °C, temperatures higher than this range usually inhibit their growth. Also at higher temperatures, thermophilic and thermotolerant bacteria including actinobacteria are well-known to remain active (Mehta and Sirari, 2018). Notwithstanding, the majority of microorganisms died above 65 °C, it can be found that further increase in temperature could be higher than 80 °C. It is apparent that the actinobacteria's enzymes (temperature-stable) mediated exothermic reactions (abiotic) might be responsible for this elevation in final temperature rather than microbial activities. Not all zones of a compost pile reach the same temperatures; therefore, it is necessary that every part of the substratum is moved to the central, hottest part of the pile by frequent turning. From a microbiological perspective, Table 2 portrays different types of microorganisms recovered from different types of wastes inside a compost pile with four main phases. In the thermophilic phase, hygienization, the process of removing/minimizing health threats associated with the application of biowaste to agricultural

Table 2 Diverse microbial species involved in the composting process of different types of wastes.

S. no.	Microbial group	Type of waste	References
1	Bacterial group		
	Staphylococcus spp., *Enterococcus* spp., *Streptococcus* spp., *Aerococcus viridians*, and *Micrococcus* spp.	Dental solid waste	Vieira et al. (2011)
	Afipia and *Hyphomicrobium* (Alphaproteobacteria), *Rhodanobacter*(Gammaproteobacteria), *Intrasporangium* (Actinobacteria), and *Bacillus* (Firmicutes)	Mixed waste	Green et al. (2010)
	Gram-negative (*Burkholderia*, *Chryseobacterium*, and *Pseudomonas*), Gram-positive (*Microbacterium* sp.)	Feather waste	Riffel and Brandelli (2006)
	Achromobacter spp., *Alcaligenes* spp., *Bacillus* spp., *Flavobacterium* spp., *Micrococcus* spp., and *Pseudomonas* spp.	Industrial water waste	Sharifi-Yazdi et al. (2001)
	Purple phototrophic bacteria	Saline wastewater	Hulsen et al. (2019)
	Thermophilic bacteria	Saline waste sludge	Gao et al. (2019)
	Candida ethanolica, *Bacillus cereus*, and *Alcaligenes faecalis*	Municipal solid waste	Pan et al. (2019)
	B. cereus PCM 2849	Pig bristles	Choinska-Pulit et al. (2019)
	Pseudomonas aeruginosa, *P. putida*, *Sphingobacterium moltivorum*, *Delftia tsuruhatensis*, *Stenotrophomonas humi*, *S. maltophilia*, *Ochrobactrum oryzae*, *O. humi*, and *Micrococcus luteus*	Waste landfills	Nourollahi et al. (2019)

Table 2 Diverse microbial species involved in the composting process of different types of wastes—cont'd

S. no.	Microbial group	Type of waste	References
2	Fungal group Ascomycota and oomycota phylum	Plastic waste	Brunner et al. (2018)
	Aspergillus niger and *A. flavus*	Wastewater	Dwivedi et al. (2012)
	Aspergillus spp., *Alternaria* spp., and *Emericela* spp.	Castor waste	Herculano et al. (2011)
	Aspergillus spp., *Penicillium* spp., *Fusarium* sp., and *Trichoderma* sp.	Garbage waste	Kumar et al. (2013)
	Aspergillus flavus, A. niger, A. terreus, Fusarium oxysporium, F. solani, Geotrichum candidum, Mucorhiemalis, Penicillium chrysogenum, Rhizopus oryzae, Trichoderma harizianum, and *Trichophyton* sp., *Aspergillus* sp.	Wastewater	Alananbeh et al. (2017)
	Trichophyton ajelloi, Aphanoascusreticulisporus, A. fulvescens, A. durus, Arthroderma quadrifidum, Chrysosporium anamorph of Arthroderma curreyi, Myceliophthoravellerea, C. keratinophilum with its teleomorph A. keratinophilus, C. europae, C. tropicum, and *Microsporum gypseum*	Sewage, municipal waste	Ulfig (2000)
	Aspergillus oryzae, Trichoderma longibrachiatum, A. fumigatus, A. niger, T. viride, Penicillium sp., and *Rhizopus* sp.	Metal finishing industry waste	Osaizua et al. (2014)
	Rhizopus spp., *Aspergillus* spp.	Poultry waste	Ramakrishnaiah et al. (2013)
	Exophiala oligosperma and *Paecilomyces variotii*	Waste gases	Estevez et al. (2005)
3	Actinomycetes group *Thermomonospora curvata*	Waste dump soil	Stutzenberger (1971)
	Streptomyces flavis 2BG and *Microbispora aerata* IMBAS-11A	Sheep wool waste	Gushterova et al. (2005)
	Kocuria palustris, Streptomyces parvus, Streptomyces griseorubens, Streptomyces rochei, Streptomyces albidoflavus, and *Streptomyces griseus*	Domestic waterwaste	Madkour et al. (2019)
	Nocardiopsis lucentensis and *Saccharomonospora azurea*	Sewage sludge	Ibraheem et al. (2017)
	Streptomycetes	Metal	Timkova et al. (2018)

land, is achieved. The extinction of numerous pathogens (human and plant), insect larvae, and weed seeds are attributable not only to the prevailing thermophilic temperature in the pile but also to the production of various antibiotics secreted by a very particular and dominant actinobacterial microflora inhabiting the compost pile (Ankidawa and Nwodo, 2012; Mehta et al., 2012). The downside of temperatures above 70°C is the extinction of the majority of mesophilic microflora accompanied by a delay in recovery once the peak temperature is achieved. But this can be avoided by effective recolonization steps.

4.3 Cooling phase

Depletion of available substrates ceases the metabolic activities of thermophilic microflora, eventually contributes to a decline in the temperature of the compost pile. At this point, the composting process enters the third stage of the composting process, i.e., the cooling phase, also known as the second mesophilic phase (Stofella and Kahn, 2001). This phase is characterized by the recolonization of mesophilic microflora which had previously been replaced by thermophilic microflora, either from protected microniches or from externally inoculated microflora, each resulting from surviving spores. Following colonization, mesophiles begin the digestion of lignin-like resistant organic materials. Other microorganisms (fungi) and macroorganisms (earthworms and sowbugs) are known to transform the coarser components into humus, too (IPTS Sevilla Spain, 2011).

4.4 Curing phase

Besides lengthy, the final curing phase, also known as maturing or aging phase, plays a critically important role in composting. The consistency of the substrates decreases during this phase, and the microbial community's structure and composition are completely altered in several succeeding steps. The proportion of fungi typically rises, while the percentage of bacteria decreases. During this phase, the compost is colonized by fungi and actinomycetes that degrade the complex compounds like cellulose, lignin, and lignin–humus complexes which are resistant to further degradation and hence become prevalent (Bernal et al., 2009; Mehta et al., 2014). Other Collembola like tiny animals (small insects), woodlice (tiny terrestrial crustaceans), and earthworms come to thrive and feed in the compost. Woodlice are tiny terrestrial crustaceans that will also feed in the compost. The organic matter is steadily mineralized by these micro and macroorganisms. Humus—the end product of composting thus obtained is comprised of numerous living entities, making it is an ideal soil amendment for enhancing/improving the growth and yield of plants (Mehta et al., 2014). A decrease in the waste heap temperature, absence of odorous odor and insect attraction, and the emergence of white or gray color owing to proliferation and prevalence of fungi and actinomycetes, ensure that composting process is complete and compost is mature as well.

The longer curing period, for instance, a year after the thermophilic stage confers the additional advantage of pathogen destruction. Numerous human pathogens usually have a short viability duration in the soil, and as long as they are exposed to the compost pile's microbiological competition, the more probable they will die quickly. Production of phytotoxins (toxic to plants) is the downside feature of immature/uncured compost (Tam and Tiquia, 1994). It may also rob the oxygen and nitrogen of soil and may contain high organic acid levels. Therefore, it is beneficial to let the compost mature fully before considering using it.

5. Determination of compost maturity

Compost maturity assessment is critical for the effective utilization of composts in the production of horticultural and agricultural produce (Cooperband et al., 2003). For the last decade, scientists have been finding a reliable and convenient way to assess compost maturity. Although numerous and great advancements have been made in determining what founds and characterizes compost as "mature," the science of compost maturity remains an art to apply. Stability and maturity are widely accepted to be separate properties of compost, though a slight difference. Stability refers to the availability of organic matter that is the constituent of composting mixture compounds that are readily available and metabolized in the presence of oxygen. Maturity is characterized as the ability of compost to sustain the growth of plants and is correlated with the presence of phytotoxic compounds, the need for oxygen, and the production of organic matter (Cooperband et al., 2003). This means the term stability is linked somewhere to the material's state/condition. It is a practical property that can be measured and quantified reasonably easily. Maturity instead depends on how to use the material and entails some judgment as well (Rynk, 2003). Generally, a "mature" compost is one where organic matter, including phytotoxic substances, is in the decomposed state and does not adversely impact it when used/applied to plant-growing conditions in this state. Various plant bioassays are used for assessing compost maturity (CCME, 2005; Ge et al., 2006). Also, the term maturity is applied to that compost that becomes ready for specific end-use (Sullivan and Miller, 2001; Ge et al., 2006). It is understood that many problems relating to the application of immature/unstable compost to the ground. The reasonably high microbial activities can cause self-heating in an unstable product (Mathur et al., 1993) which can be harmful if it heats up huge quantities of material (Brinton et al., 1995; Brinton and Evans, 2000). One of the most potent problems in bagged compost is the development of fire owing to the entrapment of flammable gases that may result in expansion and rupturing (Mathur et al., 1993; Brinton and Evans, 2000). Other problems include pathogens attraction and the production of obnoxious odors (Ge et al., 2006) and even more toxic products when the process becomes anaerobic (Brinton and Evans, 2000). Immature composts are often detrimental to plants in terms of their phytotoxic effects like

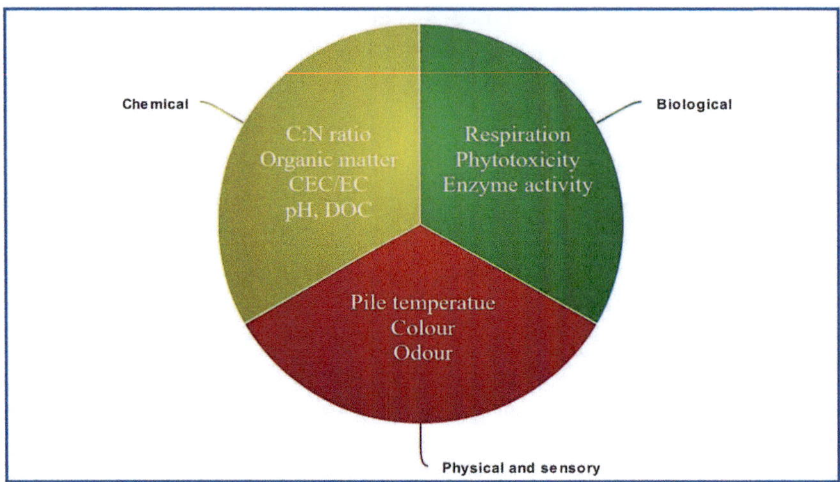

Fig. 4 Different methods (Physical, Chemical, and Biological) for determining compost maturity.

declined seed germination, restricted root growth, lessened above-ground plant growth (Butler et al., 2001), and production of high magnitude of intermediatory products like ammonia and organic acids during decomposition (Mathur et al., 1993; Sullivan and Miller, 2001; Ge et al., 2006). A number of methods have been suggested for determining the stability and (or) maturity of the compost. These methods can be of three types based on chemical, physical (including sensory), and biological characteristics (Eggen and Vethe, 2001; Sullivan and Miller, 2001; Tiquia, 2005) as depicted in Fig. 4. Notice that this is not a complete list of all investigated methods and that none of these methods are widely accepted as the "ideal" (Bio-Logic, 2001; Brewer and Sullivan, 2001).

Due to its ease, speed, inexpensiveness, and routine use, one of the most desirable methods for assessing compost maturity is "temperature monitoring" (Boulter-Bitzer et al., 2006; Weppen, 2002). The final decrease in pile temperature was well associated with a range of parameters commonly used to evaluate the maturity of the compost (Tiquia and Tam, 2002; Tiquia et al., 2002). Also reported by Boulter-Bitzer et al. (2006) that continuous monitoring of temperature variations (up and down) and oxygen availability (microbial activity indicator) presents an easy, economical, and precise method to assess the headway of compost to a mature/stable state.

The compost maturity may be assessed based on the change in its color and odor. Compost normally darkens (though, to some extent, it often relies on the feedstock material), and odors change from foul-smelling compounds like ammonia (noxious) to less noxious ones (rich and earthy). Although these sensory indicators are not so sensitive for maturity determination, they may be used to classify unstable or immature composts (Brewer and

Sullivan, 2001). These two indices, when used together, may also provide a very rough maturity estimate (TMECC, 2002). The most reliable tests for evaluating compost maturity include the rate of respiration and rewarming potential (Delgado et al., 2002; Adani et al., 2006). The rate of respiration corresponds to the current decomposition state of the waste material and is equally reliable for mature composts, irrespective of the operational conditions employed, initial feedstock material's state, and the material itself (Matteson and Sullivan, 2006). Numerous ranges of organic acids (short-chain) is produced during the initial stages of organic matter decomposition that may not only be toxic to plants but may restrict the growth of many species especially when immature compost containing medium is employed to plant them. However, as compost progresses toward maturity, these acids further decompose and phytotoxicity declines (Sullivan and Miller, 2001). Henceforth, tests pertaining to phytotoxicity like plant bioassays are regarded as "ultimate" tests to determine compost maturity by some researchers (Inbar et al., 1993; Chefetz et al., 1996), especially while a compost product is intended for use in horticulture or agriculture (Wichuk and McCartney, 2010).

Although the literature has suggested a range of compost C: N ratio limits ranging from <20:1 to <10:1 (Chefetz et al., 1996; Bio-Logic, 2001; Goyal et al., 2005), Sullivan and Miller (2001) stated that the C:N ratio should lie between 10:1 and 15:1 for mature/stable compost (Wichuk and McCartney, 2010). The outcomes of other research, however, suggest that the C:N ratio does not represent the good indicator of compost maturity (Chefetz et al., 1996; Wu et al., 2000; Brewer and Sullivan, 2001; Cooperband et al., 2003; Hutchinson and Griffin, 2008) because it has some pitfalls like (i) this ratio might vary considerably during the thermophilic stage; (ii) does not always follow or achieve the anticipated pattern; (iii) upon addition of N-rich fertilizer/feedstock in the compost, the ratio might be low even if the composting materials are not mature; and (iv) the broad inconsistency in feedstocks leads to the inconsistency in the final C:N ratios in different composts, making it difficult to impose an absolute limit on the C:N ratio that can apply to all feedstocks (Goyal et al., 2005). Another measure used to monitor compost maturity is to look for reductions in the easily and readily available organic matter (OM) that can be assessed by calculating volatile solids content (Bio-Logic, 2001; Sullivan and Miller, 2001). However, there tend to be major problems associated with OM reduction like inorganic carbonates and inactive plastic material may add to OM determinations, and can, therefore, bloat the real value (TMECC, 2002). During the composting of biosolids, no consistent trends in total volatile solid content were found, so OM was considered by Wu et al. (2000) as a poor predictor of compost maturity/stability. Some other parameters like cation exchange capacity (CEC), pH, electrical conductivity (EC), and dissolved organic carbon (DOC) are also used as maturity/stability indicators of compost by various researchers and each of these indicators has their own benefits and advantages too. Literature sources revealed that there is none of the tests can be regarded as an ideal test when examined the compost maturity/stability. If put differently, no sole test is able to accurately and reliably predict, from any number of feedstock materials, the state of maturity of composts generated under diverse process conditions (Wu et al., 2000; Bio-Logic, 2001; Brewer and Sullivan, 2001).

6. Taxonomic and metabolic microbial diversity during composting

For the composting processes, the significance of microbial communities is well known. Studies on microbial diversity including bacteria, actinobacteria, and fungi during composting have been reported extensively. Microbial diversity of compost plays an important role because they are actively involved in the breakdown of organic material. So, compost is a well-off reservoir of microbial diversity, comprising mesophilic and thermophilic bacteria, fungi, and actinomycetes. The mesophilic microflora forms the pioneer community, while thermophiles represent the climax community. It has been reported that the fast-growing Pseudomonads and Arthrobacter comprise the pioneer flora because they rapidly degrade high concentration of organic matter (Hayes et al., 1969; Stanek, 1972), while Bacilli have are dominant bacteria of compost ecosystems. During composting process, the compost maturity could be correlated with high microbial diversity in terms of microbial abundance, composition, and activity (Ryokeboer et al., 2003). To access taxonomic and metabolic microbial diversity during composting there are numerous molecular approaches, which provide a great addition to the culture-dependent techniques and have switched to the molecular and genetic level. Culture-dependent techniques are known to evaluate the isolation and characterization of only 1% of the naturally occurring microbes in the total microbial population existing in an ecosystem (Muyzer, 1999). PCR-based molecular approaches have been used for bacterial identification and its classification at the species level (Massol-Deya et al., 1995). PCR targeting the 16S rRNA gene sequencing is widely used to study the complete bacterial community composition and allows their identification as well as the prediction of phylogenetic relationships, hence increasing our knowledge about the contribution of bacterial diversity to various compost production phases. Phylogenetic information of microbes is obtained from gene sequence similarities of the 16S rRNA or 23S rRNA and 18S rRNA genes for bacteria and fungi, respectively (Chandna et al., 2013). Latest molecular approaches have revealed the identification of a wide range of uncharacterized microbes from a diversity of microorganisms which are not possible to identify with culture-dependent techniques.

Furthermore, the efficient microbes identified from the compost with these approaches can be used as compost inoculants for accelerating the composting process. However, during composting the microbial diversity may differ with the variety of organic material and nutrient supplements used during the composting. Therefore, it is important to study the diversity of microbes during composting when different agricultural byproducts like wheat bran, rice bran, rice husk, along grass clippings and bulking agents are used.

7. Techniques to analyze microbial diversity in composting

7.1 Culture-dependent phenotypic-based approaches

7.1.1 Plate count

Traditionally, diversity was determined by using the selective plate count method where viable counts of microbes were taken into consideration (Boulter et al., 2002). The morphology of a colony includes its characteristics like color, shape, margin, pigmentation, production of slime, etc. Further, the features of the microbial cell are described in terms of shape, size, Gram reaction, extracellular material like capsule, endospores, inclusion bodies, and flagella along with its location and motility, whereas microscopic fungi exist as either molds or yeasts or both. Besides, rapid and economical, this technique offers valuable information regarding the metabolic activity and culturable heterotrophic microbial populations (Devi and Soni, 2020).

7.1.2 Carbon source utilization profile/community level physiological profile (CLPP)/BIOLOG

Microbes may vary in their potential to utilize a variety of carbon sources and variations in carbon source utilization patterns form the basis of comparison of a range of microbes among different composts that in turn reflect the difference in the physiological functioning of microbial populations. CLPP has been captivated by the use of BIOLOG which is widely used to assess the functional diversity of microorganisms in the compost ecosystem. CLPP coupled with culturing the microbes for the assessment of their physiological diversity and characterization of nutritional physiology of individual microorganisms (Garland and Mills, 1991). It involves the use of BIOLOG system and BIOLOG microtiter plates in which metabolic profiling of microorganisms is done by 95 completely different carbon sources. Garland & Mills were the first to use the BIOLOG method for evaluating the functional efficiency of microbial communities. Two types of BIOLOG microtiter plates are generally used, i.e., BIOLOG GN2 and ECO plates containing 95 or 31 substrates respectively. The method is based on the redox reaction of tetrazolium violet, a redox indicator that changes color as a result of cellular respiration providing a metabolic fingerprint (Lladó and Baldrian, 2017). This method has been widely exploited in microbial diversity analysis with the aim to describe the metabolic versatility of microbial communities in many different environments including compost.

7.2 Role of culture-independent techniques

7.2.1 Fatty acid methyl ester (FAME) analysis

Fatty acid methyl ester (FAME) analysis has been extensively used as a culture-independent method for determining the wide diversity of the microbial community at the phenotypic level (Chayani et al., 2001). This method provides information on the composition of microbial communities based on different fatty acids present in their cell walls. Generally, fatty acids are the phospholipids present in the cell membrane of both prokaryotic and eukaryotic cells, comprise a relatively constant proportion of the cell biomass, and act as signature molecules hence, act as a chemotaxonomic marker. This method provides reliable information up to the genus and sometimes species level and plays an important role in studying the taxonomy of microbial communities. Sherlock MIS (MIDI Inc.) is a fully automatic system with a developed extensive database to analysis variation in fatty acid composition in different microbes. Fatty acid methyl ester (FAME) analysis represents versatility in the microbial population in an ecosystem by providing variation in the fatty acid profile (Eiland et al., 2001).

7.3 PCR-based approaches

7.3.1 Amplified ribosomal DNA restriction analysis (ARDRA)

ARDRA is appropriate approach for analyzing microbial communities and provides reliable information with respect to microbial ecology. This technique is based on DNA sequence variations present in PCR-amplified 16S rRNA genes (Smit et al., 1997). The PCR product amplified from DNA is generally digested using tetracutter restriction endonucleases like *Alu*I and *Hae*III and restricted fragments are further resolved on agarose or polyacrylamide gels. ARDRA is useful for the rapid examination of microbial communities and to compare microbial diversity present in an environment. This technique is based on the repetitive units of nuclear ribosomal DNA (rDNA) consisting of conserved coding and variable noncoding regions and hence relies on DNA polymorphism. The whole process covers; amplification of the coding and noncoding regions by PCR, digestion of amplicon by restriction endonucleases, and separation of the restriction fragments using gel electrophoresis. ARDRA is a sensitive technique and provides high resolution of genotypic characterization of microbial diversity of compost at the community level (Heyndrickx et al., 1996).

7.3.2 Enterobacterial repetitive intergenic consensus polymerase chain reaction (ERIC-PCR)

ERIC-PCR is a powerful and reliable method for DNA fingerprinting. This method is widely used in microbial diversity because it is faster, simpler, and more economical than other genomic typing methods. This method of obtaining genomic DNA fingerprints of bacteria is the repetitive extragenic palindromic-PCR (rep-PCR). The rep-PCR fingerprinting is considered a very useful molecular technique to discriminate between different species, as a variety of DNA segments present in the bacterial genome are used (Ishii and Sadowsky, 2009). The important characteristic of repetitive sequences like ERIC is that they contain repetitive, noncoding sequences which are distributed throughout the genome in a unique fashion and that help in discriminating intra-species variability among the bacterial strains (Chudzik and Stosik, 2005).

In ERIC, repeated oligonucleotides are used in DNA synthesis and the primers are designed in such a way to amplify copies of the Enterobacterial Repetitive Intergenic Consensus sequence; if the positions of copies vary among different strains, the amplification products provide each with a unique fingerprint when run on a gel.

8. Role of metagenomics in evaluating microbial diversity in compost

The term metagenomics describes the genomic analysis of those microbes, which are difficult to cultivate in a standard cultivation medium. The 16S rRNA study revealed strong evidence for the existence of uncultured microorganisms from various habitats. Metagenomic approach has provided new insights for the novel genes and gene products which include many hydrolytic enzymes, novel molecules, and antimicrobial compounds. Furthermore, metagenomic studies also reveal the role of genomic diversity in the existence of various co-operations among the microorganisms and their extraordinary potential in extreme environments. During composting, the process is generally mediated by indigenous microbial communities under aerobic conditions and in the solid state to recycle organic waste sustainably. So, compost environments are considered as one of the most significant bioreactors for renewable bioenergy on the planet as they acquire immense diversity of microbes for the efficient degradation of organic biomass (Allgaier et al., 2010). The expansion of culture-independent genome-based approaches like metagenomics for the identification of the taxonomic and metabolic functions of uncultivable microbial diversity in different environments has progressed considerably till date (Hess et al., 2011). This powerful technique allows the identification of new species and novel comprehensive sets of translated gene sequences to justify the physiology and biochemistry of degradation processes of organic waste during composting. Many culture-independent and PCR-based molecular analyses and clone libraries observed that new microbial species are enriched in compost habitats. As sequencing technologies improve, more comprehensive tools like shotgun metagenomic sequencing are applied for novel gene discovery. Functional annotation of metagenomic sequences revealed that genes involved in metabolic pathways were among the most abundant type (Yang et al., 2019). This strategy not only expands our understanding of the functions, interactions, and ecology of microbial communities related to the degradation and transformation of organic waste but also facilitates the exploitation of its generated products. The compost microbiomes, novel enzyme systems, and proteins identified by this approach could enhance the efficiency of biomass conversion in compost ecosystems and exhibit a very important role in agriculture.

Metagenomic directly analyzes the total DNA from environmental samples, provides a powerful approach in introducing the novel microbes and enzymes in microbial communities (Rosnow et al., 2016). Metagenomic applies a collection of genomic technologies and bioinformatics tools which directly access the genetic content of entire communities of organisms. Metagenomic involves sample processing sequencing

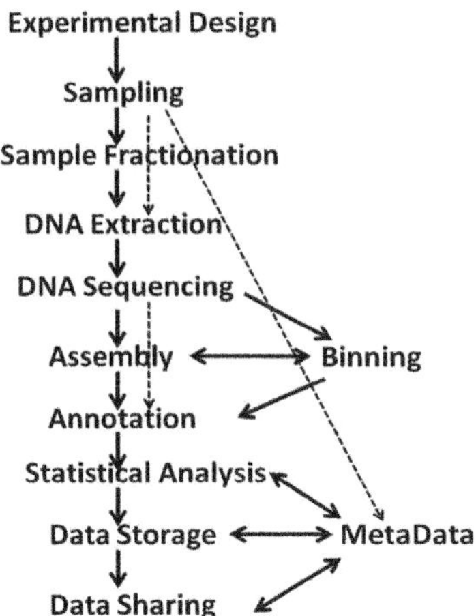

Fig. 5 Diagram depicts typical steps of metagenome and dashed arrows indicate steps that can be omitted (Thomas et al., 2012).

technology, assembly, binning, annotation, experimental design, statistical analysis, and data storage, and sharing. An overview of particular emphasis on the steps involved in a typical sequence-based metagenome approach is given in Fig. 5. Compost environments hold a huge reservoir of extensive microbes and enzymes; it is unfeasible to characterize them accurately. So, metagenomic analysis may play a very important role in accessing microbiota and microbial consortia enriched from compost ecosystems to decipher the systematic and functional significance within such a distinctive microbiome.

9. Conclusion and future outlook

Composting is the process of recycling the organic waste and possess a various role in the ecosystem including reduction of the bulk of the waste, lowering the biological oxygen demand (BOD) of waste and improving the physical characteristics of the waste, and improving the fertility of the soil that results into improving crop health and increasing agricultural productivity. Compost after application provides nutrients to the soil and maintains soil biodiversity. Composting is a self-heating, aerobic solid-phase biodegradative process of organic waste materials. A great variety of microorganisms including bacteria, fungi, and actinomycetes are involved in the process of composting. Microbial communities impart a significant role and represent an influential tool for understanding community dynamics in any complex environment like compost. Molecular approaches are very helpful in discriminating, characterizing, and classifying microbial communities up to species level. So, identification of new strains by these approaches could be very helpful in improving the quality of compost and further play a relevant role in the agriculture sector.

References

AbMuttalib, S.A., Ismail, S.N.S., Praveena, S.M., 2016. Application of effective microorganism (EM) in food waste composting: a review. Asia-Pac. J. Public Health 2, 37–47.

Adani, F., Ubbiali, C., Generini, P., 2006. The determination of biological stability of composts using the dynamic respiration index, the results of experience after two years. Waste Manag. 26, 41–48.

Adhikari, B.K., Barrington, S., Martinez, J., King, S., 2009. Effectiveness of three bulking agents for food waste composting. Waste Manag. 29, 197–203.

Alananbeh, K.M., Al-Refaee, W.J., Al-Qodah, Z., 2017. Antifungal effect of silver nanoparticles on selected fungi isolated from raw and waste water. Indian J. Pharm. Sci. 79, 559–567.

Allgaier, M., Reddy, A., Park, J.I., Ivanova, N., D'haeseleer, P., Lowry, S., Sapra, R., Hazen, T.C., Simmons, B.A., Vander Gheynst, J.S., Hugenholtz, P., 2010. Targeted discovery of glycoside hydrolases from a switchgrass-adapted compost community. PLoS One 5 (1), e8812.

Ankidawa, B.A., Nwodo, E., 2012. Recycling biodegradable waste using composting technique. J. Environ. Sci. Resour. Manage. 4, 40–49.

Awasthi, M.K., Pandey, A.K., Bundela, P.S., Khan, J., 2015. Co-composting of organic fraction of municipal solid waste mixed with different bulking waste, characterization of physicochemical parameters and microbial enzymatic dynamic. Bioresour. Technol. 182, 200–207.

Batham, M., Gupta, R., Tiwari, A., 2013. Implementation of bulking agents in composting: a review. J. Bioremed. Biodegr. 4, 205. https://doi.org/10.4172/2155-6199.1000205.

Bayer, Y., 2008. The effect of source separated on composting. MSc thesis, Institute of Science, Yildiz Technical University, Istanbul Turkey.

Bernal, M.P., Alburquerque, J.A., Moral, R., 2009. Composting of animal manures and chemical criteria for compost maturity assessment: a review. Bioresour. Technol. 100, 5444–5453.

Bio-Logic, 2001. Report on assessing compost maturity. A Final Report for the Nova Scotia Department of Environment and Labour, Bio-Logic Environmental Systems, Dartmouth, NS.

Bong, C.P.C., Lim, L.Y., Ho, W.S., Lim, J.S., Klemeš, J.J., Towprayoon, S., Ho, C.S., Lee, C.T., 2017. A review on the global warming potential of cleaner composting and mitigation strategies. J. Clean. Prod. 146, 149–157.

Boulter, I.I., Trevors, J.T., Bonland, G.J., 2002. Microbial studies of compost: bacterial identification and their potential for turfgrass pathogen suppression. World J. Microbiol. Biotechnol. 18, 661–671.

Boulter-Bitzer, J.I., Trevors, J.T., Boland, G.J., 2006. Apolyphasic approach for assessing maturity and stability in compost intended for suppression of plant pathogens. Appl. Soil Ecol. 34, 65–81. https://doi.org/10.1016/j.apsoil.2005.12.007.

Brewer, L.J., Sullivan, D.M., 2001. Maturity and stability evaluation of composted yard debris. Master of Science thesis, Oregon State University, Corvallis, OR.

Brinton, W.F., Evans, E., 2000. Compost quality standards and guidelines. Woods End Research Laboratory, prepared for New York State Association of Recyclers.

Brinton, W.F.J., Evans, E., Droffner, M.L., Brinton, R.B., 1995. Standardized test for evaluation of compost self-heating. Biocycle 36, 64–68.

Brunner, I., Fischer, M., Rüthi, J., Stierli, B., Frey, B., 2018. Ability of fungi isolated from plastic debris floating in the shoreline of a lake to degrade plastics. PLoS One 13, e0202047.

Butler, T.A., Sikora, L.J., Steinhilber, P.M., Douglass, L.W., 2001. Compost age and sample storage effects on maturity indicators of biosolids compost. J. Environ. Qual. 30, 2141–2148. https://doi.org/10.2134/jeq2001.2141.

CCME, 2005. Guidelines for Compost Quality. PN1340, Canadian Council of Ministers of the Environment, Winnipeg, Man.

Chandna, P., Nain, L., Singh, S., Kuhad, R.C., 2013. Assessment of bacterial diversity during composting of agricultural byproducts. BMC Microbiol. 13, 99.

Chang, J.I., Chen, Y.J., 2010. Effects of bulking agents on food waste composting. Bioresour. Technol. 101, 5917–5924.

Chayani, V.R., Watanbe, A., Matsuya, K., Asakawa, S., Kimura, M., 2001. Succession of microbiota estimated by phospholipids fatty acids analysis and changes in organic constituents during the composting process of rice straw. Soil Sci. Plant Nutr. 48, 143–735.

Chefetz, B., Hatcher, P.G., Hadar, Y., Chen, Y., 1996. Chemical and biological characterization of organic matter during composting of municipal solid waste. J. Environ. Qual. 25, 776–785. https://doi.org/10.2134/jeq1996.00472425002500040018x.

Choinska-Pulit, A., Laba, W., Rodziewicz, A., 2019. Enhancement of pig bristles waste bioconversion by inoculum of keratinolytic bacteria during composting. Waste Manag. 84, 269–276.

Chudzik, K.B., Stosik, M., 2005. Specific genomic fingerprints of Escherichia coli strains with repetitive sequences and PCR as an effective tool for monitoring freshwater environments. Pol. J. Environ. Stud. 14, 551–557.

Cooperband, L.R., Stone, A.G., Fryda, M.R., Ravet, J.L., 2003. Relating compost measures of stability and maturity to plant growth. Compost. Sci. Util. 11, 113–124.

Delgado, A., Garcia-Morales, J.L., Soleradel, R.R., Sales, D., 2002. Stability and maturity indexes of compost. In: Waste Manage and the Environment (First International Conference). WIT Press, Cadiz, Spain, pp. 263–268.

Devi, S., Soni, R., 2020. Relevance of microbial diversity in implicating soil restoration and health management. In: Meena, R. (Ed.), Soil Health Restoration and Management. Springer, Singapore, pp. 161–201, https://doi.org/10.1007/978-981-13-8570-4_5.

Devi, S., Sharma, C.R., Singh, K., 2012. Microbiological biodiversity in poultry and paddy straw wastes in composting systems. Braz. J. Microbiol. 43 (1), 288–296.

Duval, F.P., Quellecac, S., Trémier, A., Druilhe, C., Mariette, F., 2010. Non-destructive quantification of water gradient in sludge composting with magnetic resonance imaging. Waste Manag. 30, 610–619.

Dwivedi, S., Mishra, A., Saini, D., 2012. Removal of heavy metals in liquid media through fungi isolated from waste water. Int. J. Sci. Res. 1 (3), 181–185.

Eggen, T., Vethe, Ø., 2001. Stability indices for different composts. Compost. Sci. Util. 9 (1), 19–26.

Eiland, F., Klamner, M., Lind, A., Leth, M., Bath, E., 2001. Influence of initial C:N ratio on chemical and microbial composition during long term composting of straw. Microb. Ecol. 41, 272–280.

Estevez, E., Veiga, M.C., Kennes, C., 2005. Biofiltration of waste gases with the fungi *Exophiala oligosperma* and *Paecilomyces variotii*. Appl. Microbial. Biot. 67, 563–568.

Gao, P., Guo, L., Sun, J., Wang, Y., She, Z., Gao, M., Zhao, Y., 2019. Enhancing the hydrolysis of saline waste sludge with thermophilic bacteria pretreatment, new insights through the evolution of extracellular polymeric substances and dissolved organic matters transformation. Sci. Total Environ. 670, 31–40.

Garland, J.L., Mills, A.L., 1991. Classification and characterization of heterotrophic microbial communities on the basis of patterns of community level sole-carbon source utilization. Appl. Environ. Microbiol. 57, 2351–2359.

Ge, B., McCartney, D., Zeb, J., 2006. Compost environmental protection standards in Canada. Environ. Eng. Sci. 5, 221–234. https://doi.org/10.1139/S05-036.

Goyal, S., Dhull, S.K., Kapoor, K.K., 2005. Chemical and biological changes during composting of different organic wastes and assessment of compost maturity. Bioresour. Technol. 96, 1584–1591. https://doi.org/10.1016/j.biortech.2004.12.012.

Green, S.J., Prakash, O., Gihring, T.M., Akob, D.M., Jasrotia, P., Jardine, P.M., Watson, D.B., Brown, S.D., Palumbo, A.V., Kostka, J.E., 2010. Denitrifying bacteria isolated from terrestrial subsurface sediments exposed to mixed-waste contamination. Appl. Environ. Microbiol. 76, 3244–3254.

Gushterova, A., Vasileva-Tonkova, E., Dimova, E., Nedkov, P., Haertle, T., 2005. Keratinase production by newly isolated Antarctic actinomycete strains. World J. Microbiol. Biotechnol. 21, 831–834.

Hait, S., Tare, V., 2011. Optimizing vermistabilization of waste activated sludge using vermicompost as bulking material. Waste Manag. 31, 502–511.

Hayes, W.A., Randle, P.E., Last, F.T., 1969. The nature of the microbial stimulus effecting sporophore formation in *Agaricus bisporus* (Lange) sing. Ann. Appl. Biol. 64, 177–187.

Herculano, P.N., Lima, D.M.M., Fernandes, M.J.S., Neves, R.P., Souza-Motta, C.M., Porto, A.L.F., 2011. Isolation of cellulolytic fungi from waste of castor (*Ricinus communis* L.). Curr. Microbiol. 62, 1416–1422.

Hess, M., Sczyrba, A., Egan, R., Kim, T.W., Chokhawala, H., Schroth, G., Luo, S., Clark, D.S., Chen, F., Zhang, T., Mackie, R.I., Pennacchio, L.A., Tringe, S.G., Visel, A., Woyke, T., Wang, Z., Rubin, E.M., 2011. Metagenomic discovery of biomass-degrading genes and genomes from cow rumen. Science 331 (6016), 463–467.

Heyndrickx, M., Vauterin, L., Vandamme, P., Kersters, K., De Vos, P., 1996. Applicability of combined amplified ribosomal DNA restriction analysis (ARDRA) pattern in bacterial phylogeny and taxonomy. J. Microbiol. Methods 26, 247–259.

Hulsen, T., Hsieh, K., Batstone, D.J., 2019. Saline wastewater treatment with purple phototrophic bacteria. Water Res. 160, 259–267.

Hutchinson, M., Griffin, T.S., 2008. Evaluation of fiber content relative to other measures of compost stability. Comp. Sci. Util. 16, 6–11.

Ibraheem, I.B.M., Hammouda, O., Abdel-Raouf, N., Abdel-Tawab, M.S., Faysal, A., 2017. Assessment the ability of some bacteria and actinomycetes strains in the treatment of sewage sludge. Life Sci. J. 14 (6), 610–665.

Inbar, Y., Hadar, Y., Chen, Y., 1993. Recycling of cattle manure, the composting process and characterization of maturity. J. Environ. Qual. 22 (4), 857–863. https://doi.org/10.2134/jeq1993.00472425002200040032x.

Insam, H., de Bertoldi, M., 2007. Microbiology of the composting process. In: Diaz, L.F., de Bertoldi, M., Bidlingmaier, W., Stentiford, E. (Eds.), Compost Science and Technology. vol. 8, pp. 25–48.

IPTS Sevilla Spain, 2011. End-of-Waste Criteria on Biodegradable Waste Subject to Biological Treatment First Working Document. pp. 1–152.

Ishii, S., Sadowsky, M.J., 2009. Applications of the rep-PCR DNA fingerprinting technique to study microbial diversity, ecology and evolution. Environ. Microbiol. 11, 733–740.

Kulikowska, D., Gusiatin, Z., 2015. Sewage sludge composting in a two-stage system, carbon and nitrogen transformations and potential ecological risk assessment. Waste Manag. 38. https://doi.org/10.1016/j.wasman.2014.12.019.

Kumar, M., Ou, Y.L., Lin, J.G., 2010. Co-composting of green waste and food waste at low C/N ratio. Waste Manag. 30, 602–609.

Kumar, R., Mishra, R., Maurya, S., Sahu, H.B., 2013. Isolation and identification of keratinophilic fungi from garbage waste soils of Jharkhand region of India. Eur. J. Exp. Biol., 600–604.

Lee, J.J., Park, R.D., Kim, Y., Shim, J.H., Chae, D.H., Rim, Y.S., Sohn, B.K., Kim, T.H., Kim, K.Y., 2004. Effect of food waste compost on microbial population, soil enzyme activity and lettuce growth. Bioresour. Technol. 93, 21–28. https://doi.org/10.1016/j.biortech.2003.10.009.

Li, Y., Li, W., Wu, C., Wang, K., 2013a. New insights into the interactions between carbon dioxide and ammonia emissions during sewage sludge composting. Bioresour. Technol. 136, 385–393.

Li, Z., Lu, H., Ren, L., He, L., 2013b. Experimental and modeling approaches for food waste composting: a review. Chemosphere 93, 1247–1257.

Lim, L.Y., Chien Bong, C.P., Lee, C.T., Klemes, J.J., Sarmidi, M.R., Lim, J.S., 2017. Review on the current composting practices and the potential of improvement using two-stage composting. Chem. Eng. Trans. 61, 1051–1056. https://doi.org/10.3303/CET17611731051.

Lladó, S., Baldrian, P., 2017. Community-level physiological profiling analyses show potential to identify the copiotrophic bacteria present in soil environments. PLoS One 12 (2). https://doi.org/10.1371/journal.pone.0171638, e0171638.

Madkour, G.A., Hamed, H.M., Dar, A.M., 2019. Removal of ammonia and orthophosphate from domestic wastewater using marine actinomycetes. Egypt. J. Aquat. Biol. Fish. 23, 455–465.

Manu, M.K., Kumar, R., Garg, A., 2019. Decentralized composting of household wet biodegradable waste in plastic drums, effect of waste turning, microbial inoculum and bulking agent on product quality. J. Clean. Prod. 226, 233–241.

Massol-Deya, A.A., Odelson, D.A., Hickey, R.F., Tiedje, J.M., 1995. Bacterial community fingerprinting of amplified 16S and 16-23S ribosomal DNA gene sequences and restriction endonuclease analysis (ARDRA). In: Akkermans, A.D.L., van Elsas, J.D., de Bruijn, F.J. (Eds.), Molecular Microbial Ecology Manual. Kluwer, Dordrecht, pp. 1–8.

Mathur, S.P., Owen, G., Dinel, H., Schnitzer, M., 1993. Determination of compost biomaturity. I. Literature review. Biol. Agric. Hortic. 10, 65–85.

Matteson, T., Sullivan, D.M., 2006. Stability evaluation of mixed food waste composts. Compost. Sci. Util. 14, 170–177.

Mehta, C.M., Sirari, K., 2018. Comparative study of aerobic and anaerobic composting for better understanding of organic waste manage, a mini review. Plant. Archiv. 18, 44–48.

Mehta, C.M., Gupta, V., Singh, S., Srivastava, R., Sen, E., Romantschuk, M., Sharma, A.K., 2012. Role of microbiologically rich compost in reducing biotic and abiotic stresses. In: Satyanarayana, T., Johri, B.N., Prakash, A. (Eds.), Microorganisms in Environmental Management. Springer, New York, pp. 113–134.

Mehta, C.M., Palni, U., Franke-Whittle, I.H., Sharma, A.K., 2014. Compost, its role, mechanism and impact on reducing soil-borne plant diseases. Waste Manag. 34, 607–622.

Misra, R.V., Roy, R.N., Hiraoka, H., 2003. On-Farm Composting Methods. Food and Agriculture Organization of the United Nations.

Muyzer, G., 1999. Genetic fingerprinting of microbial communities: present status and future perspective. In: Bell, C.R., Brylinsky, M., Johnson-Green, P. (Eds.), Proceedings of the 8th International Symposium on Microbial Ecology. Atlantic Canada Society for Microbial Ecology, Halifax, Nova Scotia, pp. 1–10.

Nourollahi, A., Sedighi-Khavidak, S., Mokhtari, M., Eslami, G., Shiranian, M., 2019. Isolation and identification of low-density polyethylene (LDPE) biodegrading bacteria from waste landfill in Yazd. Int. J. Environ. Stud. 76, 236–250.

Osaizua, O.S., Olateju, K.S., Yetunde Adeola, O.O., Olanike, A.Y., 2014. Tolerance, bioaccumulation and biosorption potential of fungi isolated from metal-finishing industry waste site. J. Appl. Sci. Environ. Sanit. 9.

Ozturk, I., 1999. Anaerobic Biotechnology and Waste Water Treatment Applications. Su vakfi publishers, Istanbul Turkey, p. 320.

Pan, J., Wang, X., Cao, A., Zhao, G., Zhou, C., 2019. Screening methane-oxidizing bacteria from municipal solid waste landfills and simulating their effects on methane and ammonia reduction. Environ. Sci. Pollut. Res. 26, 37082–37091.

Ramakrishnaiah, G., Mustafa, S.M., Srihari, G., 2013. Studies on keratinase producing fungi isolated from poultry waste and their enzymatic activity. J. Microbiol. Res. 3, 148–151.

Rashad, F.M., Saleh, W.D., Moselhy, M.A., 2010. Bioconversion of rice straw and certain agro-industrial wastes to amendments for organic farming systems, 1. Composting, quality, stability and maturity indices. Bioresour. Technol. 101, 5952–5960.

Riffel, A., Brandelli, A., 2006. Keratinolytic bacteria isolated from feather waste. Braz. J. Microbiol. 37, 395–399.

Rosnow, J.J., Anderson, L.N., Nair, R.N., Baker, E.S., Wright, A.T., 2016. Profiling microbial lignocellulose degradation and utilization by emergent omics technologies. Crit. Rev. Biotechnol., 1–15.

Rynk, R., 2003. The art in the science of compost maturity. Compost. Sci. Util. 11 (2), 94–95.

Ryokeboer, J., Mergaert, J., Coosemans, J., Deprins, K., Swings, J., 2003. Microbiological aspects of bio-waste during composting in a monitored compost. J. Appl. Microbiol. 94, 127–137.

Sarkar, S., Pal, S., Chanda, S., 2016. Optimization of a vegetable waste composting process with a significant thermophilic phase. Procedia Environ. Sci. 35, 435–440.

Sharifi-Yazdi, M.K., Azimi, C., Khalili, M.B., 2001. Isolation and identification of bacteria present in the activated sludge unit, in the treatment of industrial waste water. Iran. J. Public Health 30, 91–94.

Sharma, D., Yadav, K.D., 2017. Bioconversion of flowers waste, composing using dry leaves as bulking agent. Environ. Eng. Res. 22, 237–244.

Smit, E., Leeflang, P., Wernars, K., 1997. Detection of shifts in microbial community structure and diversity in soil caused by copper contamination using amplified ribosomal DNA restriction analysis. FEMS Microbiol. Ecol. 23, 249–261.

Stanek, M., 1972. Microorganisms inhabiting mushroom compost during fermentation. Mushroom Sci. 8, 797–811.

Steiner, C., Melear, N., Harris, K., Das, K.C., 2014. Biochar as bulking agent for poultry litter composting. Carbon Manag. 2, 227–230.

Stofella, P.J., Kahn, B.A., 2001. Compost Utilization in Horticultural Cropping Systems. Lewis Publishers, Boca Raton.

Stutzenberger, F.J., 1971. Cellulase production by *Thermomonospora curvata* isolated from municipal solid waste compost. Appl. Microbiol. 22, 147–152.

Sullivan, D.M., Miller, R.O., 2001. Compost quality attributes, measurements, and variability. In: Stofella, P.J., Kahn, B.A. (Eds.), Compost Utilization in Horticultural Cropping Systems. Lewis Publishers, Boca Raton, FL, pp. 95–120.

Tam, N.F.Y., Tiquia, S.M., 1994. Assessing toxicity of spent sawdust pig-litter using seed germination technique. Resour. Conserv. Recycl. 11, 261–274.

Tanugur, I., 2009. Microbiological parameters as indicators of compost maturity. J. Appl. Microbiol. 99, 816–828. https://doi.org/10.1111/j.1365-2672.2005.02673.

Thomas, T., Gilbert, J., Meyer, F., 2012. Metagenomics—a guide from sampling to data analysis. Microb. Inform. Exp. 2 (1), 3.

Timkova, I., Sedlakova-Kadukova, J., Pristas, P., 2018. Biosorption and bioaccumulation abilities of actinomycetes/streptomycetes isolated from metal contaminated sites. Separations 5, 54.

Tiquia, S.M., 2005. Microbiological parameters as indicators of compost maturity. J. Appl. Microbiol. 99, 816–828. https://doi.org/10.1111/j.1365-2672. 2005.02673.

Tiquia, S.M., Tam, N.F.Y., 2002. Characterization and composting of poultry litter in forced-aeration piles. Process Biochem. 37, 869–880. https://doi.org/10.1016/S0032-9592(01)00274-6.

Tiquia, S.M., Wan, J.H.C., Tam, N.F.Y., 2002. Dynamics of yard trimmings composting as determined by dehydrogenase activity, ATP content, arginine ammonification, and nitrification potential. Process Biochem. 37, 1057–1065. https://doi.org/10.1016/S0032-9592(01)00317-X.

TMECC, 2002. Organic and biological properties—05.06 odor. In: Thompson, W.H., Leege, P.B., Millner, P.D., Wilson, M.E. (Eds.), Test Methods for the Examination of Composting and Compost. United States Department of Agriculture, and Composting Council Research and Education Foundation, Holbrook, NY, pp. 06–13.

Ulfig, K., 2000. The occurrence of keratinolytic fungi in waste and waste-contaminated habitats. Revista Iberoamericana DeMicologia 17, 44–50.

Vieira, C.D., de Carvalho, M.A.R., Cussiol, M.N.A., Alvarez-Leite, M.E., dos Santos, S.G., da Fonseca Gomes, R.M., Silva, M.X., Nicoli, J.R., de Macedo Farias, L., 2011. Count, identification and antimicrobial susceptibility of bacteria recovered from dental solid waste in Brazil. Waste Manag. 31, 1327–1332.

Vigneswaran, S., Kandasamy, J., Johir, M.A.H., 2016. Sustainable operation of composting in solid waste manage. Procedia Environ. Sci. 35, 408–415.

Wang, X., Pan, S., Zhang, Z., Lin, X., Zhang, Y., Chen, S., 2017. Effects of the feeding ratio of food waste on fed-batch aerobic composting and its microbial community. Bioresour. Technol. 224, 397–404.

Weppen, P., 2002. Determining compost maturity, evaluation of analytical properties. Compost. Sci. Util. 10, 6–15. https://doi.org/10.1080/1065657X.2002.10702058.

Wichuk, K.M., McCartney, D., 2010. Compost stability and maturity evaluation—a literature review. Can. J. Civ. Eng. 37, 1505–1523.

Wu, L., Ma, L.Q., Martinez, G.A., 2000. Comparison of methods for evaluating stability and maturity of biosolids compost. J. Environ. Qual. 29, 424–429. https://doi.org/10.2134/jeq 2000. 00472425002900020008x.

Yang, F., Li, G.X., Yang, Q.Y., Luo, W.H., 2013. Effect of bulking agents on maturity and gaseous emissions during kitchen waste composting. Chemosphere 93, 1393–1399.

Yang, Y., Zhang, S., Li, N., Chen, H., Jia, H., Song, X., Liu, G., Ni, C., Wang, Z., Shao, H., Zhang, S., 2019. Metagenomic insights into effects of wheat straw compost fertiliser application on microbial community composition and function in tobacco rhizosphere soil. Sci. Rep. 9, 61–68.

Zhu, N., 2007. Effect of low initial C/N ratio on aerobic composting of swine manure with rice straw. Bioresour. Technol. 98, 9–13.

Zorpas, A.A., Loizidou, M., 2008. Sawdust and natural zeolite as a bulking agent for improving quality of a composting product from anaerobically stabilized sewage sludge. Bioresour. Technol. 99, 7545–7552.

CHAPTER 13

Lichenized fungi, a primary bioindicator/biomonitor for bio-mitigation of excessive ambient air nitrogen deposition worldwide

Himanshu Rai* and Rajan Kumar Gupta
Centre of Advanced Study in Botany, Institute of Science, Banaras Hindu University, Varanasi, Uttar Pradesh, India
*Corresponding author. e-mail address: himanshurai08@yahoo.com

1. Introduction

The use of organisms to detect, monitor, and devise strategies/policies to mitigate the impact of environmental contaminants has increased exponentially in recent years (Markert and Wünschmann, 2011; Markert et al., 2012; Branquinho et al., 2015; Parmar et al., 2016; Mukhopadhyay et al., 2020). Bioindicators whereas are defined as biological processes, species, or communities used for qualitative responses to environmental stress, and the ways it changes over time, the biomonitors quantitatively determine the response of a bioindicator to a specific environmental condition (Holt and Miller, 2010). The bioindicators thus qualitatively assess biotic responses to environmental stress (e.g., the presence of the lichen *Physcia* spp. indicates polluted air), and the biomonitors quantitatively determine the response of the bioindicator (e.g., reduction of lichen chlorophyll content and PSII fluorescence or diversity indicates the presence and severity of air pollution). Biomonitors on one hand reduce the overall cost of pollution monitoring by providing an alternative to costly installation and maintenance of sophisticated instrumentation, on the other hand provide a more in-depth assessment of pollutants through increasing the survey area, the density of sampling sites, and by cumulative accumulation of data for the lifetime, representing the environmental fluctuations (Loppi, 2019).

Lichens, a symbiosis between a fungus (mycobiont) and a green (phycobiont) and/or blue-green algae (cyanobiont), are known as one of the most successful symbionts in nature, inhabiting all terrestrial domains of the planet (Galloway, 1992). The peculiar morpho-anatomy and ecophysiology of lichens (i.e., absence of root and cuticle, slow growth, long-life, large-scale absorbance of nutrients directly from the atmosphere, and tolerance to pollutants) make them extremely sensitive to environmental perturbations.

Relationship Between Microbes and the Environment for Sustainable Ecosystem Services, Volume 1
https://doi.org/10.1016/B978-0-323-89938-3.00013-X **267**

Lichens through mechanisms such as cation exchange, accumulation of tolerant biomolecules, complexation/phytochelation of pollutants, chemical conversion, gene manipulation, and through the change in proteomics and metabolomics, have been able to tolerate higher deposition of pollutants/nutrients (Silberstein et al., 1996; Sarret et al., 1998; Pawlik-Skowrońska et al., 2002; Munzi et al., 2009a,b, 2013a,b, 2017a,b; Freitag et al., 2012, Hurtado et al., 2020). The sensitivity of lichens to air pollution was first recognized in 1859–1866 (Grindon, 1859; Nylander, 1866) and since then lichens have been found as bioindicators of nearly all types of environmental pollutants (Garty, 2001; Van der Wat and Forbes, 2015; Loppi, 2019).

The sustenance of life on earth is found to be influenced by nine major factors, i.e., chemical pollution, climate change, ocean acidification, stratospheric ozone depletion, nitrogen–phosphorus biogeochemical cycle, global freshwater use, land-use change, biodiversity loss, and atmospheric aerosol loads (Fig. 1) (Rockstrom et al., 2009). Among the nine factors recognized for the sustenance of life on earth, nitrogen (N) pollution, along with biodiversity loss and climate change, had already crossed the threshold (i.e., 121 million tonnes year^{-1} against the threshold of 35 million tonnes year^{-1}) which is bound to invite deplorable environmenal changes (Fig. 1) (Rockstrom et al., 2009). Unlike other pollutants, i.e., industrial sulfur, mercury, heavy metals, and aromatic hydrocarbons, which can be curbed through technological advancement, the widespread use

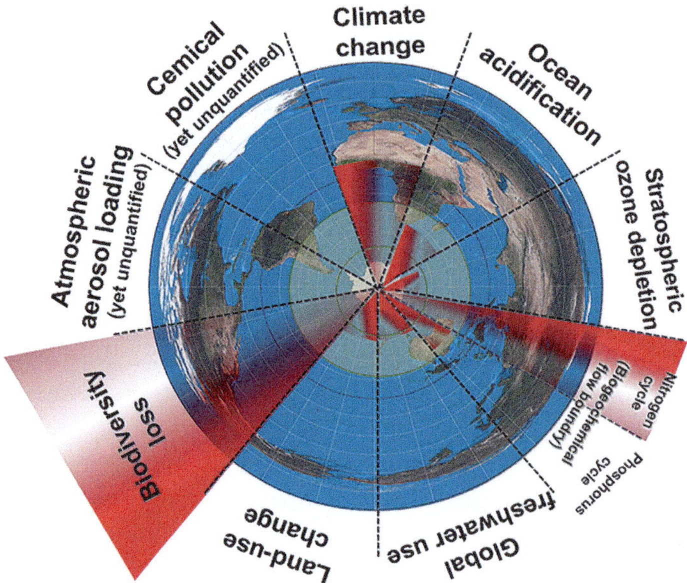

Fig. 1 The nine essential planetary systems, their proposed safe operating-represented by inner *green shading*, the current position for each variable-represented by *red wedges. (Adapted from Rockstrom et al. (2009).)*

of chemical and or organic nitrogen fertilizers in agriculture makes the perpetual release of nitrogen in the atmosphere politico-socially very complex and sensitive. The population growth on the planet has increased the food production demand leading to increased use of nitrogen fertilizers, making the release of nitrogen practically impossible to tame. Out of the whole nitrogen (N) released by zoo-anthropogenic activities, about 55% is released into the atmosphere as the reduced form of N, i.e., ammonia (NH_3), most of which is converted into ammonium (NH_4^+) (Krupa, 2003). Among the various sources of NH_3, the predominant ones are those that result from agriculture and related practices such as cattle waste, chemical fertilizers, and vegetation/stubble burning (Olivier et al., 1998; Webb et al., 2005). As per current estimation, the highest emission of ammonia occurs from east Asia-China > India, followed by Latin America, Africa, the USA, and Europe (Dianwu and Anpu, 1994; Olivier et al., 1998; Van Damme et al., 2018, 2020). After ammonia, oxidized N, i.e., Nitrates which account for about 40% of total N, are released to the atmosphere as nitric oxides-NOx (i.e., nitric oxide, NO, and nitrogen dioxide, NO_2) and are primarily of anthropogenic origins, such as fossil fuel combustion in vehicular, industrial, commercial, and domestic exhausts (Olivier et al., 1998; Gadsdon and Power, 2009).

2. Nitrogen assimilation in lichens

Lichens being a symbiont of fungi and algae needs nitrogen for their physiological functioning and biomass maintenance. The nitrogen assimilation in lichens is influenced by the intricate balance of moisture/water availability, illumination, and activity of the nitrogenase enzyme (Kershaw, 1985). Lichens probably absorb N principally in the form of ammonium, nitrate, organic nitrogen, and in cyanolichens directly from the atmosphere in gaseous form (Kershaw, 1985). Uptake of inorganic nitrogen in lichens is largely attributed to atmospheric fallout (Crittenden, 1989; Dahlman et al., 2004; Ellis et al., 2004). Though the lichen species both in natural as well as in experimental conditions have been found absorbing both reduced (NH_3) as well as oxidized (NOx) forms of atmospheric nitrogen, the majority of lichens preferably absorb ammonia in a larger quantity than oxidized forms (Díaz-Álvarez and de la Barrera, 2018; Manninen, 2018; Munzi et al., 2019; Hurtado et al., 2020).

3. Nitrogen tolerance in lichens: Probable mechanisms

Lichen species can adapt to higher ambient air nitrogen deposition through ecophysiological, biochemical, metabolic, and genetic modulation adaptations (Fig. 2). Following are the various micro to macro modulations both photobionts and mycobionts undergo to accumulate nitrogen much higher than their physiological needs (Fig. 2).

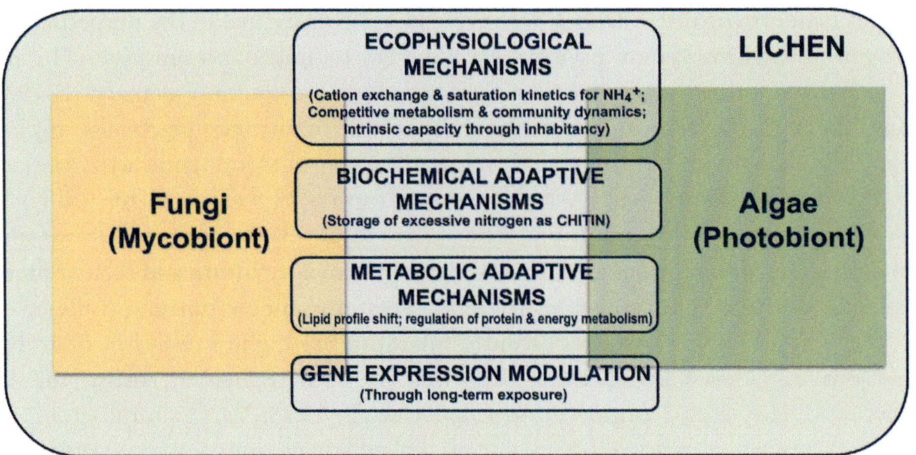

Fig. 2 Conceptual illustration of the different nitrogen tolerance mechanisms reported in lichens.

3.1 Cation exchange capacity and the saturation kinetics for NH_4^+

Munzi et al. (2009a) investigated nitrogen (N) compounds, i.e., KNO_3, NH_4NO_3, and $(NH_4)_2SO_4$, are commonly used in fertilizers that can cause chlorophyll degradation in the N-tolerant lichen *Xanthoria parietina*, and weather polyamines are responsible for the N-tolerance. The study found that excess N didn't cause chlorophyll degradation in *X. parietina* possibly due to the lower cation exchange capacity and the saturation kinetics for NH_4^+ binding sites in the plasma membrane for the N compounds (Munzi et al., 2009a). The study further found that polyamine has no significant role in N-tolerance (Munzi et al., 2009a).

3.2 Competitive metabolism and community dynamics

Hauck and Wirth (2010) based on the background studies on higher vascular plants capable of delivering carbon backbones for swift ammonium assimilation into amino acids as a criterion for tolerance to eutrophication, studied the lichens from shady habitats with known sensitivity to nitrogen pollution in Europe. The study concluded that where the shade-adapted epiphytic Licand saxicolous lichens were intolerant to nitrogen eutrophication probably due to their slower metabolism and photosynthetic rate the terricolous lichen diversity is guided by competition with other dominant ground vegetation (Hauck and Wirth, 2010). The study further suggested that the shade-growing epiphytic, corticolous, and saxicolous lichen communities are specifically sensitive to excessive nitrogen deposition.

Munzi et al. (2014) studied the photosystem II efficiency (Fv/Fm) to short-term exposure to NH_3 and distribution frequency along field NH_3 gradient on sensitive and tolerant lichen species. The study found that frequency and Fv/Fm of sensitive/ oligotrophic lichen, i.e., *Evernia prunastri* decrease abruptly above a very low NH_3 exposure (i.e., $3\,\mu gm^{-3}$) whereas tolerant lichen species, i.e., *Xanthoria parietina* retain their physiological and ecological vitality against comparatively at higher NH_3 exposure (i.e., $50\,\mu gm^{-3}$) (Munzi et al., 2014). The study concluded that the overall vitality of tolerant/nitrophytic lichens can be due to the indirect cause such as a decrease in competition due to the degradation of sensitive lichens to higher NH_3 deposition (Munzi et al., 2014). The study further highlighted the importance of establishing NH_3 pollution critical limits according to oligotrophic lichen species.

3.3 Intrinsic tolerance through inhabitancy preference and adaptation to long-term exposure

Munzi et al. (2011) analyzed the nitrogen tolerance of the nitrophilous lichen through an ex-situ experiment, to decipher the mechanisms regulating nitrogen tolerance in lichens to wet nitrogen deposition. The study involved the collection of *Xanthoria parietina* samples from habitats with different nitrogen availabilities and were treated with ammonium sulfate $(NH_4)_2SO_4$ solutions of gradient concentration (i.e., 0, 0.025, 0.05, and 0.25 M) for 5 h per day for 3 days/week. The parameters such as PSII efficiency, localization of ammonium ions, concentrations of K^+ and Mg^{2+}· and the buffer capacity were measured (Munzi et al., 2011). The study showed that lichens collected from environments with high N availability have a higher tolerance for increasing wet nitrogen treatments than the lichens collected from unpolluted areas (Munzi et al., 2011). The study shows that N tolerance to be an intrinsic character of lichens, induced by the atmosphere where the lichen inhabits.

Munzi et al. (2013a) studied the physiological mechanism of nitrogen tolerance in different lichens growing in low and high N-deposition environments. Physiological parameters such as PSII efficiency, the buffering capacity of lichen extracts, the distribution of the ions between the intra- and extracellular compartments, and potassium and magnesium concentrations were analyzed in the nitrophytic lichen *Xanthoria parietina* previously growing both in an N-poor and N-rich environment, against ammonium treatments. The physiological response of *X. parietina* was compared with the N-sensitive lichen species *Evernia prunastri* and *Usnea* sp. The study recorded similar responses between *X. parietina* from the N-poor environment and the N-sensitive species than between *X. parietina* from the N-rich environments, suggesting that *X. parietina* achieved N-tolerance after long-term exposure to N-rich environment probably through modulation of gene expressions. The study further hypothesized that N-tolerance is not an intrinsic characteristic of specific lichen species but may have been achieved by long-term exposure to high levels of N deposition.

3.4 Biochemical and metabolic adaptive tolerant mechanisms

Freitag et al. (2012) studied the metabolic profiles of *Cladonia portentosa* collected from modeled sites with differential N-deposition in Britain, using methanolic extracts through mass spectroscopy. The study found a change in the composition of lipids in lichens with a change in N-deposition where the betaine lipids showed a substantial increase, other lipids such as phosphatidylcholine decrease with increased N deposition (Freitag et al., 2012). The study found that the increase in N-deposition in *C. portentosa* changes lipid composition, as reflected by the shift from phosphatidylcholine to betaine lipids, which can potentially be used as biomarkers for increased N-deposition (Freitag et al., 2012).

Munzi et al. (2017b) studied the role of metabolomics and proteomics in the lichen *Cladonia portentosa* to elucidate the probable strategy of lichens to tolerate different forms and quantities in a controlled experiment at the Whim bog experimental site, Scotland, UK. The study found that the nitrogen forms affect diverse metabolic pathways in lichens. Specifically, the most relevant changes in protein expression were detected in the mycobiont where the oxidized nitrogen, i.e., NO_3^- primarily affected the energetic metabolism and reduced nitrogen, i.e., NH_4^+ was found to affect both transport and regulation of proteins and the energetic metabolism much more than the NO_3^-.

With background understanding from previously reported findings that increasing available nitrogen (N) increases chitin concentration in lichens, Munzi et al. (2017a) investigated the prospect of an increase in accumulation of chitin as a mechanism of N tolerance in lichens both at inter-and intra-specific levels. The study employed lichen species with different ecological N tolerances, i.e., the tolerant lichen species—*Xanthoria parietina* and *Parmotrema hypoleucinum*, and the sensitive lichens—*Evernia prunastri* and *Usnea* sp., and thalli of *X. parietina* and *P. hypoleucinum* from sites with different availabilities of nitrogen. The total nitrogen, chitin, and ergosterol contents were measured in the sample thalli of lichen species selected (Munzi et al., 2017a). The stable isotope analysis ($\delta^{15}N$) found agriculture as the predominant source of N in the study area (Munzi et al., 2017a). At inter-specific level, the nitrogen and chitin contents were higher in the tolerant species than of the sensitive species. At intra-specific level the lichen samples collected from the polluted sites recorded higher N and chitin content than in the samples collected from comparatively low polluted sites (Munzi et al., 2017a). The study concluded that chitin accumulation contributes to N stress tolerance in lichens, where the excess N is stored as chitin in the cell walls of tolerant species (Munzi et al., 2017a).

Munzi et al. (2020) studied the effect of a long time (11 years) exposure of nitrogen pollution on presumably pollution-sensitive lichen species *Cladonia portentosa*. The study tested the effect of nitrogen on photosynthetic parameters (i.e., chlorophyll *a* fluorescence, rate of photosynthesis, chlorophyll content), photobiont ultrastructure, carbon-nitrogen concentrations, and their isotopic signature, chitin concentration, surface pH, and extracellular enzymatic activity in samples exposed for 11 years to different nitrogen doses and forms (e.g., ammonia and nitrate). Except for the increase in chitin content (a probable strategy to tolerate excessive nitrogen deposition), the study found relatively no effect of long-term nitrogen deposition on the studied parameters in *C. portentosa* (Munzi et al., 2020). The study concluded that the lichen species *C. portentosa* is tolerant of long-term excessive nitrogen deposition with a low negative effect on the vitality of lichens (Munzi et al., 2020). The study concluded that both the short- and long-term responses of lichen species must be taken into consideration while designing any biomonitoring study (Munzi et al., 2020).

4. Lichens are indicators of excessive nitrogen (N) deposition along with multiple scales of diversity dynamics, biochemistry, and ecophysiology

Lichen species both in passive natural, as well as experimental settings, have been found to reflect the effect of excessive N deposition through various responses (Fig. 3). The response of lichens to N pollution is multiscale ranging from change in their diversity dynamics, biochemical profile, nutrient assimilation, and ecophysiological viability

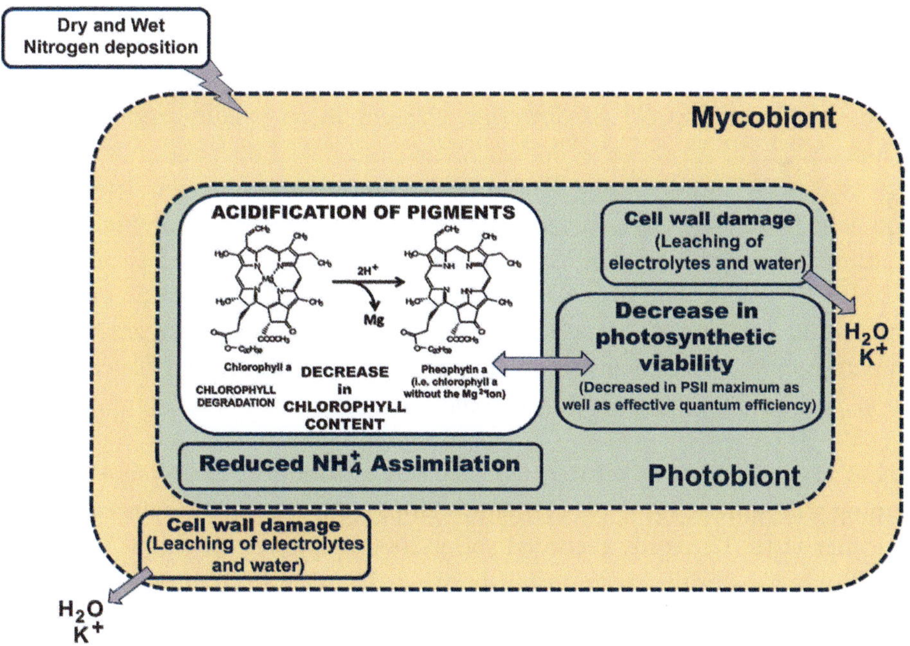

Fig. 3 Conceptual illustration of various ecophysiological effect of nitrogen eutrophication reported on lichens.

(Fig. 3). In the case of epiphytic lichens, it has been found that lichen growing on twigs represent the recent/current ambient air N deposition, lichens growing on trunks are cumulative indicative of N deposition from much earlier/distant past. The following studies showcase the ability of lichens as an appropriate indicator of N eutrophication.

van Herk et al. (2003) studied the effect of SO_2 in air, NH_4^+, NO_3^- and SO_4^{2-} on remotely situated epiphytic lichen community using regression models. Many species (i.e., *Bryoria capillaris*, *B. fuscescens*, *Cetraria pinastri*, *Imshaugia aleurites*, and *Usnea hirta*) showed a significant negative correlation with nitrogen compounds. The study found that long-distance nitrogen air pollution has a significant effect on the diversity of lichen species (van Herk et al., 2003).

Frati et al. (2006) investigated that whether the N-deposition (i.e., NO_2, NH_3) by vehicular exhaust influence the lichen diversity, lichen vitality, and the accumulation of nitrogen in epiphytic lichens and transplanted *Evernia prunastri* by measuring the parameters, i.e., accumulation of nitrogen, reduction in chlorophyll *a*, chlorophyll *b* and total carotenoids. The study found that applications of N-based fertilizers were the main factor than vehicular traffic which negatively affect the lichen community of the study area.

Geiser and Neitlich (2007) modeled the response of lichen communities to the air quality and climate gradients in western Oregon and Washington to detect the pollution gradients in the study area. The analysis found higher eutrophication in the urban-industrial and agricultural land use which was correlated with the absence of sensitive lichens and the increase of the nitrophilous lichen species. The lichen-based modeling approach was found capable of detecting changes in air quality and climate across landscapes.

Frati et al. (2007) studied the effect of high ammonia and nitrogen deposition emissions on lichens around a pig farm in central Italy by assessing the epiphytic lichen diversity, transplanted lichen species of *Xanthoria parietina* and *Flavoparmelia caperata*, along with passive NH_3 samplers. The diversity of nitrophytic species was found correlated with both ammonia and nitrogen deposition (Frati et al., 2007). *Physconia grisea* emerged as appropriate bioindicator species for NH_3 pollution, and the total N accumulation in *Xanthoria parietina* and *Flavoparmelia caperata* was correlated with ammonia emissions (Frati et al., 2007).

Davies et al. (2007) studied the diversity and distribution of epiphytes in London concerning atmospheric NO_x concentration employing atmospheric dispersion modeling at a finer scale. The study recorded about 3000 epiphytes of which 74 were lichen species, showing a positive relationship between NOx and pollution tolerant species, associated with eutrophication.

Tretiach et al. (2007) assess the comparative vitality potential of lichen *Pseudevernia furfuracea* and moss *Hypnum cupressiforme* transplanted in nylon bags. The study assessed the transmission electron microscopy observations, K cellular location, and measurements of C, N, S, and photosynthetic pigments content, CO_2 gas exchange, and chlorophyll fluorescence to analyze the exposed samples. The study found that the lichen maintained good vitality levels in comparison to the moss due to the different ecophysiology of the biomonitors.

Frati et al. (2008) studied the effects of nitrogen deposition (i.e., atmospheric NH_4^+ and nitrogen dust) in a rural area on bark pH, bark NH_4^+ and bark electric conductivity and the responses to these parameters on lichen vegetation of pine in a Mediterranean habitat. The study found that the pine vegetation was inhabited with nitrophytic lichen vegetation, suggesting diffuse eutrophication (Frati et al., 2008). Factors such as the tree circumference, distance from the stock farms, bark pH, and bark electrical conductivity were found to influence the epiphytic lichen diversity in the study area. Lichen species such as *Xanthoria parietina* was indicative of nitrogen dust pollution, and *Hyperphyscia adglutinata* occurrence reflected an increase in bark pH due to the release of high amount of ammonia (Frati et al., 2008).

Riddell et al. (2008) studied the effect of nitric acid HNO_3 gas, a secondary product of photochemical reactions of nitrogen oxides (NOx) emitting from the vehicular exhaust on *Ramalina menziesii* a native lichen of Los Angeles, USA, employing fumigation experiments in chambers for a month against control samples. The lichen exhibited a significant decline in chlorophyll content, carbon exchange capacity, and cell wall damage through increase in leachate electrical conductivity, in reference to the control samples. The study found that *R. menziesii* was able to appropriately reflect the physiological damage caused by HNO_3 exposure produced as a result of fossil fuel combustion.

Satya and Upreti (2009) studied the effect of accumulation of carbon (C), nitrogen (N), and sulfur (S) contents in an epiphytic crustose pollution tolerant lichen *Rinodina sophodes*, and its influence on the photosynthetic pigments in Gangetic plains, India. The maximum thallus C was recorded in urban land use, whereas maximum N concentration was recorded in rural habitats with dominant agricultural practices and S concentration was negligible. The thallus N, C, and S deposition were found affecting the photosynthetic vitality of *R. sophodes* causing degradation of chlorophyll, making *R. sophodes* an appropriate indicator of pollution in Gangetic plains (Satya and Upreti, 2009).

Munzi et al. (2009b) tested the suitability of cell membrane integrity in the lichen *E. prunastri* (L.) Ach. as an indicator of nitrogen (N) stress experimentally. Munzi et al. (2009a,b) incubated the lichen samples in solutions of potassium nitrate, ammonium nitrate, and ammonium sulfate, and measured the cell membrane damage, as the increase in the electrolyte leakage at the time scale of 0, 24, 48, and 96 h. Cell membrane damage was observed in *E. prunastri* at all concentrations of N compound treatments. The study found that as much of the thallus biomass is fungi, the mycobiont cell membrane damage can be used as an indicator of N stress due to ambient air N fallout.

Hogan et al. (2010) studied the phosphomonoesterase (PME) activity concerning nitrogen (N) and phosphorus (P) concentrations in *Cladonia portentosa* and assess the potential of PME as biomarkers for N pollution in the UK. *C. portentosa* collected from sites varying in wet N (NH_4^+ and NO_3^-) based on both measured and modeled data sets (Hogan et al., 2010). The PME activity in the thallus was analyzed using an enzyme-labeled fluorescent phosphatase substrate (i.e., p-nitrophenyl phosphate). The PME activity, N concentration, and N:P ratio were found to increase along the N-deposition gradient in *C. portentosa* (Hogan et al., 2010). The study hereby explores the possibility of PME as a biomarker for N deposition.

Munzi et al. (2010) studied the effect of exposure time and dose concentration on the physiological viability of lichens. The experiment was performed on N-sensitive lichen species *Evernia prunastri* and N- tolerant *Xanthoria parietina* treating the lichens for 5 weeks with solutions of NH_4NO_3 or $(NH_4)_2SO_4$. The photosystem II efficiency as chlorophyll fluorescence (Fv/Fm) was taken as a marker of lichen physiological vitality against the N manipulation (Munzi et al., 2010). The study found a deleterious effect on the photosynthetic vitality of both the lichen at different doses of the treatment, which was species-specific. The study validated that the response of lichens to N-enrichment is both time- and dose-dependent where the sensitive species respond to relatively low N-input that the tolerant species.

Wolseley et al. (2010) studied the diversity of macrolichens on trunks and twigs of trees from sites near ammonia (NH_3) monitoring stations across the UK, to study if lichens diversity was correlated to atmospheric NH_3 deposition. The results that the loss of nitrogen-sensitive species decline was prevalent even at very low N-concentration (i.e., $1\,\mu gm^{-3}$) and was correlated with the increase in N-tolerant species (Wolseley et al., 2010). Further, the study also concluded that lichens on the twigs reflected current atmospheric NH_3 concentrations whereas lichens on older trunks provide evidence of atmospheric conditions from earlier periods (Wolseley et al., 2010).

Freitag et al. (2011) studied the biochemical changes in response to nitrogen enrichment in the lichen *Cladonia portentosa* collected from sites with varying atmospheric nitrogen in Britain, using fourier transform-infrared spectroscopy (FTIR) metabolic fingerprinting. The study found an increase in the total proportions of lichen amino acids and proteins, and a decrease in the fatty acids and their esters due to higher N depositions (Freitag et al., 2011). The study provided evidence of the coupling of metabolic processes in *Cladonia portentosa*. The study thus recorded that the fingerprinting of metabolic processes in lichens can provide complementary data for mitigation of varying amounts of N deposition.

Frati et al. (2011) studied the concentration of assimilatory pigments in the lichen *Evernia prunastri* transplanted in an agricultural field with high atmospheric NH_3 concentrations in central Italy. The study found that the transplanted *E. prunastri* responded to the high NH_3 deposition with a decrease in the assimilatory pigments-chlorophylls a and b and carotenoids, and an increase of chlorophyll degradation to pheophytin (Frati et al., 2011).

Johansson et al. (2012) assess the effects of nitrogen deposition, on the epiphytic lichen community composition in an N-deficit natural boreal forest of Sweden. The experimental setup employed an irrigation-fertilization experiment where five increasing concentrations of NH_4NO_3 were applied on epiphytic lichens for four consecutive growing seasons (2006–2009), and changes in the diversity of lichens were monitored. The study recorded a reduction in the species richness of the epiphytic lichen communities, at all concentrations of N-fertilizer application (Johansson et al., 2012). The reduction in abundance was species-specific with some of the species, i.e., *Alectoria sarmentosa*, *Bryoria* spp., and *Hypogymnia physodes* increased at lower N loads and considerably decreased at high loads, with species-specific optima (Johansson et al., 2012). The study concluded that the decrease in lichen diversity along with increasing N-loads is due to physiological responses of the lichen species, increase in the photobiont: mycobiont ratio, and parasitic fungal attacks (Johansson et al., 2012).

Munzi et al. (2012) experimentally studied the physiological response of the epiphytic lichen *Evernia prunastri* to gradient exposure to different concentrations of NH_4Cl (i.e., 50, 150, and 500 μM) nitrogen compounds for about a month. The integrity of cell membranes and chlorophyll fluorescence emission (Fv/Fm and performance index-PI_{ABS}) were taken as response variables. Though no membrane damage was recorded during the period of study, a substantial decrease in photosynthetic parameters, i.e., Fv/Fm and PI_{ABS} was recorded during the experimentation. As the study involved simulation of near-natural deposition of nitrogen, it concluded that the physiological parameters of *Evernia prunastri* can be an efficient biomarker for the ambient air N deposition (Munzi et al., 2012).

Munzi et al. (2013b) studied the role of the carbon/nitrogen (C/N) balance in a pollution-sensitive lichen species *Evernia prunastri* collected from the unpolluted area of central Italy to excess N deposition, using in-vitro treatments with NH_4NO_3 and $(NH_4)_2SO_4$ solutions. The study found that the total N content of the lichen thalli increased, and the total C content decreased with increasing N experimental deposition, decreasing the C/N ratio in treated samples (Munzi et al., 2013b). The photosystem II fluorescence (Fv/Fm) ratio, measured as the vitality index was found positively correlated with thallus C content and negatively correlated with N content (Munzi et al., 2013b). The study concluded *E. prunastri* is an appropriate bioindicator of N pollution (Munzi et al., 2017b). The study further emphasized that the C/N ratio can act as a tool for early warning of N excess in different habitats (Munzi et al., 2013b).

Giordani et al. (2014) examined the nitrogen deposition quantitatively employing correlation and multiple regression models and hierarchical partitioning. The study recorded a significant variation in the diversity of growth forms of epiphytic lichens highlighting the relative importance of environmental factors. The study found macrolichens to be appropriate indicators of nitrogen deposition (Giordani et al., 2014).

Pinho et al. (2014b) found the nitrogen (N) concentration in thalli of two lichens *Parmotrema hypoleucinum* (J. Steiner) Hale, and *Xanthoria parietina* (L.) Th.Fr. in southwest Portugal and correlated it with ammonia emissions estimated from cattle information and the cover of agricultural land. The study found that the N concentrations in lichen thalli were significantly correlated with reduced-N (NH_X) emissions in the region (Pinho et al., 2014b). The study validated that, lichens can be used as a reliable biomonitoring tool for N-deposition to identify areas with excessive N-deposition and further devising mitigation strategies for such habitats.

Pinho et al. (2014a) studied the effect of atmospheric ammonia (NH_3) using the functional diversity of lichens (i.e., lichen diversity value—LDV) in a cork-oak woodland, against a point source, a cattle bark in Portugal for one year. The study found the epiphytic lichen functional diversity significantly correlated with the temporal and spatial deposition of ammonia which resulted in N-pollution tolerant species with low (oligotrophic), moderate (mesotrophic), and high (nitrophytic) tolerance (Pinho et al., 2014a). The study elucidated that the spatial modeling of the lichens can provide high spatial resolution information about the severity and the effects of atmospheric NH_3 deposition in proximity to a point- and diffuse source.

McMurray et al. (2015) using non-metric multidimensional scaling and generalized additive models analyzed the lichen community composition as an indicator of the regional climate and nitrogen (N) pollution gradients in the northern Rocky Mountains, USA. The community composition of eutrophic lichen genera such as *Physcia*, *Xanthomendoza*, and *Xanthoria* was found correlated with the high N deposition, low precipitation, and temperature extremes. The study further determined the critical N load in the area correlated with the degradation of the native lichen community. The study found that lichen community composition is strongly influenced by climatic factors, i.e., temperature regimes, relative humidity, and N-deposition. The findings were found to be useful for identifying, and sustainable management of sensitive habitats.

Yatawara and Dayananda (2019) assessed the lichen diversity and lichen species frequency by employing an index of atmospheric purity (IAP), regarding ambient air SO_2 and NO_2, measured using passive air samplers, in the tropical environment of Srilanka. The study found the lichen genus *Pyxine* a reliable pollutant tolerant indicator of all urban sites and index-based mapping as a potent tool for the assessment of pollution gradients in urban-rural tropical habitats (Yatawara and Dayananda, 2019).

Wang et al. (2019b) using experimental treatment with NH_4NO_3 solution, studied the effect of nitrogen uptake, assimilation, and the impact of excess nitrogen deposition on the symbiotic balance of dominant epiphytic lichen *Usnea* spp. in central China. The study found that the contents of chlorophyll *a* and ergosterol, which represent the photobionts and mycobionts respectively, were species-specific, and the lichens took up, assimilated, and utilized more ammonium than nitrate. Except for *Usnea longissima*, the chlorophyll contents of the other lichen species increased at first with simulated nitrogen addition, then decreased with higher nitrogen availability at a certain threshold. *U. longissima* was found to be able to regulate its nitrogen uptake, resulting in a steady photobiont-mycobiont ratio. The study suggested that epiphytic lichens in subtropics can uptake and assimilate more ammonium than the nitrate and that the balance between photobiont and mycobiont of many epiphytic lichens might change with the increasing nitrogen deposition load, ultimately impacting the diversity dynamics of lichens in the forest ecosystem.

Lichens when form an essential component of early successional stages in an ecosystem, play an important role in the soil chemistry, texture, and influence the later vegetation of the xerosere (Sparrius and Kooijman, 2013). Sparrius and Kooijman (2013) studied two inland dunes with mixed vegetation of bryophytes, lichens, and grasses in the Netherlands with low to high N deposition. The study found that higher N deposition leads to higher N availability causing lower soil pH, high microbial mass, and accelerated soil respiration and N mineralization ultimately accelerating the overall succession (Sparrius and Kooijman, 2013). Sparrius et al. (2013) studied eight inland dunes in the Netherlands, having four actives and four more stabilized drift sands with low and high N deposition respectively. The active drift sands dune's pioneer vegetation was characterized by grasses and bryophytes, whereas the stabilized drift sands were lichen rich along with moss growth. The study found that increased N deposition accelerates the succession leading to heath-forest vegetation dominated by grasses and moss, thereby a rapid change in the regional flora is attained (Sparrius et al., 2013). Sparrius et al. (2012), studied the effect of increased nitrogen (N) deposition on lichen-dominated inland dunes (along with *Polytrichum piliferum* mats, *Campylopus introflexus* mats) with low and high N deposition respectively. The study concluded that the pioneer vegetation in inland dunes is prone to N inputs where a decrease in lichen cover occurs due to increased N deposition (Sparrius et al., 2013). Thus, the increased N deposition changes the constituent of the vegetation, its pace of succession, and negatively affects the lichen cover of the affected area (Sparrius et al., 2012, 2013; Sparrius and Kooijman, 2013).

5. Lichens, the critical load assessor of N eutrophication

Root et al. (2013) estimated the nitrogen deposition in the forests of western North America using lichens employing a single regression model. The study was able to determine critical N loads employing lichens as integrative bioaccumulator of nitrogen due to their readily availability, cost-effective analysis, and accuracy of N estimation, at spatial and topographic scale (Root et al., 2013). The study highlighted the lichen-based biomonitoring approach as an appropriate method to determine the critical nitrogen load for management and mitigation purposes.

Pinho et al. (2012) determined the critical loads of nitrogen (N) deposition and long-term critical levels of the reduced form of N (i.e., ammonia) in a semi-natural

Mediterranean cork-oak woodland downwind to a known source of ammonia (i.e., cattle barn). The study used a pre-published set of lichen species categorization into N-pollution indicator functional groups such as nitrophytic, mesotrophic, and oligotrophic (Pinho et al., 2012). The study determined the annual critical loads of N-pollution by modeling the correlative relationship between lichen functional groups and annual atmospheric ammonia concentration, which can be used by policymakers to devise mitigation strategies to protect semi-natural habitats from the effects of excessive N pollution (Pinho et al., 2012).

Geiser et al. (2019) using photobiont and growth from lichen functional groups assessed the critical load of sulfur (S) and nitrogen (N) deposition in the forests of the USA employing 90% quantile regression modeled metrics. The study quantified the critical levels of S and N in the forests and the correlation observed in the change of lichen diversity, especially species of conservation concern (Geiser et al., 2019). The study thus highlights the importance of lichens in the management of pollution loads and as a reliable source of the information needed for conserving lichens and the ecosystem services. Lichens provide critical data of N eutrophication which can be utilized for the sustained management of forest ecosystems (Geiser et al., 2019).

6. Long-term lichen air N deposition monitoring: Change in pollution regime

Long-term lichen-based pollution studies mainly done in Europe has found that lichens are indicative of N pollution intensity, change in pollutant regime. Lichens respond to N pollution gradient both through community dynamics, and physiological parameters.

Sparrius (2007) analyzed the response of epiphytic lichen communities to decreasing ammonia air concentration from 1996 to 2003 in a moderately polluted area in the Netherlands. The study found an increase in nitrogen-sensitive species and a decline in the nitrogen-tolerant lichen species (Sparrius, 2007). The study recorded an overall decrease in the lichen diversity of the study area, due to the decline of N-tolerant lichens, it also recorded an increase in acidophytic lichen species such as *Bryoria fuscescens*, *Evernia prunastri*, *Hypogymnia physodes*, and *Platismatia glauca*, the native lichen flora of relatively low polluted environments (Sparrius, 2007).

Hauck et al. (2013) assessed the decline in the epiphytic lichen diversity during 150 years in the submontane and uplands broad-leaved forests of north-western Germany. The sharp species decline was attributed to forest management (i.e., the cutting of mature and decaying trees), reduction of the soil moisture, air humidity, and the deposition of acidifying nitrogen fertilizers (Hauck et al., 2013). Among these factors, excessive nitrogen deposition was one of the prominent causes of the increase of acidophilic-nitrophyte lichens (i.e., *Chrysothrichetalia candelaris*, *Physcietalia adscendentis*, *Lecanorion variae*, *Lopholoceetalia heterophyllae*) (Hauck et al., 2013).

Mayer et al. (2013) analyzed the effect of the long-term (17 years) effect of sulfur (S) and nitrogen (N) on the growth and diversity of epiphytic lichens in Austria using ordination (NMDS) and MANOVA analysis. The study found, change in species composition, decrease in total cover, and decline in species count in the epiphytic lichen community (Mayer et al., 2013). The study analyzed that the overall decrease in epiphytic lichen diversity is due to the long-term wet N deposition in the form of ammonia in precipitation and fog (Mayer et al., 2013).

Seed et al. (2013) reported open-air laboratories (OPAL) data on the diversity of selected lichens collected by citizens and modeled it with climatic and pollution data. The collected lichens reflected the gradient of N pollution in the UK. The study recognized, sensitive (*Usnea* spp., *Evernia prunastri*, *Hypogymnia physodes*, *H. tubulosa*), intermediately sensitive (*Melanelixia* spp., *Flavoparmelia caperata*, *F. soredians*, *Parmelia sulcata*, *P. saxatilis*), and N tolerant (*Xanthoria parietina*, *Xanthoria polycarpa*, *Physcia adscendens*, *P. tenella*) lichen species in different land-use types of UK. The study validated that specific lichen species can be calibrated to the gradient of N pollution with specific indicator species for lower, moderate, and high N deposition. Welden et al. (2018) reported the data of open-air laboratories (OPAL) survey data, collected by citizens of the United Kingdom during the period from 2007 to 2011, and interpreted the findings regarding nitrogen deposition. The study found that the abundance of tolerant and sensitive lichens was related to N deposition rates. The response of nitrogen-sensitive lichens to reduced nitrogen (NH_x) was more pronounced, than to oxidized nitrogen (NO_x), and the response of nitrogen-tolerant lichens was greater to oxidized nitrogen than to reduced nitrogen (Welden et al., 2018). The study further found the effect of sites in proximity to roads recorded more N-tolerant lichens than the N-sensitive ones (Welden et al., 2018). The study demonstrated the capability of citizen science to distinguish and measure the air pollution impacts over larger geographical regions and increase our understanding of lichen responses to different chemical forms of N deposition, pollution source points, and host bark pH.

Serrano et al. (2019) using *Parmotrema hypoleucinum* (Steiner) Hale, assessed the effectiveness of the European air quality directive in mitigating the nitrogen and sulfur pollution initiated in 1999–2001. The study found a reduction of sulfur emissions (2002–2011) at the ecosystem level, reflected both in observations at measurement stations and lichens alike, whereas the nitrogen deposition in lichens was found to increase in areas dominated by agricultural land-uses (Serrano et al., 2019). The study hereby highlighted the importance of biomonitors over emission estimates reached through modeling (Serrano et al., 2019). The studies herby conclude that in Europe that the mitigation policies for air pollution though have reduced sulfur dominant pollution but have been unable to check the increasing nitrogen pollution.

Herzig et al. (2020) in their recalibration studies of the Swiss lichen bioindication methods after 30 years of the launch of the studies found a change in the pollutants in the ambient air. The study recorded a major shift from sulfur- and acid-dominated pollutant regime to a more nitrogen-dominated pollution regime of NH_3/NOx and ozone, allowing recolonization and change in the epiphytic lichen vegetation both in terms of abundance and species composition. The change was characterized by an increase in nitrophytic and N-tolerant lichen species and a concurrent decrease in acidophilic lichens. The change in lichen community dynamics reflected the quantitative as well as qualitative changes in the air pollutants over the past 30 years (Herzig et al., 2020). The study hereby highlighted the crucial roles lichen biomonitoring can play in devising mitigation policies to curb air pollution.

7. Stable nitrogen isotope ratio ($\delta^{15}N$ in) lichens function of nitrogen source and their spatial distribution at the landscape level

$\delta^{15}N$ or delta-N-15 is a measure of the ratio of the two stable isotopes of nitrogen (Davies et al., 2007) $^{15}N{:}^{14}N$, in lichen thallus, and have been reported as the indicator of the spatial distribution of N emission source in landscapes with heterogeneous land use (Boltersdorf and Werner, 2013, 2014; Boltersdorf et al., 2014; Munzi et al., 2016). The $\delta^{15}N$ studies in lichens though are few and have been done primarily in Europe (Table 1), which elucidate various basic patterns of nitrogen forms primarily absorb by lichens and their influence on the lichen diversity, nutrient cycling, and physiology, which further help in differentiating the associated land-use types. The stable isotope studies in lichens, both through passive sampling and experimental setups have helped generate data on the critical limits for N deposition in landscapes and devising mitigation strategies.

(Purvis et al., 2005) studied the comparative elemental concentration and nitrogen stable isotope ratio ($\delta^{15}N$ values) in the healthy and presumably dead thallus of *Parmelia sulcata* along with their oak bark substratum, to assess the correlative accumulation of elements and indication source of nitrogen in Buckinghamshire, England. The dead samples reported a higher concentration of N and other elements than their critical loads. The higher negative $\delta^{15}N$ values in lichen samples whereas indicated that N originated mainly from ammonia, the less negative $\delta^{15}N$ values in healthy samples collected in the vicinity of roads suggested NOx as N source. The significant positive correlation between elements and the $\delta^{15}N$ values in the healthy thallus of *P. sulcata* suggested the mixed input of N both from agriculture and vehicular exhaust. The study found that the accumulation of different elements in the dead and healthy thallus was influenced by the intensity of pollution gradient from their sites of collection. The study where on one hand found *Parmelia sulcata* an appropriate bioindicator of source and quantity of N and elemental pollution, also concluded that N accumulator lichen species are also good accumulators of elements both from geological as well as pollution sources.

Table 1 The reported values of $\delta^{15}N$ with reference to the associated land use and sources of excess nitrogen deposition.

S. No.	Lichen species	Location	$\delta^{15}N$ "per mille" (‰)	Land use	Interpreted nitrogen source/concluding inferences	Reference
1.	*Xanthoria parietina* (L.) Th. Fr.	Bolu plain, Turkey, Europe	−1.5 to −8.86	Roads	Vehicular exhaust (mainly NO_x), accumulation according to the proximity of the source	Dörter et al. (2020)
			−8.86 to −18.5	Agricultural field and poultry farms	Agricultural and animal husbandry emissions (mainly NH_x), accumulation according to the proximity of the source	
2.	Lichen community of 23 Beech (*Fagus sylvatica*) forests across Europe	Europe	−1.48 to −15.04	Natural Beech Forest	Atmospheric ammonium deposition which was assimilated/accumulated by the lichen species based on their photobiont types and growth forms having varied thallus surface–volume ratio.	Hurtado et al. (2020)
3.	*Usnea lapponica* and *Letharia vulpina*	Greater Yellowstone Ecosystem, USA	−16 to −8	Natural protected area	Primarily atmospheric ammonia deposition emanating from agricultural practices with very low NO_3^- influx	Hoffman et al. (2019)

Continued

Table 1 The reported values of $\delta^{15}N$ with reference to the associated land use and sources of excess nitrogen deposition—cont'd

S. No.	Lichen species	Location	$\delta^{15}N$ "per mille" (‰)	Land use	Interpreted nitrogen source/concluding inferences	Reference
4.	*Hypogymnia physodes, Flavoparmelia caperata, Physcia stellaris*	Siberian cities of Tomsk and Prokopyevsk, Russia	−11 to +4 (mean −1.3 to −5)	Urban	NOx emitted by the vehicular and industrial exhaust	Simonova et al. (2018)
5.	*Evernia prunastri, Usnea* sp., *Parmelia sulcata, Parmelia caperata, Hypogymnia physodes*	The area of the Pyrénées-Atlantiques in southwestern France	−7.51 to −8.70 −7.75 to −8.93 −8.77 to −9.67 −8.61 to −8.99	Urban Industrial Agricultural Forest	NOx emitted by the vehicular and industrial exhaust NHx emanated from agricultural and natural ecosystem processes	Barre et al. (2018)
6.	*Anaptychia* sp.	Mexico City, Mexico	−9.4 to 5.2	Urban, rural	The isotopic ratios pointed towards mixed sources, i.e., agricultural/forested as well as urban exhaust	Díaz-Álvarez and de la Barrera (2018)
7.	*Hypogymnia physodes* *Platismatia glauca* *Pseudevernia furfuracea*	Near the metropolitan area of Helsinki, Finland	−5.07 to −6.45 −5.50 to −6.15 −3.50 to −5.25	Urban, sub-urban and associated forested area	The study supported the enhanced uptake of oxidized N (NO_2), originating from road traffic than the reduced N (NHx).	Manninen (2018)
8.	*Parmotrema hypoleucinum*	Alentejo Litoral region, Portugal, Europe	−8.27 to −10.95 −11.24 to −15.59 −9.44 to −13.16	The coastal region with oceanic influence Agricultural Urban	Mixed forms of nitrogen (NH_x, NO_x) with a relatively larger proportion of reduced form from agricultural land use than urban where oxidized nitrogen predominated	Pinho et al. (2017)

9.	*Xanthoria parietina, Parmotrema hypoleucinum, Evernia prunastri, Usnea* sp.	Companhia das Lezírias, Portugal	−5.05 to −10.21	Mediterranean cork oak woodland sites with agriculture and poultry farming practices	Agricultural activities as the main N source	Munzi et al. (2017a)
10.	*Cladia retipora*	Tasmania	−6.9 to −0.8, −7.5 (outlier)	Temperate forests, alpine, and subalpine heath, and coastal habitats	Natural as well as an anthropogenic source	Hogan et al. (2017)
11.	*Evernia prunastri, Xanthoria parietina, Cladonia portentosa*	Whim Bog, Scotland, UK	–	–	The N form is important and in the case of multiple N sources ammonium has a prominent effect on the isotopic signature of lichens than nitrate which is further guided by the fact that the species is acidophytic or nitrophytic.	Munzi et al. (2016)
12.	*Xantoria parietina*	Venafro Plain, Molise, Central Italy	−1 to 0.2	Urban	Incinerator exhaust mainly consisting of NOx with a mixture of NH_x.	Cocozza et al. (2016)
13.	*Hypogymnia physodes*	Karpacz city and their vicinity in the Karkonoski National Park area, Sudetes, SW Poland	−6.8 to −13.0	A protected area in the vicinity of an Urban to semi-urban settlements	Gaseous NO_x and/or NO_3^- ion from traffic (Vehicular exhaust)	Ciężka et al. (2016)

Continued

Table 1 The reported values of $\delta^{15}N$ with reference to the associated land use and sources of excess nitrogen deposition—cont'd

S. No.	Lichen species	Location	$\delta^{15}N$ "per mille" (‰)	Land use	Interpreted nitrogen source/concluding inferences	Reference
14.	*Usnea sphacelate, Umbilicaria decussata, Xanthomendoza borealis*	Cape Hallett, northern Victoria Land, Antarctica	−15.5 to +9.0	Antarctic habitats near an Adèlie penguin rookery	Penguin guano (NH_3)	Crittenden et al. (2015)
15.	*Parmelia sulcata, Physcia adscendens, Physcia tenella, Xanthoria parietina*	Germany	−15.2 to −1.5	Heterogenous landscape consisting of natural forests and agricultural fields	Mainly agricultural activities (NH_x)	Boltersdorf et al. (2014)
16.	*Xanthoria parietina, Physcia adscendens* and/or *Physcia tenella*	Western part of Germany	−15.2 to −1.3	Heterogeneous region with rural as well as urban landuse	Local N sources primarily agriculture and industrial livestock farming	Boltersdorf and Werner (2014)
17.	*Letharia vulpina*	Sierra Nevada, California, USA	−12.5 to −0.6	Roadside habitats	Vehicular exhaust (NO_x)	Bermejo-Orduna et al. (2014)
18.	*Hypogymnia physodes, Parmelia sulcata, Xanthoria parietina, Physcia adscendens,* and *Physcia tenella*	The western part of Germany	−14.6 to −1.9	Rural habitats	Agricultural input (NH_x)	Boltersdorf and Werner (2013)
19.	*Evernia mesomorpha*	Athabasca Oil Sands Region, north-eastern Alberta, Canada	−7.4% to +11.6%.	Industrial oil fields	Both Ammonium and Nitrate	Proemse and Mayer (2012)

20.	*Usnea aurantiaco-atra* and *Usnea antarctica*	King George Island, Antarctica	-7.67 to 4.30	Antarctic habitats	Ammonia derived from animal sources	Cipro et al. (2011)
21.	*Cladonia gracilis, Himantormia lugubris, Ochrolechia frigida, Ramalina terebrata, Sphaerophorus globosus, Stereocaulon alpinum, Usnea antarctica, Usnea aurantiaco-atra*	Barton Peninsula, south–west King George Island, Antarctica	-0.7 to 0.6; mean -7.4	Antarctic habitats	Highly influenced by the seabird guano as a source along with intrinsic physiology and biochemistry	Lee et al. (2009)
22.	Lichens	Twin Cays, Belize, Central America	0.4 to -21	Experimental open plots	The main source of N is from wet deposition primarily consisting of NH_x	Fogel et al. (2008)
23.	*Parmelia sulcata*	Buckinghamshire, England	$-8{\cdot}80$ to $-8{\cdot}04$	Urban, sub-urban to rural habitats	The mixed input of N both from agriculture (NH_x) and vehicular exhaust (NO_x)	Purvis et al. (2005)

Lee et al. (2009) along with mosses assessed the nitrogen stable isotope $\delta^{15}N$ in lichen *Cladonia gracilis*, *Himantormia lugubris*, *Ochrolechia frigida*, *Ramalina terebrata*, *Sphaerophorus globosus*, *Stereocaulon alpinum*, *Usnea antarctica*, *Usnea aurantiaco-atra* from Barton Peninsula, located in the south-west part of King George Island, Antarctica. The study found a large variation in the nitrogen isotope compositions $\delta^{15}N$, probably due to differences in physiology, biochemistry, topography, and proximity to the animal-derived sources.

Cipro et al. (2011) analyzed persistent organic pollutants as well as $\delta^{13}C$ and $\delta^{15}N$ stable isotopes in Lichens (*Usnea aurantiaco-atra* and *Usnea antarctica*), mosses, and one angiosperm species in the austral summer of 2004–2005. The lichen $\delta^{15}N$ value of $-7.67‰$ to $4.30‰$ suggested the influence of the origin from different animal–derived sources.

Proemse and Mayer (2012) analyzed the nitrogen (N) content and N stable isotope ratio ($\delta^{15}N$ values) in epiphytic lichen *Evernia mesomorpha*, to study the effect of N emission from the Athabasca Oil Sands Region, north-eastern Alberta, Canada. About 32 epiphytic samples of *E. mesomorpha* were sampled from 24 sites at different sites situated 3 to 113 km from the oil fields along with some lichen samples were also collected from the edge of the forest stand and the interior of the forest from one site at 14 km from the oil fields (Proemse and Mayer, 2012). Thallus nitrogen content was high near the oil fields. The $\delta^{15}N$ values in lichen ranged from $-7.4‰$ to $+11.6‰$ with no substantial variation in the $\delta^{15}N$ of the lichen sampled from the forest edge ($-4.8‰$) than to the interior of the stand ($-4.7‰$) (Proemse and Mayer, 2012). The study concluded that the source of N-deposition was mainly ammonia with some fraction of nitrates emanating from the oil fields.

Boltersdorf and Werner (2013) studied the comparative capability of bark, lichen tissue, and atmospheric water deposition as indicators for gradients and source of atmospheric nitrogen (N) deposition, using N estimation and stable isotope ratio-$\delta^{15}N$ in West Germany. The N content and $\delta^{15}N$ of N–tolerant lichens such as *X. parietina* and *Physcia* sp. reflect the species-specific, agriculture-related N deposition at various sites in Germany and were found to be true indicators of N-deposition gradient and N-source attributor in different habitats of Germany (Boltersdorf and Werner, 2013). Boltersdorf and Werner (2014) analyzed the nitrogen (N) content and N stable isotope ratio in epiphytic lichens *Hypogymnia physodes*, *Parmelia sulcata*, *Xanthoria parietina*, *Physcia adscendens*, and *Physcia tenella* in 24 sites of western Germany in primarily rural habitats, to assess the spatial deposition of N and their major source. The study found the species-specific accumulation of N in the lichens and the isotopic signature pointed towards agriculture practices as a major contributor of N in the form of ammonia in analyzed rural habitats. Boltersdorf et al. (2014) evaluated the lichens, mosses, and tree bark for their capability as a bioindicator of nitrogen deposition in Germany. The study estimated nitrogen content and stable isotope ratio of nitrogen $\delta^{15}N$ ($^{15}N{:}^{14}N$) in all the three bioindicators to study the species-specific variations in the spatial deposition of N in Germany

(Boltersdorf et al., 2014). Though all the considered biomonitors reflected the nation-wide variation in N deposition, the lichens were found appropriately reflecting the resident N deposition originating from agriculture-related practices through $\delta^{15}N$ assessments (Boltersdorf et al., 2014).

Bermejo-Orduna et al. (2014) studied the impact of vehicular emissions on a conifer forest, by measuring the contents of the trace elements, N, C, and their stable isotopes ($\delta^{15}N$ and $\delta^{13}C$), in the epiphytic lichen, *Letharia vulpina*. The study found that the epiphytic lichen *L. vulpina* appropriately reflected the effect of vehicular exhaust as the primary source of nitrogen pollution through thallus accumulation of N, the typical stable isotopic signature of NOx, and in a reduction of the lichen cover. The study validated the role of *L. vulpina* as a valuable biomonitor in the assessment of N deposition and source identification.

Fogel et al. (2008) in their $\delta^{15}N$ values studies in the mangrove tissues and lichens in a mangrove forest ecosystem, with experimental P-supplementary addition, found that the negative values were due to the microbial source of the nitrogen in these oligotrophic habitats.

Crittenden et al. (2015) in their studies on nitrogen (N) eutrophication by Adèlie penguin rookery at Cape Hallett, northern Victoria Land, Antarctica using *Usnea sphacelate*, *Umbilicaria decussata*, and *Xanthomendoza borealis* found that the rookery was ^{15}N-enriched. The stable isotope ratio of nitrogen $\delta^{15}N$ ($^{15}N:^{14}N$) and N:P mass ratio studies further elucidated that Antarctic lichens were capable of indicating the spatial deposition of ammonia from penguin rookeries (Crittenden et al., 2015). The study serves as an appropriate example of the impact of N transfer from marine habitats to an N-limited terrestrial ecosystem.

Cocozza et al. (2016) assessed the nitrogen stable ratio ($\delta^{15}N$ values) in the thalli of the foliose lichen, *Xantoria parietina* in plots situated on increasing distance from an incinerator plant situated in the Venafro Plain, Molise, Central Italy. The study found that the values of $\delta^{15}N$ values decreased in the samples of *X. parietina* as the distance from the incinerator plant increased (i.e., -1 to $0.2 > 3$ to $6.3 > -1.6$ to 0.2) with main input as NOx.

Ciężka et al. (2016) assessed the sulfur (S), carbon (C), nitrogen (N) content, and the stable isotopic ratio ($\delta^{13}C$, $\delta^{15}N$) in the lichen *Hypogymnia physodes* collected from 11 sites from the Karkonoski National Park in the vicinity of Karpacz city, SW Poland to assess the temporal changes in the air quality. The study recorded the influence of anthropogenic influence in sulfur, carbon, and nitrogen accumulation in the lichen, and the $\delta^{15}N$ values (i.e., -6.8 to -13.0) suggest vehicular exhaust as the main source of N deposition.

Munzi et al. (2016) assessed the efficiency of the stable isotope ratio of nitrogen $\delta^{15}N$ ($N^{15}:N^{14}$) to determine the N pollution source. The experiment exposed sensitive lichen *Evernia prunastri* and tolerant lichen species *Xanthoria parietina* for ten weeks and *Cladonia portentosa* for 11 years or 6 months, to different forms and doses of N in an experiment, and physiological parameters (Chlorophyll a fluorescence), total N, and δ^{15}N were measured (Munzi et al., 2016). The experiment found a specific correlation between the lichen N content and δ^{15}N, exhibiting the source's signature (Munzi et al., 2016). The study concluded the nitrogen stable isotopic signature in lichens as a potential parameter to be used to determine the spatial distribution of N sources at the landscape level with mixed land use.

Pinho et al. (2017) analyzed the nitrogen content and nitrogen stable isotope ratio (δ^{15}N) in lichen *Parmotrema hypoleucinum* collected from 43 sites in Alentejo litoral region, Portugal, to categorize land-use types. The study found that the sites with dominant agricultural sources emitted large amounts of N with very negative δ^{15}N values, probably as NHx (Pinho et al., 2017). Urban areas recorded substantial N-deposition with lesser δ^{15}N negative values signature of NO_x (Pinho et al., 2017). The site with a larger impact of oceanic N deposition recorded low N contentment with positive δ^{15}N values (Pinho et al., 2017). The study concluded that the quantitative N content and N stable isotope ratio from lichen thallus provide valuable information on the form and amount of atmospheric N emitted by distinct land-use types.

Hogan et al. (2017) assessed the total nitrogen and sulfur content and stable isotopic ratios (δ^{15}N, δ^{15}S) in the lichen *Cladia retipora* in Tasmania to study the effect of anthropogenic deposition of N and S. The deposition of both N and S were extremely low in all 66 samples collected. The very small negative δ^{15}N value (-6.9 to -0.8, mean -2.8) indicated the anthropogenic influence deposition which was not found correlated with the total N deposition in the lichen thallus. The study hereby concludes that N stable isotopic ratios of *C. retipora* in Tasmania are indicative of anthropogenic influences on atmospheric background N deposition and lichen can be used as an appropriate indicator for it.

With background understanding from previously reported findings that increased available nitrogen (N) increases chitin concentration in lichens, Munzi et al. (2017a) investigated the prospect of an increase in accumulation of chitin as a mechanism of N tolerance in lichens both at inter- and intra-specific levels. The study employed lichen species with different ecological N tolerances, i.e., the tolerant lichen species—*Xanthoria parietina* and *Parmotrema hypoleucinum*, and the sensitive lichens—*Evernia prunastri* and *Usnea* sp., and thalli of *X. parietina* and *P. hypoleucinum* from sites with different availabilities of nitrogen. The total nitrogen, chitin, and ergosterol contents were measured in the sample thalli of lichen species selected (Munzi et al., 2017a). The stable isotope analysis (δ^{15}N) found agriculture as the predominant source of N in the study area. The nitrogen and chitin contents were higher in the tolerant species than of the sensitive species (inter-specific level), and in the lichen, samples collected from the polluted sites than in lichen samples collected from comparatively low polluted sites (intra-specific level)

(Munzi et al., 2017a). The study concluded that chitin accumulation contributes to N stress tolerance in lichens, where the excess N is stored as chitin in the cell walls of tolerant species.

Simonova et al. (2018) studied the N-deposition in the epiphytic lichens—*Hypogymnia physodes*, *Flavoparmelia caperata*, *Physcia stellaris* in two Siberian cities Tomsk and Prokopyevsk, Russia by estimating the lichen $\delta^{15}N$ ratio. The $\delta^{15}N$ ratio in Tomsk (−5to +2; mean value −1.3) and Prokopyevsk (−11 to +4; mean value −5) were indicative of the typical isotopic signature of NOx mainly produced by anthropogenic vehicular and industrial exhaust which was higher in Tomsk than in Prokopyevsk. The study highlighted the role of the $\delta^{15}N$ ratio as an appropriate tool for detecting the N-deposition source.

Barre et al. (2018) studied the multi-element (C, N, Pb, Hg) isotope signature in both fruticose lichens (*Evernia prunastri*, *Usnea* sp.) and foliose lichens (*Parmelia sulcata*, *Parmelia caperata*, *Hypogymnia physodes*) in SW France. The $\delta^{15}N$ signature in the lichens studied were able to differentiate different sources linked with the land use of the lichen's habitat, where the higher negative ratios were found linked with NHx originating from agriculture and other forest processes and the lower negative ratios were associated with the NOx in the air emanating from the vehicular and industrial exhaust. The $\delta^{15}N$ signature was found conclusive in differentiating between the source of the origin of nitrogen pollution which was accumulated in lichens through the long-range deposition and the associated land use.

Manninen (2018) studied the effect of nitrogen (N) and sulfur (S) pollution on the acidophytic lichens growing on the conifers in the Helsinki metropolitan area. The study found that the road traffic-derived N deposition decrease the lichen diversity in the area (Manninen, 2018). Though the deposition of both oxidized and reduced N increased in lichens, the stable N isotope ratio ($\delta^{15}N$) supported the increased uptake of oxidized N originating from the vehicular exhaust as the main factor affecting the lichen diversity in the area (Manninen, 2018). The study further updated the critical loads of nitrogen and sulfur and substantiated that the tolerant lichens must be studied frequently for air quality monitoring in urban habitats.

Díaz-Álvarez and de la Barrera (2018) studied ambient air nitrogen deposition in the Mexico valley using biomonitors such as lichens along with mosses and bromeliad. The studied biomonitors reflected the spatial deposition of nitrogen content, C:N ratio, and the $\delta^{15}N$ to both season and site in the samples. The stable isotope ratio $\delta^{15}N$ concluded the mixed source of N, wet deposition (NH_X) as the main source of nitrogen pollution in rural areas along with dry deposition (NO_x) from the vehicular exhaust.

Munzi et al. (2019) studied the N stable isotope ratio (^{15}N:^{14}N) $\delta^{15}N$ in experimental manipulation with different forms (i.e., NH_3, NH_4^+, NO_3^-) and doses of N in thalli of the sensitive *Evernia prunastri* and the tolerant *Xanthoria parietina*, for ten weeks, to establish a direct correlation between lichen $\delta^{15}N$ and atmospheric $\delta^{15}N$. Different parameters, such as total N, δ^{15}N, and chlorophyll-a fluorescence, were measured against the treatments.

The δ^{15}N signature in the sensitive lichen species *E. prunastri* responded primarily to NH_3, followed by NH_4^+, with the lowest response to the NO_3^- (Munzi et al., 2019), which can be used as an appropriate tool for the interpretation of the spatial distribution of NH_3 sources (Munzi et al., 2019). Hoffman et al. (2019) studied the nitrogen (N) deposition amount, C:N ratios, and the δ^{15}N in lichens *Usnea lapponica*, *Letharia vulpina*, canopy throughfall, and bulk monitors of the montane conifer forests of the Greater Yellowstone Ecosystem. The stable isotope ratios of nitrogen were calculated relative to the atmospheric N_2 standard using delta notation. The δ^{15}N of N deposition helped to distinguish emission sources, as the δ^{15}N of NOx from combustion tends to be high while NH_x from agricultural sources was comparatively low (Hoffman et al., 2019). The study concluded that δ^{15}N can be used as a reliable tool to ascertain the source of the N–deposition for planning further mitigation strategies.

Hurtado et al. (2020) studied the effect of climate change and nitrogen deposition on the categorical functional traits, i.e., photobiont type, growth form, and quantitative functional traits such as chlorophyll a, chlorophyll *b*, and normalized phaeophytinisation index, water use, and nutrient acquisition (%C and %N; carbon and nitrogen isotopic ratio, δ^{13}C and δ^{15}N) of different lichen species along the whole of Europe. Among the photobiont categories cyano- and chlorolichens had the highest and the lowest nitrogen content respectively, whereas tripartite lichens (i.e., having green algae as primary photobiont whereas blue-green algae as secondary photobiont residing in cephalodia) had intermediate nitrogen content. According to their different nitrogen metabolisms, with chlorolichens being limited in nitrogen supply, the lowest δ^{15}N was found in chlorolichens reflecting a higher ammonium uptake. The fruticose lichens recorded the lowest δ^{15}N, probably due to the high thallus surface-volume ratio, which provides a greater number of cation exchange sites, known to be helpful to take up high amounts of ammonium from the atmosphere. The study found that lichen species responded to climate change and nitrogen deposition, which were species- and/or growth form-specific, and justify their biomonitoring potential at the continental scale (Hurtado et al., 2020).

Dörter et al. (2020) analyzed the stable isotope ratios (δ^{15}N) in lichen biomonitors to assess the source of N pollution along with other metals in Turkey. The δ^{15}N values were able to distinguish, agricultural areas, and urbanized regions. The δ^{15}N value was higher through the major roads. The more positive δ^{15}N values indicated the urban N emissions whereas lower or negative δ^{15}N values indicated agriculture and animal husbandry resulting in the emission of NH_x.

The specific values of δ^{15}N have been instrumental in differentiating both the N deposition type as well as their source (i.e., natural, agricultural, and/or fossil fuel combustion) (Table 1). The isotopic signature of δ^{15}N of nitrogen bioaccumulation in lichens helps to discern emission sources because $\delta^{15}N$ of NOx from combustion tends to be high (-5 to $+25‰$) while NHx from agricultural sources tends to be comparatively low (-40 to $-10‰$) (Table 1).

8. Effect of excessive N deposition on lichen food webs of endangered wildlife

Lichen forms a substantial part of the food of wild fauna in the tropics during the winter season. Though there are few studies from Shennongjia nature reserve in the Hubei district of central China, they indicate that the increased N pollution influences the vitality of critical staple wild food-lichens in protected forests. The impacts of nitrogen deposition on the diversity of lichens in a protected area in a forest in Shennongjia Nature Reserve of China where lichens form a considerable part of the diet of the endangered snub-nosed monkey *Rhinopithecus roxellana* (Wang et al., 2015). The study assessed the effect of wet nitrogen deposition by in-situ experiments (i.e., of the solutions of KNO_3, NH_4NO_3, $(NH_4)_2SO_4$) on four common corticolous lichens *Usnea longissima*, *U. betulina*, *Ramalina calicaris* var. *japonica*, and *Parmotrema hypoleucinum* in an in-situ experiment. The physiological parameters such as maximum quantum efficiency of the photosystem II (Fv/Fm), cell membrane integrity (measured as change in electrical conductivity-EC, of lichen dipping solution, and chlorophyll degradation (measured spectrophotometrically as ratio OD_{435}/OD_{415}) were assessed in response to the treatment time. The study found that higher nitrogen concentration led to a significant decrease of PS II quantum efficiency, enhanced chlorophyll degradation, and an increase of EC pointing to greater cell membrane damage. Though the response of lichens was species-specific a common trend of decreased vitality of lichens was observed for an increase in wet nitrogen deposition. Similar studies by Hua et al. (2016) on *Lobaria pulmonaria*, *L. orientalis*, *Usnea luridorufa*, and *U. dasopoga*, found that though the response of different lichen species for measured parameters (i.e., Fv/Fm, EC, and OD_{435}/OD_{415} was species-specific, there was an overall degradation of lichen species physiology for the increase in nitrogen deposition. Wang et al. (2019a,b) studied the effect of increasing nitrogen (N) depositions for one year on lichen species *Usnea luridorufa* a stable diet of endangered monkey species *Rhinopithecus roxellana*. The study measured the growth, survival, and phosphomonoesterase activity of *U. luridorufa* in response to increased N deposition and found a decrease in the viability of lichen species, concluding a decreased availability of lichen to the endangered monkey species (Wang et al., 2019a,b). Wang et al. (2020) highlighted the effect of higher N deposition on the diversity dynamics of lichens which ultimately affects the biodiversity of endangered primate species that consume them in their diet. The study found that higher N deposition on dominant epiphytic lichen species such as *Usnea luridorufa*, *U. longissima*, *U. betulina*, *U. dasopoga*, and *Ramalina calicaris* var. *japonica* decrease the growth rate of the thallus and negatively affects the survival of propagules (Wang et al., 2020). The nutrient analysis of lichen thallus in the experiment revealed that increased nitrogen deposition reduces the nutritional quality of the lichens, reflected as a reduced soluble protein, and soluble sugar levels and increase in the fiber content, which was found to affect the diet selection of the lichen-eating endangered primate species (Wang et al., 2020).

The studies on the effect of N pollution on food-lichen consumed by endangered wildlife highlight that increased N depositing can trigger a negative cascading effect on the lichen communities and the organisms feeding on them in the tropics. Further the studies point towards the need for the development of a proper mitigation strategy to decrease the negative effect of wet nitrogen (i.e., nitrate) deposition in protected areas having endangered fauna using lichens in their diet.

9. Lichen as a sink for nitrogen in the region of excessive deposition

Bird et al. (2019), in their 5-year experimental aerial enrichment studies, assessed the effect of elevated nitrogen (N) deposition from an oil field (Athabasca Oil Sands in Alberta, Canada) on the forests located in the vicinity. The study found that the fibric-humic (FH) material formed under terricolous lichen (i.e., *Cladonia mitis*, *C. stellaris*) mats act as a significant sink of atmospheric N deposition. The FH material restrains the release of N to mineral soil, making these mats essential for the mitigation of the N overload (Bird et al., 2019). The study further emphasized the healthy maintenance of the terricolous lichen mats in the region through unhindered natural replenishment of the necrotic/damaged portion, time to time (Bird et al., 2019).

The studies discussed in this review highlight the important role lichens can play in the early detection of excessive nitrogen deposition. The early detection of nitrogen eutrophication can be used by policymakers to devise management strategies to mitigate it using practical and sustainable methods.

Acknowledgments

The authors here like to thank many researchers who provided their articles and suggestions in preparing the manuscript. The corresponding author Dr. Himanshu Rai would like to dedicate this publication to his father Mr. Murli Rai (01.07.1945-11.05.2021), whose untimely demise though has created an unfillable void, is also a source of inspiration to follow commitments in life honestly and selflessly.

References

Barre, J.P.G., Deletraz, G., Sola-Larrañaga, C., Santamaria, J.M., Bérail, S., Donard, O.F.X., Amouroux, D., 2018. Multi-element isotopic signature (C, N, Pb, Hg) in epiphytic lichens to discriminate atmospheric contamination as a function of land-use characteristics (Pyrénées-Atlantiques, SW France). Environ. Pollut. 243, 961–971. https://doi.org/10.1016/j.envpol.2018.09.003.

Bermejo-Orduna, R., McBride, J.R., Shiraishi, K., Elustondo, D., Lasheras, E., Santamaría, J.M., 2014. Biomonitoring of traffic-related nitrogen pollution using *Letharia vulpina* (L.) hue in the Sierra Nevada, California. Sci. Total Environ. 490, 205–212. https://doi.org/10.1016/j.scitotenv.2014.04.119.

Bird, A., Watmough, S.A., Carson, M.A., Basiliko, N., McDonough, A., 2019. Nitrogen retention of Terricolous lichens in a northern Alberta Jack pine Forest. Ecosystems 22 (6), 1308–1324. https://doi.org/10.1007/s10021-019-00337-1.

Boltersdorf, S., Werner, W., 2013. Source attribution of agriculture-related deposition by using total nitrogen and δ^{15}N in epiphytic lichen tissue, bark and deposition water samples in Germany. Isot. Environ. Health Stud. 49 (2), 197–218. https://doi.org/10.1080/10256016.2013.748051.

Boltersdorf, S.H., Pesch, R., Werner, W., 2014. Comparative use of lichens, mosses and tree bark to evaluate nitrogen deposition in Germany. Environ. Pollut. 189, 43–53. https://doi.org/10.1016/j.envpol.2014.02.017.

Boltersdorf, S.H., Werner, W., 2014. Lichens as a useful mapping tool?—an approach to assess atmospheric N loads in Germany by total N content and stable isotope signature. Environ. Monit. Assess. 186 (8), 4767–4778. https://doi.org/10.1007/s10661-014-3736-3.

Branquinho, C., Matos, P., Pinho, P., 2015. Lichens as ecological indicators to track atmospheric changes: future challenges. In: Lindenmayer, D., Barton, P., Pierson, J. (Eds.), Indicators and Surrogates of Biodiversity and Environmental Change. CSIRO Publishing, Australia, pp. 77–90.

Ciężka, M., Kossowska, M., Paneth, P., Górka, M., 2016. Carbon, nitrogen and sulphur concentration and δ^{13}C, δ^{15}N values in *Hypogymnia physodes* within the montane area – preliminary data. Geosci. Rec. 2 (1), 24. https://doi.org/10.1515/georec-2016-0004.

Cipro, C.V.Z., Yogui, G.T., Bustamante, P., Taniguchi, S., Sericano, J.L., Montone, R.C., 2011. Organic pollutants and their correlation with stable isotopes in vegetation from King George Island, Antarctica. Chemosphere 85 (3), 393–398. https://doi.org/10.1016/j.chemosphere.2011.07.047.

Cocozza, C., Ravera, S., Cherubini, P., Lombardi, F., Marchetti, M., Tognetti, R., 2016. Integrated biomonitoring of airborne pollutants over space and time using tree rings, bark, leaves and epiphytic lichens. Urban For. Urban Green. 17, 177–191. https://doi.org/10.1016/j.ufug.2016.04.008.

Crittenden, P.D., 1989. Nitrogen relations of mat-forming lichens. In: Boddy, L., Marchant, R., Read, D.J. (Eds.), Nitrogen, Phosphorus and Sulphur Utilization by Fungi. Cambridge University Press, Cambridge, pp. 243–268.

Crittenden, P.D., Scrimgeour, C.M., Minnullina, G., Sutton, M.A., Tang, Y.S., Theobald, M.R., 2015. Lichen response to ammonia deposition defines the footprint of a penguin rookery. Biogeochemistry 122 (2), 295–311. https://doi.org/10.1007/s10533-014-0042-7.

Dahlman, L., Persson, J., Palmqvist, K., Näsholm, T., 2004. Organic and inorganic nitrogen uptake in lichens. Planta 219 (3), 459–467. https://doi.org/10.1007/s00425-004-1247-0.

Davies, L., Bates, J.W., Bell, J.N.B., James, P.W., Purvis, O.W., 2007. Diversity and sensitivity of epiphytes to oxides of nitrogen in London. Environ. Pollut. 146 (2), 299–310. https://doi.org/10.1016/j.envpol.2006.03.023.

Dianwu, Z., Anpu, W., 1994. Estimation of anthropogenic ammonia emissions in asia. Atmos. Environ. 28 (4), 689–694. https://doi.org/10.1016/1352-2310(94)90045-0.

Díaz-Álvarez, E.A., de la Barrera, E., 2018. Characterization of nitrogen deposition in a megalopolis by means of atmospheric biomonitors. Sci. Rep. 8 (1), 13569. https://doi.org/10.1038/s41598-018-32000-5.

Dörter, M., Karadeniz, H., Saklangıç, U., Yenisoy-Karakaş, S., 2020. The use of passive lichen biomonitoring in combination with positive matrix factor analysis and stable isotopic ratios to assess the metal pollution sources in throughfall deposition of Bolu plain, Turkey. Ecol. Indic. 113. https://doi.org/10.1016/j.ecolind.2020.106212, 106212.

Ellis, C.J., Crittenden, P.D., Scrimgeour, C.M., 2004. Soil as a potential source of nitrogen for mat-forming lichens. Can. J. Bot. 82 (1), 145–149. https://doi.org/10.1139/b03-144.

Fogel, M.L., Wooller, M.J., Cheeseman, J., Smallwood, B.J., Roberts, Q., Romero, I., Meyers, M.J., 2008. Unusually negative nitrogen isotopic compositions (δ^{15}N) of mangroves and lichens in an oligotrophic, microbially-influenced ecosystem. Biogeosciences 5 (6), 1693–1704. https://doi.org/10.5194/bg-5-1693-2008.

Frati, L., Brunialti, G., Gaudino, S., Pati, A., Rosamilia, S., Loppi, S., 2011. Accumulation of nitrogen and changes in assimilation pigments of lichens transplanted in an agricultural area. Environ. Monit. Assess. 178 (1), 19–24. https://doi.org/10.1007/s10661-010-1667-1.

Frati, L., Brunialti, G., Loppi, S., 2008. Effects of reduced nitrogen compounds on epiphytic lichen communities in Mediterranean Italy. Sci. Total Environ. 407 (1), 630–637. https://doi.org/10.1016/j.scitotenv.2008.07.063.

Frati, L., Caprasecca, E., Santoni, S., Gaggi, C., Guttova, A., Gaudino, S., Pati, A., Rosamilia, S., Pirintsos, S.A., Loppi, S., 2006. Effects of NO_2 and NH_3 from road traffic on epiphytic lichens. Environ. Pollut. 142 (1), 58–64. https://doi.org/10.1016/j.envpol.2005.09.020.

Frati, L., Santoni, S., Nicolardi, V., Gaggi, C., Brunialti, G., Guttova, A., Gaudino, S., Pati, A., Pirintsos, S.-A., Loppi, S., 2007. Lichen biomonitoring of ammonia emission and nitrogen deposition around a pig stockfarm. Environ. Pollut. 146 (2), 311–316. https://doi.org/10.1016/j.envpol.2006.03.029.

Freitag, S., Feldmann, J., Raab, A., Crittenden, P.D., Hogan, E.J., Squier, A.H., Boyd, K.G., Thain, S., 2012. Metabolite profile shifts in the heathland lichen *Cladonia portentosa* in response to N deposition reveal novel biomarkers. Physiol. Plant. 146 (2), 160–172. https://doi.org/10.1111/j.1399-3054.2012.01593.x.

Freitag, S., Hogan, E.J., Crittenden, P.D., Allison, G.G., Thain, S.C., 2011. Alterations in the metabolic fingerprint of Cladonia portentosa in response to atmospheric nitrogen deposition. Physiol. Plant. 143 (2), 107–114. https://doi.org/10.1111/j.1399-3054.2011.01484.x.

Gadsdon, S.R., Power, S.A., 2009. Quantifying local traffic contributions to NO_2 and NH_3 concentrations in natural habitats. Environ. Pollut. 157 (10), 2845–2852. https://doi.org/10.1016/j.envpol.2009.04.010.

Galloway, D.J., 1992. Biodiversity: a lichenological perspective. Biodivers. Conserv. 1 (4), 312–323. https://doi.org/10.1007/BF00693767.

Garty, J., 2001. Biomonitoring atmospheric heavy metals with lichens: theory and application. Crit. Rev. Plant Sci. 20 (4), 309–371. https://doi.org/10.1080/20013591099254.

Geiser, L.H., Neitlich, P.N., 2007. Air pollution and climate gradients in western Oregon and Washington indicated by epiphytic macrolichens. Environ. Pollut. 145 (1), 203–218. https://doi.org/10.1016/j.envpol.2006.03.024.

Geiser, L.H., Nelson, P.R., Jovan, S.E., Root, H.T., Clark, C.M., 2019. Assessing ecological risks from atmospheric deposition of nitrogen and sulfur to US forests using epiphytic macrolichens. Diversity 11 (6), 87. https://doi.org/10.3390/d11060087.

Giordani, P., Calatayud, V., Stofer, S., Seidling, W., Granke, O., Fischer, R., 2014. Detecting the nitrogen critical loads on European forests by means of epiphytic lichens. A signal-to-noise evaluation. For. Ecol. Manag. 311, 29–40. https://doi.org/10.1016/j.foreco.2013.05.048.

Grindon, L.H., 1859. The Manchester Flora. William White, London.

Hauck, M., Wirth, V., 2010. Preference of lichens for shady habitats is correlated with intolerance to high nitrogen levels. Lichenologist 42 (4), 475–484. https://doi.org/10.1017/S0024282910000046.

Hauck, M., de Bruyn, U., Leuschner, C., 2013. Dramatic diversity losses in epiphytic lichens in temperate broad-leaved forests during the last 150 years. Biol. Conserv. 157, 136–145. https://doi.org/10.1016/j.biocon.2012.06.015.

Herzig, R., Schindler, C., Urech, M., et al., 2020. Correction to: Recalibration and validation of the Swiss lichen bioindication methods for air quality assessment. Environ. Sci. Pollut. Res. 27, 28811–28812. https://doi.org/10.1007/s11356-020-09732-x.

Hoffman, A.S., Albeke, S.E., McMurray, J.A., Evans, R.D., Williams, D.G., 2019. Nitrogen deposition sources and patterns in the Greater Yellowstone ecosystem determined from ion exchange resin collectors, lichens, and isotopes. Sci. Total Environ. 683, 709–718. https://doi.org/10.1016/j.scitotenv.2019.05.323.

Hogan, C.M., Proemse, B.C., Barmuta, L.A., 2017. Isotopic fingerprinting of atmospheric nitrogen and sulfur using lichens (*Cladia retipora*) in Tasmania, Australia. Appl. Geochem. 84, 126–132. https://doi.org/10.1016/j.apgeochem.2017.06.007.

Hogan, E.J., Minnullina, G., Sheppard, L.J., Leith, I.D., Crittenden, P.D., 2010. Response of phosphomonoesterase activity in the lichen *Cladonia portentosa* to nitrogen and phosphorus enrichment in a field manipulation experiment. New Phytol. 186 (4), 926–933. https://doi.org/10.1111/j.1469-8137.2010.03221.x.

Holt, E.A., Miller, S.W., 2010. Bioindicators: using organisms to measure environmental impacts. Nat. Educ. Knowl. 3 (10), 8.

Hua, G., Chuan-hua, W., Lin, Y., 2016. Physiological response of four epiphytic lichens in Shennongjia natural reserve to different nitrogen stress. Chin. J. Ecol. 35 (3), 605–611.

Hurtado, P., Prieto, M., Martínez-Vilalta, J., Giordani, P., Aragón, G., López-Angulo, J., Košuthová, A., Merinero, S., Díaz-Peña, E.M., Rosas, T., Benesperi, R., 2020. Disentangling functional trait variation and covariation in epiphytic lichens along a continent-wide latitudinal gradient. Proc. R. Soc. B Biol. Sci. 287 (1922), 20192862. https://doi.org/10.1098/rspb.2019.2862.

Johansson, O., Palmqvist, K., Olofsson, J., 2012. Nitrogen deposition drives lichen community changes through differential species responses. Glob. Chang. Biol. 18 (8), 2626–2635. https://doi.org/10.1111/j.1365-2486.2012.02723.x.

Kershaw, K.A., 1985. Nitrogen fixation in lichens. In: Kershaw, K.A. (Ed.), Physiological Ecology of Lichens. Cambridge Studies in Ecology. Cambridge University Press, Cambridge, UK, pp. 102–140.

Krupa, S.V., 2003. Effects of atmospheric ammonia (NH_3) on terrestrial vegetation: a review. Environ. Pollut. 124 (2), 179–221. https://doi.org/10.1016/S0269-7491(02)00434-7.

Lee, Y.I., Lim, H.S., Yoon, H.I., 2009. Carbon and nitrogen isotope composition of vegetation on King George Island, maritime Antarctic. Polar Biol. 32 (11), 1607–1615. https://doi.org/10.1007/s00300-009-0659-5.

Loppi, S., 2019. May the diversity of epiphytic lichens be used in environmental forensics? Diversity 11 (3), 36. https://doi.org/10.3390/d11030036.

Manninen, S., 2018. Deriving nitrogen critical levels and loads based on the responses of acidophytic lichen communities on boreal urban *Pinus sylvestris* trunks. Sci. Total Environ. 613–614, 751–762. https://doi.org/10.1016/j.scitotenv.2017.09.150.

Markert, B., Wünschmann, S., 2011. Bioindicators and biomonitors: use of organisms to observe the influence of chemicals on the environment. In: Schröder, P., Collins, C. (Eds.), Organic Xenobiotics and Plants. Plant Ecophysiology. vol. 8. Springer, Dordrecht, pp. 217–236, https://doi.org/10.1007/978-90-481-9852-8_10.

Markert, B., Wünschmann, S., Baltrėnaitė, E., 2012. Innovative observation of the environment. Bioindicators and biomonitors: definitions, strategies and applications. J. Environ. Eng. Landsc. Manag. 20 (3), 221–239. https://doi.org/10.3846/16486897.2011.633338.

Mayer, W., Pfefferkorn-Dellali, V., Türk, R., Dullinger, S., Mirtl, M., Dirnböck, T., 2013. Significant decrease in epiphytic lichen diversity in a remote area in the European Alps, Austria. Basic Appl. Ecol. 14 (5), 396–403. https://doi.org/10.1016/j.baae.2013.05.006.

McMurray, J.A., Roberts, D.W., Geiser, L.H., 2015. Epiphytic lichen indication of nitrogen deposition and climate in the northern rocky mountains, USA. Ecol. Indic. 49, 154–161. https://doi.org/10.1016/j.ecolind.2014.10.015.

Mukhopadhyay, S., Dutta, R., Das, P., 2020. A critical review on plant biomonitors for determination of polycyclic aromatic hydrocarbons (PAHs) in air through solvent extraction techniques. Chemosphere 251. https://doi.org/10.1016/j.chemosphere.2020.126441, 126441.

Munzi, S., Branquinho, C., Cruz, C., Loppi, S., 2013a. Nitrogen tolerance in the lichen *Xanthoria parietina*: the sensitive side of a resistant species. Funct. Plant Biol. 40 (3), 237–243. https://doi.org/10.1071/FP12127.

Munzi, S., Branquinho, C., Cruz, C., Máguas, C., Leith, I., Sheppard, L., Sutton, M., 2016. Is lichen $\delta^{15}N$ an indicator of nitrogen pollution and a surrogate of nitrogen atmospheric composition? Evidence from manipulative experiments. In: Proceedings of the 2016 International Nitrogen Initiative Conference, "Solutions to Improve Nitrogen use Efficiency for the World", 4–8 December 2016, Melbourne, Australia.

Munzi, S., Branquinho, C., Cruz, C., Máguas, C., Leith, I.D., Sheppard, L.J., Sutton, M.A., 2019. $\delta^{15}N$ of lichens reflects the isotopic signature of ammonia source. Sci. Total Environ. 653, 698–704. https://doi.org/10.1016/j.scitotenv.2018.11.010.

Munzi, S., Cruz, C., Branquinho, C., Cai, G., Faleri, C., Parrotta, L., Bini, L., Gagliardi, A., Leith, I.D., Sheppard, L.J., 2020. More tolerant than expected: taking into account the ability of *Cladonia portentosa* to cope with increased nitrogen availability in environmental policy. Ecol. Indic. 119. https://doi.org/10.1016/j.ecolind.2020.106817, 106817.

Munzi, S., Cruz, C., Branquinho, C., Pinho, P., Leith, I.D., Sheppard, L.J., 2014. Can ammonia tolerance amongst lichen functional groups be explained by physiological responses? Environ. Pollut. 187, 206–209. https://doi.org/10.1016/j.envpol.2014.01.009.

Munzi, S., Cruz, C., Maia, R., Máguas, C., Perestrello-Ramos, M.M., Branquinho, C., 2017a. Intra- and inter-specific variations in chitin in lichens along a N-deposition gradient. Environ. Sci. Pollut. Res. 24 (36), 28065–28071. https://doi.org/10.1007/s11356-017-0378-3.

Munzi, S., Loppi, S., Cruz, C., Branquinho, C., 2011. Do lichens have "memory" of their native nitrogen environment? Planta 233 (2), 333–342. https://doi.org/10.1007/s00425-010-1300-0.

Munzi, S., Paoli, L., Fiorini, E., Loppi, S., 2012. Physiological response of the epiphytic lichen *Evernia prunastri* (L.) Ach. to ecologically relevant nitrogen concentrations. Environ. Pollut. 171, 25–29. https://doi.org/10.1016/j.envpol.2012.07.001.

Munzi, S., Pirintsos, S.A., Loppi, S., 2009a. Chlorophyll degradation and inhibition of polyamine biosynthesis in the lichen *Xanthoria parietina* under nitrogen stress. Ecotoxicol. Environ. Saf. 72 (2), 281–285. https://doi.org/10.1016/j.ecoenv.2008.04.013.

Munzi, S., Pisani, T., Loppi, S., 2009b. The integrity of lichen cell membrane as a suitable parameter for monitoring biological effects of acute nitrogen pollution. Ecotoxicol. Environ. Saf. 72 (7), 2009–2012. https://doi.org/10.1016/j.ecoenv.2009.05.005.

Munzi, S., Pisani, T., Paoli, L., Loppi, S., 2010. Time- and dose-dependency of the effects of nitrogen pollution on lichens. Ecotoxicol. Environ. Saf. 73 (7), 1785–1788. https://doi.org/10.1016/j.ecoenv.2010.07.042.

Munzi, S., Pisani, T., Paoli, L., Renzi, M., Loppi, S., 2013b. Effect of nitrogen supply on the C/N balance in the lichen *Evernia prunastri* (L.) ach. Turk. J. Biol. 37 (2), 165–170. https://doi.org/10.3906/biy-1205-4.

Munzi, S., Sheppard, L.J., Leith, I.D., Cruz, C., Branquinho, C., Bini, L., Gagliardi, A., Cai, G., Parrotta, L., 2017b. The cost of surviving nitrogen excess: energy and protein demand in the lichen *Cladonia portentosa* as revealed by proteomic analysis. Planta 245 (4), 819–833. https://doi.org/10.1007/s00425-017-2647-2.

Nylander, M.W., 1866. Les Lichens Du Jardin Du Luxembourg. Bull. Bot. Soc. Fr. 13 (7), 364–371. https://doi.org/10.1080/00378941.1866.10827433.

Olivier, J.G.J., Bouwman, A.F., Van der Hoek, K.W., Berdowski, J.J.M., 1998. Global air emission inventories for anthropogenic sources of NO_x, NH_3 and N_2O in 1990. Environ. Pollut. 102 (1, suppl 1), 135–148. https://doi.org/10.1016/S0269-7491(98)80026-2.

Parmar, T.K., Rawtani, D., Agrawal, Y.K., 2016. Bioindicators: the natural indicator of environmental pollution. Front. Life Sci. 9 (2), 110–118. https://doi.org/10.1080/21553769.2016.1162753.

Pawlik-Skowrońska, B., Di Toppi, L.S., Favali, M.A., Fossati, F., Pirszel, J., Skowroński, T., 2002. Lichens respond to heavy metals by phytochelatin synthesis. New Phytol. 156, 95–102. https://doi.org/10.1046/j.1469-8137.2002.00498.x.

Pinho, P., Barros, C., Augusto, S., Pereira, M.J., Máguas, C., Branquinho, C., 2017. Using nitrogen concentration and isotopic composition in lichens to spatially assess the relative contribution of atmospheric nitrogen sources in complex landscapes. Environ. Pollut. 230, 632–638. https://doi.org/10.1016/j.envpol.2017.06.102.

Pinho, P., Llop, E., Ribeiro, M.C., Cruz, C., Soares, A., Pereira, M.J., Branquinho, C., 2014a. Tools for determining critical levels of atmospheric ammonia under the influence of multiple disturbances. Environ. Pollut. 188, 88–93. https://doi.org/10.1016/j.envpol.2014.01.024.

Pinho, P., Martins-Loução, M.-A., Máguas, C., Branquinho, C., 2014b. Calibrating total nitrogen concentration in lichens with emissions of reduced nitrogen at the regional scale. In: Sutton, M.A., Mason, K.E., Sheppard, L.J., Sverdrup, H., Haeuber, R., Hicks, W.K. (Eds.), Nitrogen Deposition, Critical Loads and Biodiversity. Springer Netherlands, Dordrecht, pp. 217–227, https://doi.org/10.1007/978-94-007-7939-6_24.

Pinho, P., Theobald, M.R., Dias, T., Tang, Y.S., Cruz, C., Martins-Loução, M.A., Máguas, C., Sutton, M., Branquinho, C., 2012. Critical loads of nitrogen deposition and critical levels of atmospheric ammonia for semi-natural Mediterranean evergreen woodlands. Biogeosciences 9 (3), 1205–1215. https://doi.org/10.5194/bg-9-1205-2012.

Proemse, B.C., Mayer, B., 2012. Chapter 11: Tracing industrial nitrogen and sulfur emissions in the Athabasca oil sands region using stable isotopes. In: Percy, K.E. (Ed.), Developments in Environmental Science. Vol. 11. Elsevier, pp. 243–266, https://doi.org/10.1016/B978-0-08-097760-7.00011-1.

Purvis, O.W., Chimonides, P.J., Jeffries, T.E., Jones, G.C., Read, H., Spiro, B., 2005. Investigating biogeochemical signatures in the lichen *Parmelia sulcata* at Burnham beeches, Buckinghamshire, England. Lichenologist 37 (4), 329–344. https://doi.org/10.1017/S0024282905015288.

Riddell, J., Nash, T.H., Padgett, P., 2008. The effect of HNO_3 gas on the lichen *Ramalina menziesii*. Flora 203 (1), 47–54. https://doi.org/10.1016/j.flora.2007.10.001.

Rockstrom, J., Steffen, W., Noone, K., Persson, A., Chapin, F.S., Lambin, E.F., Lenton, T.M., Scheffer, M., Folke, C., Schellnhuber, H.J., Nykvist, B., de Wit, C.A., Hughes, T., van der Leeuw, S., Rodhe, H., Sorlin, S., Snyder, P.K., Costanza, R., Svedin, U., Falkenmark, M., Karlberg, L., Corell, R.W., Fabry, V.J., Hansen, J., Walker, B., Liverman, D., Richardson, K., Crutzen, P., Foley, J.A., 2009. A safe operating space for humanity. Nature 461, 472–475. https://doi.org/10.1038/461472a.

Root, H.T., Geiser, L.H., Fenn, M.E., Jovan, S., Hutten, M.A., Ahuja, S., Dillman, K., Schirokauer, D., Berryman, S., McMurray, J.A., 2013. A simple tool for estimating throughfall nitrogen deposition in forests of western North America using lichens. For. Ecol. Manag. 306, 1–8. https://doi.org/10.1016/j.foreco.2013.06.028.

Sarret, G., Manceau, A., Cuny, D., Van Haluwyn, C., Déruelle, S., Hazemann, J.L., Soldo, Y., Eybert-Bérard, L., Menthonnex, J.J., 1998. Mechanisms of lichen resistance to metallic pollution. Environ. Sci. Technol. 32 (21), 3325–3330. https://doi.org/10.1021/es970718n.

Satya, Upreti, D.K., 2009. Correlation among carbon, nitrogen, sulphur and physiological parameters of *Rinodina sophodes* found at Kanpur city, India. J. Hazard. Mater. 169 (1), 1088–1092. https://doi.org/10.1016/j.jhazmat.2009.04.063.

Seed, L., Wolseley, P., Gosling, L., Davies, L., Power, S.A., 2013. Modelling relationships between lichen bioindicators, air quality and climate on a national scale: results from the UK OPAL air survey. Environ. Pollut. 182, 437–447. https://doi.org/10.1016/j.envpol.2013.07.045.

Serrano, H.C., Oliveira, M.A., Barros, C., Augusto, A.S., Pereira, M.J., Pinho, P., Branquinho, C., 2019. Measuring and mapping the effectiveness of the European air quality directive in reducing N and S deposition at the ecosystem level. Sci. Total Environ. 647, 1531–1538. https://doi.org/10.1016/j.scitotenv.2018.08.059.

Silberstein, L., Siegel, B., Siegel, S., Mukhtar, A., Galun, M., 1996. Comparative studies on *Xanthoria parietina*, a pollution resistant lichen, and *Ramalina duriaei*, a sensitive species. II. Evaluation of possible air pollution-protection mechanisms. Lichenologist 28 (4), 367–383. https://doi.org/10.1006/lich.1996.0034.

Simonova, G.V., Kalashnikova, D.A., Melkov, V.N., 2018. Air quality biomonitoring with epiphytic lichens and mosses. In: 24th International Symposium on Atmospheric and Ocean Optics: Atmospheric Physics, vol. 10833, 108335Q. International Society for Optics and Photonics, https://doi.org/10.1117/12.2504559.

Sparrius, L.B., 2007. Response of epiphytic lichen communities to decreasing ammonia air concentrations in a moderately polluted area of the Netherlands. Environ. Pollut. 146 (2), 375–379. https://doi.org/10.1016/j.envpol.2006.03.045.

Sparrius, L.B., Kooijman, A.M., 2013. Nitrogen deposition and soil carbon content affect nitrogen mineralization during primary succession in acid inland drift sand vegetation. Plant Soil 364 (1), 219–228. https://doi.org/10.1007/s11104-012-1351-z.

Sparrius, L.B., Kooijman, A.M., Riksen, M.P.J.M., Sevink, J., 2013. Effect of geomorphology and nitrogen deposition on rate of vegetation succession in inland drift sands. Appl. Veg. Sci. 16 (3), 379–389. https://doi.org/10.1111/avsc.12018.

Sparrius, L.B., Kooijman, A.M., Sevink, J., 2012. Response of inland dune vegetation to increased nitrogen and phosphorus levels. Appl. Veg. Sci. 16 (1), 40–50. https://doi.org/10.1111/j.1654-109X.2012.01206.x.

Tretiach, M., Adamo, P., Bargagli, R., Baruffo, L., Carletti, L., Crisafulli, P., Giordano, S., Modenesi, P., Orlando, S., Pittao, E., 2007. Lichen and moss bags as monitoring devices in urban areas. Part I: influence of exposure on sample vitality. Environ. Pollut. 146 (2), 380–391. https://doi.org/10.1016/j. envpol.2006.03.046.

Van Damme, M., Clarisse, L., Franco, B., Sutton, M.A., Erisman, J.W., Kruit, R.W., van Zanten, M., Whitburn, S., Hadji-Lazaro, J., Hurtmans, D., Clerbaux, C., 2020. Global, regional and national trends of atmospheric ammonia derived from a decadal (2008-2018) satellite record. Environ. Res. Lett. 16. https://doi.org/10.1088/1748-9326/abd5e0, 055017.

Van Damme, M., Clarisse, L., Whitburn, S., Hadji-Lazaro, J., Hurtmans, D., Clerbaux, C., Coheur, P.-F., 2018. Industrial and agricultural ammonia point sources exposed. Nature 564 (7734), 99–103. https:// doi.org/10.1038/s41586-018-0747-1.

Van der Wat, L., Forbes, P.B.C., 2015. Lichens as biomonitors for organic air pollutants. TrAC Trends Anal. Chem. 64, 165–172. https://doi.org/10.1016/j.trac.2014.09.006.

van Herk, C.M., Mathijssen-Spiekman, E.A.M., de Zwart, D., 2003. Long distance nitrogen air pollution effects on lichens in Europe. Lichenologist 35 (4), 347–359. https://doi.org/10.1016/S0024-2829(03) 00036-7.

Wang, C.H., Hou, R., Wang, M., He, G., Li, B.G., Pan, R.L., 2020. Effects of wet atmospheric nitrogen deposition on epiphytic lichens in the subtropical forests of Central China: evaluation of the lichen food supply and quality of two endangered primates. Ecotoxicol. Environ. Saf. 190. https://doi.org/10.1016/ j.ecoenv.2019.110128, 110128.

Wang, C.-H., Munzi, S., Wang, M., Jia, Y.-Z., Tao, W., 2019a. Increasing nitrogen depositions can reduce lichen viability and limit winter food for an endangered Chinese monkey. Basic Appl. Ecol. 34, 55–63. https://doi.org/10.1016/j.baae.2018.10.006.

Wang, C.H., Yang, L., Yuan, Q.L., Munzi, S., Liu, M.J., 2015. Nitrogen sensitivity of four epiphytic lichens from habitats of *Rhinopithecus roxellana* in Shennongjia natural reserve. Chin. J. Ecol. 34 (3), 626–633.

Wang, M., Wang, C., Jia, R., 2019b. The impact of nitrogen deposition on photobiont-mycobiont balance of epiphytic lichens in subtropical forests of Central China. Ecol. Evol. 9 (23), 13468–13476. https://doi. org/10.1002/ece3.5803.

Webb, J., Menzi, H., Pain, B.F., Misselbrook, T.H., Dämmgen, U., Hendriks, H., Döhler, H., 2005. Managing ammonia emissions from livestock production in Europe. Environ. Pollut. 135 (3), 399–406. https://doi.org/10.1016/j.envpol.2004.11.013.

Welden, N.A., Wolseley, P.A., Ashmore, M.R., 2018. Citizen science identifies the effects of nitrogen deposition, climate and tree species on epiphytic lichens across the UK. Environ. Pollut. 232, 80–89. https://doi.org/10.1016/j.envpol.2017.09.020.

Wolseley, P., Sutton, M., Leith, I.D., van Dijk, N., 2010. Epiphytic lichens as indicators of ammonia concentrations across the UK. Bibl. Lichenol. 105, 75–85.

Yatawara, M., Dayananda, N., 2019. Use of corticolous lichens for the assessment of ambient air quality along rural–urban ecosystems of tropics: a study in Sri Lanka. Environ. Monit. Assess. 191 (3), 179. https://doi.org/10.1007/s10661-019-7334-2.

Further reading

Giordani, P., Malaspina, P., 2017. Do tree-related factors mediate the response of lichen functional groups to eutrophication? Plant Biosyst. 151 (6), 1062–1072. https://doi.org/10.1080/11263504.2016.1231141.

Munzi, S., Paoli, L., Fiorini, E., Loppi, S., 2012. Physiological response of the epiphytic lichen *Evernia prunastri* (L.) Ach. to ecologically relevant nitrogen concentrations. Environ. Pollut. 171, 25–29. https://doi. org/10.1016/j.envpol.2012.07.001.

Sujetovienė, G., Galinytė, V., 2016. Effects of the urban environmental conditions on the physiology of lichen and moss. Atmos. Pollut. Res. 7 (4), 611–618. https://doi.org/10.1016/j.apr.2016.02.009.

CHAPTER 14

Molecular markers and genomics assisted breeding for improving crop plants

Manish Kumar Vishwakarma[a,*]**, Punam Singh Yadav**[b]**, Ved Prakash Rai**[c]**, Uttam Kumar**[a]**, and Arun Kumar Joshi**[a]

[a]Borlaug Institute for South Asia-CIMMYT, New Delhi, India
[b]Department of Genetics and Plant Breeding, Institute of Agricultural Sciences, BHU, Varanasi, India
[c]Agricultural Research Station, Navsari Agricultural University, Tanchha, Bharuch, India
*Corresponding author. e-mail address: manishvis@gmail.com

Abbreviations

AB-QTL	advanced backcross QTL
AFLP	amplified fragment length polymorphism
CAPS	cleaved amplified polymorphic sequence
CMS	cytoplasmic male sterility
COS	conserved orthologous set
CSSLs	chromosome segment-substitution lines
DArT	diversity array technology
ESTs	expressed sequence tags
FM	functional markers
GAB	gene-assisted breeding
IBLs	inbred-backcross lines
ILs	introgression lines
InDels	insertion-deletion polymorphisms
MAB	marker-assisted backcrossing
MAGIC	multiparent advanced generation intercross
MARS	marker-assisted recurrent selection
MAS	marker-assisted selection
MTA	marker trait association
NAM	nested association mapping
NILs	near-isogenic lines
QPM	quality protein maize
QTL	quantitative trait loci
RAD	restriction site-associated DNA
RAPD	random amplified polymorphic DNA
RFLP	restriction fragment length polymorphism
RILs	recombinant inbred lines
SFP	single-feature polymorphism
SLS	MAS single large-scale MAS
SNP	single-nucleotide polymorphism
SSD	single-seed descent
SSR	simple sequence repeats

Relationship Between Microbes and the Environment for Sustainable Ecosystem Services, Volume 1
https://doi.org/10.1016/B978-0-323-89938-3.00014-1

1. Introduction

The "marker-assisted selection" term was first time cited around three decades ago in relation to its potential use by Beckmann and Soller (1986), but the theory was conceptualized earlier by Soller and Hazan in 1977. The marker-assisted selection (MAS) became popularized in reference to mapping and tagging genes with molecular markers (Xu and Crouch, 2008), in which a rather small number of non–DNA-based assays were implemented to fix favorable variants at specific loci (Koebner and Summers, 2003). The use of genetic markers as a reliable tool for the plant breeder was recognized by Sax way back in 1923 (Sax, 1923). Probably the application of MAS in plant breeding using DNA markers was first published by Concibido et al. (1996) for soybean cyst nematode (*Heterodera glycines* Ichinohe) resistance. Subsequently, the use and development of MAS in plant breeding almost for all target crops were justified by several researchers (Young and Tanksley, 1989; Ribaut and Hoisington, 1998; Joseph et al. (2004); Vishwakarma et al., 2014, 2016a,b; Yadav et al., 2015; Rai et al., 2019) in four major ways. (i) Some traits have low penetrance or complex inheritance and are expensive or time-consuming to measure, have probably difficult to manage through conventional phenotypic selection. (ii) Traits whose selection rely solely on specific environments or developmental stages that affect the target phenotype expressivity. (iii) To speed up backcross breeding and maintenance of recessive alleles during backcrossing. (iv) Pyramiding several QTL for a single target trait with complex inheritance (such as drought tolerance or other adaptive traits) or multiple monogenic traits (such as pest and disease resistances or quality traits) (Xu and Crouch, 2008). In doing so, there is consideration of four different selection schemes. From two-stage to four-stage selection, the genomic marker deployment would be extended in a stepwise manner to analyze (i) the target allele (ii) flanking markers of target (iii) all carrier chromosome markers, and (iv) all genomic marker loci.

Application of MAS for the target trait during backcrossing can be advantageous in several situations than the conventional phenotypic selection viz. (i) when a phenotypic assessment is often longer and/or more difficult and/or more costly than the molecular genotyping; (ii) when the trait is a quantitative, influenced by environment and having low heritability; (iv) the trait is semi-quantitative, e.g., plant height, adult plant resistance, etc.; (v) the trait is recessive, e.g., *Sr2* gene for stem rust and (vi) when gene pyramiding is necessary (Huang et al., 1997). Several success stories of MAS have been reported in many crops for introgression or pyramiding of major/minor genes/QTL associated with biotic (Kottapalli et al., 2010), abiotic, and quality traits by several workers from time to time (Nayak et al., 2017). The success of MAS relies on several factors, including the degree of association between the molecular marker and the target gene, the genetic base of the trait, the number of target traits to be introgressed, the number of individuals that can be analyzed, and the genetic background in which the target gene has to be transferred, etc. (Melchinger, 1990; Visscher et al., 1996; Hospital and Charcosset, 1997).

Nowadays, genotyping is being outsourced to 3rd party service providers', sampling becomes a bottleneck. To reduce the sampling error and speed up the MAS program, recently in rice, a single seed sampling strategy for MAS has been followed (Arbelaez et al., 2019). Along with this, the speed breeding strategy has also opened the avenue to speed up the different breeding programs (Watson et al., 2018). However, in this chapter, we have included every aspect of MAS in a single chapter, its updates and newly developed tools and techniques utilized in MAS nowadays, and updated about the released varieties through the application of MAS.

2. Trait choice for MAS

This is a very decisive factor for a successful MAS application. It is not mandatory to apply MAS in all cases since some traits are easy to detect in regular conventional breeding. Before application, one should consider some facts about the choice of appropriate traits.

(i) Low heritable traits

Yield is very complex traits, governed by several major and minor genes/QTLs with large and small effects. QTL detection of such traits is difficult due to low and moderate heritability, and QTL environment interaction. Albeit, MAS can be avoided if a strong QTL-environment effect is present (Li et al., 2003).

(ii) Expensive and difficult traits

Application of MAS is useful where phenotyping is not possible or less accurate. It is applicable where phenotyping is costly and time-consuming or may have low penetrance/complicated inheritance as selection for those traits relies on the developmental stage or particular environment (Xu and Crouch, 2008). Some traits where biochemical analyses are required are costly and time taking, in contrast to some traits which are easily scored like rust foliar disease resistance. A list of traits in regard to the skill needed, phenotyping cost, and time has been provided by Brennan and Martin (2007).

(iii) Simple and complex traits

Genetics of the traits also play a very crucial role in the success of the MAS program. Simple genetic traits are easier than complex traits.

(A) *Genetically complex traits*

Some traits are considered to be a very complex trait because it is governed by a small number of major and large number of minor genes/QTLs and form very complex allelic interactions and networks with the G × E interaction. Such traits create difficulty in the application of MAS. So, it is a prerequisite to apply MAS where marker trait association (MTA) is reliable and high.

(B) *Phenotyping of complex traits mapping*

For the better utilization of MAS, MTA should be robust and reliable. In this context, phenotyping should be very precise and perfect which should based on, phenotyping of a large mapping population, conduction of trials for phenotyping at multi-location and in multi environments over several years. Albeit, precision in phenotyping at multi-location and environment is practically very costly and time taking. To resolve this problem, an alternative phenotyping platform was suggested by Lippman and Zamir (2007), where all the yield component traits should be considered as one trait and trial should be organized in the different locations, seasons, and years. For handling and analysis of such data, Plabsoft would be useful, which can handle five types of data like (Germplasm and pedigree data, Phenotypic data for trait and trait complexity, Management of data from field trials, Molecular marker data for all types of common molecular markers, Project and study management data) in all important crops (Heckenberger et al., 2008).

(iv) Blending multiple traits

MAS has been proved to be a potential technique where several traits genes/ QTLs can be stacked in a single genotype. A recent successful example of genes/QTLs stacking is pyramiding of genes conferring resistance/tolerance to blast (*Pi2*, *Pi9*), gall Midge (*Gm1*, *Gm4*), submergence (*Sub1*), and salinity (Saltol) in a released rice variety CRMAS2621-7-1 as Improved Lalat (already introgressed with three BB resistance genes *xa5*, *xa13*, and *Xa21* to supplement the *Xa4* gene present in Improved Lalat). The two major difficulties which were frequently experienced with this technique are: (a) with the increase of donor parents, polymorphism in markers decreases but these problems can be solved by restoring single gene in each backcross, through selfing, or by doubled haploids (DH) production; and (b) handling populations of a suitable size.

(v) Heterosis as a trait

In 1876, Charles Darwin observed that in comparison to self-pollinated plants, cross-pollinated (hybrid) plants have good height, weight, and fertility, this observation led to the conceptual base of heterosis. Three decades later, Shull in 1914 proposed the term "Heterosis" for such stimulating effects of hybridity like increased productiveness and resistance to disease, etc. Now with the advent of molecular markers, genomic region/location of the contributory heterosis traits in plants can be easily identified and enhance the performance of hybrids by introgression of those regions into parental inbred lines through MAS. In the extension of this concept Melchinger et al. (1998) explained heterotic group, quoting same here "as a group of related or unrelated genotypes from the same or different populations, which display similar combining ability and heterotic response when crossed with genotypes from other genetically distinct germplasm groups". Application of molecular markers with the field trials database for identifying heterotic groups and introgression of exotic germplasm systematically have been successfully exemplified in rice and maize crop by Reif et al. (2003) and Aguiar et al. (2008).

3. Selection of marker type for better results through MAS

During selection of individual plants in breeding populations, selection or screening for the target gene/QTLs with the help of linked/gene-specific molecular marker is done. This procedure in MAS is referred as "foreground selection" (Hospital and Charcosset, 1997) or positive selection (Takeuchi et al., 2006). Selection of the suitable markers for the MAS program is of utmost importance, one should keep the following points in mind before marker selection: (1) reproducibility of a marker, (2) liable to automation and high throughput, (3) cost-effective, (4) nature of the marker whether it is dominant or co-dominant. The two categories of markers are functional marker (FM) and random genomic marker (RGM). The success of the MAS program largely depends on the availability of efficient molecular marker(s). For screening large populations, the use of MAS must be economical, user-friendly, and have high reproducibility across laboratories.

(A) Functional markers (FM)

Andersen and Lubberstedt (2003) coined the term functional marker and stated that the FM is gene specific, and there is no probability to lose gene during introgression program due to recombination, this is denoted as gene-assisted selection. These markers are derived through Express Sequence Tags (ESTs), functionally characterized genes, and coding genomic sequence (Rafalski, 2002). These markers can be SSR, SNP, or COS (conserved orthologous set) markers. The COS markers are derived from a highly conserved sequence so can be used, in particular, in other species (Fulton et al., 2002). With the recent availability of functional markers for the gene *Gpc-B1*, the use of MAS has allowed some success in developing wheat genotypes with enhanced grain protein content (GPC) without any yield penalty (Vishwakarma et al., 2014, 2016a,b). There will be more markers like SNPs to become routine in near future, which will aid MAS for foreground traits very efficiently, as well as association mapping studies in most of the crop plants.

(B) Random genomic marker

As described above the second type of marker are more likely to be linked to the trait of interest due to the presence of LD in the population, and known as Random Genomic Markers viz. RFLP (Restriction Fragment Length Polymorphism), RAPD (Random Amplified Polymorphic DNA), AFLP (Amplified Fragment Length Polymorphism), CAPS (Cleaved Amplified Polymorphic Sequence), SSR (Simple Sequence Repeat), SNP (Single Nucleotide Polymorphism), DArT (Diversity array technology) and InDels (insertion-deletion polymorphisms). Above all mentioned markers, the most preferable and vastly used markers are SSR and SNP, because they are feasible and cost-effective, and widely used in genotyping of a large number of segregating populations (Gupta and Varshney, 2000). In such a case, to ensure the high frequency of the target allele in the final backcross generation, adequate population size must be maintained in each backcross. High reproducibility and user-friendliness have made SSR and SNP markers as markers of choice for foreground selection during MAS. For foreground selection Cleaved amplified polymorphism sequence (CAPS) markers are also useful, but are time taking and costly in comparison to SSR or SNP. In recent years, ultra-high-throughput low-cost microarray-based molecular marker systems such as single feature polymorphisms (SFPs) are detected by hybridization of DNA or cRNA to oligonucleotide probes. DArT, and restriction site-associated DNA (RADseq) markers, have also become popular for whole-genome profiling, which generates data for multiple loci in a single assay, hence preferable for background screening (Gupta et al., 2009).

4. Types of markers and their utility in plant breeding

(A) Morphological markers

Morphological marker term are used with regard to visual difference for desired traits and are in use since the beginning of plant breeding. The selection can be done based on visible differences for traits viz. leaf shape and size, flower color, pod color, seed color, seed shape and size, hilum color, awn or panicle type and length, fruit color and shape, rind (peel) color and stripe especially in citrus fruits, flesh color, stem length, and auricle pigmentation especially in wheat crop, etc. These markers can be visualized easily and used in direct/indirect selection in the breeding program. Earlier such markers were used for the linkage mapping by classical linkage analysis (by calculation of genetic distance between the loci governing the linked traits), where genes are arranged on the chromosomes based on co-inheritance of different traits (called linked genes). A very good example of the morphological marker in the selection of high-yielding semi-dwarf wheat and rice. Likewise, Leaf tip necrosis is associated with the *Lr34* gene and Pseudo-black chaff (*Pbc*) pigmented seedling/black chaff is associated with stem rust resistance gene *Sr2* (Fig. 1). Some specific morphological markers are linked and used to identify particular varieties, like in wheat, brown glumes are specific to HD-2329 variety and crooked neck (peduncle) are specific to Kalyan Sona variety. A successful example of these markers in breeding is the introgression of *Rht10* gene (dwarf plant type) that tightly linked to *Ta1* gene (male sterility) was used to identify the male-sterile plants in breeding populations (Liu, 1991). Albeit, there are several limitations with these markers, i.e., they are very limited in numbers, if available not linked to important agronomic traits like yield and quality, sometimes show undesirable effects in plant population during growth and development. In tomatoes, about 1300 morphological and physiological results have been reported viz. fruit ripening, male sterility, fruit abscission, and disease-resistance genes but less than 400 have been mapped.

<table>
<tr><td>Leaf tip Necrosis (LTN) associated with leaf rust resistance gene Lr34 in wheat</td><td>Pseudo-black chaff (PBC) associated with stem rust resistance gene Sr2 in wheat</td></tr>
</table>

Fig. 1 Morphological markers for some traits in wheat.

(B) Protein-based or biochemical markers

These markers show the variation for protein or isozymes in electrophoresis through different molecular weights and electrophoretic mobility. It reveals the changes in amino acid sequences of concerned protein that make differences between wild-type and normal ones. It is actually differs in allele not in gene (Xu, 2010). These markers can be utilized as genetic markers to map other genes after locating on chromosomes. It is generally used in seed purity tests, and identification of specific strains, which makes its uses limited. A very good example of such a marker is the high molecular weight glutenin subunit (HMW–GS) in wheat. Payne et al. (1987) detected a correlation between HMW–GS and gluten strength and made a numeric scale to detect bread-making quality based on different subunits, which was due to different allelic variation at the Glu–D1 locus. For more detail, see Payne et al. (1987).

(C) Cytological markers

Cytological markers are the structural variation in chromosome number, shape, size, position, and banding pattern. It shows distribution and differences in euchromatin and heterochromatin. Moreover, it refers to the chromosomal banding produced by different stains; for example, G bands are produced by Giemsa stain, R bands are the reversed G bands and Q bands are produced by quinacrine hydrochloride. It is also used for linkage group identification and physical mapping, mutation, and characterization of chromosomes based on these landmarks. These physical maps are helpful in the foundation of a genetic linkage map with the assistance of a molecular marker. Albeit, the use of these cytological maps are limited in course of plant breeding. The presence of a knob at the end of the long arm of chromosome 9 in maize is a well-known example of cytological markers.

(D) **DNA markers**: DNA markers or molecular markers are categorized into three classes; hybridization-based markers (RFLP), PCR-based markers (RAPD, SSR, SNP) and both, hybridization and PCR-based markers (AFLP) (Fig. 2; modified from Staub et al., 1996)

(i) *Restriction Fragment Length Polymorphism (RFLP)*

RFLP was formulated by Botstein et al. (1980). RFLP is hybridization-based co-dominant markers. In plants, tens to thousand RFLP loci have been reported. It has two alleles at each locus and hence has a low level of polymorphism. For genotyping, restriction enzyme is used to cut DNA, fragments size is separated on agarose gel, blotted onto a membrane, hybridized, and brought to specific probes (developed from genomic or c-DNA libraries). Although. It is the first DNA marker and laid the foundation of the MAS concept. So many studies have been done with the RFLP markers (Paterson et al., 1988, 1991). Nowadays its utility is in the study of synteny of genetically close species (**Vilanova et al., 2008**).

(ii) *Random Amplified Polymorphic DNA (RAPD)*

RAPD markers are PCR-based markers which was developed by Williams et al. (1990). RAPD is 10 nucleotides long random primers, separated in the agarose gel stained with EtBr (Ethidium Bromide). Scoring of the marker is done in binary format, 1 for presence and 0 for absence of bands, and it is dominant type of marker system. The RAPD loci varies between tens to several hundred (not all polymorphic) in the plant genome. This marker system is user-friendly, but due to the low level of polymorphism, sensitivity, and repeatability, it is less preferable in the MAS program. Several studies have been conducted with the RAPD markers for different purposes (Paran and Michelmore, 1993).

(iii) *Amplified Fragment Length Polymorphism (AFLP)*

AFLP markers were formulated by Vos et al. (1995). Restriction enzyme "4 or 6 cutter" is used to cut total DNA or c-DNA. These DNA fragments are ligated with specific universal adaptors, homologous to the specific restriction sites. After that, universal primers are used to amplify these ligated flanking fragments in PCR. The amplified product is labeled with a radio-active or fluorescent dye and size separated in an acrylamide gel. AFLP is a dominant and multilocus marker easy to use in genotyping, and is reliable. The dominant mode of inheritance and low polymorphism are limitations with this marker system due to which are less suitable for MAS. The unique quality of this marker system is that it can be used in the identification of DNA methylation by the use of restriction enzymes pair (methylation resistance and sensitivity). Various studies have been performed with the AFLP markers application.

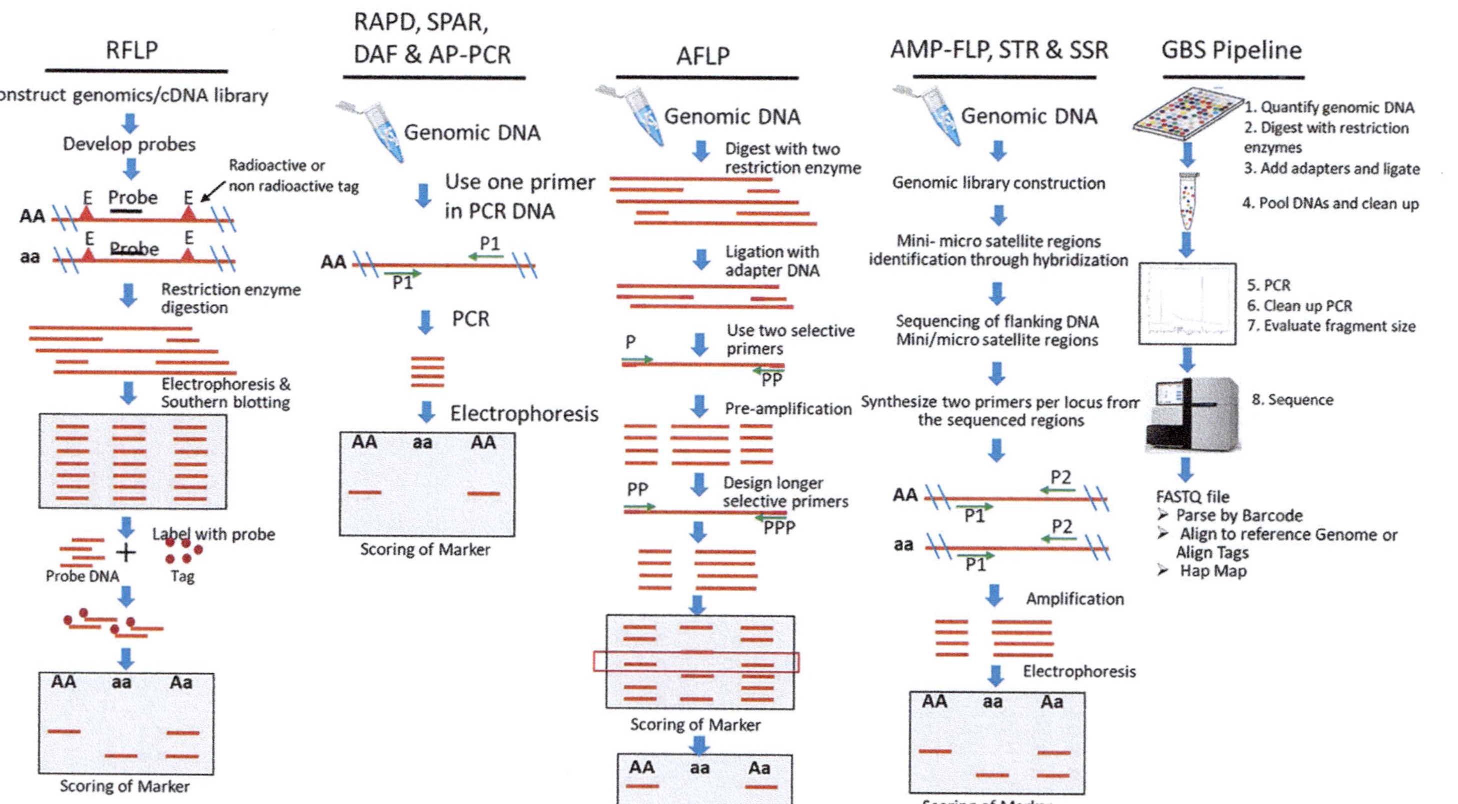

Fig. 2 Types of molecular markers based on hybridization, PCR and both. *(Modified from Staub, J.E., Serquen, F.C., Gupta, M., 1996. Genetic markers, map construction, and their application in plant breeding. Hort Sci. 31(5), 729–741.)*

(iv) *Simple sequence repeats (SSRs)*

SSRs or microsatellites were formulated by Litt and Luty (1989) but first used in plants by Akkaya et al. (1992). This marker amplifies specific loci of 1–6 nucleotide simple repeats with the set of flanking primers. For the primer identification, screen the mini or micro satellite region of genomic libraries with the probe of various combinations of 1–6 nucleotides. Then synthesize the primers homologous to the flanking sequence. This marker can be separated in agarose or poly-acrylamide gel electrophoresis with the EtBr staining. SSRs are generally co-dominant (distinguish heterozygotes from homozygotes), highly polymorphic, reliable, and user-friendly. The availability of SSR primers varies according to plant species from a few to several hundred, but lack of polymorphism of SSRs between genotypes may limit their application (Rai et al., 2013). Various studies have been performed using SSRs for different purposes viz. genetic diversity, linkage establishment, QTL mapping, MAS, etc. (Kalia et al., 2011).

(v) *Single Nucleotide Polymorphism (SNPs)*

SNPs are a single nucleotide difference in the sequence of genomic DNA which arises due to point mutations. SNPs are co-dominant in inheritance, mostly have two alleles in each locus (low level of polymorphism) highly abundant. SNPs are generally brought through point mutation, change of single nucleotide or substitute of one nucleotide with another, or insertion/deletion (InDels) of a single nucleotide. There are several SNP genotyping platforms such as Genotyping by Sequencing (GBS), Illumina Goldengate, KASPar, and SNPline genotyping systems, SNPline system, iPLEX Gold system (is based on the MassARRAY system which uses primer extension chemistry and matrix-assisted laser desorption/ionization-time of flight (MALDITOF) mass spectrometry, Affymetrix GeneChip, the Illumina Bead-Chips. SNPs have been gained special attention after several plant genome sequencing and vastly used in plants (Kumar et al., 2012; Pandey et al., 2017; Vishwakarma et al., 2016a,b). The number of studies using SNPs in plant science is rising day by day.

(vi) *Diversity Arrays Technology (DArT)*

DArT markers are microarray hybridizations based technology (Jaccoud et al., 2001). It was developed and utilized first in the study of barley crops by Wenzl et al. (2004). DArT arrays are spotted on polylysine-covered slides, rendered from genomic libraries through the amplification of candidate/random clones. Genomic DNA from the genotype samples ligated with suitable adapter after restricted with Pst1 and 4 cutters. The fragments generated by the cut are amplified. Final samples hybridized to the microarray after denaturing them and scanned with a confocal laser scanner like Tecan LS300 (Grodig,

Austria) or Affymetrix 428 (Santa clara, CA). Several researches have been done with the use of DArT and sequence-based (DArTseq) platform in crop plants (Vishwakarma et al., 2016a,b; Shasidhar et al., 2017). DArT is very useful for MAS because it provides wide genome coverage, ultra-high-throughput, and a low-cost marker system (US$0.1 per data point) (Gupta et al., 2009). Such a platform can be smartly utilized in a breeding program to select promising individuals that have the desired positive QTLs and recover the recurrent parent genome.

5. Types of selection and their procedure

(i) Marker-Assisted/Aided Selection (MAS)

Marker-assisted selection is a smart breeding procedure in which, molecular marker identification and selection are incorporated into the classical breeding program. It begins with the identification of molecular marker(s) linked with desirable gene/QTLs in a donor parent and show polymorphism with recipient parent. There is a chance of false hybrids obtained in the F1 population which can be easily eliminated by the screening of the F_1 population with a marker. If the trait-linked marker is dominant, the homozygous plant will be obtained in the F_3 generation while if it is co-dominant, the homozygous plant will be identified in the F_2 generation. The efficiency of MAS decreases as the number of genes/QTLs increases and their heritability decreases. However, MAS is relatively successful with simply inherited traits than highly complex characters governed by several genes.

The efficiency of MAS increases with an increase in the number of markers associated with a QTL. The distance between markers and the desired gene should be close (<5 cM) to ensure minimal recombination. For MAS, the preferable marker has a tight linkage with the gene of interest so that no recombination occurs between them. But, if marker and gene of interest are not tightly linked, in that case, recombination between them may decrease the proficiency of MAS as single crossover leads to alternation in linkage association. Hence, as the distance between a gene of interest and a marker increases, the efficiency of MAS decreases in the same proportion. But in case, the tightly linked marker is unavailable, the use of two flanking markers is a great choice that reduces the errors due to homologous recombination and hence, enhances the proficiency of MAS.

In classical breeding, the major challenge of gene introgression is the identification of progenies possessing the gene(s) of interest and it is a very lengthy process too. With MAS, the breeder can easily detect the marker polymorphism in the progenies of two contrasting parents at very early stages (seedling stage). Thus, MAS not only reduces the load of the population carried into the next generation by breeders but also reduces the burden of making unnecessary crosses in marker-assisted backcross breeding (MABB) and marker-assisted recurrent selection.

The success of MAS largely depends upon several factors like the number of gene(s)/QTLs to be introgressed, the distance between the marker and gene of interest, the number of progenies to be analyzed, nature of the molecular markers, and the genetic background of the recipient parent into which gene of interest has to be transferred. Hence, for marker-assisted selection following breeding strategies are used; selection of gene(s)/QTLs from breeding lines/populations, incorporation of the gene of interest from donor parent, marker-assisted backcrossing (MABC), marker-assisted recurrent selection (MARS) and pyramiding of genes.

(a) *Identification of breeding lines possessing genes of interest*

Molecular markers are equally efficient for the identification of qualitative as well as quantitative characters. Thus, MAS provides a chance of selection of that target loci whose monitoring is difficult phenotypically. Marker-assisted selection helps in the identification of genes of interest at a very early stage which is not possible in traditional breeding because visual selection for complex characters like yield is quite difficult and not possible at an early stage. Thus, MAS is a great opportunity for breeders which not only enhances early generation selection but also reduces the number of plants retained for the next generation. The breeding line/wild relatives possess a gene of interest designated as "donor parent" and it is transferred into the recipient parent by backcrossing. The traits with high heritability provide better results in comparison to characters having low heritability.

(ii) Marker-assisted backcrossing (MABC)

In general, backcross is a conventional breeding method where, improvement of an elite variety (lacking essential genes like disease resistance or quality traits), referred to as recipient/recurrent parent, is done by introgression of one or more genes/QTLs from a donor parent (have desirable gene/QTLs). The number of backcrosses for the improvement of variety depends on the behavior of gene-like if dominant gene governs the trait, six to eight backcrosses are required. While, if recessive genes govern the trait, eight to ten generations are required because after every backcross there would be a need of selfing to recover the trait. In conventional breeding, phenotyping during backcrossing is very tedious and time taking. In some cases, phenotyping is done post maturity viz. grain quality and protein; in those cases, it is very tough to select plants for the backcrossing. In this situation, the molecular marker has come up as a substitute. These markers can be used for indirect selection of the target trait. Along with that, the use of molecular markers for selection reduces the number of backcrosses, i.e., up to two to three, reducing the time of breeding program without any influence of the nature of gene (dominant or recessive). In MABC, the main objectives of the backcross are to accomplish (1) introgression of the desired trait from a donor to a recipient/recurrent parent, (2) recovery of the recipient/recurrent parent genome up to maximum, and (3) complete removal of the donor genome, leaving only the target gene/ QTL (removal of linkage drag). Molecular markers can assist selection in three ways, (1) **Foreground selection**: Markers linked to the target gene/QTL assist in indirect selection for the target gene/QTL. (2) **Background selection**: To recover the recipient/recurrent parent genome, codominant markers distributed throughout the genome are used. (3) Lastly, **Recombinant selection**: To remove the linkage drag or complete removal of the unwanted segment from the donor, require codominant markers present on the flanking region of the target gene/QTL.

Three different backcross selection strategies have been proposed by Frisch et al. (1999) in maize, i.e., two stages, three stages, and four stages based on the fastest recovery of the recurrent parent genotype (RPG), and suggested the four-stage strategies as the most proficient procedure. The two-stage selection procedure involves the selection of target allele and selection of RPG at all loci, three-stage selection strategy involves the selection of target allele, selection for recombinants at flanking markers and selection of RPG at all loci while four-stage selection procedure involves selection of target allele, selection for recombinants at flanking markers, selection of RPG on carrier chromosome and selection of RPG at all loci. The four-stage strategy is highly efficient in the fastest recovery of the recipient parent genome and reduces the chance of linkage drag.

A common question is, when does the MABC should be applied, and which parent should be used as female (donor or recurrent parent). The first answer is when an adapted, elite, superior, commercially accepted, high yielded variety in a particular environment or zone, deficient in resistance to biotic or abiotic stresses or any quality feature needed to improve for a single or few resistance/quality traits, then MABC can be applied. The second answer is using the recurrent parent as a female in the backcrossing because backcross progenies can be differentiated from selfed progenies as described by Nayak et al. (2017). Considering the flowering synchronization in mind, the recurrent parent should be planted in three to six different timing. Markers linked to desired traits should be in hand prior to starting the MABC program. While identification of polymorphic markers for background selection should also be earlier before the implication of the MABC program. Markers facilitate the recovery of recurrent parent genome and it is an utmost important part of MABC. For the calculation of recovery of recurrent parent genome Allard (1999) given formula

$$\text{Percent recovery of RPG} = \left[\left(2^{n+1} - 1\right)/2^{n+1}\right] * 100.$$

where n represents number of backcrosses. For example, it is well known in first backcross an average recovery of RPG is 75%, while practically recovery of RPG can be in the range of 50% to 100%, excluding the target gene see Nayak et al. (2017).

(a) *Foreground selection*

Marker-assisted foreground selection was proposed by Tanksley (1983) while used for the incorporation of resistance genes by Melchinger (1990). Later systematic theoretical approach of a backcross program employing foreground selection was elaborated by Hospital (2005). Identification of molecular markers either tightly linked with the gene of interest or flanked the target loci is an important step before starting foreground selection, later this marker is used for foreground selection by the screening of segregating generations for the target gene(s)/QTLs. Hence, it is commonly referred to as "foreground selection" (Hospital and Charcosset, 1997) or positive selection (Takeuchi et al., 2006). Tightly linked markers with the gene of interest provide the best results because the rate of recombination between the marker and the target gene is zero. If tightly linked markers are not available for any trait, in that case, the use of flanking markers is a better choice than single marker because flanking markers cover the target loci and only double recombination can break the bracketed association which is generally rare. It is very efficient for the incorporation of recessive genes because their transfer via classical backcross breeding requires additional recurrent selfing generations, which not only enhance the cost of the experiment but also take too much time for variety improvement. Successful applications of MAS in crops like rice and wheat have been reported (Sundaram et al., 2008; Gopalakrishnan et al., 2008; Singh et al., 2011, Vishwakarma et al., 2014, 2016a,b).

(b) *Recombinant selection*

Recombinant selection involves the selection of backcross progenies along with the gene of interest and recombinant events between the gene of interest and linked flanking markers. The main purpose of recombinant selection is the elimination of linkage drag (linkage of the gene of interest with many undesirable genes that negatively affect crop performance) by reducing the size of the donor chromosome segment possessing the gene of interest. In traditional backcross breeding, the donor segment may remain very large even after several backcrossing (Ribaut and Hoisington, 1998) but with the help of flanking markers, it can be reduced. Recombinant selection is generally used for at least two backcross generations (Frisch et al., 1999).

(c) *Background selection*

The last step of Marker-assisted backcrossing involves the identification of backcross progeny with the greatest recovery of recurrent parent genome (RPG) with the help of molecular markers that are unlinked to the gene of interest. This is known as background selection. Background selection is used to reduce the size of the incorporated region of the donor parent. The chromosome(s) that carries the gene of interest is known as carrier chromosome(s) while the other chromosomes except carrier chromosome are known as non-carrier chromosomes. The purpose of background selection is to enhance the recovery of RPG by minimizing the size of the intact chromosomal segment of donor type dragged around the gene of interest on the carrier chromosome. Which obtained by selection of those individuals which are heterozygous at the target locus but homozygous for recipient parent loci at two markers flanked the gene of interest on both sides (known as double heterozygotes) and by reducing nonrecurrent parent genome on the non-carrier chromosome (Young and Tanksley, 1989; Hospital, 2005). Neeraja et al. (2007) used a set of 18 STMS markers around the Sub1 locus which restricted the linkage drag to 2.3–3.4 Mb in the final introgressed (BC3F2) line of rice variety Swarna. In maize, Babu et al. (2005) used the flanking SSR markers to reduce the linkage drag to 8 cm around the introgressed opaque 2 allele. Conventional backcrossing requires 5–6 backcross generations to recover the RPG and there may still be a chance of linkage drag but by using markers RPG recovery may be obtained in BC_4, BC_3, or even in BC_2 (Visscher et al., 1996; Hospital and Charcosset, 1997). The recovery of RPG is obtained at a rate of $1 - (1/2t + 1)$ for each of the t generations of backcrossing (Babu et al., 2005). Thus, after six generations of backcrossing the expected recovery of RPG would be 99.2% and the end product is known as near–isogenic line with recurrent parent. In rice, Sundaram et al. (2008) reported a recovery of 96.8% of RPG in BC_4F_1 by using 50 polymorphic STMS markers with a population size of 80–145. Babu et al. (2005) identified opaque 2 allele introgressed BC_2F_1 maize line that had 95.75% of the recurrent parent genome using 72 SSR markers.

(d) *Strategy to select homozygous lines in MABC with the dominant marker*

In the MABC program, during foreground selection, we use a dominant marker that is linked to a desirable trait, but cannot differentiate homozygous and heterozygous progenies. This situation cause problem in the selection of homozygous family/plant after backcrosses. To solve this problem we suggest phenotyping assistance with the marker screening (Fig. 3). Therefore, during the selection of homozygous family, screening of ten random plants from each

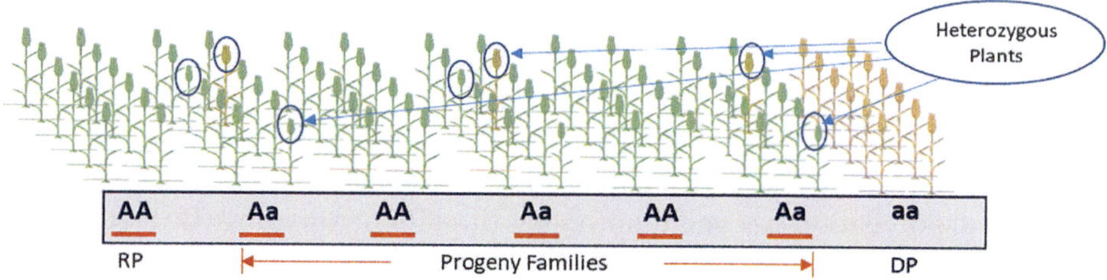

Fig. 3 Selection strategy for the dominant marker used in foreground selection, *RP*, recurrent parent; *DP*, donor parent; *AA*, homozygous family; *Aa*, heterozygous family.

family with target trait linked markers as in foreground selection, along with stringent phenotyping by planting every single family in two-row for visual similarity would be beneficial.

(iii) Marker-assisted gene pyramiding (MAGP)

Marker-assisted gene pyramiding (MAGP) is an important approach of molecular breeding which assembles multiple desirable genes from multiple parents into a single genotype. The concept of gene pyramiding has been utilized to enhance biotic resistance by incorporating two or more genes at a time (Nelson, 1978). The aim of gene pyramiding is the improvement of existing elite cultivars having few defects, for which desirable genes are identified and introgressed from the other sources. Several successful examples of gene pyramiding for bacterial blight and blast in rice has been reported (Balachiranjeevi et al., 2018). In-plant breeding, MAGP provides a potential strategy for the derivation of an ideotype that is homozygous for all the incorporated favorable alleles/genes. The multiple parents, each possesses a single target gene, used in gene pyramiding is known as founding parents. To obtain the ideal genotype, the gene-pyramiding scheme can be distinguished into two parts; the first part is known as pedigree which involves the accumulation of one copy of all target genes into a single genotype (root genotype) while the second step is known as fixation step which involves fixation of target genes into a homozygous state (Servin et al., 2004). Each node of the gene pyramiding tree is known as an intermediate genotype having two parents. After the pedigree step, many different approaches can be used for the fixation of target genes. There are three methods for the fixation of target genes in gene pyramiding: (i) production of DH population from the root genotype (ii) selfing of the root genotype directly to obtain the ideotype (iii) derive a genotype having all desirable genes in coupling phase by crossing the root genotype with a parent having no favorable genes. Some approaches like complex crossing, recurrent selection, and backcrossing can be used for gene pyramiding which largely depends on the number of founding parents and the heritability of the targeted genes. For MAGP, three breeding approaches are generally used, i.e., stepwise, simultaneous/synchronized, and convergent backcrossing. The systematic representation of genes pyramiding through stepwise, simultaneous/synchronized, and convergent backcrossing is presented in Fig. 4A, B, and C, respectively.

(a) *Stepwise backcrossing*: involves only one gene/QTL targeted or selected at a time followed by the next step of backcrossing for the second gene/QTL until all target genes/QTLs have been incorporated into the recipient parent. Thus, the benefit of this method is that it introgressed one gene at a time which results in a small population size and requires less time for genotyping. But it takes a much longer time to improve a variety (Fig. 4A).

(b) *Simultaneous/synchronized backcrossing*: involves crossing of recipient parent with each of the donor parents (suppose 4 donor parents) to generate four single-cross F_1s. Then, two double cross F_1s are obtained by crossing two

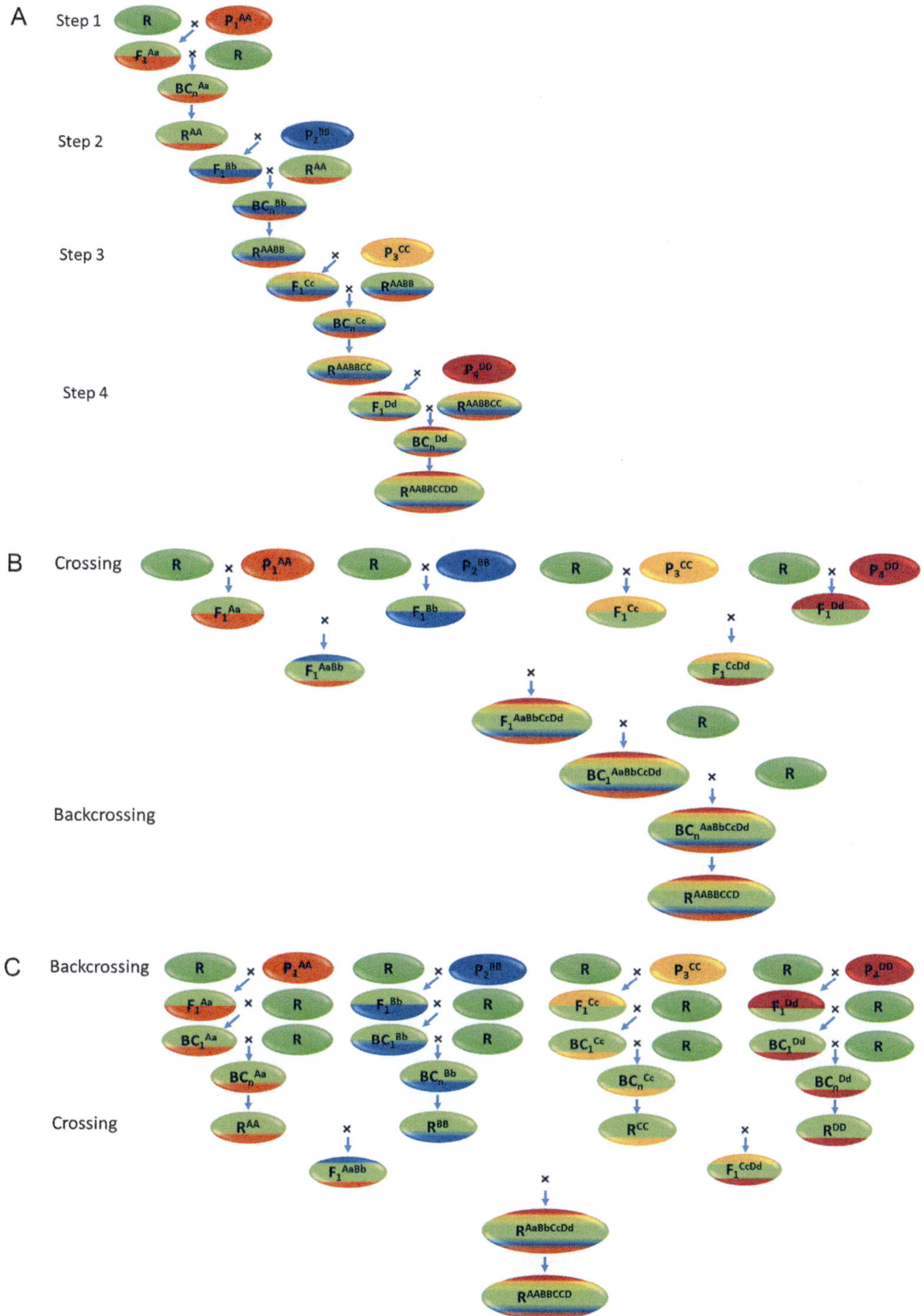

Fig. 4 (A) Stepwise backcrossing. (B) Simultaneous/synchronized backcrossing. (C) Convergent backcrossing. *R*, recurrent parent; *P1, 2, 3, and 4*, different donor for different genes; *F₁*, first filial generation; *BC*, backcross progenies.

of the four single cross F1s with each other and again these two double cross F_1s are crossed to generate a hybrid having all four target genes. For the recovery of recurrent parent genome, hybrid is now backcrossed with recurrent parent followed by one generation of selfing. This approach takes the shortest time for the improvement of a genotype but it needs a relatively larger population and more genotyping (Fig. 4B).

(c) *Convergent backcrossing*: combines the benefit of stepwise as well as simultaneous backcrossing approach. The four target genes belongs to four different donors are separately transferred into the recipient parent by single crossing, followed by backcrossing based on a marker linked to the target genes/QTLs to generate four improved lines. Out of four, two improved lines are crossed with each other and the two hybrids are then intercrossed to accumulate all four genes/QTLs together and generate the final improved line possessing all four genes. This approach requires less time as well as easy for gene fixation in MAGP (Fig. 4C). It has been successfully used for the pyramiding of blast resistance gene Piz5 and Pi5 into basmati restorer line 'PRR78' (Singh et al., 2013).

(iv) Marker–assisted recurrent selection (MARS)

Lande and Botstein (1989) suggested to identify markers linked with an important quantitative trait in a population, and utilize them for MAS in the same population. The objective of MARS is the accumulation of several favorable quantitative traits loci (QTL) by utilizing molecular markers which are linked with target traits (Bernardo, 2008). In this technique, two superior inbred genotypes that have desirable gene/QTLs for biotic/abiotic traits are used as parents. Cross these elite lines, then phenotype and genotype with genome-wide distributed markers up to $F_{3:4}$ generations. In this generation, identify the markers associated with desirable traits with the multiple regression analysis. After the establishment of the marker-trait association, select the best plants for the further crossing, which is based on the selection index that uses collective data of phenotyping and genotyping score (Selection index is a numerical score that gathers information on all the traits associated with the dependent variable like yield). Such analysis and selection can be computed with OptiMAS and QuMARS softwares. Since, the association between marker and trait loci generally changes due to recombination in each generation, hence it is re-estimated up to various generation. The final selected progenies evaluated up to $F_{3:5}$ at multilocation trials for the traits under selection. Such a program is used in the commercial industry for the improvement of Maize, Soybean, and Sunflower.

(v) Advanced backcross QTL (AB-QTL)

Advanced backcross QTL (AB-QTL) is the backcross approach that involves mapping and introgression of desirable gene/allele simultaneously from the feral/exotic germplasm into elite genotypes. This approach culminates the negative effect of traits that comes from donor parent, that creates hurdle in selection. When using traditional F_2 or RILs mapping population, even the use of molecular markers fails to select desirable alleles. Through AB-QTL, background of the recurrent parent can be enriched with only one or few desirable gene/allele leads to Near Isogenic Lines (NILs), which make ease in the identification of desirable QTLs in such population (Tanksley and Nelson, 1996). This approach has been successfully employed in several crops like Tomato (Fulton et al., 2000), Rice (Moncada et al., 2001), wheat (Kunert et al., 2007; Naz et al., 2008).

(vi) Genomic selection (GS) or Genome-wide selection (GWS)

MABC, gene pyramiding, and MARS are befitted while handling monogenic, polygenic, and quantitative trait loci (QTLs) with large effects. In general, quantitative traits expression depend upon many large and small effect QTLs. Thus, MAS is not fit and designed to handle small effects QTLs, while selection in MARS scheme based on the significant association between markers and traits, is not found suitable. To cope with the problem in MAS and MARS schemes, Meuwissen et al. (2001) conceptualize a new scheme, namely genomic selection, which uses genome-wide markers data no matter associated with the traits are significant. There are some important components of GS, viz. training population (TP), breeding population (BP), and genomic estimated breeding values (GEBV). The genomic selection scheme by using breeding population and training population has been described in Fig. 5 (Heffner et al., 2009). The step involved in a GS program

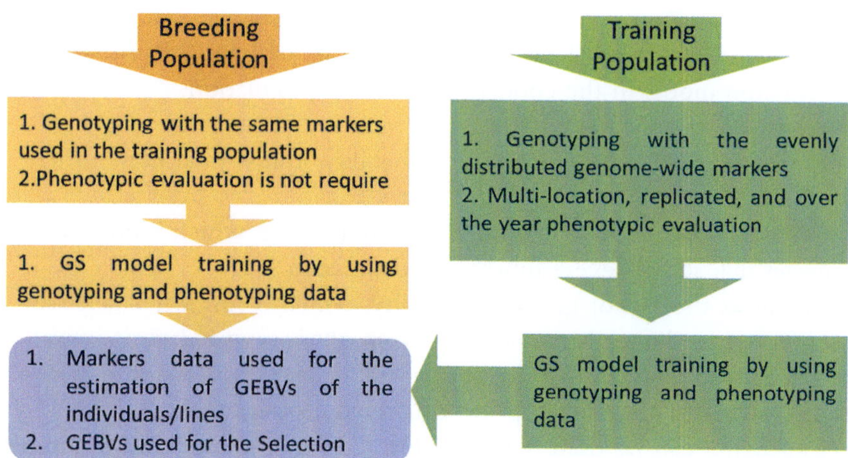

Fig. 5 Genomic selection scheme based on Heffner et al. (2009). *GEBV*, genomic estimated breeding values; *GS*, genomic selection. *Note*: GS model training is always repeated over time with the inclusion of new available lines, also included in training and breeding populations.

is to create a training population by including all possible important materials/lines suitable to improve the new breeding population. Then genotype training population's lines with the densely distributed genome-wide markers and evaluate the same TP in replication over the year in multi-location trials and multi-environment. The phenotyping and genotyping data are subjected to model training (computing the GS model parameters). Model training is continued with the addition of new markers and involvement of new traits. The GS model parameters are applicable to the breeding population, the same set of markers obtained in modeling of training population was used. Phenotypic data of the breeding population are not required at this stage. The GEBVs of individuals/lines of the breeding population are estimated based on a genotypic and phenotypic dataset of "training population". The superior/best individuals/lines are selected from the breeding population based on their GEBV estimates.

Presently, GS has been applied in several crop species such as wheat, rice, maize, sugarcane, sunflower. There are different "prediction models" (Nakaya and Isobe, 2012), are available to select the individual with the high genetic gain viz. random regression best linear unbiased predictor (RR-BLUP) (Heffner et al., 2011), Bayesian ridge regression (RR), least absolute shrinkage and selection operator (LASSO), Bayesian LASSO (Crossa et al., 2010), BayesA, BayesB, BayesCp, random forest regression (RFR), PM-RKHS pedigree information combined with molecular-reproducing kernel Hilbert space regression (Crossa et al., 2011), haplotype-based genomic best linear unbiased prediction (HGBLUP) and locally extended genomic best linear unbiased prediction (LEGBLUP) (Jiang et al., 2018).

Rewards of GS

- No QTL mapping or identification requires as the selection criterion is based on prediction model in the concerned breeding population.
- Selection accuracy and efficiency are high in GS.
- GS is very effective for low heritability traits (Dekkers, 2007; Heffner et al., 2010).
- Application of GS cut down the rate of inbreeding, which is important for those species that shows severe inbreeding depression.
- GS checks the loss of genetic variability, which is lost when selection is made based on breeding values calculated from phenotyping data.
- GS achieved high genetic gain.
- Heterosis can be predicted prior to making crosses for the hybrid.
- Cost-saving can be done with the low-cost next-generation sequencing platform, also high though put phenotyping labor cost.
- In comparison to vast conventional breeding approaches, only a few crosses are sufficient for the application of GS.

6. Success stories of MAS and perspectives

During the last three decades, several studies have been published related to QTL identification in several crops. In 2008, when a very elaborative, detailed review had been published by Collard and Mackill (2008), where they revealed the low Impact of MAS. As they mentioned (1) marker technology is still in the early stage of development; (2) MAS results may not be easily published; (3) reliability and accuracy of QTL mapping; (4) a loose association exists between marker and gene/QTL; (5) limited markers availability and limited polymorphism for markers in breeding materials; (6) effect of the genetic background; (7) QTL environment interactions; (8) high cost of MAS; (9) existence of an application gap between research laboratories and plant breeding institutes; (10) knowledge gap between plant breeders and molecular biologists or scientists in other disciplines. In this context, we have achieved remarkable growth up to some extent. From that time until now, with the fastest growing industry of low-cost sequencing technologies, we have succeeded to overcome some limitations up to some extent. There are several low-cost marker platforms available now, as discussed in above section of this chapter like SNPs, Affymetrix Axiome, and Array-based that contributed in many ways (Pandey et al., 2017). In regards to publications of the MAS work, several successful achievements and papers have been published every year see Table 1. While reliability of markers has gained wonderful heights, for example, QTL-seq (Takagi et al. (2013); Pandey et al., 2016), MutMap (Abe et al., 2012) and BSA-seq (Klein et al., 2018) are the latest fast-forward NGS-based genetic mapping approaches that identify linked gene/QTLs in the form of candidate SNPs and diagnostic markers. Along with that, with the availability of Affymetrix Axiome and KASP assay it has been possible now to genotype the desirable plants for the foreground (for the multiple desirable alleles) and background selection in one shot at a very low-cost (Leal-Bertioli et al., 2015). In addition to this, for the high-density genetic map and fine mapping, different mapping population concepts are also established viz. chromosome segment-substitution lines (CSSLs), multiparent advanced generation intercross (MAGIC), association mapping, and nested association mapping (NAM) population that are assisted with high-throughput phenotyping. With the earlier and current gene/QTLs identification studies, and newly developed technologies and approaches in the field of genomics. It has now become possible to introgress various economically and agronomically important traits in elite lines/genotypes (Janila et al., 2016; Vishwakarma et al., 2014, 2016a,b). Meanwhile, several studies have transformed in the form of released varieties in past few decades (Table 1). In conclusion, as the more rapid, reliable, and cost-effective technologies emerge, it would facilitate trait mapping and introgression programs very effectively into elite genetic backgrounds and provide a sustainable demand-driven crop improvement through genomics-assisted breeding.

Table 1 Varieties released worldwide by using Marker Assisted Selection/Gene Pyramiding.

Genotype	Improved variety/lines	Gene introgressed	Effects	Country	References
Rice					
Pusa Basmati 1	Pusa 1460, IET18990	*xa13, Xa21*	Resistance to Bacterial blight, high aroma	India	Gopalakrishnan et al. (2008)
Sambha Mahsuri (BPT-5204)	Sambha Mahsuri (IET 19046)	*xa5, xa13, Xa21*	Resistance to Bacterial blight	India	Sundaram et al. (2008)
Swarna	Swarna Sub-1A	*Sub-1A*	Resistance to Submergence	India	Nandi et al. (1997)
Pusa Basmati 6	Pusa Basmati 1728	*xa13, Xa21*	Resistance to Bacterial blight	India	ICAR IIRR (2017)
Pusa Basmati 1	Pusa Basmati 1637	*Pi9*	Resistance to Rice blast	India	ICAR IIRR (2017)
IR64	IR64 Sub 1	*Sub1*	Resistance to Submergence	India	Reddy et al. (2009)
Lalat	Improved Lalat	*Xa4 + xa5 + xa13 + Xa21*	Resistance to Bacterial blight	India	Dokku et al. (2013)
Tapaswini	Improved Tapaswini	*Xa4 + xa5 + xa13 + Xa21*	Resistance to Bacterial blight	India	Dokku et al. (2013)
BR11	BRRI dhan52	*Sub1*	Resistance to submergence	Bangladesh	Iftekharuddaula et al. (2015)
Wuyujing 3	K01, K04	*Wx-mq*	Low-amylose content gene	China	Tao et al. (2016)
	Cadet	Granule-bound Starch Synthease (GBSS)	Unique cooking and processing quality traits including amylose content Low-amylose	USA	Hardin (2000)
	Jacinato	Granule-bound Starch Synthease (GBSS)	Unique cooking and processing quality traits including amylose content Low-amylose	USA	Hardin (2000)

Mangeumbye	Improved Mangeumbye	$Xa4+xa5+Xa21$	Resistance to Bacterial blight	Republic of Korea	Suh et al. (2013)
IR64	Angke	$Xa4+xa5$	Resistance to Bacterial blight	Indonesia	Toenniessen et al. (2003)
IR64	Conde	$Xa4+xa7$	Resistance to Bacterial blight	Indonesia	Toenniessen et al. (2003)
	XieYou 218	–	Resistance to Bacterial blight	China	Cheng et al. (2007)
	MAS 26	QTL	Drought tolerance	India	Gandhi et al. (2012)
	MAS946-1	QTL	Drought tolerance	India	Gandhi et al. (2012)
	Tubigan 11		Disease resistance	Philippines	Chandrashekar (2007); Gandhi (2007)
IR64	Tubigan 7	$xa4+xa5+Xa21$	Resistance to Bacterial blight	Philippines	Chandrashekar (2007); Gandhi (2007)
Samba Mashuri	Samba Mashuri–Sub1	$Sub1$	Submergence tolerance	India	Septiningsih et al. (2009)
IR64	IR64–Sub1	$Sub1$	Submergence tolerance	India	Septiningsih et al. (2009)
Kalinga III	Birsa Vikas Dhan 111	root QTL	Drought tolerance	India	Steele et al. (2013)
	Zhongyou 6 and Zhongyou1176	$Xa21$	Bacterial blight resistance	China	Cao et al. (2003)
	MAS 946-1 (Sharada)		Drought tolerance		Chandrashekar (2007)
Pearlmillet HHB-67	HHB-67-2		Resistance to Downy Mildew	India	Hash et al. (2006)

Continued

Table 1 Varieties released worldwide by using Marker Assisted Selection/Gene Pyramiding—cont'd

Genotype	Improved variety/lines	Gene introgressed	Effects	Country	References
Soybean					
F6 derived lines from Caviness x Anand	JTN 5303		Disease resistance	USA	Arelli et al. (2007)
	Sheyenne		Tolerant to iron deficiency, Chlorosis, and lodging resistance		Helms et al. (2008)
	S99–2281		Cercospora leaf spot resistance		Shannon et al. (2009)
Barley Sloop	Sloop SA		Cyst nematode resistance	Australia	Barr et al. (2000)
	Doria	*Ryd2*	Resistance to Barley Yellow Dwarf Virus	Italy	Kosova et al. (2008)
Gairdner	Gairdner Plus		Malting quality	Australia	Elington et al. (2006)
Steptoe	Tango	APR–QTL	Resistance to stripe rust	USA	Hayes et al. (2003)
Tomato	AB2		High yield (QTL Brix9-2-5)	USA	Lippman et al. (2007)
Maize Vivek maize hybrid 9	Vivek QPM-9	*Opaque-2*	High lysine and tryptophan	India	Babu et al. (2005)
	Sunrise		Western corn rootworm resistance		Private breeding program
Wheat	Patwin	*Yr17, Lr37-Sr38,* and *Glu-D1 (5 + 10)*	Resistance to Leaf rust, stem rust, strong gluten content	USA	Helguera et al. (2003)
	Abovea, Avalanchea, Ankora	*Bdv2*	Resistance to yellow dwarf virus	USA	http://www.wheatworld.org/pdf/dubcovsky.pdf
	Ben	*Qfhs.ndsu-3AS*	Resistance to Fusarium head blight		Elias (2005)

Express	Expresso	*Yr17 + Yr15*	Resistance to Stripe rust	USA	Wheat (2009)
BW621	Lillian	*Gpc-B1*	High grain protein content	Canada	DePauw et al. (2005)
Anza	Lassik	*Lr34 + Lr37*	Resistance to Leaf rust	USA	https://fsp.ucdavis.edu/seed-catalog/wheat-varieties/Lassik
WA007869	Farnum	*Yr36 + Gpc-B1*	Resistance to Stripe rust, high grain protein content	USA	Kidwell et al. (2008)
D95-257	Westmore	*Yr36 + Gpc-B1*	Resistance to Stripe rust, high grain protein content	USA	Brevis and Dubcovsky (2008)
ProINTA Puntal	BIOINTA 2004	*Lr47*	Resistance to leaf rust	Argentina	Bainotti et al. (2009)
R4004	MS INTA 416	*Fhb1 + Lr47*	Resistance to Fusarium Head Blight and leaf rust	Argentina	Bainotti et al. (2017)
	McNeal, Reeder, Hank	*Qss.msub-3BL*	Resistance to wheat stem sawfly	USA	http://www.wheatworld.org/pdf/dubcovsky.pdf
Alturas	Cataldo	*H25*	Resistance to Hessian fly	USA	Chen et al. (2009)
	Scarlet (WA7994)	*Yr15* and *Gpc-B1*	Resistance to stripe rust and high grain protein content	USA	http://css.wsu.edu/Proceedings/2005/2005_Proceedings.pdf
94B43-BLW4	Goodeve	*Sm1*	Resistance to insect orange blossom wheat midge	Canada	DePauw et al. (2009)
	Kern	*Stb4*	Resistance to Septoria	California AES	http://www.wheatworld.org/pdf/dubcovsky.pdf
Yuma	Mace	*Wsm-1*	Resistance to wheat streak mosaic virus (WSMV)	USA	Graybosch et al. (2009)
UC1113	UC1113 (P1638741)	*Sr2 + Sr13 + Sr25*	Resistance to Ug99	USA	Yu et al. (2009)

Continued

Table 1 Varieties released worldwide by using Marker Assisted Selection/Gene Pyramiding—cont'd

Genotype	Improved variety/lines	Gene introgressed	Effects	Country	References
PBW 343	PBW723 alias Unnat PBW343	*Yr17/Lr37/ Sr38/Yr40/ Lr57*	Resistant to yellow rust	India	Sharma (2021)
Groundnut	COAN, NemaTAM		Root knot nematode resistance		Simpson et al. (2003) and Simpson and Starr (2001)
	Tifguard high O/L		Root knot nematode resistance and high O/L ratio		Chu et al. (2011)
Beans Pinto Beans Weihing	USPT-ANT-1 ABC-Weihing	$CO\text{-}4^2$ QTL	Resistance to Anthracnose Resistance to Common Bacterial Blight	USA USA	Miklas et al. (2003) Mutlu et al. (2008)
K97305	USDK-CBB-15	QTL	Resistance to Common Bacterial Blight	USA	Milkas et al. (2006)
Chase Pinto	ABCP-8	QTL	Resistance to Common Bacterial Blight	USA	Mutlu et al. (2005)
White bean	Verano	QTL	Bean golden-yellow mosaic virus (BGYMV) and common bacterial blight resistance		Beaver et al. (2008)

References

Abe, A., Kosugi, S., Yoshida, K., et al., 2012. Genome sequencing reveals agronomically important loci in rice using MutMap. Nat. Biotechnol. 30, 174–178.

Aguiar, C.G., Schuster, I., Amaral Jr., A.T., Scapim, C.A., et al., 2008. Heterotic groups in tropical maize germplasm by test crosses and simple sequence repeat markers. Genet. Mol. Res. 7 (4), 1233–1244.

Akkaya, M.S., Bhagwat, A.A., Cregan, P.B., 1992. Length polymorphisms of simple sequence repeat DNA in soybean. Genetics 132, 1131–1139.

Allard, R.W., 1999. Principles of Plant Breeding, 2nd. John Wiley & Sons, New York.

Andersen, J.R., Lubberstedt, T., 2003. Functional markers in plants. Trends Plant Sci. 8, 554–560.

Arbelaez, J.D., Tandayu, E., Reveche, M.Y., et al., 2019. Methodology: ssb-MASS: a single seed-based sampling strategy for marker-assisted selection in rice. Plant Methods 15, 78. https://doi.org/10.1186/s13007-019-0464-2.

Arelli, P.R., Pantalone, V.R., Allen, F.L., Mengistu, A., 2007. Registration of soybean germplasm JTN-5303. J. Plant Regist. 1, 69–70.

Babu, R., Nair, S.K., Kumar, A., Venkatesh, S., et al., 2005. Two-generation marker-aided backcrossing for rapid conversion of normal maize lines to quality protein maize (QPM). Theor. Appl. Genet. 111, 888–897.

Bainotti, C., Fraschina, J., Salines, J.H., et al., 2009. Registration of 'BIOINTA 2004' wheat. J. Plant Regist. 3, 165–169.

Bainotti, C.T., Lewis, S., Campos, P., et al., 2017. MS INTA 416: a new Argentinean wheat cultivar carrying Fhb1 and Lr47 resistance genes. Crop Breed Appl. Biotechnol. 17, 280–286.

Balachiranjeevi, C.H., et al., 2018. Marker-assisted pyramiding of two major, broad-spectrum bacterial blight resistance genes, Xa21 and Xa33 into an elite maintainer line of rice, DRR17B. PLoS One 13 (10). https://doi.org/10.1371/journal.pone.0201271, e0201271.

Barr, A.R., Jefferies, S.P., Warner, P., Moody, D.B., Chalmers, K.J., Langridge, P., 2000. Marker-assisted selection in theory and practice, vol. I, Adelaide, Australia, pp. 167–178.

Beaver, J.S., Porch, T.G., Zapata, M., 2008. Registration of 'Verano' white bean. J. Plant Regist. 2, 187–189.

Beckmann, J.S., Soller, M., 1986. Restriction fragment length polymorphism in genetic improvement. Oxford Surv. Plant Mol. Biol. 3, 197–250.

Bernardo, R., 2008. Molecular markers and selection for complex traits in plants: learning from the last 20 years. Crop Sci. 48, 1649–1664.

Botstein, D., White, R.L., Skolnick, M., Davis, R.W., 1980. Construction of a genetic linkage map in man using restriction fragment length polymorphisms. Am. J. Hum. Genet. 32, 314–331.

Brennan, J.P., Martin, P.J., 2007. Returns to investment in new breeding technologies. Euphytica 157, 337–349.

Brevis, J.C., Dubcovsky, J., 2008. Effect of the Gpc-B1 region from *Triticum turgidum* ssp. dicoccoides on grain yield, thousand grain weight and protein yield. In: Appels, R., Eastwood, R., Lagudah, E., Langridge, P., Mackay, M., McIntyre, L., Sharp, P. (Eds.), Proceedings of 11th International Wheat Genet Symposium, Brisbane Australia, August 24–29, 2008. Sydney University Press, pp. 1–3. http://hdl.handle.net/2123/3179.

Cao, L.Y., Zhuang, J.Y., Yuan, S.J., Zhan, X.D., Zhang, K.L., Chang, S.H., 2003. Hybrid rice resistant to bacterial leaf blight developed by marker assisted selection. Rice Sci. 11, 68–70.

Chandrashekar, M.V., 2007. UAS develops aerobic rice. Food Beverage News. www.fnbnews.com/article/detarchive.asp?articleid=22118§ionid=34. (Accessed 20 March 2009).

Chen, J., Souza, E.J., Zemetra, R.S., et al., 2009. Registration of 'Cataldo' wheat. J. Plant Regist. 3 (3), 264–268.

Cheng, S.H., Zhuang, J.Y., Fan, Y.Y., Du, J.H., Cao, L.Y., 2007. Progress in research and development on hybrid rice: a super domesticate in China. Ann. Bot. 100, 959–966.

Chu, Y., Wu, C.L., Holbrook, C.C., Tillman, B.L., Person, G., Ozias-Akins, P., 2011. Marker assisted selection to pyramid nematode resistance and the high oleic trait in peanut. Plant Genome 4, 110–117.

Collard, B.C.Y., Mackill, D.J., 2008. Marker-assisted selection: an approach for precision plant breeding in the twenty-first century. Philos. Trans. R. Soc. Lond. Ser. B Biol. Sci. 363 (557), 572.

Concibido, V.C., Young, N.D., Lang, D.A., et al., 1996. Targeted comparative genome analysis and qualitative mapping of a major partial-resistance gene to the soybean cyst nematode. Theor. Appl. Genet. 93, 234–241.

Crossa, J., Campos, G., Perez, P., et al., 2010. Prediction of genetic values of quantitative traits in plant breeding using pedigree and molecular markers. Genetics 186, 713–724.

Crossa, J., Perez, P., Campos, G., et al., 2011. Genomic selection and prediction in plant breeding. J. Crop Improv. 25, 239–261.

Dekkers, J.C.M., 2007. Prediction of response to marker assisted genomic selection using selection index theory. J. Anim. Breed. Genet. 124, 331–341.

DePauw, R.M., Knox, R.E., Thomas, J.B., et al., 2009. Goodeve hard red spring wheat. Can. J. Plant Sci. 89, 937–944.

DePauw, R.M., Townley-Smith, T.F., Humphreys, G., et al., 2005. Lillian hard red spring wheat. Can. J. Plant Sci. 85, 397–401.

Dokku, P., Das, K.M., Rao, G.J.N., 2013. Genetic enhancement of host plant-resostance of the Lalat cultivar of rice against bacterial blight employing marker assisted selection. Biotechnol. Lett. 35, 1339–1348.

Elias, E.M., 2005. Fusarium Resistant Tetraploid Wheat. United States Patent Application. 20050273875.

Elington, J., Coventry, S., Chalmers, K., 2006. Breeding outcomes from molecular genetics. In: Mercer, C. F. (Ed.), Breeding for Success: Diversity in Action, Proceedings of the 13th Australasian Plant Breeding Conference, Christchurch, New Zealand. New Zealand Grassland Association, Dunedin, pp. 743–749.

Frisch, M., Bohn, M., Melchinger, A.E., 1999. Comparison of selection strategies for marker-assisted backcrossing of a gene. Crop Sci. 39, 1295–1301.

Fulton, T.M., Grandillo, S., Beck-Bunn, T., et al., 2000. Advanced backcross QTL analysis of a *Lycopersicon esculentum, Lycopersicon parviflorum* cross. Theor. Appl. Genet. 100, 1025–1042.

Fulton, T.M., Van der Hoeven, R., Eannetta, N.T., et al., 2002. Identification, analysis, and utilization of conserved ortholog set markers for comparative genomics in higher plants. Plant Cell 14, 1457–1467.

Gandhi, D., 2007. UAS scientist develops first drought tolerant rice. The Hindu. www.thehindu.com/2007/11/17/stories/2007111752560500.htm. (Accessed 20 March 2009).

Gandhi, R.V., Rudresh, N.S., Shivamurthy, H.S., 2012. Performance and adoption of new aerobic rice variety MAS 946-1 (Sharada) in southern Karnataka. Karnataka J. Agric. Sci. 25 (1), 5–8.

Gopalakrishnan, S., Sharma, R.K., Anand Rajkumar, K., et al., 2008. Integrating marker assisted background analysis with foreground selection for identification of superior bacterial blight resistant recombinants in basmati rice. Plant Breed. 127, 131–139.

Graybosch, R.A., Peterson, C.J., Baenziger, P.S., et al., 2009. Registration of 'Mace' hard red winter wheat. J. Plant. Regist. 3, 51–56.

Gupta, H.S., Agrawal, P.K., Mahajan, V., et al., 2009. Quality protein maize for national security: rapid development of short duration hybrids through molecular marker assisted breeding. Curr. Sci. 96 (2), 230–236.

Gupta, P.K., Varshney, R.K., 2000. The development and use of microsatellite markers for genetic analysis and plant breeding with emphasis on bread wheat. Euphytica 113, 163–185.

Hardin, B., 2000. Rice breeding gets marker assists. Agric. Res. Magaz. 48, 12. USDA-ARS, USA www.ars.usda.gov/is/AR/archive/dec00/rice1200.pdf.

Hash, C.T., Sharma, A., Kolesnikova-Allen, M.A., Singh, S.D., Thakur, R.P., Raj, A.G.B., 2006. Teamwork delivers biotechnology products to Indian small-holder crop-livestock producers: pearl millet hybrid "HHB 67 Improved" enters seed delivery pipeline. J. SAT Agric. Res. 2, 16–20. http://oar.icrisat.org/id/eprint/2738.

Hayes, P.M., Corey, A.E., Mundt, C., et al., 2003. Registration of tango barley. Crop Sci. 49, 1–12.

Heckenberger, M., Maurer, H.P., Melchinger, A.E., Frisch, M., 2008. The Plabsoft database: a comprehensive database management system for integrating phenotypic and genomic data in academic and commercial plant breeding programs. Euphytica 161, 173–179.

Heffner, E.L., Jannink, J.L., Iwata, H., et al., 2011. Genomic selection accuracy for grain quality traits in biparental wheat populations. Crop Sci. 51, 2597–2606.

Heffner, E.L., Lorenz, A.J., Jannink, J.L., et al., 2010. Plant breeding with genomic selection: gain per unit time and cost. Crop Sci. 50, 1681–1690.

Heffner, E.L., Sorrels, M.E., Jannink, J.L., 2009. Genomic selection for crop improvement. Crop Sci. 49, 1–12.

Helguera, M., Khan, I.A., Kolmer, J., Lijavetzky, D., Zhong-qi, L., Dubcovsky, J., 2003. PCR assays for the *Lr37-Yr17-Sr38* cluster of rust resistance genes and their use to develop isogenic hard red spring wheat lines. Crop Sci. 43, 1839–1847.

Helms, T.C., Nelson, B.D., Goos, R.J., 2008. Registration of 'Sheyenne' soybean. J. Plant Reg. 2, 20. https://doi.org/10.3198/jpr2007.03.0146cr.

Hospital, F., 2005. Selection in backcross programmes. Philos. Trans. R. Soc. Lond Biol. Sci. 360, 1503–1511.

Hospital, F., Charcosset, A., 1997. Marker-assisted introgression of quantitative trait loci. Genetics 147, 1469–1485.

Huang, N., Angeles, E.R., Domingo, J., et al., 1997. Pyramiding of bacterial blight resistance genes in rice: marker-assisted selection using RFLP and PCR. Theor. Appl. Genet. 95, 313–320.

ICAR IIRR, 2017. Newsletter. Vol. 15. No. 1 http://www.icar-iirr.org/.

Iftekharuddaula, K.M., Ahmed, H.U., Ghosal, S., et al., 2015. Development of new submergence tolerant rice variety for Bangladesh using marker assisted backcrossing. Rice Sci. 22 (1), 16–26.

Jaccoud, D., Peng, K., Feinstein, D., et al., 2001. Diversity arrays: a solid state technology for sequence information independent genotyping. Nucleic Acids Res. 29 (4), E25.

Janila, P., Pandey, M.K., Shasidhar, Y., et al., 2016. Molecular breeding for introgression of fatty acid desaturase mutant alleles (ahFAD2A and ahFAD2B) enhances oil quality in high and low oil containing peanut genotypes. Plant Sci. 242, 203–213. https://doi.org/10.1016/j.plantsci.2015.08.013.

Jiang, Y., Schmidt, R.H., Reif, J.C., 2018. Haplotype-based genome-wide prediction models exploit local epistatic interactions among markers. G3 Genes Genomes Genetics 8, 1687–1699. https://doi.org/10.1534/g3.117.300548.

Joseph, M., Gopalakrishnan, S., Sharma, R.K., Singh, V.P., Singh, A.K., Singh, N.K., Mohapatra, T., 2004. Combining bacterial blight resistance and Basmati quality characteristics by phenotypic and molecular marker-assisted selection in rice. Mol. Breed. 13, 377–387.

Kalia, R.K., Rai, M.K., Kalia, S., et al., 2011. Microsatellite markers: an overview of the recent progress in plants. Euphytica 177 (3), 309–334.

Kidwell, K., Santra, D., deMacon, V., et al., 2008. Precision Breeding: Wheat Research Progress Report. Washington State University, Agricultural Research Center. http://arc.wsu.edu/Links/images/2008/3572Kidwell.pdf. (Accessed 20 March 2009).

Klein, H., Xiao, Y., Conklin, P.A., et al., 2018. Bulked-segregant analysis coupled to whole genome sequencing (BSA-Seq) for rapid gene cloning in maize. G3 (Bethesda) 8 (11), 3583–3592. https://doi.org/10.1534/g3.118.200499.

Koebner, R.M.D., Summers, R.W., 2003. 21st century wheat breeding: plot selection or plate detection? Trends Biotech. 21, 59–63.

Kosova, K., Chrpova, J., Sip, V., 2008. Recent advances in breeding of cereals for resistance to barley yellow dwarf virus – a review. Czeck J. Genet. Plant Breed. 44, 1–10.

Kottapalli, K., Lakshmi, N.M., Jena, K., 2010. Effective strategy for pyramiding three bacterial blight resistance genes into fine grain rice cultivar, Samba Mahsuri, using sequence tagged site markers. Biotechnol. Lett. 32, 989–996.

Kumar, S., Travis, W., Banks Cloutier, S., 2012. SNP discovery through next-generation sequencing and its applications. Int. J. Plant Genomics 2012. https://doi.org/10.1155/2012/831460, 831460.

Kunert, A., Naz, A., Dedeck, O., et al., 2007. AB-QTL analysis in winter wheat: I. Detection of favorable exotic alleles for baking quality traits introgressed from synthetic hexaploid wheat (*T. turgidum* ssp. Dicoccoides 9 *T. tauschii*). Theor. Appl. Genet. 115, 683–695.

Lande, E.S., Botstein, D., 1989. Mapping mendelian factors underlying quantitative traits using RFLP linkage maps. Genetics 121 (1), 185–199.

Leal-Bertioli, S.C.M., Cavalcante, U., Gouvea, E.G., et al., 2015. Identification of QTLs for rust resistance in the peanut wild species *Arachis magna* and the development of KASP markers for marker-assisted selection. G3 (Bethesda) 5, 1403–1413.

Li, Z.K., Yu, S.B., Lafitte, H.R., et al., 2003. QTL x environment interactions in rice I. Heading date and plant height. Theor. Appl. Genet. 108, 141–153.

Lippman, Z.B., Semel, Y., Zamir, D., 2007. An integrated view of quantitative trait variation using tomato interspecific Introgression lines. Curr. Opin. Genet. Dev. 6, 545–552.

Lippman, Z.B., Zamir, D., 2007. Heterosis: revisiting the magic. Trends Genet. 23, 60–66.

Litt, M., Luty, J.A., 1989. A hypervariable microsatellite revealed by in vitro amplification of a dinucleotide repeat within the cardiac muscle actin gene. Am. J. Hum. Genet. 44 (3), 397–401.

Liu, B.H., 1991. Development and prospects of dwarf male-sterile wheat. Chinese Sci. Bull. 36 (4), 306.

Melchinger, A.E., 1990. Use of molecular markers in breeding for oligogenic disease resistance. Plant Breed. 104, 1–19.

Melchinger, A.E., Utz, H.F., Schon, C.C., 1998. Quantitative trait locus (QTL) mapping using different testers and independent population samples in maize reveals low power of QTL detection and large bias in estimates of QTL effects. Genetics 149, 383–403.

Meuwissen, T.H.E., Hayes, B.J., Goddard, M.E., 2001. Prediction of total genetic value using genome wide dense marker maps. Genetics 157, 1819–1829.

Miklas, P.N., Kelly, J.D., Sing, S.P., 2003. Registration of anthracnose-resistant pinto bean germplasm line USPT-ANT-1. Crop Sci. 43, 1889–1890.

Milkas, P.N., Smith, J.R., Singh, S.P., 2006. Registration of common bacterial blight resistant dark red kidney bean germplasm line USDK-CBB-15. Crop Sci. 46 (2), 1005–1006.

Moncada, P., Martinez, C.P., Borrero, J., et al., 2001. Quantitative trait loci for yield and yield components in an *Oryza sativa* x *Oryza rufipogon* BC2F2 population evaluated in an upland environment. Theor. Appl. Genet. 102, 41–52.

Mutlu, N., Milkas, P.N., Steadman, J.R., et al., 2005. Registration of pinto bean germplasm line ABCP-8 with resistance to common bacterial blight. Crop Sci. 45, 806.

Mutlu, N., Urrea, C.A., Milkas, P.N., et al., 2008. Registration of common bacterial blight, rust and bean common mosaic resistant great northern bean germplasm line ABC-weighing. J. Plant Regist. 2, 120–124.

Nakaya, A., Isobe, S.N., 2012. Will genomic selection be a practical method for plant breeding? Ann. Bot. 110 (6), 1303–1316. https://doi.org/10.1093/aob/mcs109.

Nandi, S., Subudhi, P.K., Senadhira, D., et al., 1997. Mappings QTLs for submergence tolerance in rice by AFLP analysis and selective genotyping. Mol. Gen. Genet. 255, 1–8.

Nayak, S.N., Singh, V.K., Varshney, R.K., 2017. Marker-assisted selection. In: Thomas, B., Murray, B.G., Murphy, D.J. (Eds.), Encyclopedia of Applied Plant Sciences. Vol. 2. Academic Press, Waltham, MA, pp. 183–197.

Naz, A.A., Kunert, A., Lind, V., et al., 2008. AB-QTL analysis in winter wheat: II. Genetic analysis of seedling and field resistance against leaf rust in a wheat advanced backcross population. Theor. Appl. Genet. 116, 1095–1104.

Neeraja, C., Maghirang-Rodriguez, R., Pamplona, A., et al., 2007. A marker-assisted backcross approach for developing submergence-tolerance rice cultivars. Theor. Appl. Genet. 115, 767–776.

Nelson, R.R., 1978. Genetics of horizontal resistance to plant diseases. Annu. Rev. Phytopathol. 16, 359–378.

Pandey, M.K., Agarwal, G., Kale, S.M., et al., 2017. Development and evaluation of a high density genotyping 'Axiom_Arachis' array with 58K SNPs for accelerating genetics and breeding in groundnut. Sci. Rep. 7, 40577. https://doi.org/10.1038/srep40577.

Pandey, M.K., Khan, A.W., Singh, V.K., et al., 2016. QTL-seq approach identified genomic regions and diagnostic markers for rust and late leaf spot resistance in groundnut (*Arachis hypogaea* L.). Plant Biotechnol. J. 15 (8), 927–941. https://doi.org/10.1111/pbi.12686.

Paran, I., Michelmore, R.W., 1993. Development of reliable PCRbased markers linked to downy mildew resistance genes in lettuce. Theor. Appl. Genet. 85 (8), 985–993.

Paterson, A.H., Damon, S., Hewitt, J.D., 1991. Mendelian factors underlying quantitative traits in tomato: comparison across species, generations and environments. Genetics 127, 181–197.

Paterson, A.H., Lander, E.S., Hewitt, J.D., et al., 1988. Resolution of quantitative traits into mendelian factors using a complete linkage map of restriction fragment length polymorphism. Nature 335, 721–726.

Payne, P.I., Nigtingale, M.A., Krattiger, A.F., Holt, L.M., 1987. The relationship between HMW glutenin subunit composition and the bread-making quality of Britishgrown wheat varieties. J. Sci. Food Agric. 40, 51–65.

Rafalski, A., 2002. Applications of single nucleotide polymorphisms in crop genetics. Curr. Opin. Plant Biol. 5, 94–100.

Rai, V.P., Kumar, R., Kumar, S., Rai, A., Kumar, S., Singh, M., Singh, S.P., Rai, A.B., Paliwal, R, 2013. Genetic diversity in *Capsicum* germplasm based on microsatellite and random amplified microsatellite polymorphism markers. Physiol. Mol. Biol. Plants 19 (4), 575–586. https://doi.org/10.1007/s12298-013-0185-3.

Rai, A., Singh, A.M., Raghunandan, K., et al., 2019. Marker-assisted transfer of PinaD1a gene to develop soft grain wheat Cultivars. 3Biotech 9, 183.

Reddy, C.S., Babu, A.P., Swamy, B.P.M., Kaladhar, K., Sarala, N., 2009. ISSR marker based on GA and AG repeats reveal genetic relationship among rice varieties tolerant to drought, flood, salinity. J. Zhejiang Univ. Sci. B10, 133–141.

Reif, J.C., Melchinger, A.E., Xia, X.C., et al., 2003. Genetic distance based on simple sequence repeats and heterosis in tropical maize populations. Crop Sci. 43 (4), 1275–1282.

Ribaut, J.M., Hoisington, D., 1998. Marker-assisted selection: new tools and strategies. Trends Plant Sci. 3, 236–239.

Sax, K., 1923. Association of size differences with seed coat pattern and pigmentation in *Phaseolus vulgaris*. Genetics 8, 552–560.

Septiningsih, E.M., Pamplona, A.M., Sanchez, D.L., et al., 2009. Development of submergence-tolerant rice cultivars: the Sub1 locus and beyond. Ann. Bot. 102 (2), 151–160.

Servin, B., Martin, O.C., Mezard, M., et al., 2004. Toward a theory of marker-assisted gene pyramiding. Genetics 168, 513–523.

Shannon, J.G., Lee, J.D., Wrather, J.A., Sleper, D.A., Mian, M.A.R., Bond, J.P., Robbins, R.T., 2009. Registration of S99-2281 soybean germplasm line with resistance to frogeye leaf spot and three nematode species. J. Plant Regist. 3, 94–98. https://doi.org/10.3198/jpr2008.06.0307crg.

Sharma, A., Srivastava, P., Mavi, G.S., et al., 2021. Resurrection of wheat cultivar PBW343 using marker-assisted gene pyramiding for rust resistance. Front. Plant Sci. 12 (570408). https://doi.org/10.3389/fpls.2021.570408, 570408.

Shasidhar, Y., Vishwakarma, M.K., Pandey, M.K., et al., 2017. Molecular mapping of oil content and fatty acids using dense genetic maps in groundnut (*Arachis hypogaea* L). Front Plant. Sci. 8, 794. https://doi.org/10.3389/fpls.2017.00794.

Simpson, C.E., Burow, M.D., Paterson, A.H., Starr, J.L., Church, G.T., 2003. Registration of NemaTAM peanut. Crop Sci. 43, 1561. https://doi.org/10.2135/cropsci2003.1561.

Simpson, C.E., Starr, J.L., 2001. Registration of 'COAN' peanut. Crop Sci. 41, 918. https://doi.org/10.2135/cropsci2001.413918x.

Singh, A.K., Gopalakrishnan, S., Singh, V.P., et al., 2011. Marker assisted selection: a paradigm shift in basmati breeding. Indian J Genet 71, 120–128.

Singh, V.K., Singh, A.T., Singh, S.P., et al., 2013. Marker assisted simultaneous but stepwise backcross breeding for pyramiding blast resistance gene Piz5 and Pi5 into an elite basmati rice restorer line 'PRR78. Plant Breed 132, 486–495.

Staub, J.E., Serquen, F.C., Gupta, M., 1996. Genetic markers, map construction, and their application in plant breeding. Hort Sci. 31 (5), 729–741.

Steele, K.A., Price, A.H., Witcome, J.R., et al., 2013. QTLs associated with root traits increase yield in upland rice when transferred through marker assisted selction. Theor. Appl. Genet. 126 (1), 101–108.

Suh, J.P., Jeung, J.U., Noh, T.H., et al., 2013. Development of breeding lines with three pyramided resistance genes that confer broad spectrum bacterial blight resistance and their molecular analysis in rice. Rice 6, 6–11.

Sundaram, R.M., Vishnupriya, M.R., Biradar, S.K., et al., 2008. Marker assisted introgression of bacterial blight resistance in samba Mahsuri, an elite indica rice variety. Euphytica 160, 411–422.

Takagi, H., Abe, A., Yoshida, K., Kosugi, S., Natsume, S., Mitsuoka, C., Uemura, A., Utsushi, H., Tamiru, M., Takuno, S., Innan, H., Cano, L.M., Kamoun, S., Terauchi, R., 2013. QTL-seq: rapid mapping of quantitative trait loci in rice by whole genome resequencing of DNA from two bulked populations. Plant J. 74, 174–183.

Takeuchi, Y., Ebitani, T., Yamamoto, T., et al., 2006. Development of isogenic lines of rice cultivar Koshihikari with early and late heading by marker-assisted selection. Breed. Sci. 56, 405–413.

Tanksley, S.D., 1983. Molecular markers in plant breeding. Plant Mol. Biol. Rep. 1, 3–8.

Tanksley, S.D., Nelson, J.C., 1996. Advanced backcross QTL analysis: a method for simultaneous discovery and transfer of valuable QTLs from unadapted germplasm into elite breeding lines. Theor. Appl. Genet. 92, 191–203.

Tao, C., Hao, W., Ya-dong, Z., et al., 2016. Genetic improvement of japonica rice variety Wuyujing 3 for stripe disease resistance and eating quality by pyramiding Stv-bi and Wx-mq. Rice Sci. 23, 69–77.

Toenniessen, G.H., O'Toole, J.C., De Vries, J., 2003. Advances in plant biotechnology and its adoption in developing countries. Curr. Opin. Plant Biol. 6, 191–198.

Vilanova, S., Sargent, D.J., Arús, P., Monfort, A., 2008. Synteny conservation between two distantly-related Rosaceae genomes: Prunus (the stone fruits) and Fragaria (the strawberry). BMC Plant Biol. 8, 67.

Vishwakarma, M.K., Arun, B., Mishra, V.K., et al., 2016a. Marker-assisted improvement of grain protein content and grain weight in Indian bread wheat. Euphytica 208, 313–321. https://doi.org/10.1007/s10681-015-1598-6.

Vishwakarma, M.K., Mishra, V.K., Gupta, P.K., et al., 2014. Introgression of the high grain protein gene *Gpc-B1* in an elite wheat variety of Indo-Gangetic Plains through marker assisted backcross breeding. Curr. Plant Biol. 1, 60–67.

Vishwakarma, M.K., Pandey, M.K., Shasidhar, Y., et al., 2016b. Identification of two major quantitative trait locus for fresh seed dormancy using the diversity arrays technology and diversity arrays technology-seq based genetic map in Spanish-type peanuts. Plant Breed. 135, 367–375. https://doi.org/10.1111/pbr.12360.

Visscher, P.M., Thompson, R., Haley, C.S., 1996. Confidence intervals in QTL mapping by booststrapping. Genetics 143, 1013–1020.

Vos, P., Hogers, R., Bleeker, R., et al., 1995. AFLP: a new technique for DNA fingerprinting. Nucleic Acids Res. 23, 4407–4414.

Watson, A., Ghosh, S., Williams, M.J., et al., 2018. Speed breeding is a powerful tool to accelerate crop research and breeding. Nat. Plants 4, 23–29. https://doi.org/10.1038/s41477-017-0083-8.

Wenzl, P., Carling, J., Kudrna, D., et al., 2004. Diversity arrays technology (DArT) for whole-genome profiling of barley. Proc. Natl. Acad. Sci. U. S. A. 101 (26), 9915–9920.

Wheat, C.A.P., 2009. Germplasm Release. University of California, Davis. https://maswheat.ucdavis.edu/Achievements/cultivars2007.htm. (Accessed 20 March 2009).

Williams, J.G.K., Kubelik, A.R., Livak, K.J., et al., 1990. DNA polymorphisms amplified by arbitrary primers are useful as genetic markers. Nucleic Acids Res. 18, 1631–1635.

Xu, Y., 2010. Molecular Plant Breeding. CAB International.

Xu, Y., Crouch, J.H., 2008. Marker-assisted selection in plant breeding: from publications to practice. Crop Sci. 48, 391–407.

Yadav, P., Mishra, V.K., Arun, B., et al., 2015. Enhanced resistance in wheat against stem rust achieved by marker assisted backcrossing involving three independent Sr genes. Curr. Plant Biol. 2 (2015), 25–33.

Young, N.D., Tanksley, S.D., 1989. RFLP analysis of the size of chromosomal segments retained around the tm-2 locus of tomato during backcross breeding. Theor. Appl. Genet. 77, 353–359.

Yu, L.X., Abate, Z., Anderson, J.A., et al., 2009. Developing and optimizing markers for stem rust resistance in wheat. In: The Borlaug Global Rust Initiative 2009 Technical Workshop. March 17–20, 2009, Fiesta Inn, Ciudad Obrego'n, Sonora.

CHAPTER 15

Nanoherbicides: A sustainable option for field applications

Vidya Patil-Patankar[a],* **and Gaurav Sanghvi**[b]
[a]Department of Botany, PDEA's Waghire College, Pune, Maharashtra, India
[b]Department of Microbiology, Marwadi University, Rajkot, Gujarat, India
*Corresponding author. e-mail address: patankarvv.pdea@gmail.com

1. Introduction

Agriculture is the practice dealing with actions required for the production of important livestock adding nutrition to human lives and serving as an important factor for continuing the different ecological cycles. Agriculture is a vital process in maintaining the economy of the developing countries directly or by indirect means (Sekhon, 2014). With the increase in global population, urbanization, and growth of industrialization, important agricultural lands and crops are diminishing at drastic rates. On the other side, the population of the world is expected to rise and reach up to 9 billion by 2050 (PRD, 2017). Agriculture is one of the growth pillars for developing countries. as in the economy of developing countries is mainly dependent on agriculture. Agriculture serves as the main source of income and for the fulfillment of basic needs for the continuation of life all over the world. Reports from Food and agriculture organization show that approximately 50% GDP of developing countries relies on agriculture. Among the major developing countries contributors, India is one of the leading developing agriculture exporting countries. Agriculture provides 60% of livelihood to Indians. The agriculture data indicates that 17%–18% of GDP is contributed by half of the population with agriculture as the primary source of income. Apart from being the highest wheat producer, India also serves as a major producer of fruits including banana, guava, and papaya. Moreover, major production of vegetables and sprouts is also carried out in India.

India is one of the largest Asian countries with approximately fifteen thousand kilometers of land and seven thousand kilometers of coastline. Among India's total area, approximately 190 km is used for agriculture purposes. With India's population of more than 100 crores, the agricultural system of India is facing challenging issues like polluted water resources, unfertile soil, and rapid climate change at its disposal. Large diverse landscape comprising of the mountains, peninsular plateaus, forests on high altitudes, insects are also challenging for the Indian agricultural sustenance. The country also experiences diverse temperatures, heavy rainfall, and also extreme aridity and humidity. Depending

Relationship Between Microbes and the Environment for Sustainable Ecosystem Services, Volume 1
https://doi.org/10.1016/B978-0-323-89938-3.00015-3

on the soil types, physiography, and diverse ecological regions, 20 different agro regions and approximately 60 ecological sub-regions are developed in India. Further, each agro region was divided into the agro eco units at district levels for the development of sustainable agriculture practices (Gajbhiye and Mandal, 2000).

Each of the agro-ecological regions and crops has distinct problems like weeds, insects, microbes, etc. (Rao et al., 2014). India ranks second among the largest producer of livelihood stocks like rice, wheat, and sugarcane. Taking into this consideration, India needs to manage the crop deteriorating factors like the weeds and insects for better yield with resource competing (Rao and Nagamani, 2010). To achieve higher yields, root cause identification and possible solutions need to be addressed very well. For higher yields, management of the weeds is the most critical part, as weeds pose the major biological hindrance to limit crop productivity. As per the published reports, it is reported that mainly weeds are responsible for near about 30%–32% of reduction in crop yields (Bhan et al., 1999). The other reports indicate that approximately one-third of the crop losses are due to the weeds with severe effects on the quality of the crops. As per the reports by Yaduraju (2012), about US 13 billion food grains are lost due to improper weed management. The overall losses by the weeds would be too high if unintended effects of weeds are calculated on human health, nutrient quality reduction, and the nutritive content of the crops and quality of cereals.

2. Weed management

With a huge population and cheap labor, manual weeding is the most used practice for managing weeds in India. The manual weeding is further combined with the mechanical methods for weeds management. With the increase in farm wages, the combination of mechanical and manual weeds management practices has increased significantly. In the 2000s, the wages rose to 17.8% compared to the 11.6% in the 1900s (Labour Bureau, India). This has led to increased dependence on the use of herbicides on the removal of weeds (Rao et al., 2014). The most common herbicide used was 2,4-Dichlorophenoxyactetic acid (2,4-D) and isoproturon. With this as base molecules, many new generation molecules like sulfosulfron, mesosulfuron + iodosulfuron, and isproturon + 2,4-D mixture. This is a common herbicide used by wheat farmers to increase productivity. In addition, the rice farmers used to use the combination of thiobencarb, butachlor, and 2, 4-D. The use of different practices and its implications in crops along with weed management is very well reported by Chadha et al. (1997). The report has covered all the crops, fruits, medicinal important herbs, aromatic plants, and ornamental plants (Chadha et al., 1997). Most researchers of India have recommended the usage of the combination of the hand weeding practice with the synthetic herbicides (Rao and Nagamani, 2010). The dynamic evolution in weeds has also lids to the continuous redesigning of strategies to manage this weed in the field for its successful management.

3. Herbicides

Herbicides are synthetic chemicals used specifically to control or kill weed plants in an explicit, partial, or full manner (Pretty and Bharucha, 2015). Herbicides kill or suppress plants by interfering with essential plant biochemical processes such as photosynthesis. Weed management and control through biological or chemical ways possess many advantages, due to the presence/usage of herbicides in the agricultural field. Herbicide serves as a crucial function to increase the agriculture yield by guarding the crops against pests and other harmful biological entities. The combination of herbicides and pesticides has increased drastically in the agriculture system causing environmental pollution. This combination has also led to leaching, soil infertility, and runoff from the application site posing a serious challenge in the agriculture system (Pimentel, 2012). Herbicides are further categorized depending on many extrinsic and intrinsic factors.

3.1 Classification

Mode of action, site of activity, selectivity, movement, and so forth are important parameters for the classification of herbicides (Duke, 1990; Torrens and Castellano, 2014; Varshney et al., 2012). The dose of herbicide and its specificity is still an important factor to be considered for the safe use of the herbicide. In addition, the frequent usage of more concentration of the herbicides wills lids to development of resistant variety; therefore, optimum concentration of herbicides must be used for their maximum efficiency. Also, the type of applications and method used, it is critical to find some kind of harmony between these procedures and track down the ideal mode generally advantageous for the most extreme impact.

It was reported that herbicides when used in preemergent mode, they are viable against green weeds or expansive leaf weeds (Grossmann and Ehrhardt, 2007; Wright et al., 2010). Herbicides, when applied postemergent mode, they might be particular (explicit objective) or nonselective (expansive objective) (Grossmann and Ehrhardt, 2007). At the preseed treatment before the germination stage, might be the application of herbicide has been done to the seeds directly or indirectly with the dust particles. In postemergent applications, the seedlings are showered with explicit herbicides to wipe out weeds present in the field. The selection is characterized as the limit of herbicide to murder an objective plant deprived of hurting or not executing the aiming plants (Crafts, 1946). Particular herbicides are profoundly explicit and are most appropriate for the control of a particular weed-related to a particular harvest. The vast majority of the herbicides utilized in agribusiness and related ventures are exceptionally specific. Upon contact, they act by getting ingested and moved into the xylem or the phloem of the weeds, by hindering or upsetting the metabolic hardware or other biosynthetic pathways, and by harming or slaughtering the weeds (Fernández and Brown, 2013). Nonselective herbicides have restricted use in farming and other related ventures, yet they are powerful in land–recovery projects where the land is free from the weeds and the plants of interest. Further, glyphosate has been utilized globally as nonselective herbicide, however, its action is more specific when utilized in relationship with hereditarily designed harvests, which have been produced for fighting against glyphosate (Devos et al., 2008; Gustafson, 2008). The selectivity or non-selectivity of herbicides relies on different aspects, for example, plant physiology, soil geography, climate, the timing of use, the pace of utilization, and application procedure (Varshney et al., 2012). The grouping of herbicides is similarly significant for overseeing and apprehending herbicide obstruction, which keeps on being an issue in economical rural administration (Duke, 1990; Torrens and Castellano, 2014). The excess use of herbicides, actually just like different pesticides like insect poisons, may prompt expanded advancement in resistance among plants, causing injury and obliteration of helpful plants in both farming and land the board. Understanding the explanations behind ordering the herbicides depend on their methods of activity, rather than the synthetic family or the site of activity, will assist with understanding the purposes for the improvement of resistance because of their excessive use.

Plants interrelate distinctively with various herbicides dependent on factors like retention, movement, digestion, and physiological reaction. The form of herbicide activity might be pervasive at the tissue or cell levels and can also further leads to the tissue-injury for a particular cluster (Deboer et al., 2006). In addition, herbicides are very particular in their specificity and soil geography. Herbicides may either be applied straightforwardly/directly to the foliage or be added to the dirt during furrowing/plowing (Anwar et al., 2013). Herbicides may have an upright or parallel development specific to the diverse clusters (Pereira et al., 2014; Deboer et al., 2006; Martini et al., 2015). Different herbicides execute upon contact on the foliage and are powerful enough with the end goal that they don't need movement in any case (Anwar et al., 2013). Plants are flawless frameworks that comprise organs, tissues, cells, and particles, showcasing repositories of coordinated biochemical cycles. Herbicides might be reached/consumed by the plants through the roots (soil-based herbicides) or the shoots (Turgut, 2005).

The metabolic action requires the development of sap through the xylem (movement of water and supplements) and phloem (movement of sugars) (Lemoine et al., 2013; Pate et al., 1979). At the point when the herbicides enter the cell wall of the weeds, they cause tissue injury and saturate the sap, simultaneously, obstructs different biochemical pathways. Upon association with the herbicides, weeds are destroyed by the brokenness of their biochemical cycles.

Herbicides clusters are dependent on their synthetic designs of chemical molecules, which comprises a base-explicit particle encompassed by a side chain or a group(s). An adjustment inside moieties prompts a change in the movement, selectivity, and doggedness of herbicides. The more and more use of herbicides has expanded the odds of harm and injury to the non-target weeds/plants also further causing diverse ecological problems like groundwater pollution (Ronald, 2011; Piver, 1993). Majorly, the activity of herbicides depends on whether they are retarders, controllers, or hecklers of different biosynthetic pathways. The activity of herbicides for destroying a weed is surely known and needs a viability synthetic factor.

Usually, most of the herbicide's mode of activity is as per the following steps:
Contact—Must attach with the weed
Ingestion/absorption—Must be uptake by the weed
Movement/translocation—Must move to the site of activity in the weed
Harmfulness—Must have an adequate degree of fatalness or power for weed removal
Demise/Death—Injury to the weed prompting for its death.

3.2 Herbicide's mode of action

The herbicides' activity is pretty much dependent on factors like internal molecular arrangements with keeping in the focus of controlling plants through different biochemical methods. The weed-control impact at the development stage is maximum and least at the fully grown stage, as the plant goes through the phases of growing, seedling, vegetative, blossoming, and development, individually (Tehranchian et al., 2015). Reliant on the particular activity of herbicide which might include involvement of plant chemical or an organic framework that the herbicide may interfere with, subsequently harming or upsetting the standard plant development causing inevitable plant demise. Broad examination in the gap of herbicides development has prompted their characterization of its activity into different synthetic herbicides clusters. Newly found synthetic molecules and its clusters are always adding the unexplained mechanism of new generation herbicides. In addition, a few researchers may have a somewhat extraordinary strategy for arrangement in diverse herbicide clusters for better efficiency; however, those are only applicable to a few weed varieties. What is significant, in any case, is that having a top to bottom information on the method of activity of herbicides is essential in picking a particular herbicide for a particular harvest, understanding the injury manifestations, and resulting in a fitting yield. Lastly with respect to the natural destiny of herbicides (perseverance, debasement, portability) with explicit reference to the groundwater, water cleansing, soil pollution, and ecological and general wellbeing concerns with the significance of utilizing the herbicides perseveringly and as per the guidelines, rules, set by competent authorities.

4. Nanoherbicides

Nanoherbicides are herbicides made up of nanoparticles that will allow farmers to clear their fields of weeds without using toxic chemicals. Use of nanotechnology to develop nanoherbicides and these are synthesized the formula with nanosized preparations or used the nanomaterials-as base to formulate herbicides. Crop productivity can be enhanced significantly by the use of nanoherbicides applications in agriculture. Researchers have

Table 1 Comparison of nano-herbicides to conventional herbicides.

Characteristics	Nanoherbicides	Herbicides
Dispersion	Better solubility and dispersion depending on the method of nanoparticle formulation mechanism	Poor dispersion and solubility, requires combination of manual efforts for better performance
Specificity	Target specific delivery possible	Random delivery mechanism for weed removal
Release and control	Controlled Release rate and release pattern	Uncontrolled release rate and pattern
Effective duration and concentration	More effectiveness with less concentration	More effectiveness from high concentration
Value addition	Significant reduction of chemicals hazards to soil and maintains fertility of soil	Increase in chemical hazard to soil and decrease fertility

developed different ways of effective measures of delivery of herbicides. A popular example of this is encapsulating atrazine herbicide, poly (epsilon–caprolactone) nanoparticles, and so on. The most important effect showed by these nanocapsules is the strong control of the targeted species, reduction in genotoxicity level, and also known to significantly lower down the atrazine mobility in the soil. For the preparation of the nanoherbicide formulations mostly two different types of nanoparticles are used namely polymeric nanoparticles and inorganic nanoparticles. The comparison of conventional herbicides and Nano herbicides is shown in Table 1.

These nanoparticles are decomposable and with non-toxic metabolites such as polymeric substances which are mostly used in herbicide delivery systems. These formulations might increase the efficiency of the herbicide, boost the solubility, and decreases the toxicity in comparison with the conventional herbicides. Research has shown that nanoformulations assist in reducing the main disadvantage of the herbicide industry such as the evolution of herbicide–resistant plants (Manjunatha et al., 2012).

5. Nanotechnology and its implications in weed management

Nanotechnology includes the utilization and control of matter at a nano size. At the nano size, particles and atoms work unexpectedly and give different astounding and intriguing uses and applications (Fig. 1). The rise of nanotechnology has given a broad spectrum of applications by crossing different boundaries of science and shaping for better effect on all types of life forms (Sekhon, 2014). In the field of nanotechnology, nanoparticles (NPs) research is a significant approach because of its multitudinous applications. NPs are portrayed as particles having width measures not exactly or equivalent to $0.1\,\mu m$ (100 nm)

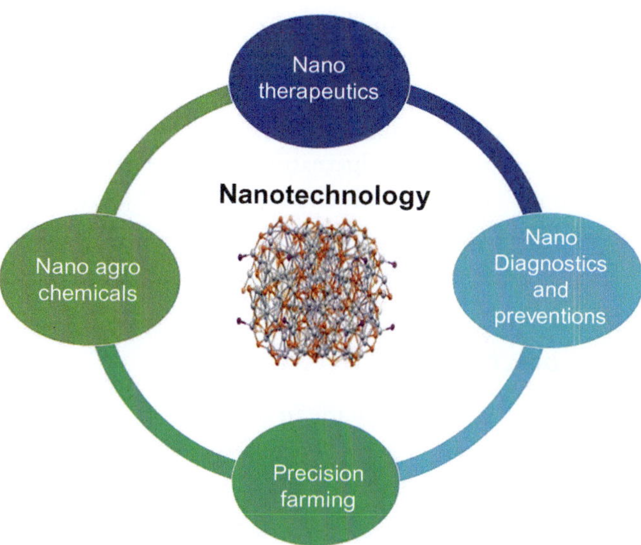

Fig. 1 Application of nanotechnology in agriculture.

and with explicit properties that rely principally upon their size (Hanemann and Szabó, 2010).

Nanotechnology as a novel branch in science leads to new agrochemical innovation and new pesticide passage to improve crop productivity, and perhaps lessen escalated pesticide use. In addition, it also opens up novel applications in farming. Nano-formulations are by and large expected to address obvious dissolvability issues of solvent, to deliver the dynamic fixing in a focused/systematic way (Kah et al., 2013). Nanostructure/nanoformulations mediated herbicide delivery, upgrades its bioavailability, improves the coordinated influx of the dynamic fixing, and empowers precise delivery systems (Sopeña et al., 2009; Preet Kaur et al., 2011). However, the use of herbicides in an improper manner has led to major consequences for ecological sites resulting in the increased levels of herbicides in the water, soil, and air pollution (Itodo et al., 2017; Koterba et al., 1993). This use of synthetic chemicals has researchers around the globe to find the substitute of chemical herbicide in form of biological or another eco-friendly alternative. With major advancements in micro and nanoscience, in comparison to the active bulk chemicals, the combination of nanotechnology in the formulation of nanoherbicides can lead to better crop yield and also significantly protect from the biological factors (Jampílek and Kráľová, 2015). In addition, the important goal of sustainable agriculture with a very low concentration of chemicals can be achieved using these nanoherbicides. The developed nanoherbicides will also decrease the environmental toxicity in the crops (Maruyama et al., 2016). The main aim of the nano herbicides is to encapsulate the herbicides' active constituents allowing them to the user for their specific concentration, reduction of heavily used synthetic chemicals having a negative impact on the environment (Sopeña et al., 2009; Hashim et al., 2014). Sustainable and controlled release of the herbicides is one of the major advantages as its use in potential weed management (Jampílek and Králová, 2015). A report on the combination of different salts like sodium chloride, sodium carbonate, and phosphates amalgamation with Zn/Aluminum layer on 2, 4-D has shown significant results in the effectiveness as the weed controller. The formulation was also displaying the simultaneous preservation of herbicidal effectiveness (Jampílek and Králová, 2015).

6. Weeds and nanotechnology

Multi-species approach with single herbicide in the variable climate brought about issues like resistance and explicit lessen control. Constant openness of plant with almost no defense system leads to the effectiveness of herbicide in one season and with another season the same formulation faces obstruction in its application and become a problem for ecological niches as well as the diverse pollution (Chinnamuthu and Kokiladevi, 2007). Foliar application of herbicide in weeds, for example, *Cynodan dactylon*, and *Solanum* sp. significantly failed in the killing of these species. The objective areas of the currently used herbicides in a plant cell are to alter or destruct the chloroplast structure, restraint of lipid biosynthesis, incursion with cell division by upsetting the mitosis process and create hindrance in cellulose biosynthesis (Böger, 2020; Wakabayashi, 1989). Contrary to the foliar application of herbicides, root assimilation has a straightforward mode of action. Roots don't have skin like leaves; albeit, developed roots are enclosed by a suberized coating. This implies that there are not many boundaries to herbicide assimilation by plant roots. Since roots are lipophilic, lipophilic herbicides will be promptly uptake/consumed. At the sub-atomic level, activities are not yet totally seen in any experiments, for some industrially accessible herbicides. As of now, no commercial herbicides exist that intrude with CO_2-fixation and sugar creation. Atomic depiction of underground plant parts for fostering a receptor-based herbicide particle having an explicit property with nano-herbicide particles like carbon nanotubes perfectly fits for destructing underground propagules of weed seeds (Zimdahl, 2018).

7. Nano-herbicides and possible actions

Weeds are a huge risk in farming; they decrease the yield up generally. So, there could be no other alternative aside from killing/removing them completely. Nanotechnology can dispose of weeds by utilizing Nano-herbicides in an eco-friendly way, without leaving any harmful synthetic chemicals in soil and climate (Pérez-de-Luque and Rubiales, 2009). Less quantity of herbicide will be utilized if dynamic formulations are joined with an effective propagation and absorption method (Fig. 2). Having size in nano

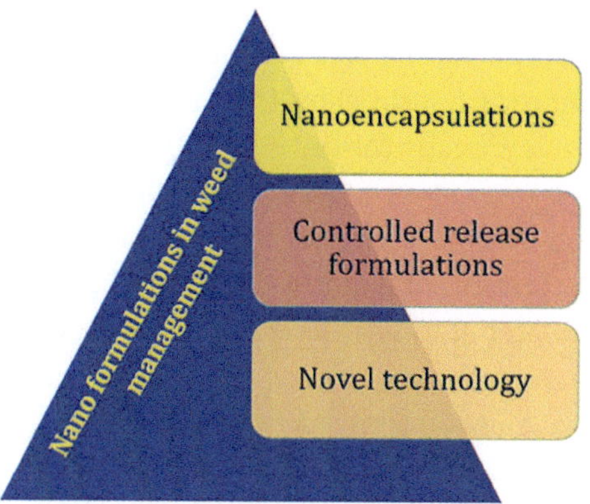

Fig. 2 Technologies for developing nano formulations in weed management.

measurements, these nanoparticles will mix with soil particles and forestall the development of weed species that have been left on the field by ordinary herbicides. Herbicides available in the market are intended to control or destroy the underground part of the weed plants. None of the herbicides hinder the action of feasible underground ground plant parts like rhizomes or tubers, which go about as a hotspot for new weeds in the resulting season. Fostering an objective of explicit herbicide embodied with nanoparticle is focused on a specific receptor found in the target weeds, which goes into roots arrangement of the weeds and moved to parts that restrain glycolysis, making the particular weed plant starve for food and gets in final death phase. Detoxification of weed deposits is vital as overuse/utilization of herbicides for a longer time deteriorates soil fertility and composition (Chinnamuthu and Boopathi, 2009). Just as nonstop utilization of the same herbicide for a steady timeframe prompts the advancement of weed resistance against specific herbicides. Up to 88% detoxification of the herbicide atrazine by Carboxy Methyl Cellulose (CMC) nanoparticles has been published (Satapanajaru et al., 2008).

8. Nanomaterial used in the synthesis

Synthesis of nonherbicides requires nanomaterials or particles which having efficiency in the effective delivery of chemicals and use of various nanomaterials-based herbicide formulations.

Nano herbicide in comparison with conventional herbicide could progress the efficacy of the herbicide, improves the solubility and reduce the toxicity. Nanoparticle-based herbicide helps to reduce the herbicide resistance potential, maintain the active ingredient's activity and prolong effects to weeds control (Manjunatha et al., 2012).

Nano particles can be act as good carrier as well as can be used in formulation of herbicide to synthesized nano herbicides. In synthesis of nonherbicide the precise herbicide molecule encapsulated with nanoparticle which act upon specific receptors present at the root of the targeted weed. Such nano particles enter inside of the roots of weeds and prevents the glycolysis of plant root system. This leads to undernourishment of plant cells and ultimately results into death. Encapsulation with nanoparticles for the control release of herbicide is another way to retain herbicide activities in soil and to control weeds in rainfed lands where dissemination of herbicides takes place through vaporization due to insufficient soil moisture. Use of adjuvant to enhance herbicidal activity is also known but presently use of nanomaterial to synthesized adjuvant in trend for to enhance potential of herbicides. Soybean micelle was applied with a nanotechnology-based surfactant to produce glyphosate-resistant crop reported presently (Manjunatha et al., 2012). Evolution of herbicide resistant plants is main drawback can be overcome with the use of nano-formulations.

Nanoparticles are used in preparation can be categorized on the basis of their size, shape and chemical nature of material used for preparation of nanoparticles. Dimensions in the macro scale of nanoparticle such as 0D includes the nanoparticles whose dimensions are all on the nanoscale, 1D—nanofibers and nanowires, 2D—nanosheets and thin films, and 3D—the materials in bulk (Singh, 2016). Classification also based on chemical nature and it is categorized in 4 main categories carbon, ceramic (metal oxides), metal, and polymeric compounds (Khan et al., 2019). The nanoparticles using cadmium oxides are shown in Fig. 3.

The biodegradable and non-toxic metabolites which are polymeric substances used for nanoherbicide formulations, these nanoparticle systems have effective herbicide delivery in targeted plants. Structures, such as fullerenes, graphene, and carbon nanotubes were comprising in Carbon-based nanomaterials (Ealias and Saravanakumar, 2017); Metal-oxide compounds such as TiO_2, ZnO, and FeO_2 are used to make inorganic solids in ceramic nanomaterial (Thomas et al., 2015); and Au, Ag, Cu, and Ni includes in metals include in nanomaterial. In polymeric compounds, dendrimers are used as Organic

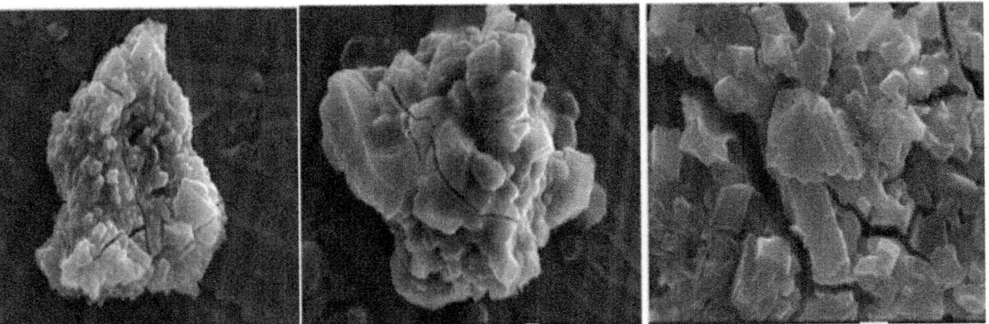

Fig. 3 Nano particle formulation using cadmium oxides. *(Adapted from Pandya, R., Patel, H., Shah, N.A., Solanki, P.S., Jani, Y.N., Keshvani, M.J., 2020. Substitutional effect of copper replacement by cadmium on structural, microstructural and electrical properties of $Cu_{1-x}Cd_xO$ oxides. Mater. Today Commun. 25, 101458.)*

nanomaterials, which are derived from organic nanoparticles that are generally symmetrical to the nucleus (Ealias and Saravanakumar, 2017).

8.1 Metal oxide nanoparticles

Metal oxide compounds have photocatalytic activities. TiO_2, ZnO, WOx, SnO_2, Fe_2O_3, CuO, ZrO_2, and MoO_3 are some examples of metal oxides used for synthesis of nano particles (Sharma et al., 2018). Photocatalytic compounds interact with light energy leads to degradation reaction (Babu et al., 2019) and to enhance the photocatalytic activity and production of reactive oxygen species, the metal oxide Nano Particles surface can be modified by metal ion doping or non-metal insertion (Bishoge et al., 2018). In general, this photocatalytic degradation with nanoparticles accelerates the reaction of degradation mostly helpful in waste treatment (Rodríguez-Méndez et al., 2017; Pérez-Sánchez et al., 2013). In agriculture, such effluents after treatment of waste can be considered as not dangerous in low concentrations (Babu et al., 2019) and they can be used for herbicide preparation as it helps to prevent the pollution of soil.

8.2 Metallic nanoparticles

Metallic Nanoparticles are comprised of metals such as Au, Pt, Ag, Zn, and Ni and alloys (Dolez, 2015). Gold nano Partials are inactive at an abundant scale become more chemically active when their size is decreased due to a larger spacing between atomic coordinates, which allows their use in catalytic applications (Saleh, 2020). Such metallic helps for catalytic activities and used for nano herbicide production.

8.3 Carbon-based nanomaterials

Nanotubes, nanofibers, fullerenes, and graphene, are some famous structures because of their various types, structures, and dimensions; these such materials are known as nanomaterials (Mathur et al., 2017; Zaytseva and Neumann, 2016). Such materials have properties like semiconductor behavior, thermal conductivity, high structural properties, and light absorption properties (Dolez, 2015). Therefore, they are used in solar cells, structural materials, electronics, and semiconductors due to the properties and applications they have. Classification of Carbon nanotubes based on the number of walls present single-wall, double-wall, or multi-wall and these walls decide the properties of Carbon nanotubes they are responsible for their structural and conductive characteristics (Pérez-Sánchez et al., 2013; Sudha et al., 2018). It can also be used for herbicide production.

8.4 Nanoparticles used in the delivery

8.4.1 Polymeric nanoparticles

Nanospheres or nanocapsules of polymeric nanoparticles are used to formulate nanoherbicides. One example of this is the polymeric nanoparticles containing atrazine herbicide.

This size, polydispersity index, pH, and encapsulation efficiency are specialized features of nanoherbicide. Also, it was found that the stability of these particles were more than 3 months. For preparation of atrazine herbicide mostly for encapsulation herbicides Poly (epsilon caprolactone) is frequently used polymer. These polymers having physico-chemical properties, also known to have enhanced bioavailability and biocompatibility. Free atrazine does not show reduced mobility of herbicide in the soil and higher herbicidal activities in comparison with nanoherbicides formulations synthesized with the help of polymeric nanoparticles (Pereira et al., 2014). Field trial of this nano-herbicide were done in Brassica sp., the polymeric nanoparticles encapsulated with atrazine were proven effective. Similarly, Grillo et al. (2012) performed another study they used the polymer for encapsulated three triazine herbicides such as atrazine, ametryn and simazine to reduce the environmental impact caused by them. Surprisingly the encapsulated polymeric nanoparticles of triazines possessed better association efficiency over 84%. Some studies shown that polymeric nanoparticles encapsulated with herbicide are safe in comparison of the traditional herbicide considering their adverse environmental impacts on human health as well on the environment.

8.4.2 Inorganic nanoparticles

Nonoparticles were prepared from inorganic nanoparticles such as Silica dioxide nanoparticles (SiNP). Silica dioxide nanoparticles were used as inorganic herbicide carriers for pH sensitive active substances. These compounds help to control optimal concentrations of herbicide and reduces rate of herbicide consumption rate. Also, these systems with easy deposit on the plant leaves reduce their wastage and protect to stabilize the herbicide. Constant release of herbicide carriers through a dynamic adsorption mechanism reported by the silica nanoparticles reported by Rani et al. (2014). Though there is need to study the control release of herbicide, leaching behavior and toxicity of SiNPS,

8.4.3 Agriculture waste-based nanoparticles

n this type nano size particles from agriculture waste are used for example rice husk waste was brought down to nano size and rice husk of nanosized was used as nanocarrier for 2,4-dichlorophenoxyacetic acid herbicide (2,4-D). Then it was used to surface adsorbed with 2,4-D so that it can be act as herbicide nanocarrier. The rice husk nanocarrier showed enhanced and reversible absorption of 2,4-D, exemplifying its individuality as good encapsulating carrier of herbicides Abigail et al. (2016).

8.4.4 Green synthesis of nanoparticles

This is nonconventional process of synthesis of nanoparticles that can be used in formulations of nano herbicides Physico-chemical approaches are used in the production of nanoparticles which leads to usage of hazardous materials, advanced equipment eventually it may have negative impact on the atmosphere. To evade such consequences 'green' synthesis will be the way which resulted from the nanotechnology which have shifted towards environmentally friendly and economically potent process. Thus, green synthesis, is great choice as part of bio-inspired protocols, which delivers consistent and justifiable methods for the biosynthesis of nanoparticles with the help of wide range of microorganisms rather than current synthetic processes. Therefore, this field is evolving fast and innovative methods in this field are continuously being developed to improve the properties of nanoparticles reported by Bahrulolum et al. (2021). Biogenesis of nano particles variety of microorganisms and plant extracts can be used successfully. Though, the use of microorganisms to harvest MtNPs is more cost-effective than that of plants. There is only concern about microorganisms is that they are able to accumulate and detoxify heavy metals due to the presence of various reductase enzymes which capable of reducing metal salts to metal nano particles consequently, microorganisms act as important nanofactories (Singh et al., 2016b). Bacteria such as *Pseudomonas deptenis* (Jo et al., 2016), *Visella oriza* (Singh et al., 2016a) *Bacillus methylotrophicus* (Wang et al., 2016) *Bhargavaea indica* and *Brevibacterium frigoritolerans* have been able to synthesize silver (Ag) and gold (Au) nanoparticles. Various genera of microorganisms such as *Bacillus, Lactobacillus, Pseudomonas, Streptomyces, Klebsiella, Enterobacter, Escherichia, Aeromonas, Corynebacterium, Weissella, Rhodobacter, Rhodococcus, Brevibacterium, Trichoderma, Desulfovibrio, Sargassum, Shewanella, Plectonemaboryanum, Pyrobaculum* and *Rhodopseudomonas* have been shown to produce such metal nanoparticles (Singh et al., 2016b). These nanofactories have enormous potential as they do not require much energy as compared to physicochemical methods and served as environmentally friendly, inexpensive, and non-toxic tools. Additional to these green synthesis of metal nanoparticles using microorganisms have several advantages compared to conventional physicochemical methods such as it offers a rapid, cost-effective, clean, non-toxic, and eco-friendly method for the synthesis which is able to give a wide range of sizes, shapes, compositions and physicochemical properties of newly synthesized metal nanoparticles (Shah et al., 2015; Ovais et al., 2018). However, the main disadvantage of microorganism-based synthesis of metal nanoparticles involves complicated steps such as microbial sampling, isolation, culturing and storage as well as the recovery of metal nanoparticles produced by this method requires downstream processing (Singh et al., 2016b).

9. Nanoherbicides formulations

9.1 Nanoencapsulations

Nanoencapsulation or encpasulation of targeted molecule with nanoparticles is a method by which synthetic compounds like pesticides and herbicides are gradually yet proficiently delivered to a specific host plant for bug management utilizing nanoparticles (Itodo et al., 2017). Taking up of the synthetic substances into the plants is currently the most encouraging novel technique for the assurance of host plants against all horticultural bugs (Prasad et al., 2014). It eases slow dispersion of a substance to the specific host for weed and creepy crawly bug control. The type of mechanism associated with nanoencapsulation of herbicides includes dissemination, disintegration, biodegradation and osmotic pressing factor having specific pH (Biswal et al., 2012). With this innovation, it is feasible to venture down the arrival of the herbicide nanoformulations under controlled conditions, consequently diminishing application measurement and improving productivity (Scrinis and Lyons, 2007). Ongoing experiments indicates that, most driving synthetic organizations center around detailing of nanoscale herbicides for efficient delivery in targeted host tissue via nanoencapsulation (Prasad et al., 2014). Herbicides, binding/encapsulation in diverse polymeric shells of nanoparticles can prompt a more secure and more helpful administration of herbicides that enhances the efficient removal of weeds with reduced environmental hazards (Jampílek and Králová, 2015). As per the reports published by Maruyama et al. (2016), the studies demonstrate that encapsulation of herbicides improves their method of activity and decreased their side effects and toxicity, which shows their appropriateness for use in the future protect environment from chemical hazards. The kinetics tests from the investigation shows that the nano-herbicides were delivered more slowly than their original form. Also, numerous experiments have given proof that nanoformulations of pesticides/herbicides can work with the augmentation of plant-based systemic resistance (SAR) against bugs. For instance, silica nanosphere can boost capacity of pesticides to infiltrate through the plant and arrive at the cell sap, subsequently applying fundamental impact on biting or sucking type bugs (Li et al., 2007). It has additionally been seen that nanoformulations can modify the mode of penetration of pesticides (Hou et al., 2016). The non-systemic use of fungicide ferbam can change and expand the infiltrating potential into tea leaves when formulated with metallic NPs (AuNPs). Such sorts of discoveries clearly will give another wilderness of advancement of pesticide definitions to get plant-based systemic resistance. Advances, for example, sustained release and controlled delivery strategies have reformed the utilization of pesticides and herbicides. Herbicides inside nanoparticles are being synthesized that can be conveniently delivered. Joined with a nanoencapsulation process and sustained delivery systems, herbicide could be applied just when essential, bringing about more prominent creation of yields and less injury to farming laborers. Numerous organizations make nanoformulations with size scopes of 100–250 nm and can break down in water more viably than current ones consequently expanding their action (Prasad et al., 2014).

9.2 Agro waste nanoparticles formulations

In this formulation, nanoparticles are made up of agricultural waste and they are acting as encapsulating herbicide that has good carrier activity. For example, as mentioned above the rice husk waste nanoparticles and was surface adsorbed with 2,4-D and it act as nanocarrier for herbicide. These rice husk, 2,4-D absorbed nanoparticles were tested for herbicidal activity when tried against target weed, *Brassica* sp. showed high herbicidal activity in comparison with free 2,4-D (Abigail et al., 2016). This herbicide delivery system has the potential benefit due to its advantages in using the waste effectively in addition to efficiently reducing the use of herbicides that are environmental contaminants. As such, weeds evolution with learned resistance due to the excess use of herbicides could be eliminated by these nanostructured herbicide carriers. The nanoparticles were found to be stable in terms of size, zeta potential, pH, and polydispersity for nearly 9 months. The triazine herbicide formulations discovered that the nano-capsules release the triazine in a precise manner through relaxing the polymeric chains were observed in vitro release experiments. It is reported in *Allium cepa* chromosome aberration assay, polymeric herbicide nanoparticles are shown to have less genotoxicity (Abigail et al., 2016). Alginate/Chitosan (Ag/Cs) were obtained from natural sources used for nanoparticle synthesis and they were preferred for encapsulation of parquet herbicide.

Formulation of this polymeric complex has simple preparation methods and the Ag/Cs nanoparticle carrier system showed significant differences in the release profiles of free paraquat and the herbicide nanoparticles. The herbicide nanocarrier has altered the negative impact of paraquat released from Ag/Cs system herbicide by interacting with soil.

In another study, the paraquat herbicide was encapsulated onto chitosan/tripolyphosphate nanoparticle. It was proved that it is efficient with this polymeric nanoparticle system as well. Cell culture viability tests and *A. cepa* chromosomal aberration tests testified to the increased safety of the polymeric herbicide systems against non-target organisms (Grillo et al., 2012).

9.3 Controlled release formulations

Controlled release formulations (CRFs) address an option in contrast to the traditional methods of herbicide application where the encapsulated molecule is accessible in a particular form in terms to accomplish the proposed effect. In this novel process, CRFs or controlled release framework is a mix of biologically active ingredients and excipient compound, generally, a polymer added to allow the mixture of all ingredients at controlled rates over a predetermined period (Jampílek and Králová, 2015; Huang et al., 2018). The herbicide CRFs may be characterized as an innovation wherein functioning active molecules/ingredients is accessible for a particular intent of fixation in a specific time to accomplish the proposed target, to arrive at ideal concentration, and to decrease any hazardous impacts (Rüegg et al., 2007). A controlled delivery deals with the mixing of agrochemicals in a quick manner to reach optimum concentration for getting better yields in various agricultural practices (Jampílek and Králová, 2015). It is presently, by and large, acknowledged truth that this innovation can contribute decidedly to human's work in battling against bugs and weeds to improve crop yields (Jampílek and Králová, 2015; Huang et al., 2018).

Table 2 Comparison of important properties of methods for synthesis of nanoparticles.

Methods	Size range (nm)	Morphology	Advantages
Co precipitation	15–200	Spherical	Fine size with higher fabrication
Microemulsion	4–12	Spherical	Highly scalable
Hydrothermal synthesis	450–520	Cubic or Spherical	Shape control
Pyrolysis	400–1100	Spherical	High production efficiency
Arc discharge	900–1100	Spherical	Shape control
Gas deposition method	5–50	Roughly spherical	Productive and useful in production of thin films
Bulk solution method	10–50	Spherical	Bulk production
Sol gel method	20–200	Spherical	Good in manufacturing of hybrid nanoparticles

In traditional application techniques, pesticides are consistently inclined to drain and vanishing which might be photolytic, hydrolytic, or microbial. The majority of this pesticide usage may lead to the elimination of pesticide fixation before they can complete their purpose (Rana and Rana, 2015). The reason for utilizing controlled delivery in innovation for agribusiness is to diminish the unreasonable effectiveness of the traditional technique and to drag out the defensive utilization of agrochemicals or biocides by keeping a compelling focus throughout a given period. By using this approach, it will lessen herbicide remains and decrease costs for laborers (Onyido et al., 2012). A photoresponsive controlled pesticide discharge film was built utilizing PEG with spiropyran as a transporter and chitosan as a film-shaping added substance. The pesticide discharge was seen from the photoresponsive controlled pesticide discharge film. Their result indicates that the work would advance the improvement of controlled-discharge pesticide innovations. In another published report by Bai et al. (2019), a novel pesticide transporter LA-NSM (lauric corrosive altered Nitraria seed feast) with controlled delivery property was set up through esterification process. The comparison of important properties with method of nanoparticle formulation is given in Table 2.

10. Applications of nano-herbicides in farming/agriculture

Nanoherbicides have the following applications and significance in agriculture sectors,

Designed for better action: Nanoscale herbicides were designed for better protection of plants from damaging influences of weeds with the application of active ingredients in a targeted manner.

Targeted action: The synthesis of specific nano herbicides molecules encapsulated with nanoparticles has precise target receptors present at the root of the targeted weed. These nanoparticles enter the roots system and are transported to achieve its action it also stops the glycolysis process of the root cells of targeted plants. Thus, this specific action leads to starvation and ultimately to the death of the weed.

Reduced phytotoxicity: Nanoencapsulated herbicides were found to reduce the phytotoxicity of herbicides on crops also regulate parasitic weeds (Pérez-de-Luque and Rubiales, 2009).

Control release of herbicides: Herbicide nanoformulations are composed of polymer coating to enable its precise and leisurely release due to likely communications between the herbicide and the polymer. CuO NPs were found to be more toxic than bulk CuO or dissolved Cu (II) ions to *Landoltia (Spirodela) punctata* and *Zea mays* (Wang et al., 2012); the action of antioxidant enzymes amplified after the treatment of *Elodea densa* plants with CuNPs (Nekrasova et al., 2011); CuO NPs induced also DNA damage in several plants.

Helps in reduction of environmental/soil contaminants: The application of nanoherbicides is associated with a less environmental impact with simultaneous preservation of herbicidal efficiency. Use of Agriculture wate formulations reduces pollution of soil but also helps effectively in controlling weeds which get adapted easily for chemical herbicides. The rice husk-based nano-formulations of herbicide its delivery system acts in multiple ways with the advantages of utilizing the waste constructively and use of environmental contamination with nano-herbicides for effective use to minimize the contaminant from the soil. Nanostructured herbicide carriers help to control evolved weeds because of the overuse of chemical herbicides in farming (Abigail et al., 2016).

Eco-friendly: Green synthesized nanoherbicides are environment-friendly in nature. Normal herbicides are made up of toxic chemicals which creates a problem in the ecosystem.

11. Conclusions and future perspectives

Chemical herbicides after overuse make the land, groundwater, and food products polluted to boost the production of the crop. Alternative to these chemically hazardous herbicides, the use of nanoherbicides can give a better eco-friendly option to increase the crop productivity and systemic removal of weed in agriculture. Even though the increase in the yield is important, but the unintended damage to the environment cannot be overlooked. Nanotechnology helps to develop nanoparticles which have unique characteristics and modes of action on the targets which can be used for the production of various agricultural products like nanoherbicides, nanopesticides, nano fertilizers and so on which could contribute to achieve sustainable and eco-friendly agricultural technology. This vision of exploiting the nanotechnological approaches in farming is still in the emerging stage. Hence, the advancement of systems that would progress the release profile of herbicides without changing their physical characteristics and novel carriers with enhanced action without substantial ecological harm is the main part that needs further focused study.

References

Abigail, M.E.A., Melvin Samuel, S., Chidambaram, R., 2016. Application of rice husk nano-sorbents containing 2,4-dichlorophenoxyacetic acid herbicide to control weeds and reduce leaching from soil. J. Taiwan Inst. Chem. Eng. 63, 318–326.

Anwar, M.P., Juraimi, A.S., Mohamed, M.T.M., Uddin, M.K., Samedani, B., Puteh, A., Man, A., 2013. Integration of agronomic practices with herbicides for sustainable weed management in aerobic rice. Sci. World J. 2013, 916408.

Babu, D.S., Srivastava, V., Nidheesh, P.V., Kumar, M.S., 2019. Detoxification of water and wastewater by advanced oxidation processes. Sci. Total Environ. 696, 133961.

Bahrulolum, H., Nooraei, S., Javanshir, N., Tarrahimofrad, H., Mirbagheri, V.S., Easton, A.J., Ahmadian, G., 2021. Green synthesis of metal nanoparticles using microorganisms and their application in the agrifood sector. J. Nanobiotechnol. 19 (1), 86.

Bai, B., Xu, X., Hai, J., Hu, N., Wang, H., Suo, Y., 2019. Lauric acid-modified Nitraria seed meal composite as green carrier material for pesticide controlled release. J. Chem. 2019, 5376452.

Bhan, V.M., Sushilkumar, Raghuwanshi, M.S., 1999. Weed management in India. Indian J. Plant Prot. 17, 171–202.

Bishoge, O.K., Zhang, L., Suntu, S.L., Jin, H., Zewde, A.A., Qi, Z., 2018. Remediation of water and wastewater by using engineered nanomaterials: a review. J. Environ. Sci. Health A Tox. Hazard. Subst. Environ. Eng. 53, 537–554.

Biswal, S.K., Nayak, A.K., Parida, U.K., Nayak, P.L., 2012. Applications of nanotechnology in agriculture and food sciences. Int. J. Sci. Innov. Discov. 2 (1), 21–36.

Böger, P., 2020. New plant-specific targets for future herbicides. In: Target Sites of Herbicide Action. CRC Press, pp. 247–282.

Chadha, K.L., Leela, D., Prabha, C., 1997. Weed Management in Horticulture and Plantation Crops. Malhotra Publishing House.

Chinnamuthu, C.R., Boopathi, P.M., 2009. Nanotechnology and agroecosystem. Madras Agric. J. 96 (1/6), 17–31.

Chinnamuthu, C.R., Kokiladevi, E., 2007. Weed management through nanoherbicides. Appl. Nanotechnol. Agric. 10, 978–981.

Crafts, A.S., 1946. Selectivity of herbicides. Plant Physiol. 21, 345–361.

Deboer, G.J., Thornburgh, S., Ehr, R.J., 2006. Uptake, translocation and metabolism of the herbicide florasulam in wheat and broadleaf weeds. Pest Manag. Sci. 62 (4), 316–324.

Devos, Y., Cougnon, M., Vergucht, S., Bulcke, R., Haesaert, G., Steurbaut, W., Reheul, D., 2008. Environmental impact of herbicide regimes used with genetically modified herbicide-resistant maize. Transgenic Res. 17 (6), 1059–1077.

Dolez, P.I., 2015. Chapter 1.1: Nanomaterials definitions, classifications, and applications. In: Dolez, P.I. (Ed.), Nanoengineering. Elsevier, Amsterdam, The Netherlands, ISBN: 978-0-444-62747-6, pp. 3–40.

Duke, S.O., 1990. Overview of herbicide mechanisms of action. Environ. Health Perspect. 87, 263–271.

Ealias, A.M., Saravanakumar, M.P., 2017. A review on the classification, characterization, synthesis of nanoparticles and their application. IOP Conf. Ser. Mater. Sci. Eng. 263, 1–15.

Fernández, V., Brown, P.H., 2013. From plant surface to plant metabolism: the uncertain fate of foliar-applied nutrients. Front. Plant Sci. 4, 289.

Gajbhiye, K.S., Mandal, C., 2000. Agro-ecological zones, their soil resource and cropping systems. In: Status of Farm Mechanization in India, cropping systems, status of farm mechanization in India, pp. 1–32.

Grillo, R., Pereira, A.E., Nishisaka, C.S., de Lima, R., Oehlke, K., Greiner, R., Fraceto, L.F., 2012. Chitosan/tripolyphosphate nanoparticles loaded with paraquat herbicide: an environmentally safer alternative for weed control. J. Hazard. Mater. 47 (8), 163–171.

Grossmann, K., Ehrhardt, T., 2007. On the mechanism of action and selectivity of the corn herbicide topramezone: a new inhibitor of 4-hydroxyphenylpyruvate dioxygenase. Pest Manag. Sci. 63 (5), 429–439.

Gustafson, D.I., 2008. Sustainable use of glyphosate in North American cropping systems. Pest Manag. Sci. 64 (4), 409–416.

Hanemann, T., Szabó, D.V., 2010. Polymer-nanoparticle composites: from synthesis to modern applications. Materials 3 (6), 3468–3517.

Hashim, N., Muda, Z., Hamid, S.A., Isa, I.M., Kamari, A., Mohamed, A., Hussein, M.Z., Ghani, S.A., 2014. Characterization and controlled release formulation of agrochemical herbicides based on zinc-layered hydroxide-3-(4-methoxyphenyl) propionate nanocomposite. J. Phys. Chem. Sci. 1, 1–6.

Hou, R., Zhang, Z., Pang, S., Yang, T., Clark, J.M., He, L., 2016. Alteration of the nonsystemic behavior of the pesticide ferbam on tea leaves by engineered gold nanoparticles. Environ. Sci. Technol. 50 (12), 6216–6223.

Huang, B., Chen, F., Shen, Y., Qian, K., Wang, Y., Sun, C., Zhao, X., Cui, B., Gao, F., Zeng, Z., Cui, H., 2018. Advances in targeted pesticides with environmentally responsive controlled release by nanotechnology. Nano 8 (2), 102.

Itodo, H.U., Nnamonu, L.A., Wuana, R.A., 2017. Green synthesis of copper chitosan nanoparticles for controlled release of Pendimethalin. Asian J. Chem. Sci. 2 (3), 1–10.

Jampílek, J., Králová, K., 2015. Applications of nanoformulations in agricultural production and their impact on food and human health. Proc. ECOpole 9 (2). https://doi.org/10.2429/proc.2015.9(2)054.

Jo, J.H., Singh, P., Kim, Y.J., Wang, C., Mathiyalagan, R., Jin, C.G., et al., 2016. *Pseudomonas deceptionensis* DC5-mediated synthesis of extracellular silver nanoparticles. Artif. Cells Nanomed. Biotechnol. 44 (6), 1576–1581.

Kah, M., Beulke, S., Tiede, K., Hofmann, T., 2013. Nanopesticides: state of knowledge, environmental fate, and exposure modeling. Crit. Rev. Environ. Sci. Technol. 43, 1823–1867.

Khan, I., Saeed, K., Khan, I., 2019. Nanoparticles: properties, applications and toxicities. Arab. J. Chem. 12, 908–931.

Koterba, M.T., Banks, W.S.L., Shedlock, R.J., 1993. Pesticides in shallow groundwater in the Delmarva Peninsula. J. Environ. Qual. 22 (3), 500–518.

Lemoine, R., La Camera, S., Atanassova, R., Dédaldéchamp, F., Allario, T., Pourtau, N., Bonnemain, J.L., Laloi, M., Coutos-Thévenot, P., Maurousset, L., Faucher, M., 2013. Source-to-sink transport of sugar and regulation by environmental factors. Front. Plant Sci. 4, 272.

Li, Z.Z., Chen, J.F., Liu, F., Liu, A.Q., Wang, Q., Sun, H.Y., Wen, L.X., 2007. Study of UV-shielding properties of novel porous hollow silica nanoparticle carriers for avermectin. Pest Manag. Sci. 63 (3), 241–246.

Manjunatha, S.B., Biradar, D.P., Aladakatti, Y.R., 2012. Nanotechnology and its applications in agriculture: a review. J. Farm Sci. 69 (1), 1–13.

Martini, L.F.D., Burgos, N.R., Noldin, J.A., de Avila, L.A., Salas, R.A., 2015. Absorption, translocation and metabolism of bispyribac-sodium on rice seedlings under cold stress. Pest Manag. Sci. 71 (7), 1021–1029.

Maruyama, C.R., Guilger, M., Pascoli, M., Bileshy-José, N., Abhilash, P.C., Fraceto, L.F., De Lima, R., 2016. Nanoparticles based on chitosan as carriers for the combined herbicides imazapic and imazapyr. Sci. Rep. 6 (1), 1–15.

Mathur, R.B., Singh, B.P., Pande, S., 2017. Carbon Nanomaterials Synthesis, Structure, Properties and Applications. CRC Press, Bora Raton, FL, USA, p. 284.

Nekrasova, G.F., Ushakova, O.S., Ermakov, A.E., Uimin, M.A., Byzov, I.V., 2011. Russ. J. Ecol. 42, 458–463.

Onyido, I., Sha'Ato, R., Nnamonu, L.A., 2012. Environmentally friendly formulations of trifluralin based on alginate modified starch. J. Environ. Prot. 3 (9), 1085.

Ovais, M., Khalilm, A.T., Ayaz, M., Ahmad, I., Nethi, S.K., Mukherjee, S., 2018. Biosynthesis of metal nanoparticles via microbial enzymes: a mechanistic approach. Int. J. Mol. Sci. 19 (12), 4100.

Pate, J.S., Atkins, C.A., Hamel, K., McNeil, D.L., Layzell, D.B., 1979. Transport of organic solutes in phloem and xylem of a nodulated legume. Plant Physiol. 63 (6), 1082–1088.

Pereira, A.E.S., Grillo, R., Mello, N.F.S., Rosa, A.H., Fraceto, L.F., 2014. Application of poly(epsilon-caprolactone) nanoparticles containing atrazine herbicide as an alternative technique to control weeds and reduce damage to the environment. J. Hazard. Mater. 2 (68), 207–215.

Pérez-de-Luque, A., Rubiales, D., 2009. Nanotechnology for parasitic plant control. Pest Manag. Sci. 65 (5), 540–545.

Pérez-Sánchez, L., Rodríguez-Méndez, A., Montufar-Reyes, I., Trejo Hernandez, R., Mayorga-Garay, M., Montoya-Lizarraga, A.C., Macías-Sámano, L.M., Reséndiz-Luján, B., Rodríguez-Morales, J., Elizalde, E., et al., 2013. Water recycling in biosystems for food production. In: Guevara-Gonzalez, R., Torres-Pacheco, I. (Eds.), Biosystems Engineering: Biofactories for Food Production in the Century XXI. Springer International Publishing, Cham, Switzerland, ISBN: 978-3-319-03879-7, pp. 77–97.

Pimentel, D., 2012. Silent Spring, the 50th Anniversary of Rachel Carson's Book.

Piver, W.T., 1993. Contamination and restoration of groundwater aquifers. Environ. Health Perspect. 100, 237–247.

Prasad, R., Kumar, V., Prasad, K.S., 2014. Nanotechnology in sustainable agriculture: present concerns and future aspects. Afr. J. Biotechnol. 13 (6), 705–713.

PRD, 2017. 2017 World Population Data Sheet with a Special Focus on Youth. Population Reference Bureau.

Preet Kaur, S., Rao, R., Hussain, A., Khatkar, S., 2011. Preparation and characterization of rivastigmine loaded chitosan nanoparticles. J. Pharm. Sci. Res. 3 (5), 1227.

Pretty, J., Bharucha, Z.P., 2015. Integrated pest management for sustainable intensification of agriculture in Asia and Africa. Insects 6 (1), 152–182.

Rana, S.S., Rana, M.C., 2015. Advances in Weed Management. Department of Agronomy, College of Agriculture, CSK Himachal Pradesh Krishi Vishvavidyalaya, Palampur, p. 183.

Rani, P.U., Madhusudhanamurthy, J., Sreedhar, B., 2014. Dynamic adsorption of a-pinene and linalool on silica nanoparticles for enhanced antifeedant activity against agricultural pests. J. Pest. Sci. 87, 191–200.

Rao, A.N., Nagamani, A., 2010. Integrated weed management in India–revisited. Indian J. Weed Sci. 42 (3&4), 123–135.

Rao, A.N., Wani, S.P., Ladha, J.K., 2014. Weed Management Research in India—An Analysis of Past and Outlook for Future.

Rodríguez-Méndez, A., Guzmán, C., Elizalde, E., Eduardo, A., Escobar-Alarcón, L., Vega, M., Alvarado-Rivera, J., Escalante, K.E., 2017. Disinfection of real wastewater by Ag–TiO$_2$ nanoparticles photocatalysis. J. Nanosci. Nanotechnol. 17, 711.

Ronald, P., 2011. Plant genetics, sustainable agriculture and global food security. Genetics 188 (1), 11–20.

Rüegg, W.T., Quadranti, M., Zoschke, A., 2007. Herbicide research and development: challenges and opportunities. Weed Res. 47 (4), 271–275.

Saleh, T.A., 2020. Nanomaterials: classification, properties, and environmental toxicities. Environ. Technol. Innov. 20, 101067.

Satapanajaru, T., Anurakpongsatorn, P., Pengthamkeerati, P., Boparai, H., 2008. Remediation of atrazine-contaminated soil and water by nano zerovalent iron. Water Air Soil Pollut. 192 (1), 349–359.

Scrinis, G., Lyons, K., 2007. The emerging nano-corporate paradigm: nanotechnology and the transformation of nature, food and agri-food systems. The Int. J. Sociol. Agric. Food 15 (2), 22–44.

Sekhon, B.S., 2014. Nanotechnology in agri-food production: an overview. Nanotechnol. Sci. Appl. 7, 31.

Shah, M., Fawcett, D., Sharma, S., Tripathy, S.K., Poinern, G.E.J., 2015. Green synthesis of metallic nanoparticles via biological entities. Materials 8 (11), 7278–7308.

Sharma, A., Ahmad, J., Flora, S.J.S., 2018. Application of advanced oxidation processes and toxicity assessment of transformation products. Environ. Res. 167, 223–233.

Singh, A.K., 2016. Introduction to nanoparticles and nanotoxicology. In: Engineered Nanoparticles. Academic Press, Cambridge, MA, USA, pp. 1–18.

Singh, B., Garg, T., Goyal, A.K., Rath, G., 2016a. Development, optimization, and characterization of polymeric electrospun nanofiber: a new attempt in sublingual delivery of nicorandil for the management of angina pectoris. Artif. Cells Nanomed. Biotechnol. 44 (6), 1498–1507.

Singh, P., Kim, Y.-J., Zhang, D., Yang, D.C., 2016b. Biological synthesis of nanoparticles from plants and microorganisms. Trends Biotechnol. 34 (7), 588–599.

Sopeña, F., Maqueda, C., Morillo, E., 2009. Controlled release formulations of herbicides based on micro-encapsulation. Cienc. Invest. Agrar. 36 (1), 27–42.

Sudha, P.N., Sangeetha, K., Vijayalakshmi, K., Barhoum, A., 2018. Chapter 12: Nanomaterials history, classification, unique properties, production and market. In: Barhoum, A., Makhlouf, A.S.H. (Eds.), Emerging Applications of Nanoparticles and Architecture Nanostructures. Elsevier, Amsterdam, The Netherlands, ISBN: 978-0-323-51254-1, pp. 341–384.

Tehranchian, P., Norsworthy, J.K., Nandula, V., McElroy, S., Chen, S., Scott, R.C., 2015. First report of resistance to acetolactate-synthase-inhibiting herbicides in yellow nutsedge (*Cyperus esculentus*): confirmation and characterization. Pest Manag. Sci. 71 (9), 1274–1280.

Thomas, S., Harshita, B.S.P., Mishra, P., Talegaonkar, S., 2015. Ceramic nanoparticles: abrication methods and applications in drug delivery. Curr. Pharm. Des. 21, 6165–6188.

Torrens, F., Castellano, G., 2014. Molecular classification of pesticides including persistent organic pollutants, phenylurea and sulphonylurea herbicides. Molecules 19 (6), 7388–7414.

Turgut, C., 2005. Uptake and modeling of pesticides by roots and shoots of parrotfeather (*Myriophyllum aquaticum*). Environ. Sci. Pollut. Res. 12 (6), 342–346.

Varshney, S., Hayat, S., Alyemeni, M.N., Ahmad, A., 2012. Effects of herbicide applications in wheat fields: is phytohormones application a remedy? Plant Signal. Behav. 7 (5), 570–575.

Wakabayashi, K., 1989. Herbicidal targets in photosynthetic light reaction and molecular design of new herbicides. Plant Prot. 43, 575–584.

Wang, C., Kim, Y.J., Singh, P., Mathiyalagan, R., Jin, Y., Yang, D.C., 2016. Green synthesis of silver nanoparticles by *Bacillus methylotrophicus*, and their antimicrobial activity. Artif. Cells Nanomed. Biotechnol. 44 (4), 1127–1132.

Wang, Z.Y., Xie, X.Y., Zhao, J., Liu, X.Y., Feng, W.Q., White, J.C., et al., 2012. Environ. Sci. Technol. 46, 4434–4441. https://doi.org/10.1021/es204212z.

Wright, T.R., Shan, G., Walsh, T.A., Lira, J.M., Cui, C., Song, P., Zhuang, M., Arnold, N.L., Lin, G., Yau, K., Russell, S.M., 2010. Robust crop resistance to broadleaf and grass herbicides provided by aryloxyalkanoate dioxygenase transgenes. Proc. Natl. Acad. Sci. 107 (47), 20240–20245.

Yaduraju, N.T., 2012. Weed management perspectives for India in the changing agriculture scenario in the country. Pak. J. Weed Sci. Res. 18, 703–710.

Zaytseva, O., Neumann, G., 2016. Carbon nanomaterials: production, impact on plant development, agricultural and environmental applications. Chem. Biol. Technol. Agric. 3, 17.

Zimdahl, R.L., 2018. Fundamentals of Weed Science. Academic Press.

CHAPTER 16

Intimate coupling of photocatalysis and biodegradation (ICPB): A viable method for removing pesticides from contaminated sites

Asha S. Raj and Preethy Chandran*
School of Environmental Studies, Cochin University of Science and Technology, Kochi, Kerala, India
*Corresponding author.

1. Introduction

Pesticides are synthetic substances used to restrain or eradicate pests from agricultural areas. Loads of pesticides are utilized *per annum*, and approximately 5% of them reach the target organism. The rest gets set down in the soil, atmosphere, and water (Pimental and Levitan 1986). Huge quantities of pesticides have been discharged into water bodies directly or indirectly without being properly treated, as a result of the tremendous development of industry and agriculture in recent times, resulting in water pollution. The chemicals contaminate water sources such as lakes, rivers, oceans, and groundwater and represent a major threat to ecological safety and human health. Pesticides can be classified based on the mode of action, target pest species, chemical nature (Table 1), and so on (Zacharia, 2011).

Europe is the largest consumer of pesticides. They consume about 45% of the total pesticides produced in the world. The United States consumes 25% of pesticides, whereas the rest of the globe consumes only 25% (De Geronimo et al., 2014). India is the world's 4th largest manufacturer of pesticides. Maharashtra (about 60,000 t) is the highest pesticide consumer within the country, followed by Uttar Pradesh (50,000 t) and Punjab (29,000 t) in the past five years. Endosulfan, DDT, Lindane, Alachlor, Aldrin etc., are some of the pesticides banned in India (Devi et al., 2017).

Several factors influence the metabolic fate of pesticides (pH, moisture, temperature, sunlight, organic matter, mobility, and persistence) (Gavrilescu, 2005). When these chemical substances enter the living organism, they will be altered by themselves to nullify the noxious effect through biodegradation, biotransformation, or mineralization (Lushchak et al., 2018). Several routes of pesticide are susceptible to human beings, mainly through biotic and abiotic factors in the environment (Anderson and Meade,

Relationship Between Microbes and the Environment for Sustainable Ecosystem Services, Volume 1
https://doi.org/10.1016/B978-0-323-89938-3.00016-5

Table 1 Classification of pesticides.

Classification of pesticides	
Based on the mode of action	Systematic and Nonsystematic (contact)
Based on target pest species	Insecticides, fungicides, herbicides, algicides, avicides
Based on the chemical nature	Organophosphates, organochlorines, carbamates, pyrethrin, and pyrethroids

2014). Dermal, oral, and inhalation are the most common routes of entry (Kim et al., 2017). Pesticides get dispersed in the human body through blood flow, and they get ejected out by way of the skin, hair, and urine (Damalas and Eleftherohorinos, 2011). The expeditious use of pesticides has devastated the environment and affected all life forms, from minute microorganisms to large animals.

2. Environmental distribution of pesticides

As the population is increasing, it is essential to cultivate a higher yield of crops. Hence, farmers prefer inorganic pesticides, which are inexpensive and easily available to negotiate the loss of crops due to pests. These harmful materials are not suitable for the environment, and they get distributed from one environmental compartment to another without being degraded. When pesticides are applied to the soil surface, they get deposited and slowly move from the top surface to the groundwater (Fig. 1).

When applied to the soil, these agrochemicals have the ability to persist for a long time in the ecosystem and tend to move towards aquatic systems (like rivers, lakes,

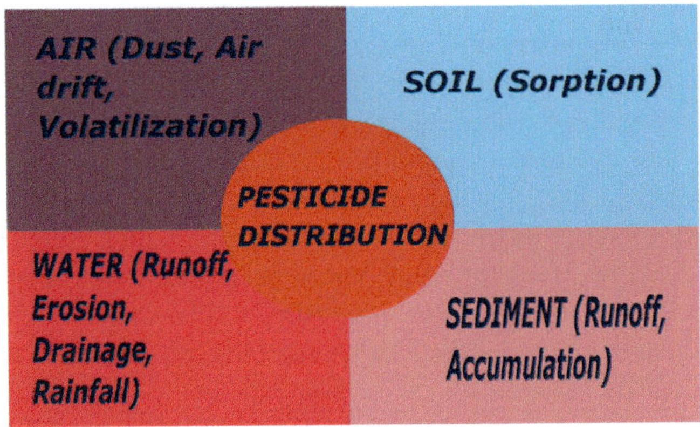

Fig. 1 Pesticide distribution in the environment.

ponds, and seas) and even percolate into groundwater. The continual application of pesticides eventually results in bioaccumulation. It causes several negative consequences, like changes in pH, soil structure, loss of soil fertility, loss of biodiversity, reduction of organic matter, and alteration of microflora (Prashar and Shah, 2016).

3. Toxicity of pesticides

Toxicity is defined as the tendency of any substance to produce damage (permanent or temporary) in living organisms. It can cause several adverse effects, like vomiting, nausea, headaches, comas, and even death. Acute and chronic are the main two types of toxicity.

Acute toxicity is due to single or multiple exposures to a toxic substance within a short period. Sweating, vomiting, nausea, headaches, blurred vision, and chest pain are signs of acute toxicity in human beings. Chronic toxicity occurs as a result of long-term exposure to contaminants and leads to the production of adverse effects.

Dermal exposure occurs when the pesticides are handled directly using hands without proper precautions and cause burns, itching, and other irritations. These chemicals are easily absorbed through the skin and lead to more adverse medical conditions like skin cancer.

Inhalation exposure: when the pesticides are sprayed or applied in powdered form over the crops, the living organisms inhale them through the nose, and ultimately they reach the lungs. In initial exposure, coughing, nausea, and dizziness are the common effects. Continuous inhalation for a long time causes lung cancer, bronchitis, and asthma.

Oral exposure: pesticides are ingested intentionally or unintentionally. Diarrhea and vomiting are common effects. However, when exposed to higher rates, it may result in the death of the person.

The LD_{50} test is a standard method for measuring acute toxicity. In this case, the chemical substance administered to the test organism should kill 50% of the organism's total population. Chronic toxicity is measured by observing detrimental (reproductive, cancer, teratogenic) effects of the test organism. Toxicity studies are carried out in vivo using different test organisms like rats, fishes, clams, algae, rabbits, earthworms, plants. Along with the studies, dose-response curves are also plotted to graphically represent the response of the test organisms towards the toxic substances (Nesheim et al., 2005).

4. Ecological effect of pesticides

Pesticides are widely used in agricultural areas, and this has an indirect influence on biological species (from a bacterium to a human). It primarily affects the soil microflora, which has a tangential effect on soil fertility and function (Imfeld and Vuilleumier, 2012) (Fig. 2).

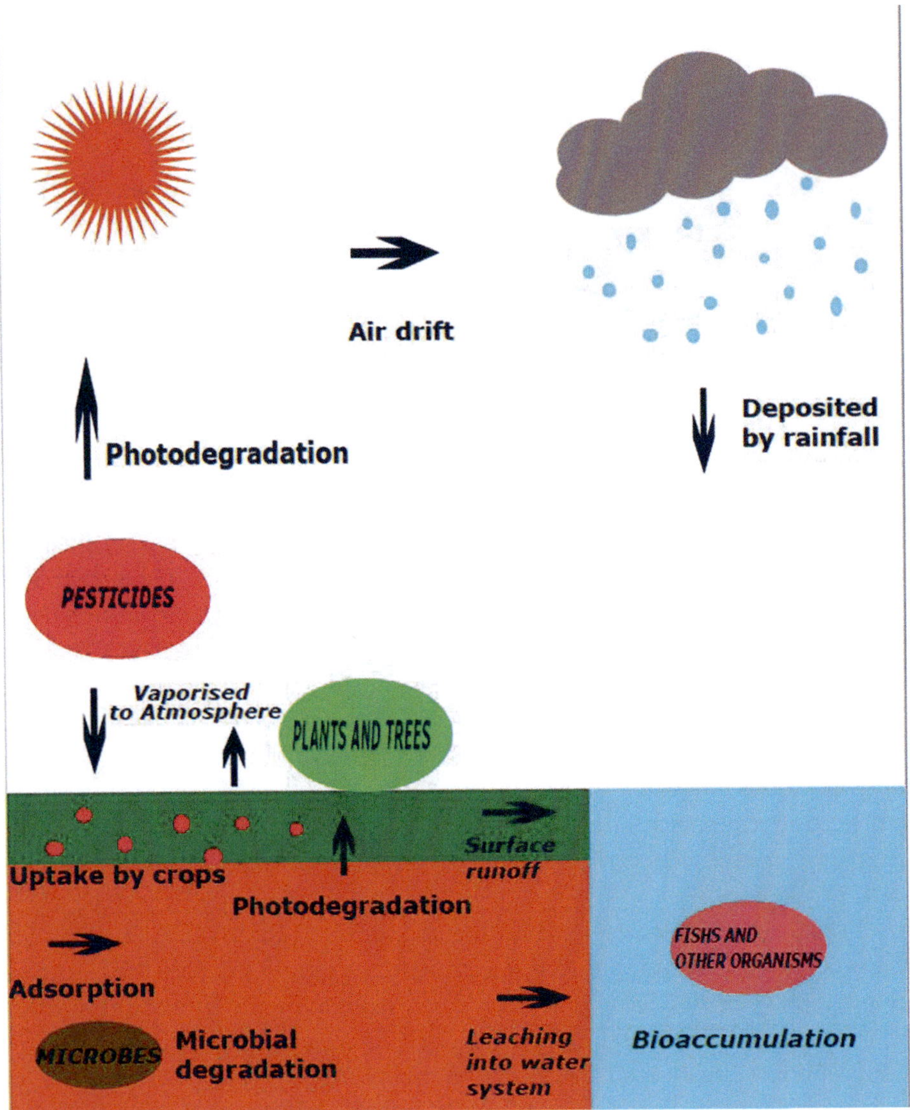

Fig. 2 Pesticide cycle in the environment.

These chemicals are noxious to earthworms, inducing reproductive and growth difficulties (Yasmin and D'Souza, 2010). Pesticides induce growth, reproduction, and teratogenic consequences in fishes through aquatic habitats, and long-term exposure leads to bioaccumulation (Mahmood et al., 2016).

5. Pesticides hazardous effects on human health

Direct exposure to pesticides or pesticide residues found in food causes a variety of health issues, including cancer, diabetes mellitus, respiratory diseases, neurological disorders, reproductive (sexual/genital) syndromes, and oxidative stress.

Photolysis (Zhao et al., 2013), photocatalysis (Chen et al., 2016), and photo–Fenton oxidation (Pereira et al., 2014) are some of the chemical oxidation approaches that have been used to investigate pesticide elimination in water. Even though these procedures are rapid and reliable, full mineralization is economically and practically impossible (Li et al., 2011a,b).

The most often used approach for treating wastewater is biodegradation, which is less expensive than chemical methods. Furthermore, assuming the organics are not excessively resistant or poisonous, biodegradation can entirely mineralize them. Complex aromatic compounds and a lack of certain enzymes in the microbial community are frequently linked to recalcitrance and toxicity (Cerniglia, 1997).

On the other hand, microbial degradation has several challenges, including the fact that it is dependent on the development and metabolism of bacteria, as well as several variables such as pH, nutrition, and temperature. Some complex contaminants are not eliminated, while others are transformed into more hazardous compounds by bacteria. To address these issues, the intimately coupled photocatalysis and biodegradation (ICPB) approach was developed and is now widely employed (Marsolek et al., 2008; Rui and Feng, 2017; Xiong et al., 2017).

6. Coupling of photocatalysis and biodegradation (ICPB): A new approach

Water bodies can be polluted by an extensive range of substances, including pesticides, domestic sewage, and toxic chemicals, which cause water pollution. Solving the drawbacks of conventional sewage treatment methods with new technology has become a recent research focus. Combining photocatalysis with biodegradation is a newly emerging approach that offers a promising way of integrating chemical and biological processes (Sarria et al., 2002). Photocatalysis converts contaminants into biodegradable intermediates, which are then decomposed and mineralized by microorganisms. The sequential coupling of photocatalysis and biodegradation is an old method that generates a variety of compounds that are hazardous and unable to biodegrade. If bacteria were present during photocatalysis, biodegradable products could be biodegraded as soon as they were generated. This might theoretically surpass the restrictions of sequential coupling. This method is known as the intimate coupling of photocatalysis and biodegradation (Figs. 3 and 4).

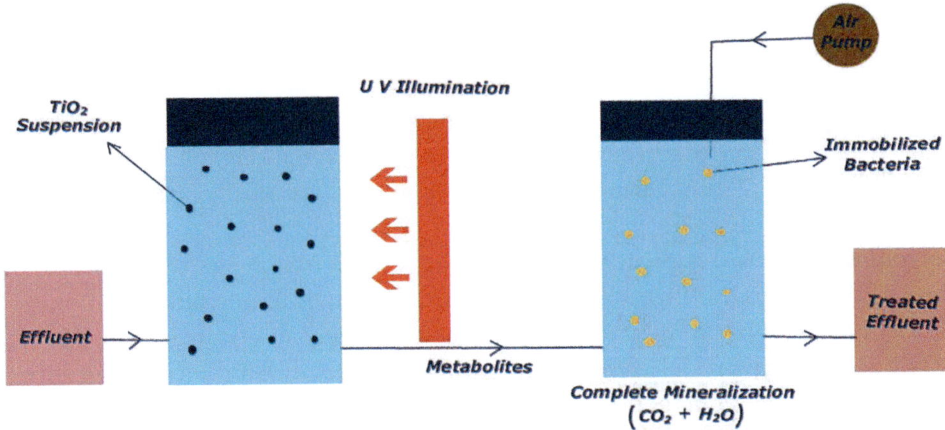

Fig. 3 Diagram of sequential coupling of photocatalysis and biodegradation.

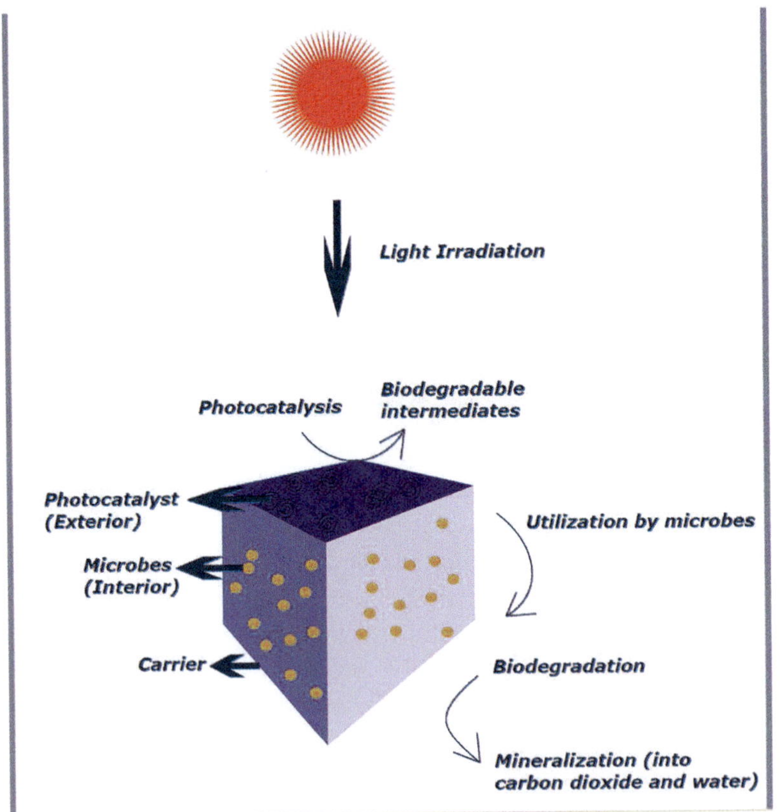

Fig. 4 Schematic diagram of ICPB.

The fundamental concept of ICPB is that photocatalysis occurs on the outer surface of the carrier material, and biofilms are protected in the interior. Rittmann (2018) was the first to reveal that ICPB could destroy 2,4,5-trichlorophenol (TCP) in a photocatalytic circulating-bed biofilm reactor. Many groups have already researched ICPB, developing a range of carriers, photocatalyst, and reactor systems. In recent years, ICPB has been extensively studied for its ability to degrade organic contaminants such as antibiotics, dyes, and PAHs.

7. Mechanism of ICPB

An illuminated reactor, porous carriers, photocatalysts, and biofilms are all common components of ICPB systems. Solar or visible light sources are recommended over intense UV radiation, harming biological components and destroying cells.

There are two steps to the operating cycle that leads to the breakdown of organic pollutants: degradation by photocatalysis and biocatalysis (Biodegradation). Photocatalyst absorbs photon energy generated by the UV or visible light during photocatalytic degradation, causing electrons (e-) to move from the valence band (VB) to the conduction band (CB) and form electron-hole pairs. Simultaneously, ROS (such as OH and O_2) targets and destroys the organic pollutants. The products produced by photocatalysis are transmitted into the carrier and biodegraded within the biofilms, which is then totally mineralized into CO_2 and water (Yu et al., 2020).

ICPB can be classified into three types based on coupling:

- UV-driven coupling supported by carriers (type-1)—The earliest and most extensively employed method. However, due to limited solar utilization and the undesired UV disinfection effect on biological processes, type-1 coupling has gained less interest in recent years.
- Visible-light-driven coupling supported by carriers (type-2)—With the advancement of visible-light-photocatalyst synthesis, type-2 coupling has received a lot of attention. The employment of a lightweight, protective carrier for microorganisms is a particularly prominent aspect of the type-1 and type-2 couplings.
- Visible/solar-light-driven direct coupling without carriers (type-3)—For *in situ* clean-up of polluted surface water, type-3 coupling appears to be preferable, which can be achieved by directly introducing a particular photocatalyst with low cytotoxicity into microorganisms under solar irradiation. However, recycling and reuse of the incorporated photocatalyst from water bodies is evident in this coupling (Zhang et al., 2021a,b).

8. Different types of photocatalyst used in ICPB

The photocatalyst is a crucial component of an ICPB system. These are substances that cause photocatalytic activities whenever exposed to illumination, which is the initial process for the degradation of contaminants. The electronic band structure and band-gap energy of a photocatalyst have a significant impact on its photocatalytic activity.

Titanium Dioxide (TiO_2): The extensively used photocatalyst, due to its robust photo activity, stability, and cheaper value and additionally, it is both safe for people and ecologically sustainable. Rittmann (2018) used Tio_2 as the photocatalyst to degrade micropollutants in a membrane biofilm reactor (MBfR). Many researchers thereafter began investigating ICPB utilizing TiO_2 as the photocatalyst. Marsolek et al. (2008) studied the degradation in a photocatalytic circulating-bed biofilm reactor using TiO_2. A novel TiO_2-coated biofilm carrier was utilized to remove 2,4,5-Trichlorophenol (Li et al., 2011a,b).

Modified TiO_2: TiO_2 has a low affinity for organic contaminants, especially hydrophobic organic pollutants when employed in water treatment applications. Hence, they are modified using different techniques like metal doping, nonmetal doping, heterogeneous composites, metal deposition, and hybrids with nanoparticles (Moma and Baloyi, 2019).

There are several studies related to the usage of modified TiO_2 as the photocatalyst. Dong et al. (2016) used a novel $Er^{3+}:YAlO_3/TiO_2$ for phenol degradation. 2,4,5-Trichlorophenol (TCP) was degraded using a mesoporous SiO_2-TiO_2 (Zhang et al., 2016). N-doped TiO_2 photocatalyst was employed to break down 4-chlorophenol. N doping has the potential to increase the carrier's photocurrent (Zhou et al., 2018). Qin et al. (2020) introduced Cu/N-codoped TiO_2 for pyrene degradation. For nitrate reduction in water, researchers used TiO_2/g-C_3N4 photocatalyst (Zhang et al., 2020).

Researchers have recently begun concentrating on nonTiO_2 photocatalysts like Mn_3O_4/MnO_2-cubicAg_3PO_4 (Chan et al., 2019). Tetracycline was degraded using the $Bi_{12}O_{17}C_{12}$ photocatalyst (Zheng et al., 2018). The carrier was coated with $BiOCl/Bi_2WO_6/Bi$ for the degradation of tetracycline hydrochloride (Fig. 5) (Liao et al., 2020).

9. Different porous carrier materials

The carrier creates a platform for the photocatalyst to be attached on the outside and permits bacteria to develop biofilm inside. Also, it protects bacteria from UV exposure and free radical attacks (Rittmann, 2018). An ideal ICPB carrier should have a suitable density (similar to water), large specific surface area and porosity, biocompatibility and excellent structural stability to resist corrosion.

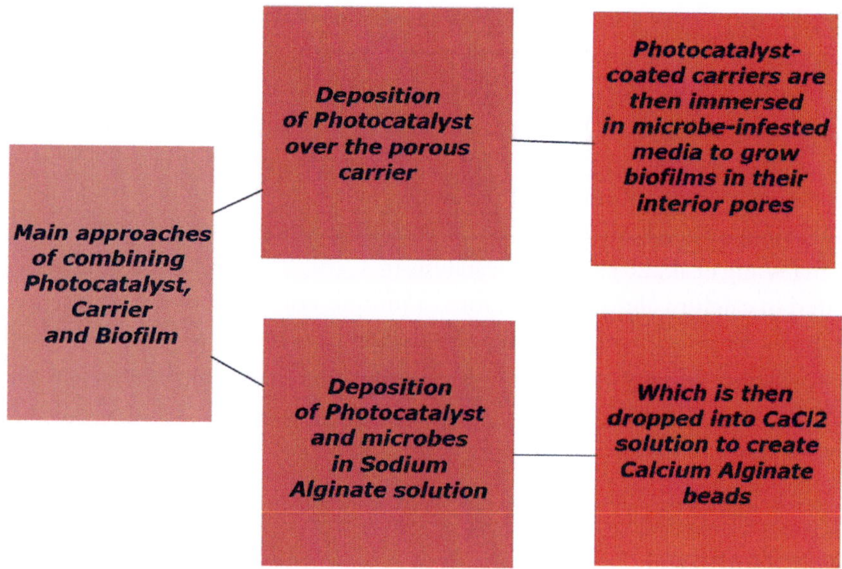

Fig. 5 Combining approach of ICPB system.

Cellulose carrier: the first researcher to introduce ICPB was Marsolek et al. (2008). In the study, cellulose carrier was utilized as the carrier material due to its porosity and low density, and it entirely assisted in the decomposition of acetate. Recently, Xiong et al. (2020) have created a novel sugarcane bagasse-based cellulose carrier. Hydrophilicity, biodegradability, high porosity, and a large specific surface area are just a few of the benefits of this carrier. Furthermore, because of its network structure, it is great to support material. However, cellulose carrier has the disadvantage of being susceptible to UV degradation and radical attack, which can degrade its physical structure. The carrier's ability to hold a dense coating is also a downside, decreasing photocatalytic activity.

Ceramic carrier: to overcome the drawbacks of the cellulose carrier, Wen et al. (2012) introduced a honeycomb–like ceramic carrier in ICPB, which biodegraded almost 100% of the phenol within the reactor. These carriers feature a rough surface that helps to maximize the amount of photocatalytic material loaded and makes them resistant to radical attack. Although ceramic carriers are resilient under UV irradiation, their adsorption capacity is low, making it difficult for internal biofilms to survive.

Polyurethane sponge: nowadays, most investigations employ commercial polyurethane sponges as ICPB carriers. Li et al. (2011a,b) became the first to propose this concept. The carrier had an average side length of 3.5 mm, a specific surface area of $4.8\,m^2/g$, high porosity (88%), and macropore sizes ranging from 50 to 500 m. The carrier had an average side length of 3.5, a specific surface area of $4.8\,m^2/g$, high porosity (88%) and a macropore size of between 50 and 500 µm. Furthermore, the biofilm generation period on polyurethane sponges was considerably lower than on porous ceramic carriers, indicating superior biocompatibility than on porous ceramics (Wen et al., 2012).

Polytetrafluoroethylene: it was also used as a carrier in ICPB systems in addition to the above-mentioned materials (Qin et al., 2020). The materials have several desirable characteristics, including high temperature and chemical resistance. Qin et al. (2020) used this carrier to degrade phenanthrene and pyrene.

Calcium alginate beads: these are nonfabricated carriers formed through a self-assembly mechanism (Wang et al., 2019). Photocatalysts (g–C_3N_4–P_{25}) and photosynthetic bacteria encapsulated in calcium alginate beads form a unique visible–light–responsive hybrid dye removal agent. The dye from contaminated water was removed with a 94% efficiency using this technique (Zhang et al., 2017a,b). Wang et al. (2019) developed a system by coupling microbe consortia and Ag_3PO_4–Fe_3O_4 nanoparticles via microcapsule using sodium alginate to degrade PAHs.

10. Microorganisms

After photocatalysis, these tiny living things form a microbial biofilm over the porous carrier and mineralize the biodegradable product into carbon dioxide and water in an ICPB system. Most of the studies use activated sludge from the wastewater treatment plants to obtain acclimatized microbes to degrade the contaminants (Marsolek et al., 2008; Li et al., 2020; Zhang et al., 2021a,b). As the inoculum, researchers utilized activated sludge from the Water Reclamation Plant and grew it onto TiO_2–coated carriers for ICPB (Li et al., 2012a,b) (Fig. 6).

Later on, researchers started using pure strains with the capacity to metabolize certain recalcitrants. Zhang et al. (2017a,b) used pure strains of *Rhodospirillum* sp. for reactive dye removal. Phenolic compounds were degraded using microalgae biofilm (*Scenedesmus obliquus*) (Zhong et al., 2021). Zhang et al. (2021a,b) appointed *Geobacter* and *Thauera* for the degradation of 2,4,6-trichlorophenol.

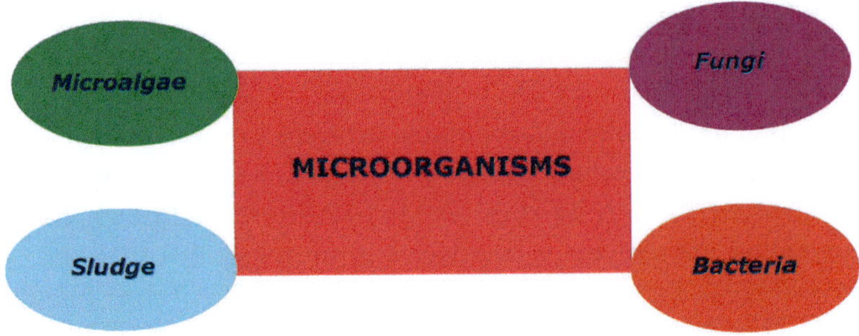

Fig. 6 Microbes used in ICPB.

11. Reactors

The photocatalytic circulating-bed biofilm reactor (PCBBR) is the first studied reactor in ICPB. This is the operational unit that provides the most suitable conditions for the ICPB system to operate functionally. 2,4,5-trichlorophenol (TCP) and COD were eliminated from the reactor. Unfortunately, one disadvantage was that the carrier could not carry enough photocatalyst, which reduced the circulation in the reactor. Li et al. (2012a,b) used a manufactured sponge carrier to resolve the issue. Researchers later switched to visible light instead of UV in PCBBR, which gave better biofilm protection and accelerated the disintegration rate (Xiong et al., 2017).

12. Applications of ICPB

ICPB, a novel wastewater treatment approach, has shown significant promise as a low-cost, environmentally friendly, and potential long-term treatment. The technology has been used to degrade a variety of organic contaminants, including antibiotics, dyes, and PAHs since it was initially created by Prof. Rittmann's group (Rittmann, 2018).

Dye: the textile sector generates a significant volume of dye-containing wastewater, which pollutes water and land and has negative consequences for human health, animals, and plants. The degradation of reactive dyes RB5 (Li et al., 2012a,b), reactive dyes RB5, RY86, and RR120 (Li et al., 2012a,b), and methyl red (Waghmode et al., 2019) have been studied in ICPB. Xiong et al. (2020) recently demonstrated that the ICPB system efficiently removed 92.08% of the dye within 6 h.

Antibiotics have recently been identified as an emergent class of environmental toxins, owing to their widespread use in humans and animals. Antibiotics spread in the environment, particularly in natural water systems, leading to the global establishment and spread of antibiotic resistance. ICPB has recently been used to remove antibiotics from wastewater. Tetracycline was degraded to 94.3% in six cycles using visible-light-induced ICPB (Li et al., 2020) and several other antibiotics, including oxytetracycline (Ding et al., 2018), tetracycline (Xiong et al., 2018), and oxycycline (Xiong et al., 2018), have been successfully removed using ICPB.

PAHs (polycyclic aromatic hydrocarbons): PAHs (polycyclic aromatic hydrocarbons) are a broad group of persistent contaminants that are found throughout the environment. PAH pollution in the aquatic environment is gaining attention due to its carcinogenic and mutagenic potential (Zhao et al., 2014). Many studies have shown that ICPB systems have great potential for degrading PAHs in wastewater. Cai et al. (2019) degraded 96% of phenanthrene in 20 min by combining visible light and Mn_3O_4/MnO_2-Ag_3PO_4. Qin et al. (2020) reported the complete degradation of pyrene using the ICPB system.

Pesticides: pesticides are among a wide spectrum of organic micropollutants with environmental consequences. The ICPB system has been tested as a new sewage treatment method for pesticide-contaminated water and found to be very effective. There are only a few studies related to the removal of pesticides. Li et al. (2011a,b) reported that 92% of 2,4,5-Trichlorophenol was degraded by a novel TiO_2-Coated Biofilm Carrier.

13. Conclusion

Pesticide contamination is one of the emerging problems in the environment. This complex mixture of chemicals in water offers a long-term hazard to aquatic life, fish, birds, and humans because many of these compounds are toxic, persistent, and bioaccumulative. ICPB is a revolutionary technology that combines photocatalysis with biocatalysis benefits. This method can be used to degrade emerging contaminants such as pesticides, hormones, and cosmetics, which are resistant to degradation in the natural environment and pose a severe threat to the environment and human health.

References

Anderson, S.E., Meade, B.J., 2014. Potential health effects associated with dermal exposure to occupational chemicals. Environ. Health Insights 8. EHI-S15258.

Cai, H., Sun, L., Wang, Y., Song, T., Bao, M., Yang, X., 2019. Unprecedented efficient degradation of phenanthrene in water by intimately coupling novel ternary composite Mn_3O_4/MnO_2-Ag_3PO_4 and functional bacteria under visible light irradiation. Chem. Eng. J. 369, 1078–1092.

Cerniglia, C.E., 1997. Fungal metabolism of polycyclic aromatic hydrocarbons: past, present and future applications in bioremediation. J. Ind. Microbiol. Biotechnol. 19 (5–6), 324–333.

Chan, C.Y., Chan, H.S., Wong, P.K., 2019. Integrated photocatalytic-biological treatment of triazine-containing pollutants. Chemosphere 222, 371–380.

Chen, F., Yang, Q., Sun, J., Yao, F., Wang, S., Wang, Y., Wang, X., Li, X., Niu, C., Wang, D., Zeng, G., 2016. Enhanced photocatalytic degradation of tetracycline by $AgI/BiVO_4$ heterojunction under visible-light irradiation: mineralization efficiency and mechanism. ACS Appl. Mater. Interfaces 8 (48), 32887–32900.

Damalas, C.A., Eleftherohorinos, I.G., 2011. Pesticide exposure, safety issues, and risk assessment indicators. Int. J. Environ. Res. Public Health 8 (5), 1402–1419.

De Geronimo, E., Aparicio, V.C., Bárbaro, S., Portocarrero, R., Jaime, S., Costa, J.L., 2014. Presence of pesticides in surface water from four sub-basins in Argentina. Chemosphere 107, 423–431.

Devi, P.I., Thomas, J., Raju, R.K., 2017. Pesticide consumption in India: A spatiotemporal analysis. Agric. Econ. Res. Rev. 30 (1), 163–172.

Ding, R., Yan, W., Wu, Y., Xiao, Y., Gang, H., Wang, S., Chen, L., Zhao, F., 2018. Light-excited photoelectrons coupled with bio-photocatalysis enhanced the degradation efficiency of oxytetracycline. Water Res. 143, 589–598.

Dong, S., Dong, S., Tian, X., Xu, Z., Ma, D., Cui, B., Ren, N., Rittmann, B.E., 2016. Role of self-assembly coated Er^{3+}: $YAlO_3/TiO_2$ in intimate coupling of visible-light-responsive photocatalysis and biodegradation reactions. J. Hazard. Mater. 302, 386–394.

Gavrilescu, M., 2005. Fate of pesticides in the environment and its bioremediation. Eng. Life Sci. 5 (6), 497–526.

Imfeld, G., Vuilleumier, S., 2012. Measuring the effects of pesticides on bacterial communities in soil: a critical review. Eur. J. Soil Biol. 49, 22–30.

Kim, K.H., Kabir, E., Jahan, S.A., 2017. Exposure to pesticides and the associated human health effects. Sci. Total Environ. 575, 525–535.

Li, F., Lan, X., Wang, L., Kong, X., Xu, P., Tai, Y., Liu, G., Shi, J., 2020. An efficient photocatalyst coating strategy for intimately coupled photocatalysis and biodegradation (ICPB): powder spraying method. Chem. Eng. J. 383, 123092.

Li, G., Park, S., Kang, D.W., Krajmalnik-Brown, R., Rittmann, B.E., 2011b. 2, 4, 5-Trichlorophenol degradation using a novel TiO_2-coated biofilm carrier: roles of adsorption, photocatalysis, and biodegradation. Environ. Sci. Technol. 45 (19), 8359–8367.

Li, G., Park, S., Rittmann, B.E., 2012a. Degradation of reactive dyes in a photocatalytic circulating-bed biofilm reactor. Biotechnol. Bioeng. 109 (4), 884–893.

Li, G., Park, S., Rittmann, B.E., 2012b. Developing an efficient TiO_2-coated biofilm carrier for intimate coupling of photocatalysis and biodegradation. Water Res. 46 (19), 6489–6496.

Li, Q., Guo, B., Yu, J., Ran, J., Zhang, B., Yan, H., Gong, J.R., 2011a. Highly efficient visible-light-driven photocatalytic hydrogen production of CdS-cluster-decorated graphenenanosheets. J. Am. Chem. Soc. 133 (28), 10878–10884.

Liao, Q., Rong, H., Zhao, M., Luo, H., Chu, Z., Wang, R., 2020. Interaction between tetracycline and microorganisms during wastewater treatment: a review. Sci. Total Environ. 757, 143981.

Lushchak, V.I., Matviishyn, T.M., Husak, V.V., Storey, J.M., Storey, K.B., 2018. Pesticide toxicity: a mechanistic approach. EXCLI J. 17, 1101.

Mahmood, I., Imadi, S.R., Shazadi, K., Gul, A., Hakeem, K.R., 2016. Effects of pesticides on environment. In: Plant, Soil and Microbes. Springer, Cham, pp. 253–269.

Marsolek, M.D., Torres, C.I., Hausner, M., Rittmann, B.E., 2008. Intimate coupling of photocatalysis and biodegradation in a photocatalytic circulating-bed biofilm reactor. Biotechnol. Bioeng. 101 (1), 83–92.

Moma, J., Baloyi, J., 2019. Modified titanium dioxide for photocatalytic applications. In: Photocatalysts-Applications and Attributes, p. 18.

Nesheim, O.N., Fishel, F.M., Mossler, M., 2005. Toxicity of pesticides. EDIS 2005 (8).

Pereira, J.H., Queirós, D.B., Reis, A.C., Nunes, O.C., Borges, M.T., Boaventura, R.A., Vilar, V.J., 2014. Process enhancement at near neutral pH of a homogeneous photo-Fenton reaction using ferricarboxylate complexes: application to oxytetracycline degradation. Chem. Eng. J. 253, 217–228.

Pimental, D., Levitan, L., 1986. Pesticides: amounts applied and amounts reaching pests. Biosciences 36, 86–91.

Prashar, P., Shah, S., 2016. Impact of fertilizers and pesticides on soil microflora in agriculture. In: Sustainable Agriculture Reviews. Springer, Cham, pp. 331–361.

Qin, Z., Zhao, Z., Jiao, W., Han, Z., Xia, L., Fang, Y., Wang, S., Ji, L., Jiang, Y., 2020. Coupled photocatalytic-bacterial degradation of pyrene: removal enhancement and bacterial community responses. Environ. Res. 183, 109135.

Rittmann, B.E., 2018. Biofilms, active substrata, and me. Water Res. 132, 135–145.

Rui, D., Feng, Z., 2017. Intimate coupling of photocatalysis and biodegradation to synchronously degrade pollutants. Prog. Chem. 29 (9), 1154.

Sarria, V., Parra, S., Adler, N., Péringer, P., Benitez, N., Pulgarín, C., 2002. Recent developments in the coupling of photoassisted and aerobic biological processes for the treatment of biorecalcitrant compounds. Catal. Today 76 (2–4), 301–315.

Waghmode, T.R., Kurade, M.B., Sapkal, R.T., Bhosale, C.H., Jeon, B.H., Govindwar, S.P., 2019. Sequential photocatalysis and biological treatment for the enhanced degradation of the persistent azo dye methyl red. J. Hazard. Mater. 371, 115–122.

Wang, Y., Chen, C., Zhou, D., Xiong, H., Zhou, Y., Dong, S., Rittmann, B.E., 2019. Eliminating partial-transformation products and mitigating residual toxicity of amoxicillin through intimately coupled photocatalysis and biodegradation. Chemosphere 237, 124491.

Wen, D., Li, G., Xing, R., Park, S., Rittmann, B.E., 2012. 2, 4-DNT removal in intimately coupled photobiocatalysis: the roles of adsorption, photolysis, photocatalysis, and biotransformation. Appl. Microbiol. Biotechnol. 95 (1), 263–272.

Xiong, H., Zou, D., Zhou, D., Dong, S., Wang, J., Rittmann, B.E., 2017. Enhancing degradation and mineralization of tetracycline using intimately coupled photocatalysis and biodegradation (ICPB). Chem. Eng. J. 316, 7–14.

Xiong, J., Guo, S., Zhao, T., Liang, Y., Liang, J., Wang, S., Zhu, H., Zhang, L., Zhao, J.R., Chen, G., 2020. Degradation of methylene blue by intimate coupling photocatalysis and biodegradation with bagasse cellulose composite carrier. Cellulose 27 (6), 3391–3404.

Xiong, W., Zeng, G., Yang, Z., Zhou, Y., Zhang, C., Cheng, M., Liu, Y., Hu, L., Wan, J., Zhou, C., Xu, R., 2018. Adsorption of tetracycline antibiotics from aqueous solutions on nanocomposite multi-walled carbon nanotube functionalized MIL-53 (Fe) as new adsorbent. Sci. Total Environ. 627, 235–244.

Yasmin, S., D'Souza, D., 2010. Effects of pesticides on the growth and reproduction of earthworm: a review. Appl. Environ. Soil Sci. 2010. https://doi.org/10.1155/2010/678360, 678360.

Yu, M., Wang, J., Tang, L., Feng, C., Liu, H., Zhang, H., Peng, B., Chen, Z., Xie, Q., 2020. Intimate coupling of photocatalysis and biodegradation for wastewater treatment: mechanisms, recent advances and environmental applications. Water Res. 175, 115673.

Zacharia, J.T., 2011. Identity, physical and chemical properties of pesticides. In: Pesticides in the Modern World-Trends in Pesticides Analysis, pp. 1–18.

Zhang, C., Fu, L., Xu, Z., Xiong, H., Zhou, D., Huo, M., 2017a. Contrasting roles of phenol and pyrocatechol on the degradation of 4-chlorophenol in a photocatalytic–biological reactor. Environ. Sci. Pollut. Res. 24 (31), 24725–24731.

Zhang, C., Li, Y., Shen, H., Shuai, D., 2021a. Simultaneous coupling of photocatalytic and biological processes: a promising synergistic alternative for enhancing decontamination of recalcitrant compounds in water. Chem. Eng. J. 403, 126365.

Zhang, H., Liu, Z., Li, Y., Zhang, C., Wang, Y., Zhang, W., Wang, L., Niu, L., Wang, P., Wang, C., 2020. Intimately coupled TiO_2/g-C_3N_4 photocatalysts and in-situ cultivated biofilms enhanced nitrate reduction in water. Appl. Surf. Sci. 503, 144092.

Zhang, H., Lu, Y., Li, Y., Wang, C., Yu, Y., Zhang, W., Wang, L., Niu, L., Zhang, C., 2021b. Propelling the practical application of the intimate coupling of photocatalysis and biodegradation system: system amelioration, environmental influences and analytical strategies. Chemosphere, 132196.

Zhang, L., Xing, Z., Zhang, H., Li, Z., Wu, X., Zhang, X., Zhang, Y., Zhou, W., 2016. High thermostable ordered mesoporous SiO_2–TiO_2 coated circulating-bed biofilm reactor for unpredictable photocatalytic and biocatalytic performance. Appl. Catal. B Environ. 180, 521–529.

Zhang, X., Wu, Y., Xiao, G., Tang, Z., Wang, M., Liu, F., Zhu, X., 2017b. Simultaneous photocatalytic and microbial degradation of dye-containing wastewater by a novel g-C_3N_4-P_{25}/photosynthetic bacteria composite. PLoS One 12 (3), e0172747.

Zhao, C., Pelaez, M., Duan, X., Deng, H., O'Shea, K., Fatta-Kassinos, D., Dionysiou, D.D., 2013. Role of pH on photolytic and photocatalytic degradation of antibiotic oxytetracycline in aqueous solution under visible/solar light: kinetics and mechanism studies. Appl. Catal. B Environ. 134, 83–92.

Zhao, Z., Zhang, L., Cai, Y., Chen, Y., 2014. Distribution of polycyclic aromatic hydrocarbon (PAH) residues in several tissues of edible fishes from the largest freshwater lake in China, Poyang Lake, and associated human health risk assessment. Ecotoxicol. Environ. Saf. 104, 323–331.

Zheng, J., Chang, F., Jiao, M., Xu, Q., Deng, B., Hu, X., 2018. A visible-light-driven heterojuncted composite WO_3/$Bi_{12}O_{17}C_{12}$: synthesis, characterization, and improved photocatalytic performance. J. Colloid Interface Sci. 510, 20–31.

Zhong, N., Yuan, J., Luo, Y., Zhao, M., Luo, B., Liao, Q., Chang, H., Zhong, D., Rittmann, B.E., 2021. Intimately coupling photocatalysis with phenolics biodegradation and photosynthesis. Chem. Eng. J., 130666.

Zhou, T., Ma, R., Zhou, Y., Xing, R., Liu, Q., Zhu, Y., Wang, J., 2018. Efficient N-doping of hollow core-mesoporous shelled carbon spheres via hydrothermal treatment in ammonia solution for the electrocatalytic oxygen reduction reaction. Microporous Mesoporous Mater. 261, 88–97.

Index

Note: Page numbers followed by *f* indicate figures and *t* indicate tables.

A

Acyl-homoserine lactone (AHL), 132–133
Advanced backcross QTL (AB-QTL), 318–319
AGPs. *See* Antibiotic growth promoters (AGPs)
Agriculture, 215–216
    alliance, 71–72
Agro-ecosystem, 205
Allelochemicals, 155
1-Aminocyclopropane-1-carboxylase (ACC), 35
Ammonia (NH$_3$), 268–269
Ammonification, 37
Ammonium (NH^{+4}), 268–269
Amplified fragment length polymorphism (AFLP),
    305–306
Amplified ribosomal DNA restriction analysis
    (ARDRA), 258
AMPs. *See* Antimicrobial peptides (AMPs)
Antibiotic growth promoters (AGPs), 71
Antibiotics, 69–70
Antimicrobial
    activity, 75
    clinical, 71–72
    compounds, 77
    conventional, 71–72
    feeding, 71
    resistant genes, 71
    therapeutic, 72–73
    transferable, 73
Antimicrobial peptides (AMPs), 167
Arachidonic acid (AA), 20–21
Arbuscular mycorrhizal fungi/Arbuscular
    mycorrhizal parasites (AMF), 11–12, 181
Auxin hormone/indole-3-acetic acid, 34

B

Bacillus species, 202
Bacteria
    agriculture, 9–10
    agrochemicals, 10–11, 11*t*
Bacteriocins, 77
Baculovirus (BV), 221

BCA. *See* Biocontrol agents (BCA)
Benzene, toluene, ethyl benzene, xylene (BTEX)
    compounds, 38
Betaine lipids, 272
Bioaugmentation, 145–146
Biocontrol activity, endophytes, 156
Biocontrol agents (BCA), 216
Biocontrol efficiency, enhancing strategies, 164–166
    integrated biocontrol strategies, 165–166
    management, indigenous endophytic microflora,
        166
    optimized formulation, inoculation techniques,
        165
Biocontrol mechanisms, 156–164, 229–231
    antibiosis, 160–161
    competition, 161
    detoxification, 162
    exudates, roots, 229–230, 230*t*
    foliar microbiome, detoxifying enzymes, 231
    induced resistance, 163–164
        induced systemic resistance, 163–164
        systemic acquired resistance, 163
    iron-chelating compounds, use, 231
    parasitism, lysis, 161–162
    production, siderophores, 156–160, 157–159*t*,
        160*f*
    substrate, biocidal volatiles, use, 230–231
Biofilmmediated bioremediation, 123–124
Biofilms, 128–142
    bioremediation, 142–145, 144*f*
    detection, characterization, assaying, 134–142,
        135–140*t*
        chemical methodologies, 141–142
        microbiological, molecular methodologies,
            134–140
        physical methodologies, 140–141
    different pollutants, applications, 145–149
    diversity, 129
    formation, stages, 129–133, 131*f*
    matrix, composition, 133–134, 133*f*
    reactors, 146

Biological control (CB), 215–216
Biological oxygen demand (BOD), 239–240
Biological soil quality, 30
Biostimulation, 145–146
Biosurfactants, 124–125
Biotransformation, nutrients, 34
Biotropic mycoparasitism, 228–229
BOD. *See* Biological oxygen demand (BOD)
Bulking agents, 241

C

Camellia oleifera cake (COC), 146–147
CAPS. *See* Cleaved amplified polymorphic
    sequence (CAPS)
Carbon dioxide (CO_2), 36–37, 143–145, 206
Carbon source utilization profile/community level
    physiological profile (CLPP)/BIOLOG, 257
Cation exchange capacity (CEC), 255–256
Cell wall degrading enzyme (CWDE), 228–229
CFU. *See* Colony forming unit (CFU)
Chemical oxygen demand (COD), 149
Chitin, 272
Chromosome segment-substitution lines (CSSLs),
    321
Cleaved amplified polymorphic sequence (CAPS),
    305–306
Climate-smart agriculture (CSA), 204
COD. *See* Chemical oxygen demand (COD)
Colony forming unit (CFU), 134–140
Colorimetric methodologies, 142
Composting, 240–246, 242–245*t*, 246*f*
  cooling phase, 252
  curing phase, 252–253
  maturity, determination, 253–256, 254*f*
  mesophilic phase, 249
  metagenomics, evaluating microbial diversity,
    259–260, 260*f*
  phases, 248–253, 249*f*
  process, types, 246–247
    aerobic, 247
    anaerobic, 247, 248*f*
  taxonomic, metabolic microbial diversity, 256
  techniques, analyze microbial diversity
    culture-dependent phenotypic-based
      approaches, 257
    fatty acid methyl ester (FAME) analysis,
      257–258
    PCR-based approaches, 258–259

role, culture-independent techniques,
    257–258
thermophilic phase (35–65°C), 249–252,
    250–251*t*
Coral Probiotic Hypothesis (CPH), 97–98
Cordyceps militaris, edible medicinal fungus, 21–22
Crystal violet (CV), 141–142
CSA. *See* Climate-smart agriculture
Culture medium, 79
CWDE. *See* Cell wall degrading enzyme (CWDE)
Cytokinins (CK), 201–202

D

DArT. *See* Diversity array technology (DArT)
Dichlorodiphenyltrichloroethane (DDT), 143–145
Disease suppressive soils, 191
Dissolved organic carbon (DOC), 255–256
Diversity array technology (DArT), 305–306
Double heterozygotes, 314–315

E

Earthworm-microorganisms, crop productivity
  microflora role, 110–112, 111*f*
    drilosphere role, 110–111
    rhizosphere role, 111–112
  nature's ploughman, 109–112
  vermicompost
    biofertilizer, 112
    crop growth, productivity, 112–113, 113–114*t*
    nutrient uptake, effect, 114
    soil, physical, chemical biological properties,
      115
  vermitechnology, co-treatment of OFMSW,
    115–117, 116*f*
Edible microbial biomass
  biotechnological tools, generation, 19–22
    Cas9 CRISPR system, 21–22
    genetic modification, 19–21, 20*f*
    RNA interference (RNAi), 21
  microorganisms, possible routes, 13–17
    bacteria, 16
    filamentous fungi, 16
    microalgae, 14–15, 15–16*t*
    potential use as food, 13–14, 14*t*
    yeast, 16–17, 16*t*
  nutritional value, contextual approaches, 2–4
  probiotics-prebiotics-symbiotics, functional
    properties, 5–6

production, 17–19
   advantages, disadvantages, 17
   CO2 conversion, 17–18, 17–18*t*
   process, 18–19, 19*f*
profit dynamics, 4
safety, 6–13
   food safety, plant-microorganism-soil
      relationship, 9–13
   microorganism, plant-microorganism-soil
      relationship, 8–9
EIS. *See* Electrochemical impedance spectroscopy
      (EIS)
Eisenia fetida, 112
Electrical conductivity (EC), 31, 255–256
Electrochemical impedance spectroscopy (EIS),
      140–141
Endophytes, 155
Enterobacterial repetitive intergenic consensus
      polymerase chain reaction (ERIC-PCR),
      258–259
Entomopathogenic fungi, 221–222
Environmental pollution, 125–127, 126–128*t*
Environmental stress, 85–86
Eudrilus eugeniae, 112
European Consensus (EC), 3
Express Sequence Tags (ESTs), 305
Extracellular DNA (eDNA), 134–140
Extracellular polymeric substances (EPSs), 124–125

F

Fatty acid methyl ester (FAME), 257–258
Fourier transform-infrared spectroscopy (FTIR),
      147–148, 276
Free-living rhizobacteria (ePGPR), 34–35
Fungicides, 53–54*t*

G

Gene pyramiding, 302–303
"Generally Recognized as Safe" status, 73
Genetically complex traits (GXE), 303
Genetic engineering, antimicrobial peptides, 166–168
GHG. *See* Greenhouse gases (GHG)
Gibberellins (GA), 201–202
Gram-negative bacteria, 77
Gram-positive bacteria, 77
Granulovirus (GV), 221
Greenhouse gases (GHG), 94, 204, 246
GXE. *See* Genetically complex traits (GXE)

H

Heavy metals (HM), 12, 122–123
Herbicides, 335–337
   classification, 335–337
   mode of action, 337
Herpes simplex virus 1 (HSV-1), 8
High molecular weight glutenin subunit
      (HMW-GS), 307
HM. *See* Heavy metals (HM)
Holobiont, 85–86
Hydraulic retention time (HRT), 149
Hyperparasites/parasiting pathogen, 228–229

I

IFIC. *See* International Food Information Council
      (IFIC)
IHNV. *See* Infectious hematopoietic necrosis virus
      (IHNV)
Immunocompromised patients, 73
Indole acetic acid (IAA)/auxin, 35, 201–202
Induced systemic resistance (ISR), 163–164,
      203–204
Infectious hematopoietic necrosis virus
      (IHNV), 79
Inorganic metals/trace element, 122–123
Inorganic nitrogen ions, 148–149
Insertion-deletion polymorphisms (InDels),
      305–306
International Food Information Council (IFIC), 3
Intimate coupling, photocatalysis and
      biodegradation (ICPB), 359–361, 360*f*
   applications, 365–366
   mechanism, 361
   microorganisms, 364, 364*f*
   photocatalyst types, 362, 363*f*
   porous carrier materials, 362–364
   reactors, 365
ISR. *See* Induced systemic resistance (ISR)

K

Key engines, microbes, 35–36

L

Lactic acid bacteria (LAB), 4
Lichenized fungi, primary bioindicator/biomonitor,
      267–269, 268*f*
   air N deposition monitoring, pollution regime,
      280–282

Lichenized fungi, primary bioindicator/biomonitor
  (Continued)
  critical load assessor, N eutrophication, 279–280
  excessive N deposition, lichen food webs,
    293–294
  indicators, excessive nitrogen (N) deposition,
    273–279, 273*f*
  lichen, sink, 294
  nitrogen assimilation, 269
  nitrogen isotope ratio ((δ15N in), spatial
    distribution, 282–293, 283–287*t*
  nitrogen tolerance, mechanisms, 269–273, 270*f*
    biochemical, metabolic adaptive tolerant
      mechanisms, 272–273
    cation exchange capacity, saturation kinetics,
      NH, 270
    competitive metabolism, community
      dynamics, 270–271
    intrinsic tolerance, long-term exposure, 271
Linked genes, 306

M
Malic enzyme (ME), 20–21
Man Ragosa and Sharp (MRS) medium, 80
Marker assisted backcross breeding (MABB),
  311–312
Marker-assisted backcrossing (MABC), 312
Marker-assisted recurrent selection (MARS), 312
Marker-assisted selection (MAS), 302
Membrane biofilm reactor (MBfR), 362
Methane (CH₄), 143–145, 206
MFC. *See* Microbial fuel cells (MFC)
Microbes, 198–204, 199*f*
  antibiotics production, 202
  combating climate change, 204–207, 205*f*
  defined, 123–124
  enzyme production, 203
  induced systemic resistance, 203–204
  nitrogen fixation biological, 200
  phosphate solubilization, 200–201
  phytohormones production, 201–202
  plant stress management, 204
  siderophore production, 203
Microbial biocontrol, crop protection, food
    security, 217–223
  biofertilizers, 217–220, 217*f*
    mycorrhiza fungi association, 220
    nitrogen fixing microbes, 218, 218*f*
    phosphorus solubilizing microbes, 219
    plant growth-promoting microbes, 219
  biofungicides, 223
  bioherbicides, 222–223
  biopesticides, 220–222
    bacteria, 220, 221*f*
    fungi, 221–222
    nematodes, 222
    viruses, 221
  postharvest, management, applications,
    223–224
Microbial biofilms, 128–129
Microbial community, 85–86
  climatic factors, 86–96
    drought stress, 91–94
    increased levels, CO2, lands, 92–93*t*, 94
    ocean acidification, 94–95
    temperature elevation, thermotolerance,
      87–91, 87–91*t*
    water soluble oil fractions, 95–96
  future, applications, 97–99, 97*f*
    genetic engineering, 98
    metagenomics, 98
    next-generation sequencing, others, 98–99
    probiotics, 97–98
Microbial control activity, 225–229
  antagonism, 225–226
    antibiotic production, 226, 226–227*f*
    competition, 227, 228*t*
    parasitism, 228–229, 229*f*
Microbial fuel cells (MFC), 123–124
Microbial organic volatile compounds (mVOC),
  207
Microbial transformation, atmospheric nitrogen, 37
Microbiology, 9
Microbiome, stimulants
  application process, 185–190, 229
    phytoremediation, 189–190
    plant growth, crops, 187–188
    stress management, 188–189, 189*f*
    sustainable agriculture, 185–187, 186*f*
  crop plants, 180–181, 180*f*
  major crops, 181–185
    potato, 185
    rice, 182–183
    sugarcane, 185
    tomato, 181–182
    wheat, 183–184, 184*t*
Microflora, 70–71

Microorganisms/free-floating microbial
        cells, biocontrol agents, sustainable
        agriculture, 45–46, 128–129
    bacteria, fungi, 46
    biocontrol agents mechanisms, 48–63, 49–54*t*
      actinomycetes, 61–63, 61–63*t*
      antibiosis, 56
      antobosis, 57–60, 58–60*t*
      competition, 55–56
      hyperparasitis, 55
      induced resistance, 57
      metabolite production, 56–57
      plant growth promotion, 57
      secretion lytic enzymes, 56
    desired product production, 46–48
      evaluation (vitro,glasshouse), 47
      formulation, desired product, 47
      isolation, desired microorganism, 46–47
      mass production, 47
      registration, 48
      testing, isolates field conditions, 47
Molecular
    breeding, 316–318
    markers, genomics assisted breeding, 302–303
      selection, markers, MAS, 305–306
      traits choice, MAS, 303–304
      types, markers, plant breeding, 306–311, 307*f*,
        309*f*
      types, selection, procedure, 311–320, 315*f*,
        317*f*, 319*f*, 322–326*t*
Multiparent advanced generation intercross
    (MAGIC), 321
Mycorrhizae, 11–13, 12*t*, 220

N

Nano-herbicides, 337–338, 338*t*
  applications, agriculture, 348–349
  formulations, 345–348, 348*t*
    agro waste, 346–347
    controlled release formulations (CRFs),
      347–348
    nanoencapsulations, 345–346
Nanomaterial, synthesis, 341–345, 342*f*
  carbon-based nanomaterials, 343
  metallic nanoparticles, 343
  metal oxide nanoparticles, 343
Nanoparticles, delivery, 343–345
  agriculture waste-based, 344
  green synthesis, 344–345
  inorganic, 344
  polymeric, 343–344
Natural attenuation, 145–146
Near isogenic lines (NILs), 318–319
Nested association mapping (NAM), 321
Nicotinamide adenine dinucleotide phosphate
    (NADPH), 20–21
Nitric oxide (NO), 268–269
Nitrogen dioxide (NO$_2$), 268–269
Nitrogen fixation, 37
Nitrous oxide, 206
NMR. *See* Nuclear magnetic resonance (NMR)
    imaging
Non-phosphate solubilizing bacteria (NPSB),
    200–201
Norovirus (NV), 8
Nuclear magnetic resonance (NMR) imaging,
    140–141
Nucleopolyhedrovirus (NPV), 221

O

Obligate parasites, 222
Open-air laboratories (OPAL), 281
Organic matter (OM), 255–256
Outer membrane proteins (OMPs), 130–132

P

PAH. *See* Polycyclic Aromatic Hydrocarbons
    (PAH)
Pathogens, 225
Pedigree, 316–318
Persistent organic pollutants (POPs), 143
Pesticides, 349–350, 355–356, 356*t*
  ecological effect, 357–358, 358*f*
  environmental distribution, 356–357, 356*f*
  hazardous effects, 359
  toxicity, 357
PGPB. *See* Plant growth promoting bacteria (PGPB)
PGPR. *See* Plant growth promoting rhizobacteria/
    plant development advancing rhizobacterias
    (PGPR)
Phosphate solubilizing bacteria, 200–201
Phospholipid fatty acid (PLFA), 115
Phosphomonoesterase (PME), 276
  activity, 293–294
Photocatalytic circulating-bed biofilm reactor
    (PCBBR), 365

Phytohormones, 34
Phytoimmunity, 155
Phytomicrobiome, 33
Phytopathogens, 155
Phytostabilization, 12
Pig veterinary Society, 71–72
Plagues/natural enemies, 215–216
Plant growth promoting bacteria (PGPB), 187–188
Plant growth promoting rhizobacteria/plant development advancing rhizobacterias, (PGPR), 34–35, 181
Plant microbiome, 178–179, 179f
   language, 229
PME. *See* Phosphomonoesterase (PME)
Poliovirus (PV-1), 8
Polychlorinated biphenyls (PCBs), 143–145
Polycyclic aromatic hydrocarbon (s) (PAH/PAHs), 38, 95–96, 122
Polyunsaturated fatty acids (PUFA), 20–21
POPs. *See* Persistent organic pollutants (POPs)
Predators, 215–216
Priority lethal pollutants, 122, 123f
Probiotics, 4, 69–70
   adverse effects, 74
   animal health, 70–71
   antiviral effect, 79
   aquaculture, applications, 80
   competitive exclusion, 75–76
   feed antibiotics, viable substitute, 71–73
   immune response, improvement, 79
   mechanisms, actions, 76–77
   nutrients, enzymatic attributed, digestion, 78
   risk assessment protocol, 73–74
   selection principle, 74–75, 75–76f
   separation, methods, 79–80
   water quality, impact, 78
Propidium monoazide (PMA), 134–140
Pseudomonas, 202

Q

Quantitative polymerase chain reaction (qPCR), 134–140
Quantitative trait loci (QTLs), 319–320
Quorum sensing (QS)/density sensing, 132–133

R

Random amplified polymorphic DNA (RAPD), 305–306
Reduce pathogen infection, 190–191, 190t

Restriction fragment length polymorphism (RFLP), 305–306, 308
Restriction site-associated DNA (RADseq), 305–306
Rheology, 3–4
Rhizomicrobiome, 33
Root exudates, 32

S

SAR. *See* Systemic acquired resistance (SAR)
Sequencing batch moving bed biofilm reactor (SBMBBR), 149
Simple sequence repeat (SSR), 305–306
Single feature polymorphisms (SFPs), 305–306
Single nucleotide polymorphism (SNP), 305–306
Small-angle X-ray/neutron scattering, 140–141
Soil, 29–30
   microbes, enzymes, restoration, reclamation, 38–39
   microbial influence, biogeochemical cycles, applications, 35–37, 36t
      carbon, microbial decomposers, 36–37
      nitrogen cycle, 37, 37t
   microbiome, plant-soil-microbial interactions, 32–34
      plant productivity, health, 33
      plant root, microbes, interactions, 34
      plants, effect, 32–33
   pollution, 31
   quality, assessment, 30, 30f
   rhizobacteria, physiology, plant-growth promoting, 34–35
Soil carbon (C) transformation, 36–37
Soil organic carbon (SOC), 31
SR-lXRF. *See* Synchrotron radiation micro-X-ray fluorescence microscopy (SR-lXRF) analysis
Sulfate reducing bacteria (SRB), 146–147
Sustainable production agroecosystems, 12–13
Symbiotic bacteria (iPGPR), 34–35
Synchrotron radiation micro-X-ray fluorescence microscopy (SR-lXRF) analysis, 146
Synthetic minimal microbiome, 98
Systemic acquired resistance (SAR), 163

T

Titanium dioxide (TiO$_2$), 362
Total Petroleum Hydrocarbons (TPH), 95–96
Trinitrotoluene (TNT), 143–145

U

Ultrasonic time-domain reflectometry, 140–141

V

Vermicasts, 112
Veterinary Feed Directive (VED), 71
Viable but non-culturable (VBNC) biofilm cells, 134–140
Volatile organic compounds (VOCs), 230–231

W

Wastewater treatment, 359
Water (H$_2$O), 143–145
Water soluble oil fraction (WSF), 95–96

Weed management, 334
  nanotechnology, 338–340, 339*f*, 341*f*
World Health Organization, 2

X

X-ray absorption near-edge structure (XANES) spectroscopy analysis, 146
X-ray computed tomography, 140–141

Y

Yeast Extract Agar (YEA), 80

Z

Zero hunger, 2